Fundamentals Of Mechanics

◆·◆·◆·◆

University Physics Volume 1

BY

Samuel J. Ling
Truman State University

Printed By
Createspace, Charleston
An Amazon.com Company

Samuel J. Ling
Department of Physics
Truman State University
Kirksville, MO 63501, USA.
sling@truman.edu

ISBN-10: 1-98527-463-9
ISBN-13: 978-1-98527-463-1

©2018, Samuel J. Ling.

All rights reserved. Printed in the United State of America. Except as permitted under the Copyright Act of 1976 and Amendments, no part of this publication may be reproduced or distributed in any form or by any means without the prior written permission of the Author. To obtain permission to use the material from this work, please submit a written request to Samuel J. Ling, Professor of Physics, Truman State University, 100 E. Normal St., Kirksville, MO 63501, USA. For information regarding permission, contact sling@truman.edu.

Cover Photo:

STS-64 Shuttle Mission Imagery (9 Sept. 1994) — With a crew of six NASA astronauts aboard, the space shuttle Discovery heads for its nineteenth Earth-orbital mission. Onboard were astronauts Richard N. Richards, L. Blaine Hammond, Carl J. Meade, Mark C. Lee, Susan J. Helms and Jerry M. Linenger. Photo credit: National Aeronautics and Space Administration (NASA)

Typed in LaTeX
Distributed by CreateSpace.com

Dedicated to the loving memory of my parents

Preface

A solid foundation in physics is necessary for many disciplines in the pure and applied sciences. This requirement has been traditionally satisfied by a one-year long course in general physics, which covers mostly topics from classical physics. In order to make the first year physics more accessible and appealing to students textbooks for these courses have attempted to present the same material in various formats, more colorful graphics, systematic problem solving strategies, online homework access, etc. These changes have come at a considerable cost - while the content of these books have become less rigorous their prices have doubled or tripled.

The loss of rigor has been very unfortunate and detrimental for students. For instance, it is well-known that many students now do not develop understanding of basic concepts such as an vectors and kinematics. Further, the difference between the magnitude and the components of vectors is muddled in the minds of students, and you are very likely to find students quoting numbers such as -100 N force and spring force equal to $-kx$, acceleration equal to -9.81 m/s^2, just to mention a few problems.

Many more students do not learn to think physics from a conceptual perspective, instead, they learn to rely on formula-manipulations without understanding where those formulas come from. As a result, after successfully completing a one-year course in physics, some of even the best students do not know how to derive simple results - some others do not understand why they need to learn how to derive basic results.

Often the wide range of the coverage material is blamed for these deficiencies. However, an examination of the current books reveals that they present various topics in essentially unrelated way and do not emphasize enough the few basic ideas which are essential for understanding physics. This type of presentation does not lend itself to teaching students the important of derivations and places undue emphasis on the use of final results in mostly numerical problems.

This book is based on a first-year course that I have taught for a number of years at Truman State University. It has a standard selection of topics suitable for the first-year physics course but differs from the existing books in the degree of emphasis placed on the understanding of the fundamental ideas and the care with which the derivation of various results are presented.

My general approach in writing this book has been to present concepts in as broad a context as possible at the mathematical level of the student. For instance, I avoid the vector-killing presentation of 1-d motion; instead we dive directly into 2- and 3-dimensional treatment and present 1-d as a special and very useful case. The approach - general leading the specific

- makes it possible to develop ideas and concepts starting from few basic principles. The level of the rigor has been carefully managed so that the book should be accessible to those students who are concurrently enrolled in the Calculus sequence.

Various detailed guided exercises and challenging problems in this book will help students develop their own approach to problem solving rather than simply learn to follow a set of rules. Many books try to encourage students to memorize a step-by-step process of solving certain problems. However, this approach to physics is contrary to physics thinking as it emphasizes "getting" the right answer rather than learning to identify the first principles applicable to the given situation and then reasoning correctly from those first principles.

Although learning to follow a set recipe can show better performance on tests that test whether students know how to follow those recipes, this approach turns physics into a subject of word problems rather than a subject where a student learns how to examine real situations through simple models. To help students grasp the approach unique to physics, the examples in this book contain detailed analysis of various ideas, approximations and their limitations. It is hoped that by studying these examples a student will develop his/her own sense of the "physics" reasoning that he/she can bring to bear when confronting new situations.

The four major topics of the classical physics, Mechanics, Thermodynamics, Electrodynamics, and Optics, are spread over five volumes so that each volume would be easy to carry around in a back pack. Selection from Modern Physics makes up the sixth volume. These six volumes should serve as a strong base for the future work in physics and engineering.

I am sure there are many typographical errors still left in the text despite numerous revisions. Some students, especially Mr. Brian McClain and Mr. Robert Ashcraft, have read earlier drafts of this text and were instrumental in fixing typographical errors. I am very thankful to several students for their feedback in other courses where parts of the present and other volumes were used. I would appreciate hearing about any typos and mistakes in the book so that I can correct them for the next edition. I would also welcome any other observations or remarks you may have regarding the content and approach of this book.

During the writing I had much support from my family and am truly grateful for that. Without their understanding and encouragement I would not have been able to complete this task.

Samuel Ling
Kirksville, Missouri, USA.

TO THE TEACHER

University Physics is organized in six volumes:

1. Fundamentals of Mechanics
2. Applications of Mechanics
3. Thermodynamics
4. Electricity and Magnetism
5. Optics
6. Modern Physics

1. Fundamentals of Mechanics

After the introductory chapter which presents the basic approach of physics, vectors are presented in great detail. Every student is encouraged to master the concepts associated with vectors before proceeding any further in the text.

The important topics of kinematics is presented in an approach that emphasizes the general definition before the particular applications. This order is different from the "normal" practice in other textbooks at this level. I believe that the normal order of doing the one-dimensional problems first is counter-productive to student-learning, and may even be harmful. One-dimensional problems de-emphasize the directional aspect of vectors, thereby introducing many long-lasting misconceptions such as the inability to tell the difference between the magnitude and the component of a vector. Many examples of this deficiency can be cited. For instance, students tend to apply the constant acceleration equations even when they are not applicable, e.g., in the simple harmonic motion. Too many problems done early on in one-dimensional constant acceleration has a way of leaving a lasting impression in the minds of students!

For instance, the magnitude of the spring force is learned as $F = -kx$ rather than $F = k\Delta l$. This becomes problematic when a student encounters problems of two or more masses connected by springs; the definition $F = -kx$ does not work any more but $F = k\Delta l$ is still useful, and students have to unlearn the bad habit. This type of unforced error should be avoided from the start.

Although, the one-dimensional constant acceleration case is an important problem, working on 1-dimensional problems for two or three weeks cements a message to students that "everything is 1-d". A student becomes so used to the constant acceleration formulas to the extent that everything, even the Simple Harmonic Motion, appears to be a constant acceleration problem to the student. Every physics teacher has encountered this problem in some form or other. In my teaching experience, an exposure to the one-dimensional first is responsible for students having difficulties with the concept of direction of vectors. Here, I try to avoid the problem from developing in the first place by presenting vectors first.

Many teachers feel that students would be far better off by first mastering the correct definitions and then using those definitions in applications such as one-dimensional problems. However, the books seem to go the other way. In this book I have spread the topic of kinematics over two chapters - the first contain the fundamental definitions and the second the applications. A teacher may wish to treat the the two chapters as one and lecture by going back and forth between the two chapters as you develop the concepts of displacement, velocity and acceleration.

The subject of kinematics is followed by a long chapter on forces. Various common forces are introduced and studied in the context of the static equilibrium. This is important because students should learn that forces exist independent of acceleration. When students encounter forces first in the context of $\vec{F} = m\vec{a}$, many confuse the $m\vec{a}$ as being a force. It is difficult for students to separate the two sides of the equality as different physical entities - actually, I have encountered physic professors who have told me that $m\vec{a}$ is a force! Studying statics, where you study $\vec{F} = 0$, before tacking the dynamics gives an opportunity to learn about forces themselves. I have introduced the concept of torque in this chapter so that I could discuss the static equilibrium situation for macroscopic objects more fully. This also helps develop the concept that forces act at different particles of an extended body. For some teachers, this may be an early introduction of torque. The topics of torque are separated enough that a teacher can design his/her course to skip the section(s) on the torque in this chapter and return to this topic after reaching the chapter on rotation. I would recommend that you do not skip torque at this stage as it will help your students understand the distributive nature of forces such as Normal and frictional forces.

The chapter on forces introduces students to a multitude of common forces. The distributive nature of forces on the extended bodies are illustrated and emphasized. To help with the concept of force as interaction between bodies I have introduced Newton's third law of motion at this stage before presenting Newton's second law. This chapter also provides an important learning ground for the techniques of the free-body diagrams.

Next, Newton's laws of motion are presented systematically, and the dynamics of single particles and multiparticles are studied extensively. The dynamics is continued in the next chapter using the concept of impulse. The conservation of momentum is presented, both in the context of the third law of motion and then again in the chapter on momentum.

We continue the discussion on Newton's laws of motion by introducing impulse in the next chapter. This chapter shows the emergence of the principle of conservation of momentum. Important applications to collisions and changing mass systems such as rocket is discussed here.

The concept of work is developed from a fundamental perspective. Rather than use Fd to introduce the concept of work, we go directly to the dot product and the infinitesimal work, $dW = \vec{F} \cdot d\vec{r}$. We develop the main ideas of the conservation of energy from the ground up based on the work-energy theorem. By separating the work on multiparticle systems into the work by the internal forces and the external forces a student is encouraged to see how energy conservation can be applied more broadly. Various potential energy formulas are derived from the first principles and the role of the reference energy and reference point in space are made plain to the student.

After the chapter on the energy, there is a rather long chapter on rotation. A teacher should set aside a good deal of time for a thorough treatment of rotation. The main emphasis in rotation is on the angular momentum in the context of fixed-axis rotation. The conservation of angular momentum is discussed here. The rolling motion of an extended body is developed from the mechanics of the individual particles.

Finally, to complete the foundational topics in mechaincs, a discussion of the modification of Newton's second law in a non-inertial frame is presented. This is necessary for a treatment of the Coriolis force.

To keep the present volume to a manageable length the other topics, which I have labeled as Applications of Mechanics, such as gravitation, oscillations, mechanical waves, and fluid mechanics have been presented in the second volume.

2. Applications of Mechanics

This volume continues the discussions on mechanics. This volume and Volume 1 should be thought of as one book on Mechanics.

Volume 2 starts with the chapter on Newton's law of universal gravitation which includes an extensive discussion on the planetary motion and Kepler's laws. I have laid out the basis for the effective potential energy and discussed the fundamental approach for studying the central force problem. The orbit equation is discussed in considerable detail.

The two chapters on the vibrations and waves introduce students to the basic aspects of the Simple Harmonic Motion, coupled oscillators and waves. The concept of normal modes are introduced and serve as the basis for a discussion of other topics. I have tried to take the mystery out of the wave motion. Learning the basic aspects of coupled motion helps a student extend his/her understanding to the vibrations in a system where the motion of a large number of particles are coupled.

A chapter on statics introduces internal stress and strain in a solid body. This chapter would be a review for students who have completed the chapter on force presented in the first volume. Finally, the fluid statics and dynamics are presented over two chapters, where the differences between the dynamic pressure and the static pressure are highlighted and clarified.

3. Thermodynamics

The subject of thermodynamics is presented here from the traditional perspective. First, the effects of heat on substances is described in significant detail. The wrong definition of temperature in terms of the kinetic energy of molecules, quoted often in most textbooks, is avoided here. Instead, we rely on the notion of the thermal equilibrium to introduce temperature. The only magic that is thrust on the students without any explanations is the ideal gas law. This shortcoming is remedied in the chapter on the kinetic theory. The ideal gas provides a good system in which various aspects of thermodynamics can be explored more fully by a student. the first chapter of this volume explains students the meaning of heat and carefully defines the concept of temperature based on the thermal equilibrium.

Next we discuss the first law of thermodynamics as an expanded view of the conservation of energy. Various applications of the first law, including Calorimetry, are presented to solidify the understanding of the special nature of thermal interaction between a system and the surroundings. The second law of thermodynamics is next presented in the classic forms of Kelvin-Planck and Clausius, and illustrated by applications to the Carnot engine and other practical engines. Clausius's theorem is also proved in this chapter. An entire chapter is then devoted to the concept of Entropy where reversible and irreversible processes are also defined. Finally, there is a chapter on the kinetic theory in which we go into fair amount of detail of looking at macroscopic systems from constituents. I have also included simple models of gases in this chapter.

4. Electricity and Magnetism

This volume introduces classical Electricity and Magnetism. Various aspects of classical Electricity and Magnetism are compactly combined in Maxwell's equations. In this volume we systematically build towards Maxwell's equations.

First, we address phenomena displayed in the vicinity of static charges, i.e., electric charges that are not moving. The explanation of these phenomena is in terms of electric field ($\vec{E}$). Then, we handle magnetic phenomena that appear when charges are moving steadily. The new phenomena is understood in terms of magnetic field ($\vec{B}$) associated with moving charges. The electric field associated with static charges is static, meaning independent of time. The magnetic field associated with steady current of charges is also static. Finally, we address the rich phenomena that occur when electric and magnetic field can change with time. A good understanding of multi-variate calculus can help you greatly when studying theories of electricity and magnetism. I have developed some of the mathematics where it is used, but our treatment is not rigirous, instead I have appealed to the conceptual understanding.

5. Optics

The volume on optics is self-contained containing introductions to both geometric and physical optics. Some of the concepts of waves from Volume 1 has been presented once again when discussing the wave optics. A repeat of these ideas in the context of light often gives students a second exposure which is often necessary to learn them well.

This volume begins with a discussion of the fundamental aspects of light. The next two chapters are devoted to geometric optics. The second chapter is entirely concerned with the image formation by reflection and refraction. Various basic optical elements are discussed here. The applications of geometric optics to instruments are presented next, where a detail picture of optics of the eye is also given.

The rest of the book is devoted to the wave or physical optics. In physical optics I have tried to reinforce everywhere the concept of the addition of amplitudes which is central to the understanding of wave phenomena. Some of the sections in these chapters are mathematically demanding, but it would be worthwhile for a student to work though the calculations in the book and attempt the more challenging problems found in the exercises and problems.

6. Modern Physics

Physics since around year 1900 AD is called Modern Physics. The ideas of Modern Physics differ fundamentally from classical ideas based on Newtonian mechanics and Maxwell's electricity and magnetism. Some of the changes in approach to Physics were made nececessary by the inability of classical theories to explain strange new phenomena observed at microscopic scales while other changes arose due to mathematical and physical inconsistencies among various parts of physics. In this volume we will introduce ideas from Relativity, Quantum Mechanics, Material Science, Nuclear Physics, Fundamental Particles, Gravity, and Cosmology.

TO THE STUDENT

This book was written to provide you with a solid foundation of introductory physics. The book is expected to be accessible to students who are at least concurrently enrolled in a Calculus course. The emphasis in this book is on generating a deeper understanding of physics rather than creating a superficial familiarity with the subject. You cannot gain a good understanding of physics without a lot of effort on your part.

Dear student, you should realize that reading physics textbook requires time and patience. Rushing through the textbook will not be enough to learn physics. You are expected to sit down with the book and really study the book to get the material presented here. To help you take notes down while you are studying I have left a healthy margin on the side of each page. The fonts have been chosen to make the reading less straining to the eye. You would realize that the more time you put into your studies, the more you would get out of your studies, and you will derive much satisfaction from learning the subject.

The chapters build on each other. You should not skip over any problems you encounter in the early chapters since those problems are more likely to show up again in the later chapters.

The exercises and problems are key parts of the textbook. The exercises and problems at the end of chapters have been selected carefully to improve your understanding and challenge you as well. The exercises should help you improve your understanding of the basic definitions of single concepts, and problems should help you think about multiple concepts together.

Many of the problems are non-trivial and you are expected to spend a good deal of time solving them. These problems would prepare you well for the advanced courses in your major and lead to a fruitful career in the pure and applied sciences. Some of the exercises have multiple parts so that you can explore a physical situation more fully and develop a sense of a more complete understanding.

Contents

1 **INTRODUCTION**   1
    1.1 THE WORLD OF PHYSICS 2
    1.2 FUNDAMENTAL UNITS 4
       1.2.1 Length 4
       1.2.2 Time 5
       1.2.3 Mass 7
       1.2.4 Relation Between Metric and Imperial Units 8
       1.2.5 Other Metric Units 9
       1.2.6 The SI AND CGS Systems of Units 9
       1.2.7 Conversion of Units 10
    1.3 UNCERTAINTY IN MEASUREMENTS 12
       1.3.1 Uncertainty and Precision 12
       1.3.2 Uncertainty and Significant Figures 13
       1.3.3 The Scientific Notation For Numbers 15
       1.3.4 Rounding Off 15
       1.3.5 Uncertainty and Accuracy 16
    1.4 PROPAGATION OF UNCERTAINTIES 16
    1.5 ORDER OF MAGNITUDE ESTIMATES 19
    1.6 DIMENSIONAL ANALYSIS 20
    1.7 EXERCISES 22
    1.8 PROBLEMS 24

2 **VECTORS**   25
    2.1 MOTIVATION FOR VECTORS 26
    2.2 GEOMETRICAL VIEW OF VECTORS 27
       2.2.1 What is a Vector? 27
       2.2.2 Multiplication of a Vector by a Scalar 28
       2.2.3 Unit Vector 28

		2.2.4	Addition of Vectors 29
		2.2.5	Subtraction of Vectors 31
		2.2.6	Vector Equations and Polygons 32
		2.2.7	Multiplication of a Vector with Another Vector 32
		2.2.8	Scalar Product Or Dot Product 33
		2.2.9	Vector or Cross Product 35
	2.3	ANALYTICAL VIEW OF VECTORS 37	
		2.3.1	Base Vectors 37
		2.3.2	Magnitude of Vectors 39
		2.3.3	Directions of Vectors 40
		2.3.4	Adding and Subtracting Vectors Analytically 44
		2.3.5	Scalar Products Analytically 46
		2.3.6	Vector Product Analytically 50
	2.4	EXERCISES 51	
	2.5	PROBLEMS 56	
3	**KINEMATICS**		**59**
	3.1	INTRODUCTION 60	
	3.2	POSITION VECTOR 60	
	3.3	DISPLACEMENT VECTOR 61	
	3.4	AVERAGE VELOCITY 69	
		3.4.1	Average Speed 73
	3.5	VELOCITY AND SPEED 74	
		3.5.1	Velocity in One-dimensional Motion 74
		3.5.2	Velocity - General 82
		3.5.3	Speed 87
	3.6	ACCELERATION 88	
	3.7	MOTION USING POLAR COORDINATES 98	
		3.7.1	Polar Coordinates 98
		3.7.2	Unit Vectors $\hat{u}_r$ and $\hat{u}_\theta$ 99
		3.7.3	Velocity and Acceleration in Polar Coordinates 101
		3.7.4	Circular Motion 102
	3.8	EXERCISES 109	
	3.9	PROBLEMS 122	
4	**APPLICATIONS OF KINEMATICS**		**123**
	4.1	CONSTANT VELOCITY 124	

4.2	CONSTANT SPEED		125
4.3	CONSTANT ACCELERATION		125
	4.3.1	Planar Motion	126
4.4	ONE DIMENSIONAL MOTION WITH CONSTANT ACCELERATION		128
	4.4.1	Free Fall: Application of Constant Acceleration	131
4.5	TWO DIMENSIONAL MOTION WITH CONSTANT ACCELERATION		137
	4.5.1	Projectile Motion	137
4.6	VARIABLE ACCELERATION		145
4.7	RELATIVE MOTION		149
	4.7.1	Observers Moving at Uniform Velocity	150
	4.7.2	Observers Moving at Uniform Acceleration	152
4.8	EXERCISES		154
4.9	PROBLEMS		162

5 FORCES AND STATIC EQUILIBRIUM   165

5.1	MEASURING FORCES		166
	5.1.1	Distinction Between Mass and Weight	166
	5.1.2	A Method of Measuring Forces	167
5.2	WHAT IS A FORCE?		168
	5.2.1	Forces and Newton's Third Law of Motion	168
5.3	FORCES AS VECTORS		171
	5.3.1	Resultant force	171
	5.3.2	Component of a Force in a Given Direction	172
5.4	SOME COMMON FORCES		174
	5.4.1	Weight and Gravitational Force	175
	5.4.2	Tension Force and Hooke's Law	177
	5.4.3	Spring force and Hooke's Law	178
	5.4.4	Normal Forces at a Contact Surface	179
	5.4.5	Static Frictional Force	183
	5.4.6	Kinetic or Sliding Frictional Force	186
	5.4.7	Rolling Friction	187
	5.4.8	Viscous drag	188
5.5	TORQUE OR MOMENT OF A FORCE		189
5.6	STATIC EQUILIBRIUM		195
	5.6.1	Static Equilibrium and Newton's First Law of Motion	195

		5.6.2 Problems of Static Equilibrium	197
	5.7	EXERCISES	207
	5.8	PROBLEMS	215

6 NEWTON'S LAWS OF MOTION   219

- 6.1 NEWTON'S FIRST LAW OF MOTION 220
- 6.2 NEWTON'S SECOND LAW OF MOTION 223
  - 6.2.1 Definition of Mass 223
  - 6.2.2 Definition of Momentum 223
  - 6.2.3 The Second Law 225
  - 6.2.4 Operational Definition of Force and Mass 228
- 6.3 NEWTON'S THIRD LAW OF MOTION 230
- 6.4 PROBLEMS CONCERNING MOTION OF PARTICLES .. 232
- 6.5 PROBLEMS CONCERNING MOTION OF EXTENDED BODIES 238
  - 6.5.1 General Considerations 238
  - 6.5.2 Examples of Translational Motion of one Body .. 240
  - 6.5.3 Coupled Systems 244
- 6.6 EXERCISES 254
- 6.7 PROBLEMS 260

7 IMPULSE AND COLLISION   263

- 7.1 IMPULSE 264
  - 7.1.1 Impulse of a Constant Force 264
  - 7.1.2 Impulse of a Time Dependent Force 265
  - 7.1.3 Area Under the Curve Method 268
- 7.2 IMPULSE AND CHANGE OF MOMENTUM 270
- 7.3 TRANSLATIONAL MOTION OF MULTIPARTICLE SYSTEMS 271
  - 7.3.1 Two-Particle System 271
  - 7.3.2 Generalization 273
  - 7.3.3 Motion of Center of Mass 273
- 7.4 CALCULATIONS OF CENTER OF MASS 275
  - 7.4.1 Examples of Discrete Masses 275
  - 7.4.2 Examples of Continuous System 278
- 7.5 ISOLATED SYSTEMS AND CONSERVATION OF MOMENTUM 282
- 7.6 COLLISIONS 286
- 7.7 COLLISIONS IN THE CM FRAME 291
- 7.8 SYSTEMS WITH TIME VARYING MASS 294

	7.8.1	Rocket Motion With no External Force 295
	7.8.2	Rocket Motion With Constant External Force 299

7.9 EXERCISES 300

7.10 PROBLEMS 306

8 WORK, ENERGY AND POWER 309

8.1 WORK 310

    8.1.1 General Formula for Work 313

8.2 WORK-ENERGY THEOREM 319

    8.2.1 Integrating Equation of Motion of One Particle in One Dimension 319

    8.2.2 Integrating Equation of Motion of One Particle in Three Dimensions 321

    8.2.3 Work-Energy Theorem Applied to Multiparticle Systems . 322

    8.2.4 Uses of Work-Energy Theorem 324

    8.2.5 Work-Energy Theorem For Conservative Forces 325

8.3 POTENTIAL ENERGY FUNCTIONS 327

    8.3.1 Potential Energy Associated with Gravity 328

    8.3.2 Potential Energy Associated with the Spring Force 331

    8.3.3 Potential Energy Associated With the Universal Gravitational Force 333

    8.3.4 Force and Potential Energy Function 336

8.4 CONSERVATION OF ENERGY 336

8.5 POWER 339

8.6 ELASTIC COLLISIONS 340

8.7 EXERCISES 343

8.8 PROBLEMS 348

9 FIXED-AXIS ROTATION 351

9.1 KINEMATICS OF PURE ROTATION 352

9.2 ANGULAR MOMENTUM 363

    9.2.1 Angular Momentum of a Point Particle 363

    9.2.2 Angular Momentum of an Extended Body and the Moment of Inertia 366

    9.2.3 Calculations of Moments of Inertia 369

9.3 DYNAMICS OF FIXED-AXIS ROTATION 379

    9.3.1 Rotational Dynamics of a Single Particle 380

    9.3.2 Rotational Dynamics of Extended Bodies 381

	9.3.3	The Law for Fixed Axis Rotation 382
	9.3.4	Practice With Torque Calculations 382
	9.3.5	Example Problems - Single Rigid Bodies 383
	9.3.6	Example Problems - Coupled Systems 386
9.4	CONSERVATION OF ANGULAR MOMENTUM 391
	9.4.1	For One Particle 391
	9.4.2	Conservation of Angular Momentum For Extended Bodies . 393
9.5	ROTATIONAL WORK AND KINETIC ENERGY 396
9.6	ROLLING MOTION IN A STRAIGHT LINE 397
9.7	EXERCISES 403
9.8	PROBLEMS 411

10 NONINERTIAL FRAMES 415

- 10.1 ACCELERATING FRAME 416
  - 10.1.1 Kinematics in Accelerating Frame 416
  - 10.1.2 Newton's Second Law in Accelerating Frame 417
- 10.2 ROTATING FRAME 420
  - 10.2.1 Kinematics in a Uniformly Rotating Frame 420
  - 10.2.2 Newton's Second Law in Uniformly Rotating Frame 425
  - 10.2.3 Newton's Second Law in Earth's Frame 425
- 10.3 CORIOLIS FORCE 430
- 10.4 EXERCISES 432
- 10.5 PROBLEMS 436

Chapter 1

INTRODUCTION

Contents

1.1	THE WORLD OF PHYSICS	2
1.2	FUNDAMENTAL UNITS	4
	1.2.1 Length	4
	1.2.2 Time	5
	1.2.3 Mass	7
	1.2.4 Relation Between Metric and Imperial Units	8
	1.2.5 Other Metric Units	9
	1.2.6 The SI AND CGS Systems of Units	9
	1.2.7 Conversion of Units	10
1.3	UNCERTAINTY IN MEASUREMENTS	12
	1.3.1 Uncertainty and Precision	12
	1.3.2 Uncertainty and Significant Figures	13
	1.3.3 The Scientific Notation For Numbers	15
	1.3.4 Rounding Off	15
	1.3.5 Uncertainty and Accuracy	16
1.4	PROPAGATION OF UNCERTAINTIES	16
1.5	ORDER OF MAGNITUDE ESTIMATES	19
1.6	DIMENSIONAL ANALYSIS	20
1.7	EXERCISES	22
1.8	PROBLEMS	24

1.1 THE WORLD OF PHYSICS

Welcome to physics - the most fundamental science.

Physics is devoted to the understanding of all natural phenomena. In physics we try to understand physical phenomena at all scales, from the world of subatomic particles to the entire universe. Despite the breadth of the subject, various subfields of physics share a common core. The same basic training in physics will prepare you to work in any area of physics and related areas of science and engineering.

Introductory courses in physics follow the historical development of the subject with the **classical physics** or the pre-1900 physics, followed by the modern physics or the post-1900 physics. Newtonian mechanics is the foundation of classical physics and we will begin with a thorough treatment of the subject. You will learn to examine the world from the mechanical viewpoint of Sir Isaac Newton (1642-1727), the great architect of mechanics and Calculus, among other things. After you have mastered the basics of Newtonian mechanics, you will learn how to apply Newton's laws of motion to such diverse areas as gravitation, fluids, waves and sound.

Part 1 of this textbook is devoted entirely to these topics. Often heat is also included in a course on the classical physics. Part 2 gives an introductory treatment of the laws of thermodynamics. Part 3 covers the subject of electricity and magnetism. Finally, in part 4 of this textbook you will study the fundamentals of optics. A course on Modern physics usually covers special relativity and introductory quantum mechanics and you will be taught these topics in a future course.

A physicist's understanding of the nature is usually expressed in **conjectures, theories, and laws**, which are usually expressed mathematically. Expressing the laws of physics mathematically allows one to investigate and deduce their implications more fully. Therefore, a solid training in mathematics is essential for a student to appreciate and understand the laws of physics. Classical physics uses calculus extensively. This book will use calculus to express the ideas of physics more precisely.

Nature is quite complicated. Over the past several centuries physicists have discovered a number of laws that have universal applicability but they are not necessarily infallible. The laws of physics summarize the current understanding of the nature. A particular theory's merits are decided on how well its predictions hold up against reproducible experiments or observations. This makes physics an **empirical science**, although physicists do spend much time developing models and theories. In physics, experiments and theory complement one another, and together, they provide the most powerful tools at our disposal for unlocking the mysteries of nature.

1.1. THE WORLD OF PHYSICS

Figure 1.1: The image of the oldest recorded supernova RW86 observed by Chinese in 185 A.D. The bluish color are in the X-ray from NASA's Chandra X-ray Observatory and the European Space Agency's XMM-Newton Observatory and the yellow and red are infrared data from NASA's Spitzer Space Telescope, as well as NASA's Wide-Field Infrared Survey Explorer (WISE). The Chinese had recorded that the guest star remained in the sky for eight months.Credits: X-ray: NASA/CXC/SAO & ESA; Infared: NASA/JPL-Caltech/B. Williams (NCSU).

The **theories** in physics range from the empirical to the abstract. Often a theory is simply a model or a mental image that helps grasp the experimental observations better. Theories may also suggest new experiments to render the physical situation clearer. Physicists also explore theories that seek to unearth the abstract principles that can explain a number of phenomena. Although the words - theory and model - are often used interchangeably, theories are usually understood to mean more general ideas while models usually refer to a particular phenomenon.

When many predictions of a general theory have been tested and found to be correct, the theory gains the status of a **law of physics**. The testing of the predictions of laws is usually limited by the development in technology, which itself is guided by a progress in physics, and/or the human ingenuity. With increased precision, or an access to some previously inaccessible physical conditions, or some hitherto unimagined experiment, the old accepted laws are often tested in new ways.

Sometimes, new tests show flaws in the accepted laws. Since many such exciting discoveries have been made in the past, physicists keep an open mind about their subject and treat the presently accepted laws as tentative. If a law is inconsistent with reliable observations, then either the law is modified or a new theory is invented to replace the old one. In this sense, physics is quite dynamic, changing continuously as our understanding of the world improves. Whether this process of correcting and replacing the old laws with new ones will ever end up with the eventual discovery of "the final laws" is unclear at this time since we do not know if such immutable final laws even exist.

1.2 FUNDAMENTAL UNITS

In order to understand a physical phenomenon, we make careful measurements of the relevant physical quantities. To facilitate communication among people and to build devices that would work with each other, the measurements of physical quantities are expressed in terms of agreed-upon standards. Length, time, and mass are fundamental quantities in mechanics. Other fundamental quantities will be added to this list when we study heat, electricity and optics.

1.2.1 Length

Length is a measure of the distance between two points in space. Before, the French Revolution (1790), different standards of length were used in different countries, and even different localities in the same country; for instance, a Greek foot was approximately 1.012 times the English foot, a Roman foot was approximately 0.97 of an English foot, etc. Furthermore, to make matters worse, the multipliers between different units in common use were not uniform; for instance, there are three feet in a yard, and twelve inches in a foot.

In 1792, the government of France after the French revolution adopted a new system of weights and measures with meter as the fundamental unit of length. The name meter comes from the Latin word metrum and the Greek word metron, both meaning "measure". The meter was defined as 10^{-7} times or one ten-millionth of the distance on the meridian from the North Pole to the equator passing through Paris. The factor was chosen to get a size close to the "human scale". Later it was found that prototypes based on earth-based definition were 0.2 mm too short due to the flattening of earth due to its rotational motion.

In 1889, the first meeting of International Committee for Weights and Measures (CIPM for Comité International des Poids et Mesures) replaced the earth-based meter to the distance between two fine markings on a Platinum-Iridium rod kept at zero degrees Celsius temperature and standard pressure at the International Bureau of Weights and Measures (BIPM, Bureau International des Poids et Mesures) near Paris, France. Accurate copies of the original rod were made and distributed to other standard-keeping laboratories throughout the world. These secondary standards were used to produce more accessible copies such as meter rulers for the general public.

1.2. FUNDAMENTAL UNITS

With the advancement in optical technology, it became possible to measure lengths more precisely than the fine markings on the standard meter rod. In 1960 a new standard for meter based on the wavelength of orange-red light emitted by Krypton-86 in a gas discharge tube was adopted. The meter was redefined to be equal to $1,650,763.73$ wavelengths of this light.

To reduce further uncertainty in measurements, in 1983 General Conference on Weights and Measures (CGPM, Conférence Générale des Poids et Mesures) replaced the definition of meter based on Krypton-86 to the one based on the measurement of the speed of light. An exact value of $299,792,458$ m/s for the speed of light is assumed which is then used to define the meter.

One **meter** is the length of the path travelled by light in vacuum during a time interval of $1/299,792,458$ of a second.

Note that this way of defining a meter uses the unit of time (second) to define the unit of length. With this definition of the unit of length, the precision of length is now tied with the precision of time measurements. We will see below that the time measurements have become extremely accurate and therefore a better unit to serve as a base unit for other units.

1.2.2 Time

To define time we think of either a natural phenomenon that repeats itself or an experiment that can be performed repeatedly. For instance, oscillations of a pendulum provides a basis for time in an experimental setting while earth's rotation provides a naturally occurring phenomenon that has been used for time measurements.

The SI unit of time is one second. How one second came to be a standard of time is a fascinating story unto itself in the history of science. It is known that ancient civilizations used the apparent motion of celestial bodies across the sky - Sun, Moon, planets and stars - for keeping track of the passage of time and seasons. It is known that Egyptians made time keeping devices and used a similar system as our own.

For instance, the current division of a year into 365 days seems to have come from Egyptian calendar as far back as 3100 BCE (Before Common Era) based on the rising of the "Dog Star" in Canis Major, now called Sirius, next to the sun every 365 days, which coincided with the flooding of Nile. Egyptians had built obelisks (slender, tapering, four-sided monuments) as far back as 3500 BCE whose shadow was used to determine time during the day. The obelisks were like primitive sundials and had markings at the base to indicate the shadows corresponding to the shortest and longest days of the year.

Figure 1.2: Sundial in thyme garden at Minnesota Landscape Arboretum. Photographed June 17, 2007 at 12:21 solar time. Photo-credit: S. E. Wilco, via Wikimedia Commons.

Around 1500 BCE the Egyptians invented the sundial that divided the day from sunrise to sunset into ten parts plus two "twilight hours". Similarly they also divided nighttime into 12 hours thus making a total of 24 hours in a full day. The divisions of an hour into sixty minutes and a minute into sixty seconds is said to have come from the Sumerian culture, which had a sexagesimal system that was based on number 60.

Egyptians also invented water clock or clepsydra before 1500 BCE, the earliest time keeping device not dependent on the motion of celestial objects. Greeks started using water clocks around 325 BCE and built even more impressive and elaborate water clocks. The complexities were added to make the flow of water as steady as possible. Despite these efforts it was very difficult to control the water flow with high accuracy and new mechanical clocks were needed.

Little progress after the Egyptian inventions seems to have been made in time keeping until Galileo Galilei (1564 - 1642) who suggested using the natural period of a pendulum. Although Galileo sketched a design of a pendulum clock, he never constructed it. The first pendulum clock was built by Christian Huygens of Netherlands in 1656. It had an error of less than 1 minute a day and was the most accurate clock to date. Christian Huygens also invented the balance wheel and spring assembly in 1675, which led to the construction of more accurate clocks. The oscillations of the balance wheel, which oscillates at around 5 cycles per second, provide the time standard for the mechanical watch. In 1889 Sigmund Riefler's clock was made that kept time with an error of less than a hundredth's of a second a day and became a standard fixture for astronomers.

With the discovery of piezoelectricity in 1880 by Pierre and Jacques Curie, the Curie brothers, it was found that piezo-crystals of quartz vibrate at a definite frequency when one applies voltage upon them. A vibrating quartz crystal generates an oscillating current of constant frequency that can be determined quite accurately with appropriate electrical circuitry. In 1927 a Canadian-born telecommunications engineer Warren Morrison (1896-1980) invented the quartz watch. Morrison and others demonstrated that the accuracy of time based on quartz crystals far exceeded clocks based on balance wheel and spring assembly. Today inexpensive electronic clocks based on quartz vibration are commonplace.

Figure 1.3: A quartz watch by Seiko.

Despite a better performance of quartz crystals, they are no match for atomic clocks developed in 1940's and 50's. The possibility of atomic clock based on atomic beam magnetic resonance was first suggested in 1945 by I. Rabi of Columbia University (New York). In 1949 the National Bureau of Standards of United States (now called the National Institute of Standards and Technology or NIST) developed the first atomic clock using ammonia molecule. However, the atomic clock based on ammonia molecule was not much better than the existing quartz clocks. In 1955 Louis Essen at the National Physical Laboratory in United Kingdom constructed the

1.2. FUNDAMENTAL UNITS

Figure 1.4: Atomic clock NIST-F1 at National Institute of Standards and Technology, USA.

world's first atomic clock based on the atomic transitions of cesium atoms that had an accuracy of 1 sec in 300 years. The measured time using the atomic clock was compared with the time based on the rotation of earth and found to be much more accurate and stable. Therefore, in 1967 the 13th General Conference of Weights and Measures decided to replace the definition of a second by the following.

One **second** is the duration of $9,192,631,770$ periods of the radiation corresponding to the transition between the two hyperfine levels of the ground state of the cesium 133 atom.

Although there have been improvements in atomic clocks, the definition of a second today is the one adopted in 1967. Atomic clocks are getting better everyday, and in 2010 the reported uncertainty for NIST-F1 clock shown in Fig. 1.4 was merely 3×10^{-16} second in one second, i.e. 1 second in 3 million years - a fantastic precision!

1.2.3 Mass

Mass is a measure of mechanical response of an object. Two objects of equal mass, regardless of their chemical content, shape or size, are accelerated equally when subjected to the same force, and two objects of different masses have different accelerations when subjected to the same force.

The SI unit of mass, the kilogram (kg), is the only base quantity now that is still defined by a physical artifact. The original sample was an alloy made up of 90% platinum and 10% iridium by mass in the shape of a cylinder of height 39 mm and diameter 39 mm in 1879 by George

Matthey of Johnson Matthey and stored at the atmospheric pressure in a special triple-bell jar at BIPM, the International Bureau of Weights and Measures near Paris. The alloy was chosen for its non-corrosive properties and the shape was chosen to correspond to the minimum area for a given volume of a cylinder; the spherical shape would be better for minimizing the surface exposed, but since spheres roll off easily it was decided that a cylinder would serve better. The definition of a kilogram can be given as follows.

One **kilogram** is the mass of the prototype of the kilogram kept at the International Bureau of Weights and Measures.

Forty copies of the original prototype were made in 1882 and distributed to various countries. Copies of these secondary standards were made widely available to the tradesmen and general public. Thus all 1-kg samples are traceable to the international prototype kept at BIPM.

The unit of kilogram defined by an artefact has some intrinsic problems. For instance, the prototype may be damaged or corrode due to oxidation or other wear and tear due to the environment. As a result of these problems, the prototype is said to be gaining approximately 1 micro-gram per year. Therefore, it is hard to keep the standard kilogram constant to a very high degree of precision. Presently, several new methods for defining kilogram are being investigated. A particularly attractive possibility is to define the kilogram based on the mass of a fixed number of molecules of a substance that can be made with high purity. In another method developed at National Institute of Standards and Technology (NIST), force of gravity on a standard kilogram is balanced by magnetic force between two coils in a **Watt balance** (Fig. 1.5).

1.2.4 Relation Between Metric and Imperial Units

Although metric units are preferred in scientific circles, the Imperial units, also called the English units, are also used commonly at least in the United States. Therefore, it is important to note their relations.

The Imperial unit of length such as inch (in), foot (ft), yard(yd) and mile(mi) do not have independent standards. Instead they are now defined in terms of the metric unit of meter. The unit of inch is taken to be exactly 2.54 cm, and the conversion of other units into the metric units are made by first converting them into inches, and then using this exact conversion of inches into centimeters.

$$\boxed{1 \text{ in} = 2.54 \text{ cm} \quad (Exactly).} \tag{1.1}$$

The other units of length in the English system have the following relations

1.2. FUNDAMENTAL UNITS

Figure 1.5: The Watt balance at National Institute of Standards and Technology, USA. (Photo by Richard Steiner)

to inch.

$$1 \text{ ft} = 12 \text{ in}$$
$$1 \text{ yd} = 3 \text{ ft}$$
$$1 \text{ mi} = 1760 \text{ yd}$$

The English unit of mass is a pound (lb) or pound-mass, which is approximately 1 lb = 453.59237 grams ≈ 453.6 grams. The English unit of time is same as the metric unit of time, namely, second.

1.2.5 Other Metric Units

Often a meter, a kilogram, or a second is too large or too small a unit for the physical phenomena under study. In the metric system, multiples of positive and negative powers of 10 are then used to simplify the numerical values to ordinary sizes and unit names are given to various multiples of the fundamental units meter, kilogram, and second in a uniform way by adding prefixes to the names of units (Table 1.1). For instance, one hundredth of a meter is called a centimeter, one thousand meters is a kilometer, and a millionth of a second is called a microsecond.

1.2.6 The SI AND CGS Systems of Units

There are two systems of units based on the metric units that are commonly used in science - the International System (SI) and the centimeter-gram-second (cgs) systems. In the SI system all units are expressed in

Table 1.1: SI prefixes

Factor	Prefix	Symbol	Factor	Prefix	Symbol
10^1	deka-	da	10^{-1}	deci-	d
10^2	hecto-	h	10^{-2}	centi-	c
10^3	kilo-	k	10^{-3}	milli-	m
10^6	mega-	M	10^{-6}	micro-	μ
10^9	giga-	G	10^{-9}	nano-	n
10^{12}	tera-	T	10^{-12}	pico-	p
10^{15}	peta-	P	10^{-15}	femto-	f
10^{18}	exa-	E	10^{-18}	atto-	a
10^{21}	zetta-	Z	10^{-21}	zepto-	z
10^{24}	yotta-	Y	10^{-24}	yocto-	y

meter, kilogram and second while in the cgs system they are in centimeter, gram and second. The SI system is dominant in textbooks while cgs is more convenient for laboratory experiments. The SI system for mechanical properties is also called MKS or meter-kilogram-second system.

All mechanical properties can be expressed in terms of length, mass and time. Therefore, their units can be related to the units of length, mass and time. For this reason, the units meter, kilogram and second are called <u>fundamental units</u> while other units are said to be derived from them and are called <u>derived units</u>. There are other properties that are not related to motion of objects, such as an electric charge or temperature of an object and therefore, the set of fundamental units or base units contains additional quantities. Presently, there are seven base standards in the SI system of units as shown in Table 1.2. In the initial part of the book we will be using only the units of length, time and mass.

Table 1.2: Base SI units

Quantity	SI Unit Name	Unit Symbol
Length	meter	m
Time	mecond	s
Mass	kilogram	kg
Electric current	ampere	A
(Thermodynamic) Temperature	kelvin	K
Amount of substance	mole	mol
Luminous intensity	candela	cd

1.2.7 Conversion of Units

Conversion of units is frequently needed in physics and you should make sure you have a method of calculation that works consistently for you. A

1.2. FUNDAMENTAL UNITS

particularly useful method is to multiply the given number by a fraction whose value is 1. The fraction is chosen so that numerator and denominator are different units for the same physical quantity. For instance, if we need to change minutes to hours, we will need a fraction of hours to minutes or minutes to hours depending on whether minute to be changed is in the numerator or the denominator respectively. We now illustrate this method of changing units with examples.

Example 1.2.1. Changing one unit only. How many seconds are in 20 minutes?

Solution. You can start with 20 minutes and then multiply with a fraction of second [the unit desired] to minute [the unit to be converted] whose value is 1. The fraction is obtained by the conversion of 60 seconds in 1 minute.

$$
\begin{aligned}
20 \text{ min} &= 20 \text{min} \times \frac{60 \text{ sec}}{1 \text{ min}} \\
&= 20\cancel{\text{min}} \times \frac{60 \text{ sec}}{1 \cancel{\text{min}}} \\
&= 1200 \text{ sec}.
\end{aligned}
$$

Example 1.2.2. Changing product of units. A rectangular board is 5 in by 2 in. The are of the board is 10 square inches. How many square centimeters in 10 square inch?

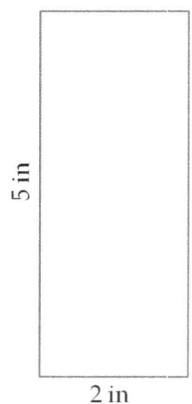

Solution. A square inch is a unit of area. Note that square inches stands for inches times inches. It is usually helpful to write out square inches this way before become proficient at conversions and can skip this step. We need a fraction for each inch to be converted. The fraction needed has *cm* at the numerator and *in* at the denominator, which we can construct from the fact that there are 2.54 *cm* in 1 *in*.

The calculation is presented below. You should note that when I used the calculator I got way too many digits than the three digits in 2.54 that went into the calculation, so I put in another step where the final answer was rounded off to three digits. This happens a lot in calculations and you should learn to round off numbers to correct number of digits. I will show you later a consistent way of rounding off that is commonly practised.

$$
\begin{aligned}
10 \text{ in}^2 &= 10 \text{ in} \times \text{in} \\
&= 10 \text{ in} \times \text{in} \times \frac{2.54 \text{ cm}}{1 \text{ in}} \times \frac{2.54 \text{ cm}}{1 \text{ in}} \quad \text{(no change)} \\
&= 10 \cancel{\text{in}} \times \cancel{\text{in}} \times \frac{2.54 \text{ cm}}{1 \cancel{\text{in}}} \times \frac{2.54 \text{ cm}}{1 \cancel{\text{in}}} \\
&= 10 \times 2.54 \text{ cm} \times 2.54 \times \text{ cm} \\
&= 64.516 \text{ [from calculator] cm}^2 \\
&= 64.5 \text{ cm}^2.
\end{aligned}
$$

Example 1.2.3. Changing product of units. Convert 1.5 lb mi/h^2 into kg m/s^2?

Solution. This example is definitely messier than the previous ones. We will need four fractions, one for each factor of the units; we will combine the two factor of hours into one to save space. We will also need to round off the final answer. Here is the gory details of the calculation.

$$\begin{aligned} 1.5 \text{ lb mi/h}^2 &= 1.5 \text{ lb mi/h}^2 \times \frac{0.4536 \text{ kg}}{1 \text{ lb}} \times \frac{1.6093 \text{ km}}{1 \text{ mi}} \times \frac{1 \text{ h}^2}{3600^2 \text{ s}^2} \\ &= 1.5 \text{ l\!b mi\!/\!h}^2 \times \frac{0.4536 \text{ kg}}{1 \text{ l\!b}} \times \frac{1609 \text{ m}}{1 \text{ m\!i}} \times \frac{1 \text{ h}^2}{3600^2 \text{ s}^2} \\ &= 8.4488... \text{ [from calculator] kg m/s}^2 \\ &= 8.4 \times 10^{-5} \text{kg m/s}^2. \text{ [Rounding off to two digits.]} \end{aligned}$$

Note that I rounded off the final number to two digits - I did that because the lease precise number was 1.5 which has two significant digits.

1.3 UNCERTAINTY IN MEASUREMENTS

1.3.1 Uncertainty and Precision

Recall that the time measured by the NIST-F1 atomic clock is highly precise, the uncertainly being only 3×10^{-16} sec in 1 sec. Even though the NIST atomic clock is highly precise, it is not 100% precise. It is a fact of every measuring device that no measurement is 100% precise. The uncertainties in measurements results from either the difficulties in characterizing the sample perfectly or in the precision of the measuring instruments, or both.

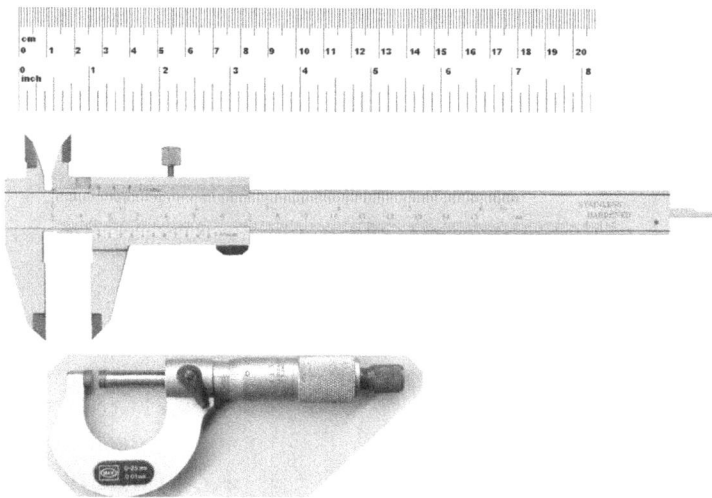

Figure 1.6: The ruler, vernier caliper and micrometer measure lengths to increasing precisions.

The uncertainties are reported in one of the two equivalent ways: **absolute uncertainty** and **relative uncertainty**. The absolute uncertainty

1.3. UNCERTAINTY IN MEASUREMENTS

is also called the **absolute error** since the absolute uncertainty is indicated by the absolute value of the error in measurements. Suppose we measure the time of swing of a pendulum and find that most values fall between 9 sec and 11 sec with the average value of 10 sec and a spread of 0.4 seconds symmetrically around the average value. The spread here, for example, may come from the standard deviation of the values or some other way you are able to estimate how the observed values are distributed. We write our observation as:

$$10 \text{ sec} \pm 0.4 \text{ sec}.$$

The spread written this way is called absolute uncertainty(error). The problem with the absolute uncertainty(error) is that we do not know how bad the uncertainty is relative to the main value. *The relative uncertainty(error) expresses the absolute uncertainty as a percentage of the main value, which is also called* **the percentage relative error**. Thus, in our example, the relative uncertainty(error) would be

$$0.4/10, \text{ or, } 4\%$$

. Therefore, the reading of the period can be expressed in two alternate ways:

Absolute uncertainty(error) notation: $10 \text{ sec} \pm 0.4 \text{ sec}$ *or* (10 ± 0.4) sec.

Same reading in relative uncertainty(error) notation: 10 sec, percent relative error 4%.

Although there is an uncertainty in every measurement, I will not be indicating any uncertainty in values given in problems or tables in this book for the sake of brevity. When an uncertainty is not specified, the general practice is to regard the last digit to be uncertain by one or two units. For example, if the length of a rod is given to be 55.7 cm then it is usually assumed that the length is between 55.6 cm and 55.8 cm using one unit of the last digit as uncertainty. I will follow this practice when truncating the digits in numerical problems.

Uncertainty not specified: 55.7 cm will mean $55.5 \text{ cm} \pm 0.1$ cm.

1.3.2 Uncertainty and Significant Figures

The uncertainty in the value of a physical quantity refers to a lack of complete knowledge about that quantity. For instance, a reading of 2.5 cm $\pm$ 0.1 cm for the length of an object says that we are uncertain about the value 2.5 cm by an amount 0.1 cm. Clearly, a claim better than 0.1 cm in this case is not meaningful. We cannot have a reading of the main value more precise than the uncertainty will allow. For instance, a reading of 2.51 cm $\pm$ 0.1 cm is meaningless since it claims that the average value expected is known up to 0.01 cm while it is uncertain by 0.1 cm - now, think about it, how can a quantity be known to a precision of 100th place and uncertain in the 10th place? A normal practice is to round off the uncertainty to one non-zero digit and use that absolute uncertainty figure as a guide to round off the main number noting that the main number can be no more precise than what is allowed by the uncertainty number. **Significant figures** or significant digits refers to the number of digits allowed in the main number. Thus, 2.5 cm $\pm$ 0.1 cm has two significant digits and 2.54 cm $\pm$ 0.02 cm has three.

In numerical calculations where uncertainty is not explicitly stated it will be understood that the last digit displayed is uncertain. We will pay particular attention in our calculations so that the final result we report as our answer does not have too many insignificant digits or that we are not rounding off at too few significant figures. It is also advisable that you keep one or two extra digits in the intermediate steps of a calculation since they tend to effect the final outcome sometimes, and only in the end, round off the final number to the appropriate significant figures. The calculation of the uncertainty in a quantity that depends on other quantities is a complicated matter and you will find simple examples later in the chapter that illustrate how uncertainties propagate from measured to derived quantities. Here, we make do with a simple rule that captures the spirit of significant figures.

<u>The least precise value in an expression tends to control the maximum number of significant figures left at the end of a calculation.</u>

Example 1.3.1. Illustration of simple rule of significant figures
The length, width and height of a rectangular parallelepiped are given to be 1.5 cm, 0.600 cm, and 0.105 cm. What is the volume of the parallelepiped up to the correct number of significant figures using the simple rule of significant figures?

Solution. To apply the simple rule of significant figures in a calculation, we need to identify the least precise number in the input numbers. Here the length of 1.5 cm has two significant figures, the width of 0.600 cm and height of 0.105 cm each have three. Thus, the least precise number has two significant digits. Therefore, the volume will have no more than two significant digits.

$$\begin{aligned} \text{Volume} &= 1.5 \text{ cm} \times 0.600 \text{ cm} \times 0.105 \text{ cm} \\ &= 0.0945 \text{ cm}^3 \text{ (from calculator; incorrect number of digits)} \end{aligned}$$

Now we need to round off the number obtained above to the correct number of significant digits, which is two here. The final number 0.0945 has one digit to the left of the decimal and four digits to the right of the decimal. The decimal is readily moved to get rid of the leading zeros if we write the number using a power of ten as a multiplier. Therefore, the only digits which matter are the contiguous non-zero digits, i.e., 9, 4 and 5 here.

We seek a two-digit approximation of number 945: the digit 4 is called

1.3. UNCERTAINTY IN MEASUREMENTS

the least significant digit. Now, we need to decide about rounding off the digits to the right of the left significant digit: will 945 be rounded off to 940 or 950? Since we have a 5 after the least significant digit 4 here, we can round up the number to 950.

$$\text{Volume} = 0.095 \text{ cm}^3 \text{ or } 9.5 \times 10^{-2} \text{ cm}^3$$

1.3.3 The Scientific Notation For Numbers

In the scientific notation, we use powers of 10 to write a decimal number. This helps us avoid any ambiguity in the reporting of the precision in the data. For example, 36000 is written as 3.6×10^4 and 36000. as 3.6000×10^4, which clearly shows that 36000 has two significant digits and 36000. has five.

In the scientific notation, it is a normal practice to force as many powers of ten (positive or negative) as necessary to make the multiplier of ten contain only one non-zero digit from 1 and 9 in the non-decimal part, and place the required trailing zeros on the right side of decimal to display appropriate number of significant figures.

A major advantage of the scientific notation is that it permits us to easily read off the significant figures in the decimal number. Thus, 1.5×10^{-3} m has two significant digits, 9.05×10^6 s has three, and 9.050×10^3 kg has four. Note that in the scientific notation, the decimal number multiplying a power of 10 is greater or equal to 1 and less than 10. The following examples illustrate the standard scientific notation and the same numbers not in standard scientific notation.

Scientific	1.5×10^{-3} m	9.05×10^6 s	9.050×10^3 kg
Non-scientific	0.0015 m	90.5×10^5 s	9050. kg

1.3.4 Rounding Off

In numerical calculations, we must often round off numbers to display the result up to the correct number of significant digits. This requires making decision about the least significant digit. We use the following rules for whether or not to increment the least significant digit when eliminating the insignificant part of a number.

- If the fraction to the right of the least significant digit is greater than $\frac{1}{2}$, then we increment the least significant digit by one.

- If the fraction to the right of the least significant digit is less than $\frac{1}{2}$, then we do not increment the least significant digit.

- If the digit to the right of the least significant digit is equal to 5, then we increment the least significant digit only when it is odd.

This leads to incrementing of least significant digit in a group of numbers only half the time, and therefore reduces the chance of the introduction of a systematic error by either incrementing all the time or not incrementing all the time. Thus, 3.5 may be rounded up to 4 and 4.5 may rounded down to 4. As long as you practice this consistently, then over a large group of numbers you would not introduce any systematic error which might creep in by rounded up a borderline case all the time.

1.3.5 Uncertainty and Accuracy

The **accuracy** of a measurement does not refer to the precision of the measurement, but instead, to the closeness of the measured value to the "true value", also called the accepted value. The accuracy is often expressed in a similar way to the percentage error, except that our objective now is to compare the best experimental value to the accepted value. The deviation of the experimental value from the accepted value is often written as a percentage, which we may call the **percentage deviation from the accepted value**.

Percent deviation from the accepted value.

$$Percent\ deviation\ from\ the\ accepted\ value = \frac{Accepted\ Value - Experimental\ Value}{Accepted\ Value} \times 100\%.$$

Thus, if you measured the width of a metal plate to be 5.7 cm while it was manufactured to the "exact" specification of 6.0 cm. Then, your percent error will be 5% as obtained from the following calculation.

$$Percent\ deviation = \frac{|6.0 - 5.7|}{|6.0|} \times 100\% = 5\%.$$

The causes of percent deviation may be a defect in the ruler or a faulty procedure of measurement.

1.4 PROPAGATION OF UNCERTAINTIES

In physics, many physical quantities are derived from other quantities that are directly measured in an experiment. How do uncertainties in the measured quantities propagate into derived quantities? Several different situations occur routinely.

Case 1: The derived quantity is a constant times the measured quantity. For instance, if you measure the diameter of a spherical ball to be $D \pm \Delta D$, what would be the radius R of the ball? We know that radius is $1/2$ of the diameter. Therefore, we simply divide the diameter by 2.

$$R \pm \Delta R = \frac{D \pm \Delta D}{2} = \frac{D}{2} \pm \frac{\Delta D}{2}. \qquad (1.2)$$

1.4. PROPAGATION OF UNCERTAINTIES

Therefore, the uncertainty in radius is 1/2 of the uncertainty in the diameter.

Case 2: Two measured quantities are added to get the composite quantity. For instance, suppose you measure the length and width of a rectangle to be $L \pm \Delta L$ and $W \pm \Delta W$ respectively. What would be the uncertainty in the perimeter? The perimeter P of a rectangle is given as 2 times the length plus width.

$$P \pm \Delta P = 2\left(L \pm \Delta L + W \pm \Delta W\right) = 2\left(L + W\right) \pm 2\left(\Delta L + \Delta W\right). \tag{1.3}$$

Therefore, the uncertainty in the perimeter would be two times the sum of the uncertainties in length and width.

Case 3: A measured quantity is raised to a power. For instance, suppose you measure the diameter of a circle to be $D \pm \Delta D$. What would be the area A?

$$A \pm \Delta A = \frac{\pi}{4}\left(D \pm \Delta D\right)^2 = \frac{\pi}{4}D^2 \pm \frac{\pi}{2}D\Delta D + \frac{\pi}{4}(\Delta D)^2. \tag{1.4}$$

Normally, the uncertainty will be much smaller than the main number. Therefore, the term containing the square of the uncertainty in the expansion on the right side can be ignored in comparison with other terms to yield the following.

$$A \pm \Delta A = \frac{\pi}{4}D^2 \pm \frac{\pi}{2}D\Delta D. \tag{1.5}$$

Therefore, the uncertainty in the area of the circle will be $\Delta A = \frac{\pi}{2}D\Delta D$, which shows that the uncertainty in area of a circle not only depends on the uncertainty in the diameter but also the diameter itself.

Case 4: Two or more measured quantities are multiplied together. This is the case when we try to calculate area of a rectangle or volume of a parallelepiped. Suppose you measure the length, width and height of a parallelepiped to be $L \pm \Delta L$, $W \pm \Delta W$ and $H \pm \Delta H$, and you wish to calculate its volume V with uncertainty in the volume ΔV, how would that work?

$$\begin{aligned} V \pm \Delta V &= (L \pm \Delta L) \times (W \pm \Delta W) \times (H \pm \Delta H) \\ &= LWH \pm (WH\Delta L + LH\Delta W + LW\Delta H), \end{aligned} \tag{1.6}$$

where, once again, we have assumed that the uncertainties are much smaller than the main readings so that we can neglect terms that have product of uncertainties with each other, and kept terms that have only one uncertainty factor in them. The largest deviation from the main value occur when all the $\pm$ are $+$ or $-$, that is $(WH\Delta L + LH\Delta W + LW\Delta H)$. Thus, the value of the volume ranges between $V_{min} = LWH - (WH\Delta L + LH\Delta W + LW\Delta H)$ and

$V_{max} = LWH + (WH\Delta L + LH\Delta W + LW\Delta H)$. In practice, we add each of the contributions to the uncertainty in the composite quantity in quadrature, i.e. square them first and then take the square root of the sum as illustrated next. Let ΔV_L is the uncertainty in volume due to uncertainty in length, and similarly for ΔV_W and ΔV_H. Then,

$$\Delta V_L = WH\Delta L.$$
$$\Delta V_W = LH\Delta W.$$
$$\Delta V_H = LW\Delta H.$$

Since each individual source of uncertainty is independent of other uncertainties, the net uncertainty in the volume is constructed by using Pythagorean theorem, giving a diagonal in the space of uncertainties.

$$\Delta V = \sqrt{(\Delta V_L)^2 + (\Delta V_W)^2 + (\Delta V_H)^2}. \quad (1.7)$$

Example 1.4.1. Propagation of uncertainties. The length, width and height of a metal rectangular parallelepiped were measured to be 2.54 cm ± 0.01 cm, 0.500 cm ± 0.005 cm, and 0.208 cm ± 0.001 cm respectively. Find the average volume, and absolute and relative uncertainties in volume.

Solution. The average volume is obtained by simply multiplying the average length, width and height and making sure to keep the appropriate significant figures.

$$\begin{aligned} \text{Volume} &= 2.54 \text{ cm} \times 0.500 \text{ cm} \times 0.208 \text{ cm} \\ &= 0.264 \text{ cm}^3 = 2.64 \times 10^{-1} \text{ cm}^3. \end{aligned}$$

The uncertainty in volume will be constructed from uncertainty in volume due to uncertainty in each of the measured quantities. Thus the uncertainty in volume due to uncertainty in length is

$$\begin{aligned} \Delta V_L &= WT\Delta L \\ &= 0.500 \text{ cm} \times 0.208 \text{ cm} \times 0.01 \text{ cm} \\ &= 0.001 \text{ cm}^3 \ (keeping\ only\ the\ most\ significant\ digit). \end{aligned}$$

Similarly, the uncertainty in volume due to the uncertainty in width is 0.003 cm^3, and the uncertainty in volume due to the uncertainty in thickness is 0.001 cm^3.

Now we add the three sources of uncertainties in the volume in quadrature to obtain the net uncertainty.

$$\Delta V = \sqrt{(\Delta V_L)^2 + (\Delta V_W)^2 + (\Delta V_T)^2} = 0.003 \text{ cm}^3.$$

Hence, the volume of the block is $(2.64 \pm 0.03) \times 10^{-1}$ cm^3.

The relative uncertainty in volume is obtained from the ratio of absolute uncertainty to the average value.

$$\text{Relative uncertainty} = \frac{\Delta V}{V_{ave}} = \frac{0.03}{2.64} = 1\%.$$

1.5 ORDER OF MAGNITUDE ESTIMATES

It is a common misconception that physics is an exact science. One of the most important skills in developing an understanding of a physical phenomenon is the ability to understand it qualitatively, and figure out roughly how things might work. The process of determining a reliable rough estimate usually involves identification of correct principles and a good guess about the relevant variables. Estimating based on physical principles is very useful in developing physics intuitions. Note that estimating does not mean randomly guessing a number or a formula, but it means employing reasonable physics to relate variable which ought to be related based on sound physical reasoning. Enrico Fermi of the University of Chicago was famous for making rough estimates in the so-called "back of the envelop" calculations, and therefore these types of estimations are also called Fermi Problems.

Often, it is possible to estimate the value of an unknown quantity within a factor of 10 of the exact answer. A factor of 10 is also called one order of magnitude. A factor of 100 corresponds to two orders of magnitude and so forth.

To make some progress in estimating, you need to have some definite ideas about how variables may be related. The following strategies can help you in practising the "art of guessing".

- When dealing with an area or a volume of a complex object, introduce a simple model of the object such as a sphere or a cube.

- Guess the linear dimension, such as the radius of the sphere first, and then use your guess to obtain the volume or area.

- There is no need to go beyond one significant figure.

- Finally, check to see if your answer is reasonable. If you get some wacky answer, check to see if your units are right.

Example 1.5.1. Counting marbles in a jar. Estimate the number of marbles in a jar that is 10 cm in diameter and 20 cm tall as shown.

Solution. Counting the marbles in the Fig. 1.7 across we find that there are approximately 10 marbles in the diameter of the jar. Now, since the jar is 10 cm in diameter, we approximate the diameter of one marble to be 1 cm.

Gauss proved that when spheres are packed as well as it can be, they occupy 74% of the space [You need this information from geometry to make a better guess. But if you didn't know this, you could still make a reasonable guess about the percentage of space occupied by the marbles, and that would be OK.] Now we know how to estimate the number of marbles in the jar: divide 74% of the volume of the jar by the volume

Figure 1.7: Marble Jar

Figure 1.8: Provervial spherical cow.

of one marble. The actual number of marbles is probably less than this number since you would not get the optimal packing assumed here. I leave the final number for you to work out.

Example 1.5.2. The spherical cow. Estimate the area of cowhide on the body of one thousand cows.

Solution. The area of the surface of a cow is very difficult to figure out exactly. But it is possible to approximate a cow by a sphere of diameter about 1 meter for the purposes of surface area as indicated in Fig. 1.8. We could also assume a cow as a cube of 1-meter side with more or less the same result. Multiplying the surface area of one cow by 1000 we get an estimate of the area of the cowhide of 1000 cows.

$$A = \frac{4\pi (1 \text{ m})^2}{\text{cow}} \times 1000 \text{ cows} \longrightarrow 10^4 \text{ m}^2.$$

1.6 DIMENSIONAL ANALYSIS

Figure 1.9: James Clerk Maxwell (1831 - 1879) made invaluable contributions to the theories of electricity and magnetism.

Dimensional analysis is a powerful tool for studying the dependence of a physical quantity on the dimensions and properties of the system. In the most basic form, dimensional analysis is about checking the units in a physics equation: **you should get the same units on both sides of an equation if the equation is valid**. But, you can often go beyond just checking the units, and discover interdependence among physical quantities if you apply dimensional analysis to a given physical situation. In this application, dimensional analysis often helps you guess the right physics as we will see in the example below.

Following the Scottish physicist James Clerk Maxwell (1831 - 1879), we denote the dimension of a physical quantity by enclosing its symbolic name in square brackets. Thus, dimension of time is denoted by $[T]$, length by $[L]$, and mass by $[M]$. There are five base dimensional quantities corresponding to five base units: length $[L]$, time $[T]$, mass $[M]$, temperature $[\theta]$, and electric current $[I]$. The angle in radian does not have any dimensions, as you can easily see from the angle subtended by an arc of a circle: the angle in radian subtended by an arc is equal to the ratio of arc length to the radius of the circle.

The dimensions of other physical quantities are determined by first expressing their units in base units, and then by replacing the unit names by the corresponding base physical quantities. For instance, the unit of speed is meter per second, which means that its dimension is $[L]/[T]$, i.e length over time. Similarly, the dimensions of density are $[M]/[L]^3$, and electric charge is $[I][T]$. The dimensions of some commonly encountered mechanical quantities are listed in Table 1.3. You do not need to memorize these dimension yet. You will encounter all of these quantities in future chapters. Table 1.3 is included here as a reference that can be used to solve problems in this chapter to get a feel for how dimensional analysis

1.6. DIMENSIONAL ANALYSIS

works in calculations.

Table 1.3: Dimensions and SI units of common mechanical quantities

Quantity	Dimension	SI unit
Mass	[M]	kg
Time	[T]	s
Velocity	$[L][T]^{-1}$	m.s^{-1}
Acceleration	$[L][T]^{-2}$	m.s^{-2}
Angle	Dimensionless	rad*
Angular velocity	$[T]^{-1}$	s^{-1}, or, rad.s^{-1}
Density	$[M][L]^{-3}$	kg.m^{-3}
Momentum	$[M][L][T]^{-1}$	kg.m.s^{-1}
Force	$[M][L][T]^{-2}$	kg.m.s^{-2} = Newton = N
Work, Energy	$[M][L]^2[T]^{-2}$	kg.m^2.s^{-2} = Joule = J
Torque	$[M][L]^2[T]^{-2}$	kg.m^2.s^{-2} = N . m
Power	$[M][L]^2[T]^{-3}$	kg.m^2.s^{-3} = Watt = W
Pressure, Stress	$[M][L]^{-1}[T]^{-2}$	kg.m^{-1}.s^{-2} = Pascal = Pa

*rad or radian is not a unit since it comes from the ratio of two lengths, the arc length and the radius, the result of which is a unitless quantity we call radian.

Example 1.6.1. Guessing the formula for frequency of a pendulum. The frequency of the pendulum is the number of cycles it makes in unit time. How does the frequency of a pendulum depend upon its length, mass, angle of swing and force of gravity?

Solution. The dimensional analysis can be used to find a formula for the frequency f of a pendulum of length l, and mass m, which swings between angles $\pm\theta$ radians of the vertical axis. Because pendulum swings as a result of gravity, we need to include the acceleration due to gravity g as one of the possible variables. We assume that the mass of the string is negligible compared to the mass of the pendulum bob. Our task is to find the frequency, f, as a function of m, l, θ, and g.

$$frequency,\ f = f(m, l, \theta, g).$$

Now, we anticipate that frequency would go as some power of each of the physically relevant variables.

$$[f] = h(\theta) \times [l]^a \times [m]^b \times [g]^c. \qquad (1.8)$$

where $h(\theta)$ is dimensionless since angle θ is dimensionless, and exponents a, b, and c are to be determined. Dimensional analysis would not help us with the form of the function $h(\theta)$ since θ is dimensionless. Now, let us write out the dimensions of all physical quantities I have listed.

$[f] = 1/[T];\ [l] = [L];\ [m] = [M];\ [g] = [L]/[T]^2;\ [\theta] = Dimensionless.$

Now, putting the dimensions in Equation 1.8 we find

$$\frac{1}{[T]} = h(\theta) \times [L]^a \times [M]^b \times \frac{[L]^c}{[T]^{2c}}.$$

Equating the exponents of $[L]$, $[T]$ and $[M]$ on the two sides of the equation gives us the following relations among a, b and c. If a particular dimension is missing on one side of the equation, e.g. $[M]$ is missing on the left side, then the exponent of that quantity would be zero on that side of the equation.

$$a + c = 0; \ b = 0; \ 2c = 1, \implies a = -\frac{1}{2}; \ b = 0; \ c = \frac{1}{2}.$$

Therefore, we find that the frequency of a pendulum is

$$f = h(\theta)\sqrt{\frac{g}{l}}.$$

To appreciate the power of dimensional analysis, compare the following exact answer for the frequency of a pendulum obtained from a more difficult calculation based on Newton's second law of motion and small angle approximation.

$$f = \frac{1}{2\pi}\sqrt{\frac{g}{l}}.$$

The dimensional analysis gave us the important part of physics that frequency of a pendulum does not depend on its mass, and it is inversely proportional to the square root of length of the pendulum with very little effort on our part.

1.7 EXERCISES

Unit Conversion Practice

Ex 1.7.1. A car travels at a speed of 65 miles per hour (mph). What is the speed in meters/second?

Ex 1.7.2. In Astronomy, a common unit of distance is a light-year (ly), which is equal to the distance that light travels in one year, which is taken to be 365.25 days. How much is one ly in meters?

Ex 1.7.3. The density of aluminum is 2.7 grams per cubic-centimeter. (a) What is the density in kilogram/cubic meter? (b) What is the density in lb/ft^3?

Ex 1.7.4. Astronomical unit (AU) is a measure of distance that is equal to the average distance between the Earth and the Sun. One AU is approximately equal to 150 million kilometers. Evaluate the average distances between the Sun and following planets, Mercury, Mars, and Jupiter, in AU if these distance in meter are 5.79×10^{10} m, 2.28×10^{11} m and 7.78×10^{11} m respectively.

Significant Figures and Uncertainties

Ex 1.7.5. Evaluate the following to the correct number of significant figures/digits. (a) $2.52 + 1.008$, (b) 3.01×5, (c) $2.51/\pi$, (d) $\pi(1.5)^2$, (e) $3.5^{1/3}$, (f) $\sqrt{5.34}$, (g) $15.5 \times (1.1)^{1/4}$, (h) π^2 (Trick question!).

Ex 1.7.6. The side of a square plate is measured to be 2.1 cm $\pm$ 0.1 cm. (a) What are the absolute and relative uncertainties in the measured value of the side? (b) Find the perimeter and the area of the square to the approprtiate significant figures. (c) Find the absolute and relative uncertainties in the perimeter and the area of the square.

Ex 1.7.7. The radius of a circle was measured to be 3.55 cm $\pm$ 0.05 cm. (a) Find the circumference and the area of the circle to the approprtiate significant figures. (b) Find the absolute and relative uncertainties in the circumference and the area of the circle.

Ex 1.7.8. Use a ruler with mm markings to measure the side of the given square in Fig. 1.10, and find the average values and uncertainties in the values of of the perimeter and the area. (Note: Your measurements will have an average value and an uncertainty, which you will use to deduce the average values and uncertainties in the perimeter and the area.)

Figure 1.10: Exercise 1.7.8

Ex 1.7.9. Use a ruler with mm markings to measure the diameter of the given circle in Fig. 1.11, and calculate the average and the uncertainties of circumference and area. (Note: Your measurements will have an average value and an uncertainty.)

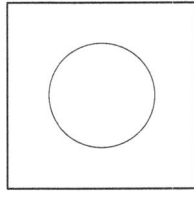

Figure 1.11: Exercise 1.7.9

Ex 1.7.10. The diameter of a spherical ball of mass 20.0 ± 0.1 grams was measured by a micrometer and found to be 16.582 mm $\pm$ 0.002 mm. (a) Find the average volume of the sphere. (b) Find the absolute and relative uncertainties of the volume. (c) Find the density of the ball, giving both the average value and the uncertainty.

Ex 1.7.11. The diameter of a spherical steel ball was measured by a Vernier caliper to be 10.35 mm $\pm$ 0.01 mm. Assume the density of steel to be 7.8 g/cm^3. (a) Find the average volume of the sphere. (b) Find the absolute and relative uncertainties in the volume. (c) Find the mass of the sphere, giving both the average value and the uncertainty.

Ex 1.7.12. The height and diameter of a platinum cylindrical rod are found to be 39.000 mm $\pm$ 0.001 mm and 390 mm $\pm$ 1 mm respectively. Find its volume, giving both the average value and the uncertainty.

Order of Magnitude

Ex 1.7.13. Estimate the amount of blood pumped by your heart in a day.

Ex 1.7.14. Estimate the total number of hair on your head.

Ex 1.7.15. Estimate the total mass of all the water in the oceans of the Earth.

Ex 1.7.16. Estimate the amount of gasoline used in cars each year in the United States of America.

Dimensional Analysis

Ex 1.7.17. A physics student claims that he has found a new force that depends on the density D, the velocity v, and the acceleration a as given by the following relation.
$$F = D\, v^6/a^2.$$
(a) Check the dimensions to decide if the equation makes sense. (b) If this force depends only D, v, and a, predict the form of the relation by finding the correct powers for each of these quantities in the magnitude of the force.

Ex 1.7.18. Hooke's law gives the magnitude of the force of a spring by $F = kx$, where x is the stretching or compression and k the spring constant. Find the dimensions of the spring constant.

Ex 1.7.19. By using dimensional analysis, find a formula for the oscillation frequency of a mass m attached to a spring of spring constant k. Note that the magnitude of the spring force is $F = k\Delta l$, where Δl is the stretching or compression of the spring.

1.8 PROBLEMS

Problem 1.8.1. In Astronomy, a unit of distance, called parsec (pc), is in common use. It is defined to be the distance at which an object of size 1 Astronomical Units (AU) ($\approx 150 \times 10^6$ km) will subtend an angle of 1 arc-second, which is equal to 1/3600 of one degree. Two stars in a binary star system separated by 8.6×10^{14} m are seen to subtend an angle of 0.2 arcsec. How far away are the stars (a) in pc, and (b) in meters.

Problem 1.8.2. A boat is seen to be 2 meters under water when it is 4 km from the shore. Use this data to estimate the radius of earth.

Problem 1.8.3. The British physicist G. I. Taylor argued that the radius R of a spherically symmetric nuclear explosion must depend on the energy E, the initial density of air D, and time t since explosion. Using dimensional analysis, find a formula for the radius at time t after the explosion. [Challenging problem]

Problem 1.8.4. By dimensional analysis, find a formula for the oscillation frequency of a star of radius R and density D. Note that you will also need the dimensions of Newton's gravitational constant, which is $[G_N] = [L]^3[T]^{-2}[M]^{-1}$. Skip this problem if you do not know Newton's gravitational force and G_N. You will learn about this force in a coming chapter.

Chapter 2

VECTORS

Contents

2.1	MOTIVATION FOR VECTORS		26
2.2	GEOMETRICAL VIEW OF VECTORS		27
	2.2.1	What is a Vector?	27
	2.2.2	Multiplication of a Vector by a Scalar	28
	2.2.3	Unit Vector	28
	2.2.4	Addition of Vectors	29
	2.2.5	Subtraction of Vectors	31
	2.2.6	Vector Equations and Polygons	32
	2.2.7	Multiplication of a Vector with Another Vector	32
	2.2.8	Scalar Product Or Dot Product	33
	2.2.9	Vector or Cross Product	35
2.3	ANALYTICAL VIEW OF VECTORS		37
	2.3.1	Base Vectors	37
	2.3.2	Magnitude of Vectors	39
	2.3.3	Directions of Vectors	40
	2.3.4	Adding and Subtracting Vectors Analytically	44
	2.3.5	Scalar Products Analytically	46
	2.3.6	Vector Product Analytically	50
2.4	EXERCISES		51
2.5	PROBLEMS		56

2.1 MOTIVATION FOR VECTORS

Imagine trying to follow the motion of a car on a flat planar surface. Suppose the car starts from some place marked O and goes to point A, 300 m to the north, as given in Fig. 2.1. The car then turns right, and goes 400 m in that direction and arrives at point B. The final point B is

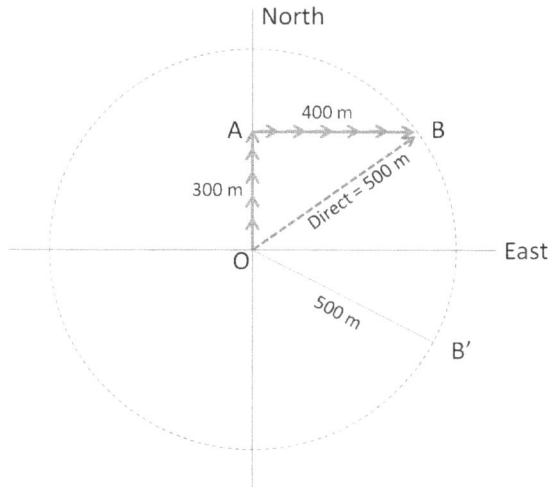

Figure 2.1: The displacement of the car from O to B is give by the direct distance from O to B and the direction from O to B. Note that the direct distance $\overline{OB}$ is not equal to the sum of distances $\overline{OA}$ and $\overline{AB}$.

clearly not 700 m from the starting place O as far as the direct distance $\overline{OB}$ is concerned. Of course, B is 700 m from O on the road $O - A - B$. Using the Pythagoras theorem, we find that the point B is only 500 m away from O. But, if someone said to you that the final place of the car is 500 m away from the starting place, you wouldn't know if the car is at B or at some other place such as B' on the circle of radius 500 m. You know instinctively that you also need a direction in addition to the distance. In a planar motion just one angle with a reference axis is sufficient to nail down the direction. By using elementary trigonometry, we find that B is in the direction of approximately 37° North of East from the starting place O, i.e. $\angle$ East-O-B is approximately 37°.

The direct distance along with the direction of the net movement is called **displacement**. Geometrically, we represent a displacement by an arrow in space whose size represents the direct distance using a definite scale and the direction corresponds to the direction of the net movement. Thus, the displacement of the car shown in Fig. 2.1 is represented by an arrow from O to B and denoted symbolically by drawing an arrow over segment name OB, i.e. by $\overrightarrow{OB}$. This displacement has a magnitude given by the length of the segment $\overline{OB}$ and a direction given by the direction of the arrow from point O to point B.

Notice that the successive displacements $\overrightarrow{OA}$ and $\overrightarrow{AB}$ make up the entire displacement $\overrightarrow{OB}$. Although the lengths $\overline{OA}$ and $\overline{AB}$ do not add to give the length of $\overline{OB}$, we introduce special rule of addition between quantities like displacement so that the displacements $\overrightarrow{OA}$ and $\overrightarrow{AB}$ do add to give the displacement $\overrightarrow{OB}$. The special rule needed to add the displacements is called the parallelogram law of addition. This law can be described as follows.

- Represent a displacement by an arrow of appropriate size.

- If you put the arrow for the second displacement at the end of the arrow for the first displacement, then the arrow from the start to finish is the sum of the two displacements. This way of adding physical quantities that have both a magnitude and a direction is called the **parallelogram law of addition**.

2.2 GEOMETRICAL VIEW OF VECTORS

2.2.1 What is a Vector?

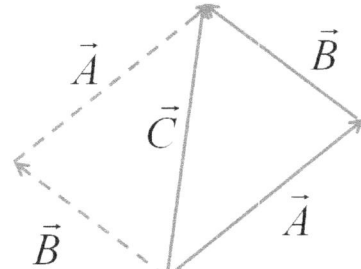

Figure 2.2: The parallelogram law of addition. Here the diagonal $\vec{C} = \vec{A} + \vec{B}$. The other diagonal gives the difference of the two vectors as we will see below.

Vectors are physical quantities that have a direction and a magnitude, and add according to the parallelogram law of addition. Common examples of vectors in mechanics are: (1) the position vector - location of an object in relation to a reference point, (2) the displacement vector - change in position in the course of time, (3) the velocity vector - rate of change in the position vector, (4) the acceleration vector - rate of change of the velocity vector, (5) the force vector, (6) the momentum vector, and (7) the angular momentum vector.

From the list of vectors, it would appear that everything in mechanics is a vector, but that is not the case. Indeed, many important physical quantities are not vectors. For instance time, speed, energy and mass are not vectors. They are scalar quantities. Since vectors obey special rules of addition and multiplication, which are very different than the rules for scalars, you will need to keep track of what is a vector and what is not, so that you will use the appropriate rule. If you use scalar arithmetic rules for vectors, you will definitely get wrong results.

We will use an arrow over letters to denote a vector, for example $\vec{F}$ for a force vector, $\vec{v}$ for velocity vector, $\vec{a}$ for acceleration vector, etc. Occasionally, we will denote a vector by a bold-faced letters, e.g. in **A**. The magnitude of a vector will be denoted by either vertical bars around the vector symbol or by the same symbol without the arrow over it. For instance, the magnitude of the vector $\vec{F}$ will be denoted by either $|\vec{F}|$ or simply F.

Notation: Vectors will have an arrow over the symbol and their magnitude will be denoted by the same symbol without the arrow.

An arrow, whose length corresponds to the magnitude of the vector and whose direction refers to the direction of the vector in space, represents a vector geometrically. The length of the arrow requires a scale so that two arrows for the same physical quantity can be compared as shown in Fig. 2.3.

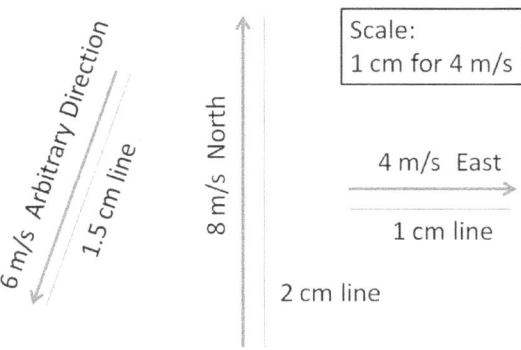

Figure 2.3: Vectors of different magnitudes are drawn using the same scale regardless of their directions.

If two vectors have the same magnitude and the same direction, they are equal no matter where on the paper you happen to draw the two vectors. If two vectors have the same magnitude but different directions, they are not equal. For instance, a displacement of 200 m towards West and a displacement of 200 m East are not equal even though they have the same magnitude since they point in different directions.

2.2.2 Multiplication of a Vector by a Scalar

A scalar is any real number, positive, negative, or zero. The effect of multiplying a vector by a scalar is very different for positive and negative scalars. Multiplication of a vector $\vec{A}$ by a scalar s gives a new vector $\vec{B} = s\vec{A}$. If s is a positive scalar, then $\vec{B}$ has the same direction as $\vec{A}$ but its magnitude is s times the magnitude of $\vec{A}$. That is, $|\vec{B}| = s|\vec{A}|$. If $s > 1$ then $|\vec{B}|$ is s times greater and if $s < 1$ then $|\vec{B}|$ is s times smaller. Note that dividing by the scalar s is equivalent to multiplying by $1/s$, so the division is not a different operation than multiplication. We can obtain a vector of any length in the same direction as the original vector by multiplying it by an appropriate positive real number.

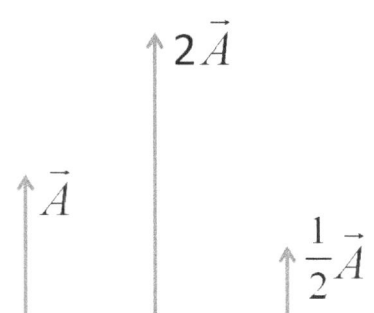

Figure 2.4: Multiplying a vector by positive real numbers gives vectors of different lengths and same direction as the original vector.

Multiplication of a vector $\vec{A}$ by a negative number $(-s)$ [minus s] gives a vector $\vec{C}$ that is not only s times the original vector but also its direction is opposite to the direction of $\vec{A}$. For instance, when velocity vector of 10 m/s pointed North is multiplied by -2, you get another velocity vector that has a magnitude 20 m/s and pointed South.

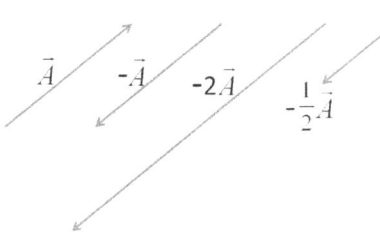

Figure 2.5: Multiplying a vector by a negative numbers gives vectors in the opposite direction.

2.2.3 Unit Vector

Evidently dividing a vector $\vec{A}$ by its magnitude $|\vec{A}|$ one should obtain a vector whose length is 1 with no units and the same direction as the direction of vector $\vec{A}$. This vector is called **unit vector** in the direction of $\vec{A}$. We will denote unit vectors obtained this way by placing a karat over the symbol of the original vector, $\hat{A}$. Thus, unit vector $\hat{A}$ is

$$\boxed{\hat{A} = \frac{\vec{A}}{|\vec{A}|}.} \qquad (2.1)$$

Unit vectors are very useful for constructing vectors of arbitrary lengths. For instance, if we want a displacement vector of length 5 cm in the direction of some unit vector $\hat{u}$, then all we need to do is to multiply $\hat{u}$ by 5 cm, giving the desired vector $(5 \text{ cm})\hat{u}$. Now, a force vector of length 10 N in the same direction as $\hat{u}$ will be $(10 \text{ N})\hat{u}$. You can see that unit vectors have universal applicability regardless of how you construct them. They are holders of the information about directions in space. As you might have guessed it, there are infinite number of unit vectors, one for each direction in space.

2.2.4 Addition of Vectors

Vectors add differently than simple numbers as we have seen in the addition of two displacement vectors above. For instance, two vectors of magnitude 1 units each can add to yield a vector whose magnitude can be any number between zero and two units depending on the relative orientation of the two. That is, $1 + 1$ for vectors can be anywhere between 0 and 2. That is why you need to be extra careful when dealing with vectors.

Let us take another example of addition of vectors. This time, I will use forces on a ring to illustrate the addition process of vectors. Suppose you pull a rigid ring by two forces, $\vec{F_1}$ and $\vec{F_2}$, in opposite directions and with equal strengths [represented in the figure with equal-length arrows] applied at the two opposite ends of the ring as shown in Fig 2.6. It would not come as surprise to you that the ring is not pulled in either direction. This is because the net force on the ring is zero because the vector sum also takes into account the directions of the vectors.

Now, suppose you apply three forces on the ring as in Fig. 2.7(a). This time, you will find that, if the three forces form sides of a triangle as shown in the Fig. 2.7(b), then the ring is not pulled in any direction. On the other hand, if the forces do not form a triangle, then there will be a net pull on the ring in some direction.

We interpret the experiment depicted in Fig. 2.7 as saying that the sum of two of the forces has exactly the same magnitude as the third force but acts in the opposite direction to it. Only then, the sum of the three will give a net zero force on the ring. The triangle diagram tells us about the addition rule of two vectors: if vectors $\vec{F_1}$ and $\vec{F_2}$ are non-collinear, they form adjacent sides of a parallelogram when the tail of one is placed at the tip of the other, and the sum of vectors $\vec{F_1}$ and $\vec{F_2}$ will be the vector

Figure 2.6: Two forces of equal magnitude but opposite directions acting on a rigid ring result in net zero force.

Check for yourself if the sum of vectors $\vec{F_1}$ and $\vec{F_3}$ cancels vector $\vec{F_2}$. Also check if the sum of vectors $\vec{F_2}$ and $\vec{F_3}$ cancels vector $\vec{F_2}$.

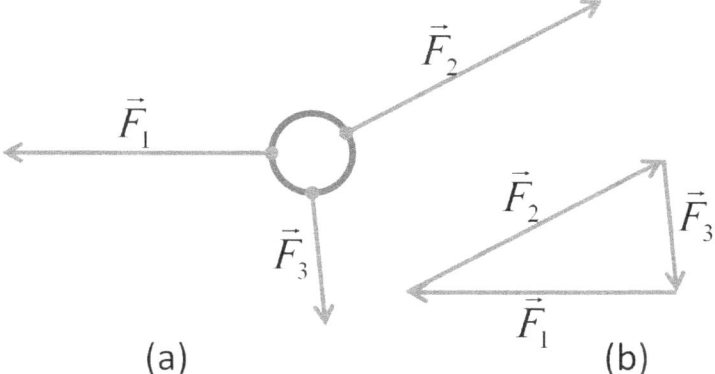

(a) (b)

Figure 2.7: (a) Forces on a ring whose net force is zero. (b) When the three forces are balanced, they form sides of a triangle and vectors make a round trip in the triangle.

along the diagonal of that parallelogram as shown in Fig 2.8.

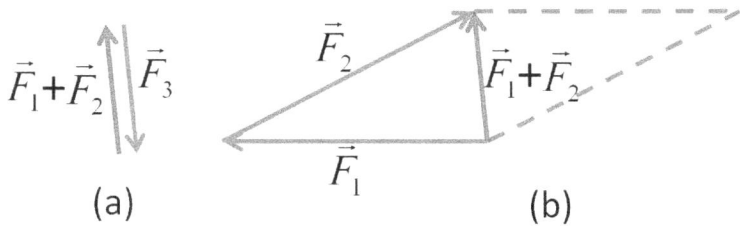

(a) (b)

Figure 2.8: Balanced forces on the ring in Fig. 2.7. (a) To cancel force $\vec{F}_3$, the sum of forces $\vec{F}_1$ and $\vec{F}_2$ must be in the opposite direction to $\vec{F}_3$ and have the same magnitude. (b) The sum of two forces according to the parallelogram law of addition - the tail of vector $\vec{F}_2$ is placed at the tip of vector $\vec{F}_1$ and the sum is read from the tail of vector $\vec{F}_1$ to the tip of $\vec{F}_2$.

Our procedure of addition of vectors parallels the procedure Isaac Newton gave in his masterpiece Principia Mathematica regarding the addition of forces:

"A body, acted on by two forces simultaneously, will describe the diagonal of a parallelogram in the same time as it would describe the sides by those forces separately."

The same rule of addition applies to all vectors. Geometrically it amounts to drawing the second vector at the tip of the first vector and the method is known as **tip-to-tail method** of adding vectors. For instance, to add vector $\vec{B}$ to vector $\vec{A}$, i.e. to obtain $\vec{A}+\vec{B}$, draw vector $\vec{A}$ and then, draw vector $\vec{B}$ by placing the tail of vector $\vec{B}$ at the tip of the arrow for vector $\vec{A}$. The arrow from the tail of $\vec{A}$ to the tip of $\vec{B}$ is the sum. This rule is called the parallelogram law of addition because the two vectors being added form the sides of a parallelogram and the sum is one of the diagonals of the parallelogram.

You will find that, vector $\vec{B}+\vec{A}$ obtained by placing the tail of vector $\vec{A}$ at the tip of vector $\vec{B}$ is equal to the vector $\vec{A}+\vec{B}$ obtained by placing the tail of $\vec{B}$ to the tip of $\vec{A}$ as shown in Fig. 2.9. Thus, adding two vectors can be done in any order, i.e., vector addition is commutative.

$$\vec{A}+\vec{B}=\vec{B}+\vec{A}.$$

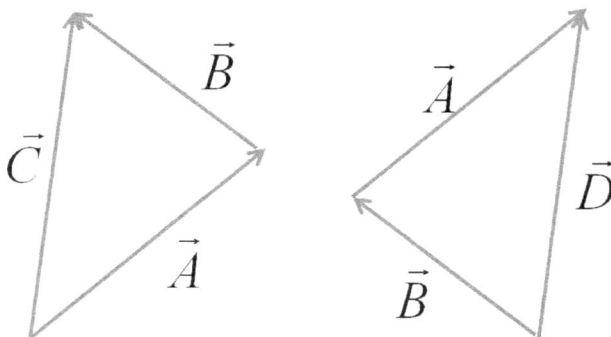

Figure 2.9: $\vec{C} = \vec{A}+\vec{B}$ and $\vec{D} = \vec{B}+\vec{A}$ are equal.

The parallelogram law of addition can be extended to the addition of more than two vectors by simply placing each additional vector starting from the tip of the last vector added as illustrated in Fig. 2.10.

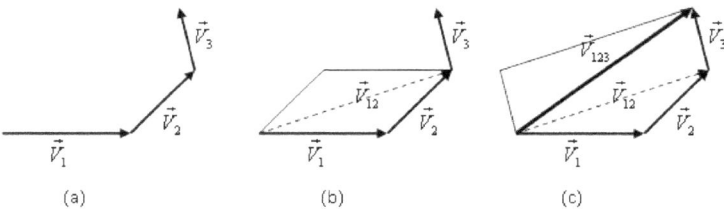

Figure 2.10: Addition of more than one vector according to the parallelogram law of addition id done by placing each successive vector at the tip of the last vector added. (a) Vectors to be added, (b) Sum $\vec{V}_{12}$ of $\vec{V}_1$ and $\vec{V}_2$, and (c) Sum $\vec{V}_{123}$ of $\vec{V}_{12}$ and the third vector V_3.

2.2.5 Subtraction of Vectors

Subtracting a vector $\vec{B}$ from another vector $\vec{A}$ can be turned into adding the negative of vector $\vec{B}$ to $\vec{A}$.

$$\vec{A}-\vec{B}=\vec{A}+\left(-\vec{B}\right).$$

Therefore, the geometrical procedure to obtain $(\vec{A}-\vec{B})$ will be to draw vector $\vec{A}$ and at the tip of vector $\vec{A}$ draw the vector $(-\vec{B})$, which is obtained from vector $\vec{B}$ by reversing the direction of the later. From the tail of $\vec{A}$ to the tip of $(-\vec{B})$ is the vector $(\vec{A}-\vec{B})$. You would obtain the same result for $(\vec{A}-\vec{B})$ by drawing $\vec{B}$ such that its tip meets the tip of $\vec{A}$. Then $(\vec{A}-\vec{B})$ is from the tail of $\vec{A}$ to the tail of $\vec{B}$.

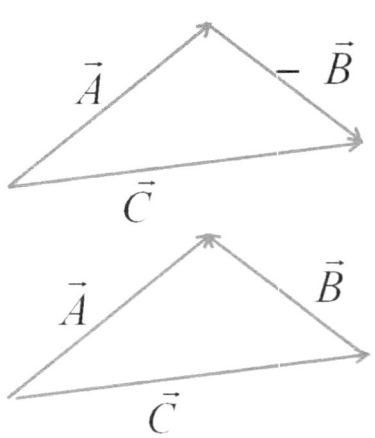

Figure 2.11: Two ways of obtaining $(\vec{A}-\vec{B})$: (a) From tail of $\vec{A}$ to the tip of $(-\vec{B})$, and (b) From tail of $\vec{A}$ to the tail of $\vec{B}$.

2.2.6 Vector Equations and Polygons

Recall that when you walk around a polygon and end up at the starting place, then your direct distance from the starting place to ending place is zero. This means that the net displacement must be a zero vector, also called the null vector. Thus, if you add vectors by placing the tail of one vector at the tip of another vector and the result is a closed polygon, the net sum of all vectors will be zero vector as illustrated in Fig. 2.12.

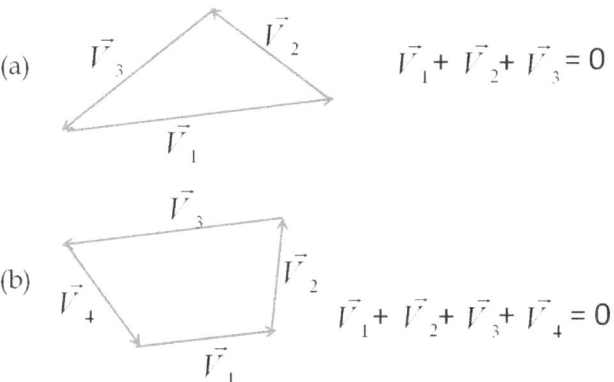

Figure 2.12: Polygon of vectors arranged tip-to-tail add up to zero or null vector.

Therefore, for any polygon of vectors we can write a vector equation. In Fig. 2.12(a), the three vectors give rise to the vector equation

$$\vec{V}_1 + \vec{V}_2 + \vec{V}_3 = 0,$$

which is the same as

$$\vec{V}_1 + \vec{V}_2 = (-\vec{V}_3),$$

which shows that the sum of the two vectors $\vec{V}_1$ and $\vec{V}_2$ is equal to the vector $(-\vec{V}_3)$, i.e. another vector that has the same magnitude as vector $\vec{V}_3$ but has an opposite direction to that of vector $\vec{V}_3$. Similarly, Fig. 2.12(b) corresponds to the vector equation

$$\vec{V}_1 + \vec{V}_2 + \vec{V}_3 + \vec{V}_4 = 0.$$

You will find many vector equations in this textbook. For example, the second law of motion is a vector equation $\vec{F} = m\vec{a}$, which says that the force vector obtained by vector addition of all the forces on the object of mass m is equal to the vector obtained by multiplying the acceleration vector with the mass.

2.2.7 Multiplication of a Vector with Another Vector

You have already learned what happens when you multiply a vector by a real number (also called a scalar). How do you multiply a vector by another vector? A vector has a numerical part in the magnitude and a

2.2. GEOMETRICAL VIEW OF VECTORS

non-numerical part in the direction. For instance, a velocity vector can be 3 m/s East and another velocity can be 4 m/s Up. Granted you can think of a way to multiply 3 and 4, but how would you multiply East and Up?

It turns out that two types of multiplication are found in physics. (1) In one type of multiplication, called the scalar product, the product of two vectors is actually a scalar number that depends upon the magnitudes of the two and the angle between them. We call this type of multiplication a scalar product. Examples of this type of multiplication are work and energy. (2) In the other type of multiplication between two vectors, called the vector product, the product is another vector. Examples of this vector products are torque and angular momentum.

2.2.8 Scalar Product Or Dot Product

The **scalar product** of two vectors $\vec{A}$ and $\vec{B}$, often called the **dot product**, is defined as the product of three quantities: (1) magnitude of the vector $\vec{A}$, (2) magnitude of the vector $\vec{B}$ and (3) cosine of the angle θ between the two vectors when they are drawn tail-to-tail. The angle θ is the smaller of the two angles that the two vectors make with each other. We denote the scalar product by placing a dot between the two vectors as in $\vec{A} \cdot \vec{B}$. The result of multiplying the three numbers can be positive, zero or negative real number.

$$\vec{A} \cdot \vec{B} = |\vec{A}||\vec{B}|\cos\theta. \quad (2.2)$$

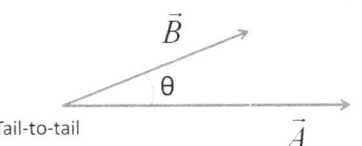

Figure 2.13: Arranging vectors to be multiplied together.

Geometrically, you can see that $|\vec{B}|\cos\theta$ is the length of the projection of vector $\vec{B}$ on vector $\vec{A}$ as shown in Fig. 2.14. The projection of $\vec{B}$ over $\vec{A}$ is obtained by drawing the two vectors so that their tails are at the same point, and then drawing a perpendicular line from the tip of $\vec{B}$ onto the line of $\vec{A}$. From the right-angled triangle so-formed, you can immediately see that the projection of $\vec{B}$ on $\vec{A}$ will have the value $|\vec{B}|\cos\theta$.

Therefore, you could state the scalar product in another way: the scalar product is equal to the product of $|\vec{A}|$ and the projection of vector $\vec{B}$ on vector $\vec{A}$. To draw a projection, sometimes, you may need to extend the vector in one or the other direction. If the projection of vector $\vec{B}$ falls on the opposite side of the direction of $\vec{A}$, then we say that the projection has a negative value, otherwise projection will have a positive value. If vector $\vec{B}$ is perpendicular to vector $\vec{A}$, then the projection will be zero.

Equivalently, you can draw a projection of vector $\vec{A}$ on vector $\vec{B}$. The length of this projection will be $|\vec{A}|\cos\theta$, which we can multiply with the length of vector $\vec{B}$ to get the value of the scalar product. Therefore, we can state the scalar product in yet another way: the scalar product is equal to the product of $|\vec{B}|$ and the projection of vector $\vec{A}$ on vector $\vec{B}$.

When the vector upon which a projection is being sought needs to be extended backwards as shown in Fig. 2.15, you must be careful with signs.

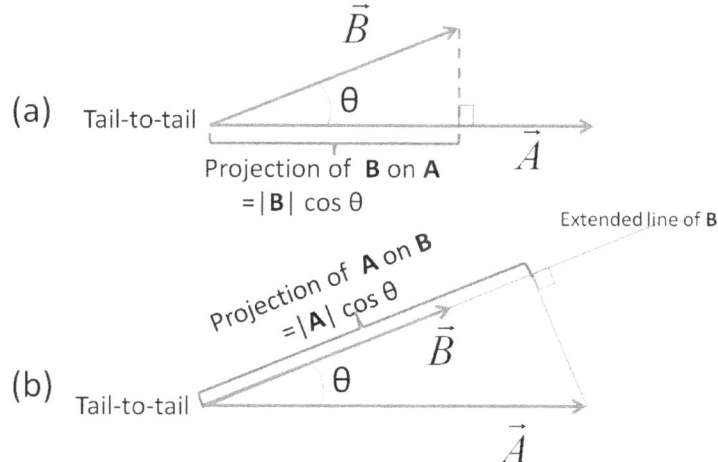

Figure 2.14: Projection of one vector over another. For two vectors, you have two projections: (a) the projection of $\vec{B}$ over $\vec{A}$ and (b) the projection of $\vec{A}$ over $\vec{B}$. Sometimes, you need to extend the line of one vector to draw a projection as shown in (b). If the projection from one vector falls on the extension on the opposite side, the projection has a negative value as shown in Fig. 2.15.

It is easier to work with the supplementary angle ϕ in the triangle formed by the projection of $\vec{B}$ onto the backward extended line of $\vec{A}$, but in the end, we are interested in the cosine of the other angle, the angle between the vectors, which is here θ. Therefore, we need to place a negative in front of the value $|\vec{B}|\cos\phi$.

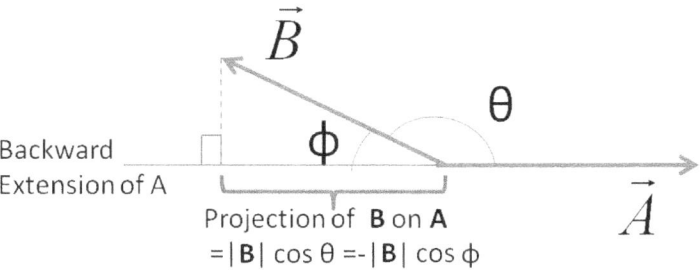

Figure 2.15: Projection of one vector over another sometimes requires extension of the line of action of the vector. If the projection from one vector falls on the extension on the opposite side, the projection has a negative value.

Note that if the dot product between two vectors is zero, then, either one or both of the vectors has a zero length, or the angle between them in ninety degrees since $\cos 90° = 0$.

$$\text{If } \vec{A}\cdot\vec{B} = 0, \text{ either } |\vec{A}| = 0, \text{ or } |\vec{B}| = 0, \text{ or } \theta = 90°.$$

Another useful property of the scalar product occurs when you take the scalar product of a vector with itself. Since $\cos\theta = 1$ here, the scalar product is equal to the square of the magnitude of the vector.

$$\vec{A} \cdot \vec{A} = |\vec{A}|^2.$$

In particular, the scalar product of a unit vector $\hat{u}$ is equal to 1 since the magnitude of a unit vector is 1 by definition.

$$\hat{u} \cdot \hat{u} = 1^2 = 1.$$

Based on the definition of the scalar product given above, you can prove the following algebraic properties of the scalar product.

1. Linearity: $\vec{A} \cdot (s\vec{B}) = s\vec{A} \cdot \vec{B}$, where s is scalar.
2. Distributive: $\vec{A} \cdot (\vec{B} + \vec{C}) = \vec{A} \cdot \vec{B} + \vec{A} \cdot \vec{C}$.
3. Commutative: $\vec{A} \cdot \vec{B} = \vec{B} \cdot \vec{A}$.

2.2.9 Vector or Cross Product

The other product of interest to us is the vector product, which is often called the cross product. Unlike the result of the scalar product, the result of a vector product is another vector, whose magnitude depends upon magnitudes of the two vectors and the angle between them, and whose direction depends on the orientations of the two vectors being multiplied. So, to define the vector product, we need rules for the magnitude as well as the direction for the product.

The **vector product** or the **cross product** between two vectors $\vec{A}$ and $\vec{B}$ is denoted by $\vec{A} \times \vec{B}$, which we will represent by the symbol $\vec{C}$.

$$\vec{C} \equiv \vec{A} \times \vec{B}.$$

Rule for the magnitude of $\vec{C}$:

Let θ be the angle between the two vectors $\vec{A}$ and $\vec{B}$ when they are drawn with their tails at the same point, just as we have done when we defined the scalar product.

$$\boxed{|\vec{C}| = |\vec{A}||\vec{B}|\sin\theta.} \quad (2.3)$$

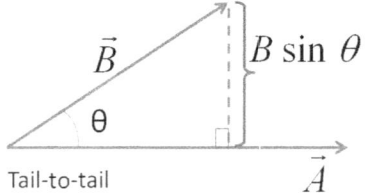

Figure 2.16

Note that $|\vec{B}|\sin\theta$ is the length of the perpendicular when you draw the projection of vector $\vec{B}$ onto vector $\vec{A}$ as shown in Fig. 2.16. Therefore, we see that $|\vec{A}||\vec{B}|\sin\theta$ is equal to the area of the parallelogram formed by the two vectors drawn so that they come out of the same point as illustrated in Fig. 2.17.

Rule for the direction of $\vec{C}$:

When vectors $\vec{A}$ and $\vec{B}$ are drawn so that their tails are at the same point, they define a plane if they are non-collinear. [If the vectors are

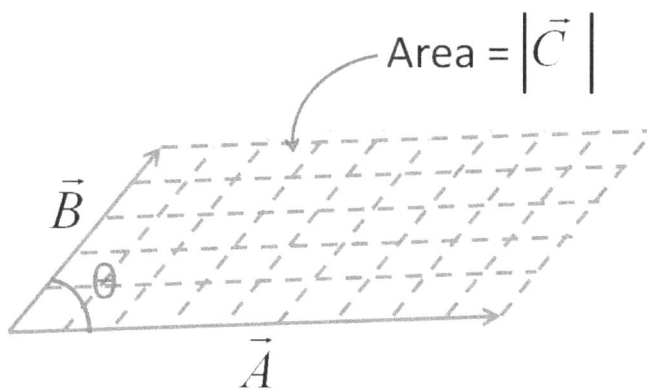

Figure 2.17: The magnitude of a vector product is equal to the area of the parallelogram formed by placing the two vectors on the sides of the parallelogram.

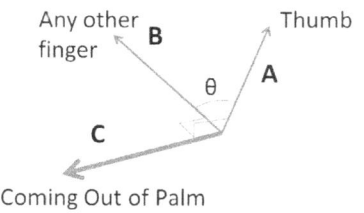

Figure 2.18: Directions of vectors in the Right Hand Rule.

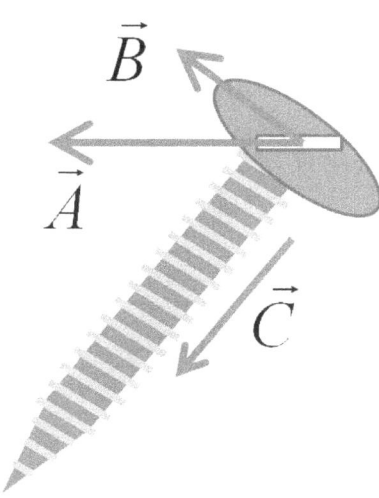

Figure 2.19: Right Hand Rule illustrated by the advancing right-handed screw.

collinear, their vector product will be zero since θ is either zero or 180° making $\sin\theta = 0$, so you do not need a rule for direction.] The direction of vector $\vec{C}$ is perpendicular to this plane. Now, there are two directions for perpendicular to a plane. The direction of $\vec{C}$ is obtained by a **Right Hand Rule**: if the thumb of your right hand points towards vector $\vec{A}$ and any of the other fingers point in the direction of vector $\vec{B}$, then the vector product $\vec{C}$ points in the direction coming out of palm as shown in Fig. 2.18. Another useful way to determine the direction of $\vec{C}$ is that of a right-handed screw (Fig. 2.19). Place vectors $\vec{A}$ and $\vec{B}$ in the plane of the head of the screw so that when you turn the screw so that the screw advances forward when $\vec{A}$ rotates towards $\vec{B}$. Then the direction of the screw advancing is the direction of the vector product $\vec{C}$.

From the definition of a vector product you can conclude that the vector product of a vector with itself will be zero since the angle with itself is zero and $\sin\theta = \sin 0 = 0$. Another useful result of cross product is the cross product of two vectors that are perpendicular to each other. In that case $\sin\theta = \sin 90° = 1$. Therefore, the magnitude of a vector product of two mutually perpendicular vectors is simply the product of the magnitudes of the two vectors.

$$\vec{A} \times \vec{A} = 0$$
$$|\vec{A} \times \vec{B}| = |\vec{A}||\vec{B}|, \text{ if } \vec{A} \text{ is perpendicular to } \vec{B}.$$

By using the definition for the vector product you can prove the following algebraic properties of the vector product.

1. Linearity: $\vec{A} \times (s\vec{B}) = s\vec{A} \times \vec{B}$, where s is scalar.

2. Distributive: $\vec{A} \times (\vec{B} + \vec{C}) = \vec{A} \times \vec{B} + \vec{A} \times \vec{C}$.

3. Anti-commutative: $\vec{A} \times \vec{B} = -\vec{B} \times \vec{A}$.

Of particular importance is the anti-commutative property: it shows that $\vec{A} \times \vec{B}$ is not equal to $\vec{B} \times \vec{A}$. That is, the order matters.

2.3 ANALYTICAL VIEW OF VECTORS

The analytic view of vectors or the algebraic method for vectors is based on the decomposition of vectors along perpendicular axes. It is easily seen that an arbitrary vector can be constructed by adding appropriate vectors along x, y, and z-axes.

For simplicity of drawing let us look at an example of vectors in the xy-plane. In Fig. 2.20 you can see how an arbitrary vector $\vec{A}$ can be written as a sum of two vectors $\vec{A}_1$ and $\vec{A}_2$, one along each Cartesian axis. Similar arguments can be applied to a vector in the three-dimensional space. Therefore an arbitrary vector can be constructed from three vectors $\vec{A}_1$, $\vec{A}_2$ and $\vec{A}_3$ along the x, y and z-axes respectively.

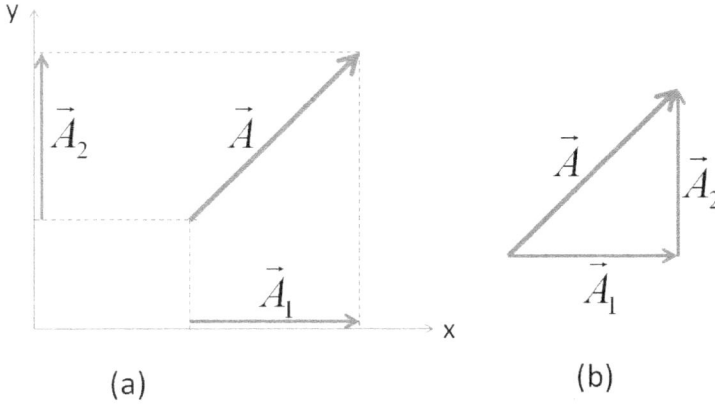

Figure 2.20: An arbitrary vector can be constructed by a sum of appropriate vectors parallel to the Cartesian axes. Here vector $\vec{A}$ is sum of vectors $\vec{A}_1$ and $\vec{A}_2$. (a) The vector $\vec{A}_1$ along the x-axis is formed by the projections of the two ends of the vector $\vec{A}$ on the x-axis. Similarly for $\vec{A}_2$ along the y-axis. (b) The triangle shows that $\vec{A} = \vec{A}_1 + \vec{A}_2$.

2.3.1 Base Vectors

Since we need vectors along the axes, either pointed towards the positive axes or in the opposite directions, we define unit vectors pointed towards the positive x, y and z-axes for convenience of writing them. Let $\hat{u}_x$, $\hat{u}_y$, and $\hat{u}_z$ denote unit vectors in the direction of the positive x, y and z-axes respectively. These unit vectors are also called **base vectors**. Often these vectors are also written as $\hat{i}$, $\hat{j}$, and $\hat{k}$ respectively. We will use $\hat{u}_x$, $\hat{u}_y$, and $\hat{u}_z$ notation to make the role of the coordinates explicit. If you are more comfortable with $(\hat{i}, \hat{j}, \hat{k})$ notation, you may continue to use them since they are less cumbersome to write and could save time.

By definition, these vectors have unit magnitude. Since these vectors are perpendicular to each other, their dot products are also zero. Furthermore, a calculation of their cross-products will show that cross-product of

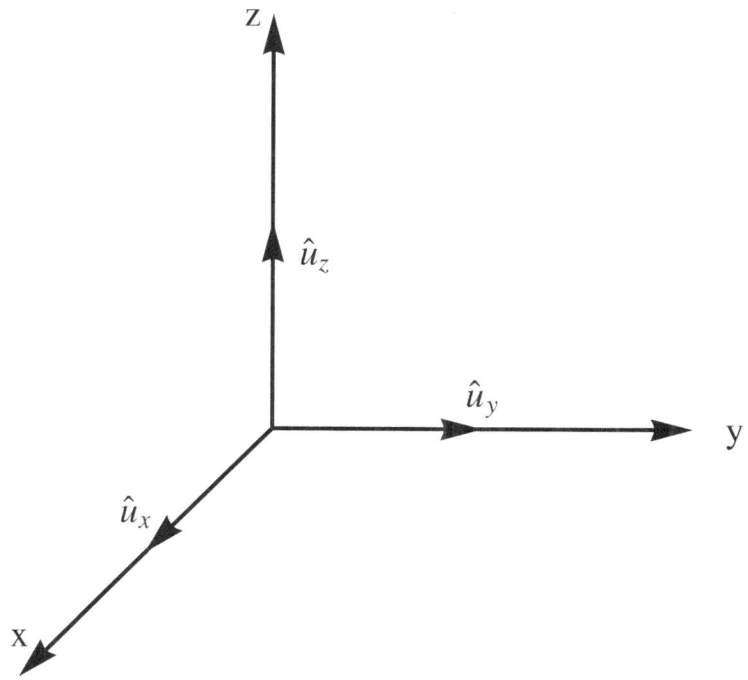

Figure 2.21: The unit vectors $\hat{u}_x$, $\hat{u}_y$, and $\hat{u}_z$ in the directions of the positive x, y and z-axes.

two of them takes into plus or minus of the third.

$$\hat{u}_x \cdot \hat{u}_x = \hat{u}_y \cdot \hat{u}_y = \hat{u}_z \cdot \hat{u}_z = 1 \text{ (since unit vectors)} \quad (2.4)$$

$$\hat{u}_x \cdot \hat{u}_y = \hat{u}_x \cdot \hat{u}_z = \hat{u}_y \cdot \hat{u}_z = 0 \text{ (since perpendicular)} \quad (2.5)$$

$$\hat{u}_x \times \hat{u}_y = \hat{u}_z \quad (2.6)$$

$$\hat{u}_y \times \hat{u}_z = \hat{u}_x \quad (2.7)$$

$$\hat{u}_z \times \hat{u}_x = \hat{u}_y \quad (2.8)$$

Now, we can see that vector $\vec{A}_1$ along the x-axis in Fig. 2.20 can be written as a real number A_x times the unit vector $\hat{u}_x$. When the real number A_x is positive the vector $\vec{A}_1$ is pointed towards the positive x-axis and when A_x is negative, the vector $\vec{A}_1$ is pointed towards the negative x-axis.

Similarly, we can write vector $\vec{A}_2$ along the y-axis as a real number A_y times the unit vector $\hat{u}_y$. If we had a vector $\vec{A}_3$ along z-axis, then we would write the vector as a real number A_z times the unit vector $\hat{u}_z$.

$$\vec{A}_1 = A_x \hat{u}_x \text{ (vector along } x \text{ axis)} \quad (2.9)$$

$$\vec{A}_2 = A_y \hat{u}_y \text{ (vector along } y \text{ axis)} \quad (2.10)$$

$$\vec{A}_3 = A_z \hat{u}_z \text{ (vector along } z \text{ axis)} \quad (2.11)$$

The real numbers A_x, A_y and A_y are called the components of vector $\vec{A}$, which can be written as the sum of the vectors $\vec{A}_1$, $\vec{A}_2$, and $\vec{A}_3$ along the axes. Fig. 2.22 shows how vector $\vec{A}$ can be constructed by adding the

2.3. ANALYTICAL VIEW OF VECTORS

vectors along the Cartesian axes. Therefore, we can write any vector in terms of the base vectors.

$$\vec{A} = A_x \hat{u}_x + A_y \hat{u}_y + A_z \hat{u}_z, \quad (2.12)$$

where the sum of $A_x \hat{u}_x$, $A_y \hat{u}_y$ and $A_z \hat{u}_z$ is a vector sum. A simpler notation for the representation of a vector is to just list the components in order, as in $\vec{A} = (A_x, A_y, A_z)$. We will refrain from this notation since it ignores the explicit role of unit base vectors. However, if you are comfortable with this notation, it can save you some time since you do not have to write down the unit vectors all the time, especially when you are doing long calculations.

Vectors in terms of components: $\vec{A} = A_x \hat{u}_x + A_y \hat{u}_y + A_z \hat{u}_z = (A_x, A_y, A_z)$

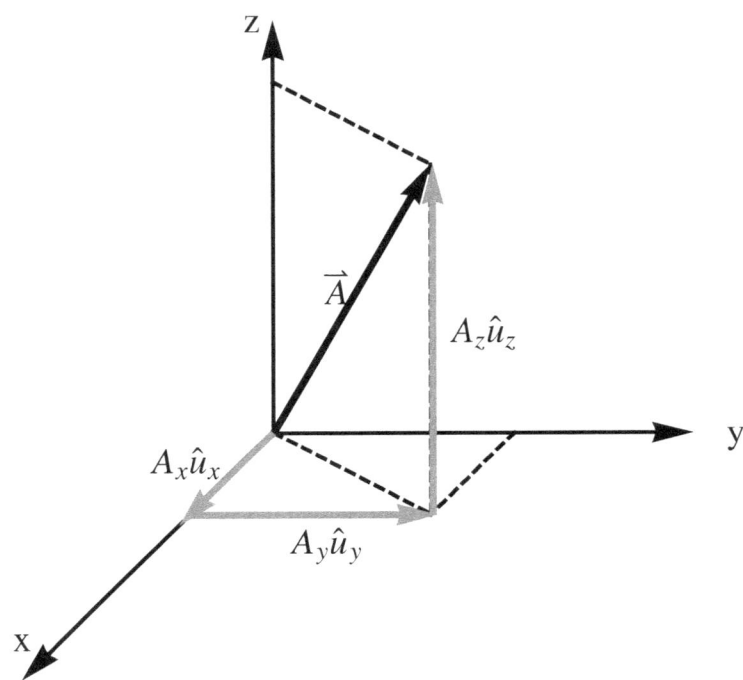

Figure 2.22: Placing the tail of vector $A_y \hat{u}_y$ at the tip of vector $A_x \hat{u}_x$ and then placing the tail of vector $A_z \hat{u}_z$ at the tip of $A_y \hat{u}_y$ give the vector $\vec{A}$ as the sum of the three: $\vec{A} = A_x \hat{u}_x + A_y \hat{u}_y + A_z \hat{u}_z$.

2.3.2 Magnitude of Vectors

You can find the magnitude of a vector from its components by using Pythagoras theorem in Fig. 2.22. It is immediately seen that the magnitude of a vector is equal to the square root of the sum of the squares the components.

$$|\vec{A}| = \sqrt{A_x^2 + A_y^2 + A_z^2}. \quad (2.13)$$

2.3.3 Directions of Vectors

The directions of vectors is automatic in the graphical or geometric picture of vectors. You always draw the complete vectors when working geometrically. In the analytic or algebraic picture we work with components when performing algebraic manipulations. In the end, we must put together the information contained in the components into the magnitudes and directions of vectors. The specification of a physical direction in space very much depends upon whether the vector has only one non-zero component, or two non-zero components, or three non-zero components. These will be called one-dimensional, two-dimensional and three-dimensional situations respectively.

1. Direction for 1-D situations

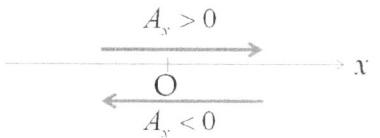

Figure 2.23: One-dimensional vector $\vec{A} = A_x \hat{u}_x$. The direction of the vector is towards the positive x-axis if $A_x > 0$ and towards the negative x-axis if $A_x < 0$.

If a vector $\vec{A}$ has only one non-zero component, say the x-component, the vector will be $A_x \hat{u}_x$ analytically. Depending upon the sign of the x-component A_x, the vector $\vec{A}$ will be pointed either towards the positive x-axis if $A_x > 0$ or towards the negative x-axis if $A_x < 0$. Therefore, if only one component of a vector is non-zero, the sign of the value of the component will have sufficient information to deduce the direction of the vector with respect to the coordinate axis. Due to this reason, sometimes the sign itself is mistakenly taken to be the direction of the vector. Note that direction of a vector is the physical direction in space and cannot be plus or minus something.

2. Direction for 2-D situations

If a vector has two non-zero components, the vector will fall in a plane. For instance, if only the x and y-components of a vector are non-zero, the vector will fall in the xy-plane. Similarly, if only the x and z-components are non-zero, the vector will be in the xz plane, etc. In these cases, the direction in the plane is specified by an angle with one of the axes: the vector is drawn, or re-drawn if needed, so that the tail of the vector is at the origin, then the angle that the arrow makes with one of the axis is sufficient to tell the direction of the vector in the plane.

To be concrete, consider a vector $\vec{A}$ that falls entirely in the xy-plane such that we have the following analytic representations.

$$\vec{A} = A_x \hat{u}_x + A_y \hat{u}_y.$$

Depending upon the signs of A_x and A_y, the vector arrow may point in the first, second, third, or fourth quadrant as shown in Fig. 2.24. Beware of the common formula based on the trigonometry of the vector in the first quadrant in Fig. 2.24. The direction of the vector in the first quadrant with respect to the positive x-axis direction is given by angle θ_1. From the right-angled triangle $\triangle OPQ$ in Fig. 2.24(I) it is seen that the tangent of this angle is related to the components A_x and A_y as follows.

$$\tan(\theta_1) = A_y/A_x \implies \theta_1 = \arctan(A_y/A_x). \tag{2.14}$$

2.3. ANALYTICAL VIEW OF VECTORS

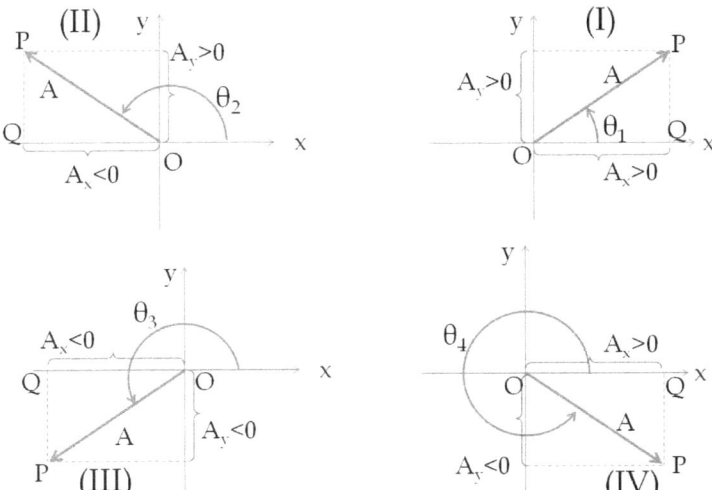

Figure 2.24: Vector in a plane. The direction of a vector in a plane can be in one of the four quadrants and can be assigned with respect to the positive x-axis direction. The common formula $\theta = \arctan(A_y/A_x)$ gives all four angles if correction is made for the particular quadrant: $\theta_1 = \theta$, $\theta_2 = \theta + 180°$, $\theta_3 = \theta + 180°$, and $\theta_4 = \theta + 360°$.

When the vector falls in the second quadrant, the right-angled triangle $\triangle OPQ$ in Fig. 2.24(II) gives the supplementary angle to the angle θ_2, that is the tangent of angle $\angle QOP$ is equal to A_y/A_x. Therefore, we must add 180° to the value obtained by arc-tangent of the ratio A_y/A_x.

$$\theta_2 = 180° + \arctan(A_y/A_x). \tag{2.15}$$

If a vector falls in the third quadrant, the direction is opposite to the direction in the first quadrant with the signs of both A_x and A_y reversed. Therefore, the angle from the positive x-axis is 180° more than θ_1, which is equal to $\arctan(A_y/A_x)$.

$$\theta_3 = 180° + \arctan(A_y/A_x). \tag{2.16}$$

Finally, the direction in the fourth quadrant is exactly opposite to that of the direction in the second quadrant when we reverse the signs of both A_x and A_y. Therefore,

$$\theta_4 = 360° + \arctan(A_y/A_x). \tag{2.17}$$

Warning! Beware of the value of arctangent from a calculator. You need to interpret the value from your calculator based on the quadrant in which the vector is pointed. Think, before you commit to the display on the calculator screen.

Example 2.3.1. Direction of a vector in a plane A velocity vector of a projectile is given in the xy-plane of a Cartesian coordinate system with the right horizontal direction for the positive x-axis and the direction up for the positive y-axis. At a particular instant the velocity has the following value: $\vec{v} = 3\hat{u}_x - \sqrt{3}\hat{u}_y$ in units of $[m/s]$, where I have kept the unit separate for convenience. What is the direction of the velocity vector?

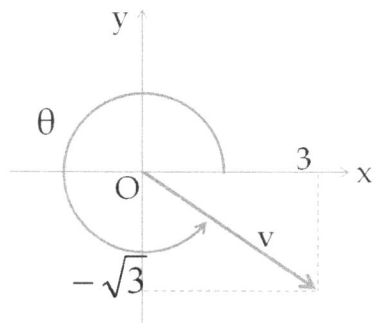

Figure 2.25: Example 2.3.1

Solution. Geometrically, we could draw the vector and read off the angle with respect to either of the two axes. That would be sufficient for specifying the direction.

The discussion above has shown that, alternately, we could obtain the angle from the given components without drawing the vector in the xy-plane. Here the components of the vector tell us that the vector is pointed in the fourth quadrant since $v_x > 0$ and $v_y < 0$. Therefore, the larger angle of the vector going counter-clockwise from the positive x-axis is given by the angle θ_4, which we will write simply as θ (see Fig. 2.25).

$$\theta = 360° + \arctan\left(-\sqrt{3}/3\right) = 360° - 30° = 330°.$$

The calculation reveals that using analytic method can save a lot of work.

Other Choices for Specifying Directions in a Plane

There is no reason why you should have to refer the direction of a vector in the xy-plane with respect to the direction of the positive x-axis. Often it is more convenient to provide the direction of a vector with respect to the closest axis. The negative axis directions are usually labeled with a "bar" over the axis symbol. For instance, while the positive x-axis is Ox, the negative x-axis is written as $O\bar{x}$, and similarly for other axes. Then, we have a number of convenient choices for giving directions of a vector $\overrightarrow{OP}$ in a plane drawn from the tail at the origin: $\angle xOP$, $\angle \bar{x}OP$, $\angle yOP$, and $\angle \bar{y}OP$.

For instance, the direction of a vector in the third quadrant does not have to be specified by the large angle θ_3 but rather by the small angle $\angle QOP$, which is $\angle \bar{x}OP$ the vector makes with the negative x-axis.

Third quadrant: $\angle QOP = \angle \bar{x}OP = \arctan\left(A_y/A_x\right)$. (2.18)

Similarly, the direction of the vector in the fourth quadrant is more conveniently specified with the smaller angle $\angle QOP$ with respect to the x-axis, that is, going in the clockwise direction from the x-axis rather than going counterclockwise direction. This is easily seen to equal $\arctan\left(A_y/A_x\right)$.

Fourth quadrant: $\angle QOP$ in $\triangle QOP = \angle xOP = \arctan\left(A_y/A_x\right)$. (2.19)

3. Direction for 3-D situations

The directions for vectors with all three of its components non-zero require two angles. Typically, we use the polar and azimuthal angles of a spherical coordinate system. The commonly used symbols in physics are θ for the polar angle and ϕ for the azimuthal angle, which is exactly opposite of the convention in math books for these angles. Since we are studying physics here we will stick to the physics notation.

The polar angle θ for a vector $\vec{A}$ is the angle the vector makes with the z-axis when the tail of the vector is placed at the origin as shown in Fig.

2.3. ANALYTICAL VIEW OF VECTORS

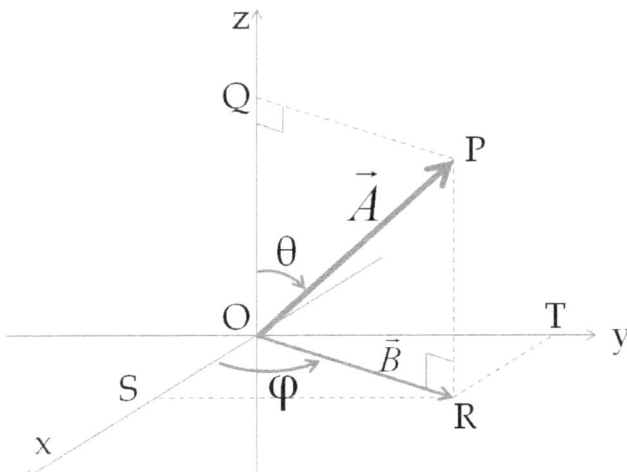

Figure 2.26: Vector in space can be given by using spherical coordinates given by the polar angle θ and the azimuthal angle ϕ. The polar angle is the angle between the vector and the positive z-axis, and the azimuthal angle is the angle between the positive x-axis and the direction of the projection of the vector on the xy-plane.

2.26. The range of polar angle is between 0 and $180°$ or π radians. The zero degree direction is towards positive z-axis and the $180°$ direction is towards the negative z-axis.

Azimuthal angle is somewhat complicated: we first draw the vector in space with its tail at the origin, and then project the vector on the xy-plane by drawing a normal from the tip of the vector to the xy-plane as shown in the figure. The projection on the xy-plane defines another vector $\vec{B}$, which may be called projected vector, which has only x and y-components non-zero. The x and y-components of the projected vector $\vec{B}$ are equal to the x and y-components of the original vector $\vec{A}$ as you can verify from the given figure: $B_x = A_x$ and $B_y = A_y$. The direction of the projected vector $\vec{B}$ in the xy-plane with respect to the positive x-axis is given by the angle the projected vector makes with the positive x-axis direction. This angle is called the azimuthal angle of the original vector $\vec{A}$ as shown in the figure.

Azimuthal angle has a range of 0 to $360°$ or 2π radians, with the direction taken from the positive x-axis towards the positive y-axis, also called counter-clockwise as seen from the positive z-axis, as shown in Fig. 2.26. Sometimes, the range of azimuthal angle is given between $-180°$ and $180°$ or $-\pi$ radians to π radians where the negative angle is for clockwise direction from the positive x-axis when seen from the perspective of the positive z-axis.

The right-angled triangles $\triangle OPQ$ and $\triangle OSR$ in Fig. 2.26 can be used to obtain the following relations for the polar and azimuthal angles

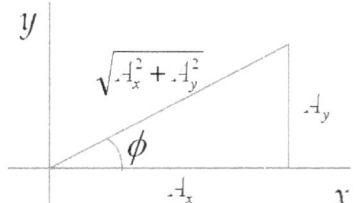

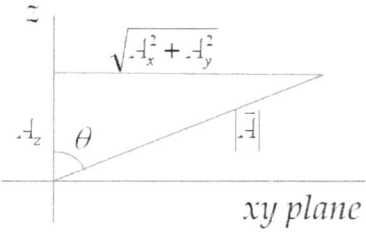

Figure 2.27: Angles ϕ and θ.

in terms of the Cartesian components (A_x, A_y, A_z) of the vector.

$$\boxed{\tan\theta = \frac{\sqrt{A_x^2 + A_y^2}}{A_z} \quad \text{(Add 180° if } A_z < 0.\text{)}} \quad (2.20)$$

$$\boxed{\tan\phi = \frac{A_y}{A_x} \quad \text{(Be mindful of the quadrant.)}} \quad (2.21)$$

The triangles in Fig. 2.26 can also be used to obtain the components of a vector from its magnitude and direction given by polar and azimuth angles. I will give the answer and leave the exercise of proving them to the student as an exercise. Let the magnitude of a vector be A and the direction in three-dimensional space be given by polar angles θ and ϕ. The Cartesian components of the vector will be:

$$\boxed{A_x = A\sin\theta\cos\phi} \quad (2.22)$$

$$\boxed{A_y = A\sin\theta\sin\phi} \quad (2.23)$$

$$\boxed{A_z = A\cos\theta} \quad (2.24)$$

Example 2.3.2. Vector in space The displacement vector from the center of the Earth to St. Louis, Missouri has a length of 6.4×10^6 m. A Cartesian axis is chosen with its origin at the center of the Earth, z-axis pointed towards polar North and x and y-axes in the equatorial plane. With a particular choice of x and y-axes, the Cartesian components of the vector are (-14780 m, -4234000 m, 4800000 m). What are the values of polar and azimuthal angles for the direction of the vector?

Solution. This is a straightforward application of formulas given above. We only need to be careful in using arctangent and make appropriate choice for the quadrant. The data shows that the angle will be in the third quadrant in the xy-plane.

$$\tan\theta = \frac{\sqrt{A_x^2 + A_y^2}}{A_z} = \frac{4234025}{4800000} = 0.882 \implies \theta = 41.4°.$$

$$\tan\phi = \frac{A_y}{A_x} = 286.5$$

$$\implies \text{Since, third quadrant, } \phi = 180° + \arctan(286.5) = 269.8°.$$

2.3.4 Adding and Subtracting Vectors Analytically

Adding and subtracting vectors is a rather simple task if done analytically. Each vector is first written as a sum of vectors along the three Cartesian axes using their components. Then, the component vectors along each Cartesian axis are summed to obtain the components of the resultant vector. Let $\vec{A} = A_x\hat{u}_x + A_y\hat{u}_y + A_z\hat{u}_z$ and $\vec{B} = B_x\hat{u}_x + B_y\hat{u}_y + B_z\hat{u}_z$ be two vectors we want to add. The Cartesian components of their sum

2.3. ANALYTICAL VIEW OF VECTORS

$\vec{C} = \vec{A} + \vec{B}$ is

$$\begin{aligned}\vec{C} &= \vec{A} + \vec{B} \\ &= (A_x \hat{u}_x + A_y \hat{u}_y + A_z \hat{u}_z) + (B_x \hat{u}_x + B_y \hat{u}_y + B_z \hat{u}_z) \\ &= (A_x + B_x)\hat{u}_x + (A_y + B_y)\hat{u}_y + (A_z + B_z)\hat{u}_z\end{aligned}$$

Therefore, the components of the sum are

$$\boxed{C_x = A_x + B_x}$$
$$\boxed{C_y = A_y + B_y}$$
$$\boxed{C_z = A_z + B_z}$$

Similarly, subtracting vector $\vec{B}$ from vector $\vec{A}$ gives the following result.

$$\begin{aligned}\vec{D} &= \vec{A} - \vec{B} \\ &= (A_x - B_x)\hat{u}_x + (A_y - B_y)\hat{u}_y + (A_z - B_z)\hat{u}_z\end{aligned}$$

Example 2.3.3. Calculation of a net force A pendulum bob has two forces on it - gravity of Earth and tension in the string. At a particular instant the pendulum is at an angle 30° from the vertical. At that instant it has a tension of 5 N (unit N called Newton will be explained later) pointed towards the point of suspension, and a force of gravity equal to 10 N pointed down. Find the magnitude and direction of the net force, defined as the sum of all forces.

Solution. It is helpful to make a drawing of the physical situation as shown in Fig. 2.28. The figure also shows a choice of axes for analytical

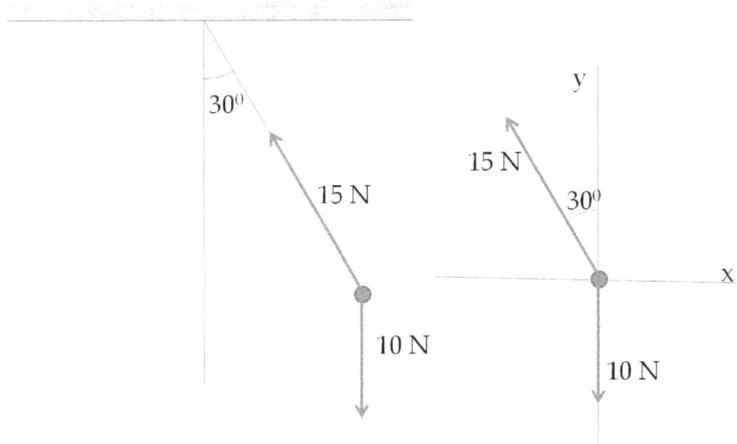

Figure 2.28: Example 2.3.3. Normally, we draw two figures. The figure on the left shows the physical setting of the problem and the figure on the right collects all the forces emanating from the same point where the origin of coordinate system is placed. The choice of axes are shown in the diagram. The components of the vectors are obtained from their projections on the axes.

calculations. The magnitude of the forces and their directions are also shown on the figure for convenience. Let us denote force of gravity by $\vec{F}_g$ and the force of tension by $\vec{F}_T$. From the figure, we find the following decomposition of the vectors, where we keep track of units on the right side of the equation. The force of gravity has only the y-component and the force of tension has a negative x-component and a positive y-component with respect to the axes chosen in the figure. To work out the components I have redrawn the vector in Fig. 2.28 with tails at the origin of the axes so that simple projections onto the axes gives the corresponding components. I will also place the units separately so that you can focus on the other aspects of the calculations.

$$\vec{F}_g = -10\hat{u}_y \quad [\text{N}]$$
$$\vec{F}_T = -15\sin 30°\hat{u}_x + 15\cos 30°\hat{u}_y \quad [\text{N}]$$

Now, the sum of these two forces will have the following components.

$$\begin{aligned} \vec{F}_{net} &= \vec{F}_g + \vec{F}_T \\ &= -7.5\hat{u}_x + 3.0\hat{u}_y \quad [\text{N}] \end{aligned}$$

We are not done yet. We have found the representation of the net force in the chosen coordinate system. The magnitude and direction of the net force can now be obtained from the components.

Magnitude of the net force: $\left|\vec{F}_{net}\right| = \sqrt{(-7.5)^2 + (3.0)^2} = 22.5$ N.

The direction is in the second quadrant given by the angle θ with respect to the negative x-axis, $O\bar{x}$.

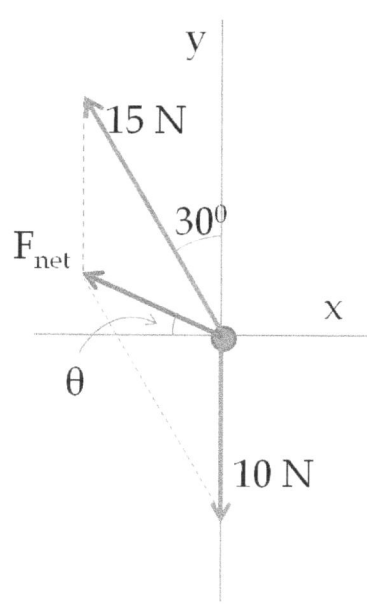

Figure 2.29: Example 2.3.3. The direction of the net force.

$$\theta = \arctan(3/(-7.5)) = -22°.$$

The negative sign says that the angle is clockwise from the negative x-axis into the second quadrant. We could, of course, specify the counterclockwise angle with respect to the positive x-axis, which will be 158°.

2.3.5 Scalar Products Analytically

Using the dot products between the base vectors one can write the scalar product between two vectors $\vec{A}$ and $\vec{B}$ in terms of their components. The calculation is left as an exercise for the student.

$$\boxed{\vec{A} \cdot \vec{B} = A_x B_x + A_y B_y + A_z B_z,} \quad (2.25)$$

which must be identical to the geometric definition of the dot product. Therefore,

$$|\vec{A}||\vec{B}|\cos\theta = A_x B_x + A_y B_y + A_z B_z. \quad (2.26)$$

Applying the result in Eq. 2.25 to the dot product of a vector $\vec{A}$ with itself gives

$$|\vec{A}|^2 = A_x^2 + A_y^2 + A_z^2. \quad (2.27)$$

2.3. ANALYTICAL VIEW OF VECTORS

Since the amplitude of a vector is a positive quantity, only the positive root is physical. Taking the positive root of both sides, we see that the amplitude of a vector can be computed from its components.

$$\boxed{|\vec{A}| = \sqrt{A_x^2 + A_y^2 + A_z^2}.} \quad (2.28)$$

Projections

From our geometrical discussions we know that the scalar product between two vectors also gives information about the projection of one vector onto another.

$$\vec{A} \cdot \vec{B} = AB \cos\theta, \quad (2.29)$$

which gives the projection of A on B, namely $A\cos\theta$, as:

$$A \cos\theta = \frac{\vec{A} \cdot \vec{B}}{B}. \quad (2.30)$$

If vector $\vec{B}$ is a unit vector then we can use Eq. 2.30 to find the component of the vector along the direction of the unit vector. For instance if $\vec{B}$ is the unit vector along the x-axis, we obtain the x-component of $\vec{A}$ as the following calculation demonstrates.

$$\begin{aligned}
\vec{A} \cdot \hat{u}_x &= (A_x \hat{u}_x + A_y \hat{u}_y + A_z \hat{u}_z) \cdot \hat{u}_x \\
&= A_x \hat{u}_x \cdot \hat{u}_x + A_y \hat{u}_y \cdot \hat{u}_x + A_z \hat{u}_z \cdot \hat{u}_x \\
&= A_x \times 1 + A_y \times 0 + A_z \times 0 \\
&= A_x
\end{aligned}$$

This shows that to obtain any component of a vector we just need a dot product of the vector with the corresponding base vector.

$$\vec{A} \cdot \hat{u}_x = A_x \quad (2.31)$$
$$\vec{A} \cdot \hat{u}_y = A_y \quad (2.32)$$
$$\vec{A} \cdot \hat{u}_z = A_z \quad (2.33)$$

Equations 2.31 to 2.33 are very useful, not only for finding components of a vector, but also the angles a vector makes with the Cartesian axes as we see by explicitly writing out the scalar product of the left side of these equations. Let α, β, and γ be the angles that vector $\vec{A}$ makes with the positive x, y and z-axes respectively as shown in Fig. 2.30 . That is, the angles between the vector $\vec{A}$ and the unit vectors $\hat{u}_x$, $\hat{u}_y$ and $\hat{u}_z$ are α, β, and γ, respectively. We find

$$|\vec{A}||\hat{u}_x|\cos\alpha = A_x$$
$$|\vec{A}||\hat{u}_y|\cos\beta = A_y$$
$$|\vec{A}||\hat{u}_z|\cos\gamma = A_z$$

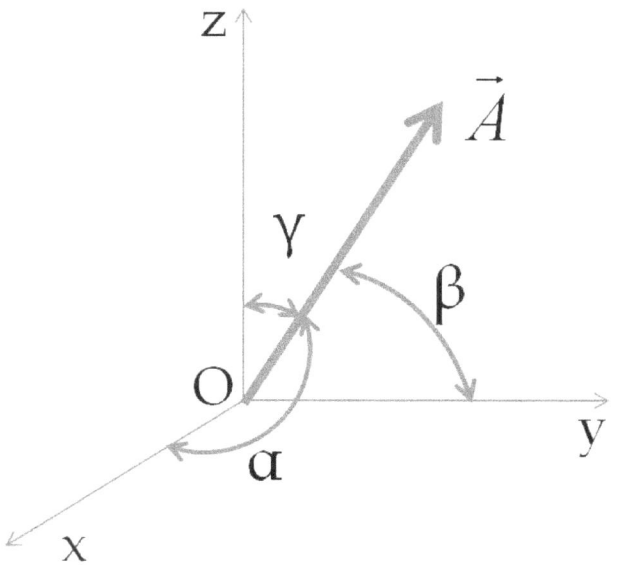

Figure 2.30: Angles for Direction Cosines.

Since the magnitudes of the unit vectors are all 1, we obtain the following formulas for the cosines of the angles. These cosines are called direction cosines of the vector.

$$\boxed{\cos\alpha = \frac{A_x}{|\vec{A}|} = \frac{A_x}{\sqrt{A_x^2 + A_y^2 + A_z^2}}} \quad (2.34)$$

$$\boxed{\cos\beta = \frac{A_y}{|\vec{A}|} = \frac{A_y}{\sqrt{A_x^2 + A_y^2 + A_z^2}}} \quad (2.35)$$

$$\boxed{\cos\gamma = \frac{A_z}{|\vec{A}|} = \frac{A_z}{\sqrt{A_x^2 + A_y^2 + A_z^2}}} \quad (2.36)$$

Example 2.3.4. Angle between two vectors. Two vectors have the following Cartesian representations: $\vec{A} = 3\hat{u}_x + 4\hat{u}_y + 5\hat{u}_z$ and $\vec{B} = -6\hat{u}_x + 7\hat{u}_y + 8\hat{u}_z$. What is the angle between the the vectors when they are drawn so that their tails are at the same point?

Solution. This example is an application of Eq. 2.26, which is a very useful formula for finding angles between any two vectors. We need the magnitudes of the vectors and their components to compute the angle between them. From the given components, the magnitudes of the two vectors are:

$$A = \sqrt{3^2 + 4^2 + 5^2} = 7.07.$$
$$B = \sqrt{(-6)^2 + 7^2 + 8^2} = 12.2.$$

Therefore, the cosine of the angle between these vectors is

$$\cos\theta = \frac{\vec{A}\cdot\vec{B}}{|\vec{A}||\vec{B}|} = \frac{A_xB_x + A_yB_y + A_zB_z}{AB} = \frac{-3\times 6 + 4\times 7 + 5\times 8}{7.07\times 12.2} = 0.58,$$

which gives the following angle between the two vectors.

$$\theta = \arccos(0.58) = 55°.$$

Example 2.3.5. Use of direction cosines. Find the angle between the body diagonal of a parallelepiped of dimensions $a \times b \times c$ and the edges. Use the result to find the angle the body diagonal of a cube makes with the edges.

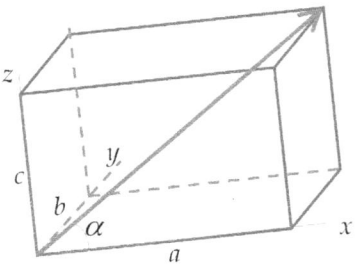

Figure 2.31: Body diagonal of a parallelepiped.

Solution. Let us place a Cartesian coordinate system so that the body diagonal from one corner to farthest corner be given by the following vector $\vec{A} = a\hat{u}_x + b\hat{u}_y + c\hat{u}_z$. Therefore, the direction cosines with the edges are

$$\cos\alpha = \frac{a}{\sqrt{a^2 + b^2 + c^2}}$$

$$\cos\beta = \frac{b}{\sqrt{a^2 + b^2 + c^2}}$$

$$\cos\gamma = \frac{c}{\sqrt{a^2 + b^2 + c^2}}$$

If the parallelepiped is a cube, then all the sides are equal and the three angles will also be equal. Setting $a = b = c$ in these equations results in

$$\cos\alpha = \cos\beta = \cos\gamma = \frac{1}{\sqrt{3}}.$$

Therefore, the angle the body diagonal makes with the edges is $54.7°$.

Example 2.3.6. Law of cosines using the scalar product. Some standard trigonometric results can be obtained rather easily with the use of vectors. One such result is the law of cosines which can be derived using the scalar product since the scalar product involves cosine of the angle between two vectors. Let $\vec{A}$ and $\vec{B}$ be vectors along two adjacent sides of a triangle drawn with their tails at the vertex. Let the angle between $\vec{A}$ and $\vec{B}$ be denoted by θ. We now place a third vector from the tip of $\vec{B}$ to the tip $\vec{A}$ on the third side of the triangle. Then, it is seen that vector $\vec{C}$ is equal to $\vec{A} - \vec{B}$.

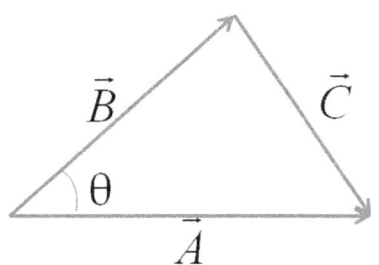

Figure 2.32: Diagram for law of cosines.

$$\vec{C} = \vec{A} - \vec{B}.$$

We now take the scalar product of each side with itself to obtain.

$$\vec{C} \cdot \vec{C} = \left(\vec{A} - \vec{B}\right) \cdot \left(\vec{A} - \vec{B}\right).$$

After expansion on the right side we find

$$C^2 = A^2 + B^2 - 2\vec{A} \cdot \vec{B},$$

where we have used the fact that the order of the vectors in a dot product does not matter, i.e. $\vec{A} \cdot \vec{B} = \vec{B} \cdot \vec{A}$. We can now use the dot product between $\vec{A}$ and $\vec{B}$ to get the law of cosines for a triangle.

$$C^2 = A^2 + B^2 - 2AB\cos\theta.$$

2.3.6 Vector Product Analytically

A vector product is also relatively easy to work out by making use of the results of the vector products among the base vectors given in Eqs. 2.6-2.8. Let us rewrite them here also:

$$\boxed{\hat{u}_x \times \hat{u}_y = \hat{u}_z}$$

$$\boxed{\hat{u}_y \times \hat{u}_z = \hat{u}_x}$$

$$\boxed{\hat{u}_z \times \hat{u}_x = \hat{u}_y}$$

$$\boxed{\hat{u}_x \times \hat{u}_x = \hat{u}_y \times \hat{u}_y = \hat{u}_z \times \hat{u}_z = 0.}$$

Now, to find the cross product of two vectors we will first express each vector in terms of the base vectors and then expand the product out by using distributive, associative, and linearity properties of the vector product operation.

$$\vec{A} \times \vec{B} = (A_x\hat{u}_x + A_y\hat{u}_y + A_z\hat{u}_z) \times (B_x\hat{u}_x + B_y\hat{u}_y + B_z\hat{u}_z) \quad (2.37)$$
$$= A_x\hat{u}_x \times (B_x\hat{u}_x + B_y\hat{u}_y + B_z\hat{u}_z)$$
$$+ A_y\hat{u}_y \times (B_x\hat{u}_x + B_y\hat{u}_y + B_z\hat{u}_z)$$
$$+ A_z\hat{u}_z \times (B_x\hat{u}_x + B_y\hat{u}_y + B_z\hat{u}_z) \quad (2.38)$$

The calculations for the three lines in Eq. 2.38 are similar. Let us work out the first line in detail.

$$A_x\hat{u}_x \times (B_x\hat{u}_x + B_y\hat{u}_y + B_z\hat{u}_z)$$
$$= A_xB_x(\hat{u}_x \times \hat{u}_x) + A_xB_y(\hat{u}_x \times \hat{u}_y) + A_xB_z(\hat{u}_x \times \hat{u}_z)$$
$$= A_xB_y\hat{u}_z - A_xB_z\hat{u}_y.$$

The second and third lines in Eq.2.38 work out similarly with the following results.

$$A_y\hat{u}_y \times (B_x\hat{u}_x + B_y\hat{u}_y + B_z\hat{u}_z) = -A_yB_x\hat{u}_z + A_yB_z\hat{u}_x$$
$$A_z\hat{u}_z \times (B_x\hat{u}_x + B_y\hat{u}_y + B_z\hat{u}_z) = A_zB_x\hat{u}_y - A_zB_y\hat{u}_x.$$

Summarizing, the vector product in terms of components gives the following.

$$\boxed{\vec{A} \times \vec{B} = \hat{u}_x(A_yB_z - A_zB_y) + \hat{u}_y(A_zB_x - A_xB_z) + \hat{u}_z(A_xB_y - A_yB_x).}$$
$$(2.39)$$

The result of the vector product can also be obtained by evaluating the following determinant, which serves as a good way of organizing terms in the cross product.

$$\vec{A} \times \vec{B} = \begin{vmatrix} \hat{u}_x & \hat{u}_y & \hat{u}_z \\ A_x & A_y & A_z \\ B_x & B_y & B_z \end{vmatrix} \quad (2.40)$$
$$= \hat{u}_x(A_yB_z - A_zB_y) + \hat{u}_y(A_zB_x - A_xB_z) + \hat{u}_z(A_xB_y - A_yB_x).$$

Once cross product has been calculated, the magnitude and direction of the resulting vector can be determined in the usual way described above.

Example 2.3.7. Numerical example of vector product. Evaluate the vector product of the following vectors given in a particular Cartesian coordinate system: $\vec{A} = 2\hat{u}_y + 3\hat{u}_z$ and $\vec{B} = 4\hat{u}_x + 5\hat{u}_y + 6\hat{u}_z$.

Solution. The determinant notation of a vector product helps organize the information well. Note that in $\vec{A}$ there is no component vector along the x-axis, which will cause an entry of zero in the corresponding place in the determinant.

$$\vec{A} \times \vec{B} = \begin{vmatrix} \hat{u}_x & \hat{u}_y & \hat{u}_z \\ 0 & 2 & 3 \\ 4 & 5 & 6 \end{vmatrix}$$
$$= \hat{u}_x (2 \times 6 - 3 \times 5) + \hat{u}_y (3 \times 4 - 0) + \hat{u}_z (0 - 2 \times 4)$$
$$= -3\hat{u}_x + 12\hat{u}_y - 8\hat{u}_z.$$

Example 2.3.8. Unit normal to the plane of two vectors. Two non-collinear vectors $\vec{A}$ and $\vec{B}$ define a plane in space since they can always be drawn from the same point. In a particular coordinate system suppose the two vectors have the following representations $\vec{A} = 2\hat{u}_y + 3\hat{u}_z$ and $\vec{B} = 4\hat{u}_x + 5\hat{u}_y + 6\hat{u}_z$. Find a unit vector normal to the plane.

Solution. We know that a vector product of two vectors is normal to the plane of the two vectors. Therefore, we can determine the unit normal vector from the cross product of the two vectors. We have already worked out the cross product of these vectors in Example 2.3.7.

$$\vec{A} \times \vec{B} = -3\hat{u}_x + 12\hat{u}_y - 8\hat{u}_z.$$

Now, we must divide this vector by its magnitude to obtain the unit normal, $\hat{n}$.

$$\hat{n} = \frac{\vec{A} \times \vec{B}}{|\vec{A} \times \vec{B}|} = \frac{-3\hat{u}_x + 12\hat{u}_y - 8\hat{u}_z}{\sqrt{3^2 + 12^2 + 8^2}} = -0.20\hat{u}_x + 0.82\hat{u}_y - 0.54\hat{u}_z.$$

Note that the magnitude of $\hat{n}$ is equal to 1.00 for two decimal places. Beyond that the rounding error in the calculation makes the magnitude deviate from 1. Note also that there is another unit vector in addition to $\hat{n}$ that is normal to the plane of vectors $\vec{A}$ and $\vec{B}$: the other vector is $-\hat{n}$ which is pointed in exactly the opposite direction to $\hat{n}$.

2.4 EXERCISES

Geometric Picture of Vectors

Ex 2.4.1. In this exercise you will draw two vectors to scale on a graph paper, and then add them graphically. Two forces act on an object: a force of 3 Newton(N) in some direction and 4 N force in the direction 90° from the direction of the 3 N force such that the two forces fall in a plane. (a) Choose a scale for drawing, such as a 1 cm line for 1 N, and draw the vectors on the same graph paper using your scale. Make sure that you give your scale with the figure. (b) Draw or redraw the vectors so that the tail of the second vector is at the tip of the first vector. Use tip-to-tail method to draw the vector that is equal to the sum of the two vectors. (c) Use the scale for the drawing and convert the length of the sum vector on the graph to determine the magnitude of the sum. (d) Use a protractor and read off the direction of the sum vector with respect to the force whose magnitude is 3 N.

Ex 2.4.2. Three forces act at three points on a ring such that their directions are in one plane. We will use one of the forces to indicate the directions of the other two. The forces are: 3 N force in some direction in the plane, 6 N force in a direction that is 60° counterclockwise from the direction of the 3-N force and 4.5 N force that is acting 90° clockwise from the direction of the 3-N force. Graphically add the given force vectors by drawing them in a tip-to-tail way, and find the magnitude and direction of the sum force, also called the net force.

Ex 2.4.3. Three forces act at three points on a ring such that their directions are in one plane. We will use one of the forces to indicate the directions of the other two. The forces are: 20 N force in some direction in the plane, 60 N force in a direction that is 135° counterclockwise from the direction of the 20-N force and 45 N force that is acting 30° clockwise from the direction of the 20-N force. By adding these forces graphically find out the fourth force needed to make the net force on the ring zero.

Ex 2.4.4. Write vector equations from the vector diagrams given in Fig. 2.33.

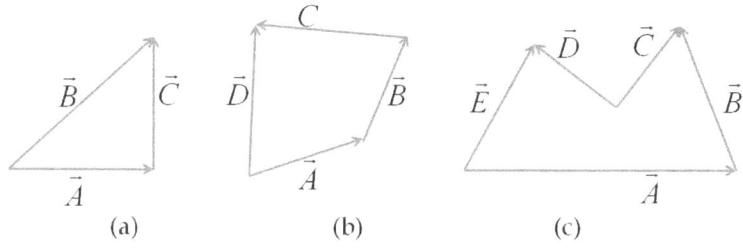

Figure 2.33: Exercise 2.4.4.

Ex 2.4.5. Draw vector diagrams for the following vector equations. (a) $\vec{A} + \vec{B} - \vec{C} = 0$; (b) $\vec{A} - \vec{B} - \vec{C} = 0$; (c) $\vec{A} - \vec{B} - \vec{C} + \vec{D} = 0$; (d) $\vec{A} + \vec{B} + \vec{C} - \vec{D} = 0$; (e) $\vec{A} + \vec{B} + \vec{C} + \vec{D} - \vec{E} = 0$.

Ex 2.4.6. A force $\vec{F}$ of magnitude 10 N and a displacement $\vec{s}$ of magnitude 10 cm are shown to their corresponding scales in Fig. 2.34. To work graphically on the given vectors, it may be helpful to transfer the figure to another paper using the same scale. (a) Determine the scales for the two vectors that was used for the given drawings, i.e., state the length of the force arrow that represents 1 N of force and the length of the displacement arrow that represents 1 cm of displacement. Since the figure shows two different properties we have two different scales, one for each property. (b) Draw the projection of the force vector on the displacement vector, and determine the value of the projection of the force vector onto the displacement vector in the unit of N. (c) Draw the projection of the displacement vector on the force vector, and determine the value of the projection of the displacement vector onto the force vector in the unit of cm. (d) Verify that the product of the projection of $\vec{F}$ on $\vec{s}$ with the magnitude of $\vec{s}$ is equal to the product of the projection of $\vec{s}$ on $\vec{F}$ with the magnitude of $\vec{F}$ within the margin of error of your measurements of the projections.

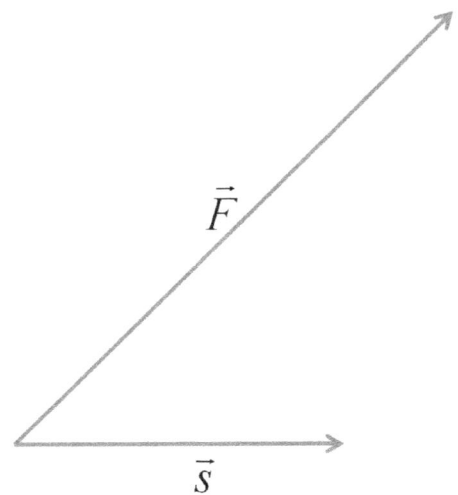

Figure 2.34: Exercise 2.4.6.

Ex 2.4.7. Repeat the exercises given in Exercise 2.4.6 for the force and displacement vectors given in Fig. 2.35 with the magnitude of force equal to 40 N and the magnitude of displacement equal to 3 m.

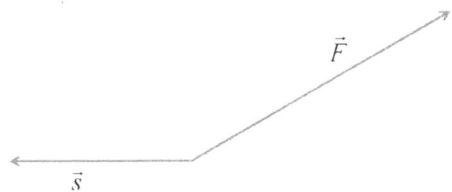

Figure 2.35: Exercise 2.4.7.

Ex 2.4.8. Determine graphically the magnitude and direction of the vector product $\vec{s} \times \vec{F}$ of the two vectors given in Fig. 2.34.

Ex 2.4.9. Determine graphically the magnitude and direction of the vector product $\vec{s} \times \vec{F}$ of the two vectors given in Fig. 2.35.

Analytic Picture of Vectors

Note: Although I have included units when giving values of physical vectors that have dimensions, it is cumbersome to write units all the time in calculations. A regular practice is to convert all quantities in the same system of units before using those numbers in the calculations. If you follow this advice, you do not need to write units all the time in your calculations. You can do the calculations without the units, and then put the units at the end in your final answer.

Ex 2.4.10. Draw the following vectors written in component form. Here m is the unit meter. (a) $(2 \text{ m}) \, \hat{u}_x + (4 \text{ m}) \, \hat{u}_y$. (b) $(3 \text{ m}) \, \hat{u}_x + (-5 \text{ m}) \, \hat{u}_z$. (c) $(-1 \text{ m}) \, \hat{u}_y + (4 \text{ m}) \, \hat{u}_z$.

Ex 2.4.11. (a) Find the components of the vector given in Fig. 2.36 with respect to the axes $Oxyz$ and $Ox'y'z'$. The z and z'-axes are perpendicular to the given figure. Assume the given vector is entirely in the xy-plane of the two coordinate systems. (b) Write the vector using the unit vectors along axes in the two different coordinates - you should have two answers, one for each coordinate system. (c) Since there are two different representations for the same vector in the two different coordinates, and since the choice of coordinate system is arbitrary, what physical content do the values of components have for the vector?

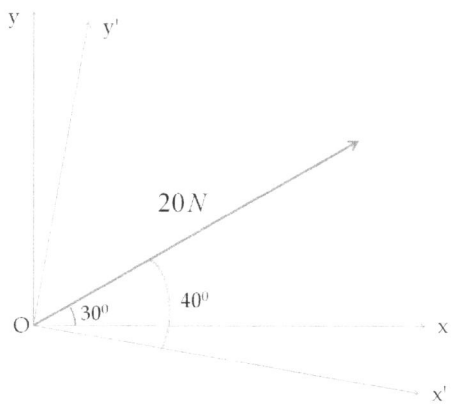

Figure 2.36: Exercise 2.4.11.

Ex 2.4.12. Determine the x and y-components of the following vectors in the xy-plane of a particular coordinate system. Here, the terms counterclockwise and clockwise refer to the rotation direction as you look from the side of the positive z-axis. (a) A force of magnitude 10 N in the direction of $30°$ counterclockwise from the positive x-axis direction. (b) A force of magnitude 10 N in the direction of $30°$ clockwise from the negative x-axis direction. (c) A force of magnitude 10 N in the direction of $30°$ counterclockwise from the positive y-axis direction. (d) A force of magnitude 10 N in the direction of $30°$ clockwise from the negative y-axis direction.

Ex 2.4.13. Determine the magnitudes and directions of the following vectors given in the component form with respect to a particular Cartesian

2.4. EXERCISES

coordinate system, where the positive x-axis points to the East, the positive y-axis points to the North and the positive z-axis points vertically up. Make sure you give the units for the magnitudes of the vectors if appropriate. From any angle(s) you calculate, state the physical direction in space for each vector. Here m is the unit meter. (a) $(3\text{ m})\,\hat{u}_x + (4\text{ m})\,\hat{u}_y$. (b) $(-3\text{ m})\,\hat{u}_x + (4\text{ m})\,\hat{u}_y$. (c) $(-3\text{ m})\,\hat{u}_x + (-4\text{ m})\,\hat{u}_y$. (d) $(3\text{ m})\,\hat{u}_x + (-4\text{ m})\,\hat{u}_y$.

Ex 2.4.14. Determine the magnitudes and directions of the following vectors given in component form with respect to a particular Cartesian coordinate system, where the positive x-axis points to the East, the positive y-axis points to the North and the positive z-axis points vertically up. Make sure you give units for the magnitudes of the vectors if appropriate. From any angle(s) you calculate, state the physical direction in space for each vector. Here m is the unit meter. (a) $(3\text{ m})\,\hat{u}_x + (4\text{ m})\,\hat{u}_y + (12\text{ m})\,\hat{u}_z$. (b) $(-3\text{ m})\,\hat{u}_x + (4\text{ m})\,\hat{u}_y + (12\text{ m})\,\hat{u}_z$. (c) $(3\text{ m})\,\hat{u}_x + (-4\text{ m})\,\hat{u}_y + (12\text{ m})\,\hat{u}_z$. (d) $(3\text{ m})\,\hat{u}_x + (4\text{ m})\,\hat{u}_y + (-12\text{ m})\,\hat{u}_z$. (e) $(-3\text{ m})\,\hat{u}_x + (-4\text{ m})\,\hat{u}_y + (-12\text{ m})\,\hat{u}_z$.

Ex 2.4.15. Three velocity vectors fall in the xy-plane of a coordinate system and have the following representations: $\vec{V}_1 = (3\text{ m/s})\hat{u}_x + (2\text{ m/s})\hat{u}_y$, $\vec{V}_2 = (9\text{ m/s})\hat{u}_x + (11\text{ m/s})\hat{u}_y$, and $\vec{V}_3 = (-8\text{ m/s})\hat{u}_y$. (a) Find the x and y-components of the sum of the three vectors. (b) Determine the magnitude and direction of the sum.

Ex 2.4.16. Three force vectors fall in the yz-plane of a coordinate system and have the following representations: $\vec{V}_1 = (100\text{ N})\hat{u}_y + (200\text{ N})\hat{u}_z$, $\vec{V}_2 = (-500\text{ N})\hat{u}_y + (100\text{ N})\hat{u}_z$, and $\vec{V}_3 = (-400\text{ N})\hat{u}_y + (500\text{ N})\hat{u}_z$. (a) Find the y and z-components of the sum of the three vectors. (b) Determine the magnitude and direction of the sum.

Ex 2.4.17. (a) Find the x and y-components of the two forces given in Fig. 2.37 with respect to the coordinates given in the figure. (b) Add the two vectors and find the magnitude and direction of the resultant vector.

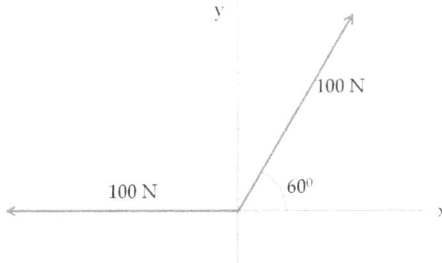

Figure 2.37: Exercise 2.4.17.

Ex 2.4.18. Calculate the scalar product between the following pairs of vectors: (a) $(10\text{ N})\,\hat{u}_x + (-5\text{ N})\,\hat{u}_y$ and $(3\text{ m})\,\hat{u}_x + (4\text{ m})\,\hat{u}_y$, (b) $(40\text{ N})\,\hat{u}_x + (-10\text{ N})\,\hat{u}_y$ and $(3\text{ m})\,\hat{u}_x + (4\text{ m})\,\hat{u}_z$, (c) $(100\text{ N})\,\hat{u}_x + (-10\text{ N})\,\hat{u}_y$ and $(30\text{ m})\,\hat{u}_y + (4\text{ m})\,\hat{u}_z$, (d) $(2\text{ N})\,\hat{u}_x + (3\text{ N})\,\hat{u}_y + (4\text{ }N)\,\hat{u}_z$ and $(4\text{ m/s})\,\hat{u}_x + (3\text{ m/s})\,\hat{u}_y + (2\text{ m/s})\,\hat{u}_z$, (e) $(100\text{ N})\,\hat{u}_x + (-10\text{ N})\,\hat{u}_y + (100\text{ }N)\,\hat{u}_z$ and $(30\text{ m/s})\,\hat{u}_y + (4\text{ m/s})\,\hat{u}_z$.

Ex 2.4.19. Find the angle between each pair of vectors given in Ex. 2.4.18.

Ex 2.4.20. Find the angles each vector makes with positive x, y and z-axes. (a) $3\hat{u}_x + 2\hat{u}_y + 4\hat{u}_z$, (b) $-3\hat{u}_x + 2\hat{u}_y + 4\hat{u}_z$, (c) $3\hat{u}_x - 2\hat{u}_y + 4\hat{u}_z$, (d) $3\hat{u}_x + 2\hat{u}_y - 4\hat{u}_z$, (e) $3\hat{u}_x - 2\hat{u}_y - 4\hat{u}_z$, (f) $-3\hat{u}_x - 2\hat{u}_y - 4\hat{u}_z$,

Ex 2.4.21. Find unit vectors in the directions of the following vectors. (a) $(10\ N)\ \hat{u}_x + (-5\ N)\ \hat{u}_y$, (b) $(3\ m)\ \hat{u}_x + (4\ m)\ \hat{u}_z$, (c) $(30\ m)\ \hat{u}_y + (4\ m)\ \hat{u}_z$, (d) $(2\ N)\ \hat{u}_x + (3\ N)\ \hat{u}_y + (4\ N)\ \hat{u}_z$, (e) $(100\ N)\ \hat{u}_x + (-10\ N)\ \hat{u}_y + (100\ N)\ \hat{u}_z$.

Ex 2.4.22. For each vector in the following list find two unit vectors in the xy-plane perpendicular to it. (a) $\hat{u}_x$, (b) $\hat{u}_y$, (c) $\hat{u}_x + \hat{u}_y$, (d) $\hat{u}_x - \hat{u}_y$, (d) $\frac{1}{2}\hat{u}_x - \frac{\sqrt{3}}{2}\hat{u}_y$, (f) $a\hat{u}_x + b\hat{u}_y$, (g) $\cos(\theta)\hat{u}_x + \sin(\theta)\hat{u}_y$.

Ex 2.4.23. Evaluate the vector products:

(a) $[(3\ m)\ \hat{u}_x + (4\ m)\ \hat{u}_y] \times [(10\ N)\ \hat{u}_x + (-5\ N)\ \hat{u}_y]$, (b) $[(3\ m)\ \hat{u}_x + (4\ m)\ \hat{u}_z] \times [(40\ N)\ \hat{u}_x + (-10\ N)\ \hat{u}_y]$, (c) $[(30\ m)\ \hat{u}_y + (4\ m)\ \hat{u}_z] \times [(100\ N)\ \hat{u}_x + (-10\ N)\ \hat{u}_y]$, (d) $[(4\ m/s)\ \hat{u}_x + (3\ m/s)\ \hat{u}_y + (2\ m/s)\ \hat{u}_z] \times [(2\ N)\ \hat{u}_x + (3\ N)\ \hat{u}_y + (4\ N)\ \hat{u}_z]$, (e) $[(30\ m/s)\ \hat{u}_y + (4\ m/s)\ \hat{u}_z] \times [(100\ N)\ \hat{u}_x + (-10\ N)\ \hat{u}_y + (100\ N)\ \hat{u}_z]$.

2.5 PROBLEMS

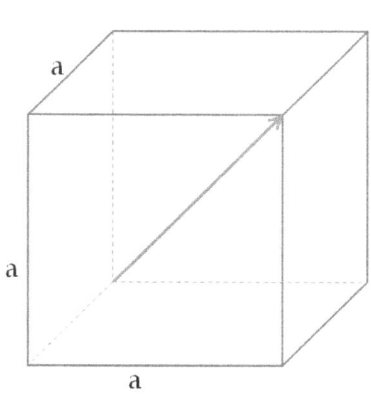

Figure 2.38: Problem 2.5.1.

Problem 2.5.1. Find a unit vector that is perpendicular to the body diagonal of a cube shown in Fig. 2.38.

Problem 2.5.2. Find the angle between the lines from the center of one face to the corners on the opposite face of a cube.

Problem 2.5.3. Two body diagonals of a cube cross at the center of the cube. Find the angle between them.

Problem 2.5.4. Three vectors $\vec{A}$, $\vec{B}$, and $\vec{C}$ are placed on the adjacent edges of a parallelepiped with their tails at the common vertex. Prove that $\vec{A} \cdot \left(\vec{B} \times \vec{C} \right)$ is equal to the volume of the parallelepiped.

Problem 2.5.5. Prove the following identity for arbitrary vectors $\vec{A}$, $\vec{B}$, and $\vec{C}$:

$$\vec{A} \cdot \left(\vec{B} \times \vec{C} \right) = \vec{B} \cdot \left(\vec{C} \times \vec{A} \right) = \vec{C} \cdot \left(\vec{A} \times \vec{B} \right).$$

Problem 2.5.6. Prove that, if the magnitude of the two vectors $\vec{A}$ and $\vec{B}$ are equal, then $\vec{A} + \vec{B}$ and $\vec{A} - \vec{B}$ are perpendicular to each other.

Problem 2.5.7. Prove that, if $|\vec{A} - \vec{B}| = |\vec{A} + \vec{B}|$, then vector $\vec{A}$ is perpendicular to vector $\vec{B}$.

Problem 2.5.8. Suppose $\vec{A} \cdot \vec{B} = \vec{A} \cdot \vec{C}$. Is $\vec{B} = \vec{C}$? Why or why not? Give a graphical interpretation of the given statement also.

Problem 2.5.9. Suppose $\vec{A} \times \vec{B} = \vec{A} \times \vec{C}$. Is $\vec{B} = \vec{C}$? Why or why not? What is the most we can say about $\vec{B}$ and $\vec{C}$? Give a graphical interpretation of the given statement also.

2.5. PROBLEMS

Problem 2.5.10. Suppose $\vec{A}\cdot\vec{B} = \vec{A}\cdot\vec{C}$ and $\vec{A}\times\vec{B} = \vec{A}\times\vec{C}$. Is $\vec{B} = \vec{C}$? Why or why not?

Problem 2.5.11. Prove the following identity for vector cosines, $\cos^2\alpha + \cos^2\beta + \cos^2\gamma = 1$, where α, β, and γ are angles a vector makes with the positive x, y and z-axes respectively.

Problem 2.5.12. We proved the law of cosines in the text using the scalar product of vectors placed on the sides of a triangle. Prove the law of sines by using the vector product between vectors on the sides of a triangle. The law of sine says that if the sides of a triangle have lengths A, B, and C and the angles opposite to the sides are $\angle A$, $\angle B$, and $\angle C$ respectively, then

$$\frac{\sin \angle A}{A} = \frac{\sin \angle B}{B} = \frac{\sin \angle C}{C}$$

Problem 2.5.13. Prove that an arbitrary vector $\vec{A}$ can always be written as a sum of a vector parallel to a vector $\vec{B}$ and a vector perpendicular to $\vec{B}$. That is, show that

$$\vec{A} = a\vec{B} + b\vec{B}_\perp,$$

where $\vec{B}\cdot\vec{B}_\perp = 0$, and a and b are some scalars. Hint: Use the projection of $\vec{A}$ on $\vec{B}$ to construct the vector you need for the vector parallel to $\vec{B}$. Let $\vec{A}_1$ be the vector parallel to $\vec{B}$ that you need. Then, show that $\vec{A} - \vec{A}_1$ is a vector that is perpendicular to vector $\vec{B}$.

Chapter 3

KINEMATICS

Contents

3.1	INTRODUCTION	**60**
3.2	POSITION VECTOR	**60**
3.3	DISPLACEMENT VECTOR	**61**
3.4	AVERAGE VELOCITY	**69**
	3.4.1 Average Speed	73
3.5	VELOCITY AND SPEED	**74**
	3.5.1 Velocity in One-dimensional Motion	74
	3.5.2 Velocity - General	82
	3.5.3 Speed	87
3.6	ACCELERATION	**88**
3.7	MOTION USING POLAR COORDINATES .	**98**
	3.7.1 Polar Coordinates	98
	3.7.2 Unit Vectors $\hat{u}_r$ and $\hat{u}_\theta$	99
	3.7.3 Velocity and Acceleration in Polar Coordinates	101
	3.7.4 Circular Motion	102
3.8	EXERCISES	**109**
3.9	PROBLEMS	**122**

3.1 INTRODUCTION

Kinematics deals with the basic physical quantities necessary for a description of motion. We use a properly marked three-dimensional Cartesian coordinate system and a clock to define position, displacement, velocity and acceleration of an object in precise mathematical language. The coordinate system is called the **reference frame** or simply the frame. The choice of origin and the directions of axes of the coordinate system are arbitrary. We will see that true vectors such as displacement and velocity are independent of this choice. The choices are made to simplify the mathematical description of the motion.

If we assume that the three-dimensional space we inhabit consist of geometrical points, then each point of space will have unique values for Cartesian coordinates (x, y, z) in a particular frame. The coordinates of the geometrical points in space can be used to indicate the location of an object if the object occupies only one space point at a time. Fundamental particles are assumed to be point-like in this sense, and we refer to them as **point particles**. However, we would also like to study extended objects such as boxes, people, cars, planets, stars and galaxies, which occupy some volume of space containing infinitely many points of space.

How can we follow the motion of extended, non-point-like, objects? We will demonstrate in a later chapter that there exists a special geometrical point for each object, called the center of mass, whose motion is like that of a point particle. We will find that the motions of various parts of an extended object naturally separate into a motion of its center of mass and another motion, such as rotation or deformation, about the center of mass.

Until we have a demonstration of this separation, you can assume that, whenever we discuss the motion of a large object by assigning one point to it, we are looking at the motion of the center of mass. If an object does not change shape or orientation, then all points on the object would have the same motion as the center of mass. Therefore, for non-rotating rigid bodies, you could choose a convenient point of the object for describing its motion. For instance, for a moving train, the position of the train can be said to be given by the front end or back end of the train as long as the train does not rotate.

3.2 POSITION VECTOR

In order to locate the position of a particle, we need to measure the distance of the particle from some reference point and a way to tell the direction from the reference point. A properly marked three-dimensional Cartesian coordinate system with origin at the reference point adequately specifies the location of points in space. Therefore, the values of the three coordinates (x, y, z) of a point can completely specify the position of a

3.3. DISPLACEMENT VECTOR

particle as shown in Fig. 3.2.

The vector from the origin of the coordinate system to the location of the particle is called the **position vector**. The position vector is denoted by $\vec{r}$. Note that the coordinates (x, y, z) of P do not represent any vector, it is the arrow from the origin to the point P that is the position vector. We say that position of the particle is the space point of coordinates (x, y, z) and the position vector is the vector $\vec{r}$ as shown in Fig. 3.2. The

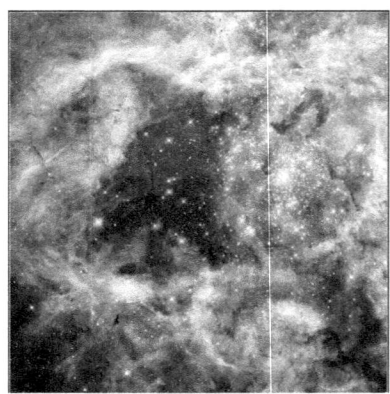

Figure 3.1: The Hubble telescope took this picture of a stellar group called R136 in the Large Magellanic Cloud. These stars are "young", only a few million years old. The region is a hotbed for new star births. How would you assign positions to these stars? Photo credits: NASA/ESA.

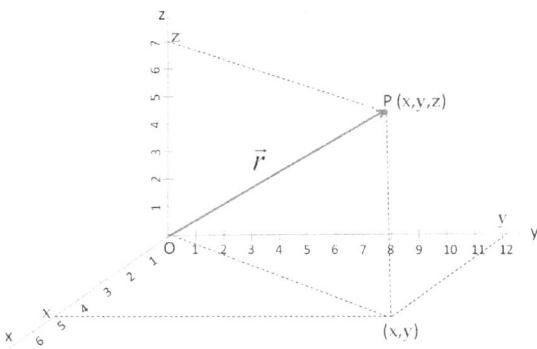

Figure 3.2: Position vector $\vec{r}$ is pointed from the reference point O to the location of the point P in space. Choice of a different origin will change the position vector, which will be from the new origin to the same location P.

position vector can be written using the unit base vectors of the Cartesian coordinates, which we have chosen to denote by $(\hat{u}_x, \hat{u}_y, \hat{u}_z)$.

$$\vec{r} = x\,\hat{u}_x + y\,\hat{u}_y + z\,\hat{u}_z. \tag{3.1}$$

Since, one end of the position vector is at the arbitrarily chosen reference point, the position vector does depend on the choice of the coordinate system. If the origin of the coordinate system was moved to another place, the position vector from the new origin will be different than the one from the old origin. We will see below that although position vector itself may not be a "true" vector, the change in position, being independent of the choice of a coordinate system, is a true vector.

3.3 DISPLACEMENT VECTOR

One of the characteristics of a moving object is that its position changes with time. The position of a point particle traces out a trajectory in space as it moves from one point to the next. To describe this change, we define a quantity called the **displacement vector** or simply displacement. Let the positions of a particle be $P_1(x_1, y_1, z_1)$ at time t_1 and $P_2(x_2, y_2, z_2)$ at time t_2. Then, the vector $\vec{s}_{12}$ from point P_1 to point P_2 is called the displacement vector as illustrated in Fig. 3.3. The magnitude of $\vec{s}_{12}$ is the direct distance between points P_1 and P_2 and the direction is in the direction of the arrow from P_1 to P_2. We will use the symbol $\vec{s}$ for the

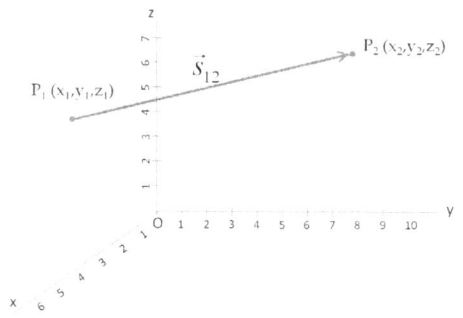

Figure 3.3: The displacement vector $\vec{s}_{12}$ from point P_1 to point P_2 in space.

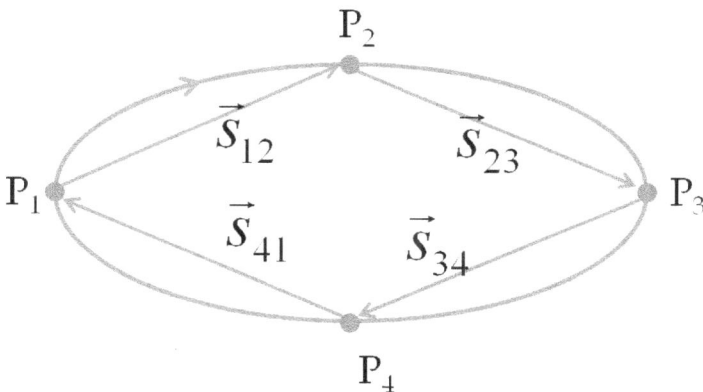

Figure 3.4: Displacement vectors between four successive points in the motion of a planet in an elliptical path around the sun.

displacement vector rather than the symbol $\vec{d}$ since we will soon take the derivative of displacement and other functions and the letter d will be used for that purpose.

In Fig. 3.4 shows four displacement vectors in the motion of a planet around the Sun. The planet moves in an elliptical orbit, but the displacements between points on the trajectory are given by straight segments.

Note that the definition of the displacement vector does not mention the origin or the reference point: **the displacement vector is the vector from the initial position to the final position**, period. We illustrate this fact in Fig. 3.4 by drawing four displacement vectors in the motion of a planet. It is clear that you do not need any coordinate system to define a displacement vector.

Although we do not need any reference point or coordinate system for a definition of the displacement vector, it is often helpful to write the displacement vector in terms of the position vectors of the particle, which does require a reference point, which we will place at the origin of a coordinate system.

Let $\vec{r}_1$ be the position vector of the particle when it is at P_1 and $\vec{r}_2$ be

3.3. DISPLACEMENT VECTOR

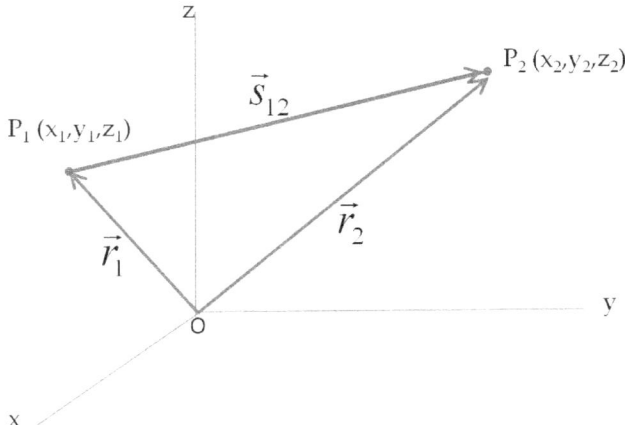

Figure 3.5: The displacement vector $\vec{s}_{12}$ from point P_1 to point P_2 in space can be written in terms of position vectors of the initial and final points as $\vec{s}_{12} = \vec{r}_2 - \vec{r}_1$ as the triangle of vectors shows here.

the position vector when it is at P_2, as shown in Fig. 3.5. Then, it is easy to see that vectors $\vec{r}_1$, $\vec{s}_{12}$ and $(-\vec{r}_2)$ form a closed triangle of vectors. In other words, the sum of vectors $\vec{r}_1$ and $\vec{s}_{12}$ is equal to $\vec{r}_2$. Therefore, the displacement from P_1 to P_2 is equal to the vector obtained by subtracting the initial position vector from the final position vector.

$$\boxed{\vec{s}_{12} = \vec{r}_2 - \vec{r}_1.} \tag{3.2}$$

Since, each position vector can be written in terms of the coordinates of the corresponding points, we can write the displacement vector in terms of the coordinates of the two points also.

$$\begin{aligned}\vec{s}_{12} &= \vec{r}_2 - \vec{r}_1 \\ &= (x_2\hat{u}_x + y_2\hat{u}_y + z_2\hat{u}_z) - (x_1\hat{u}_x + y_1\hat{u}_y + z_1\hat{u}_z) \\ &= (x_2 - x_1)\hat{u}_x + (y_2 - y_1)\hat{u}_y + (z_2 - z_1)\hat{u}_z. \end{aligned} \tag{3.3}$$

An example of the resolution of a displacement vector in the xy-plane is displayed in Fig. 3.6. Note that regardless of the placement of the displacement vector, the projections of the beginning and ending of the displacement vector on the Cartesian axes give the length of vectors along the Cartesian axes that make up the original displacement vectors. You can say that $(x_2 - x_1)\hat{u}_x$ along the x-axis and $(y_2 - y_1)\hat{u}_y$ along the y-axis add up to the displacement vector shown in the figure.

Example:

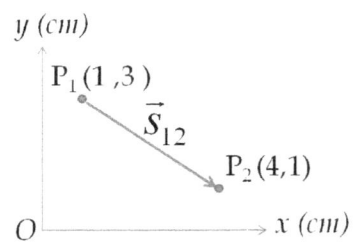

$\vec{r}_1 = 1\hat{u}_x + 3\hat{u}_y$ [cm]
$\vec{r}_2 = 4\hat{u}_x + 1\hat{u}_y$ [cm]
$\vec{s}_{12} = (4 - 1)\hat{u}_x$
$\quad + (1 - 3)\hat{u}_y$ [cm]

The factors $(x_2 - x_1)$, $(y_2 - y_1)$, $(z_2 - z_1)$ multiplying the unit vectors for the Cartesian axes in Eq. 3.3 are the x, y and z-components of the displacement vector from P_1 to P_2. As explained in the last chapter, components of a vector can be used to determine the magnitude and direction of the vector.

Magnitude of $\vec{s}_{12}$:

$$\boxed{|s_{12}| = \sqrt{(x_2 - x_1)^2 + (y_2 - y_1)^2 + (z_2 - z_1)^2}.}$$

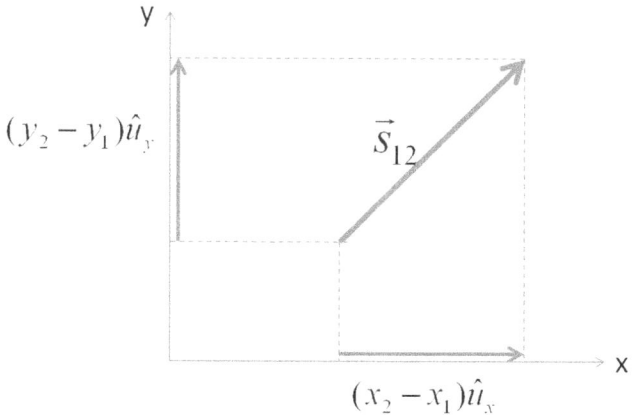

Figure 3.6: Displacement vector in terms of the base vectors shown for a displacement in the xy-plane.

Direction of $\vec{s}_{12}$:

To determine the direction of $\vec{s}_{12}$ either you draw the vector in a properly marked axis system, or give angles with respect to reference directions. Three different cases arise:

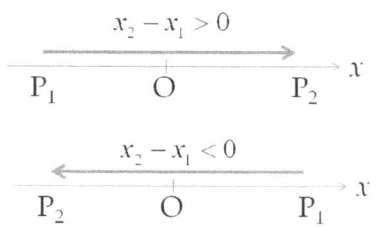

Figure 3.7: Directions along one axis.

1. If a displacement happens along one of the Cartesian axes, the sign of the component determines the direction with respect to the axis. For instance, if the displacement falls on the x-axis, then $\vec{s}_{12} = (x_2 - x_1)\hat{u}_x$. The x-component $(x_2 - x_1)$ will be positive if $x_2 > x_1$, giving a displacement in the direction of the unit vector $\hat{u}_x$, that is, towards the positive x-axis. On the other hand, if $x_2 < x_1$, the component $(x_2 - x_1)$ will be negative, which will give direction opposite to the unit vector $\hat{u}_x$ since multiplication of a vector by a negative number gives a vector that is in the opposite direction.

2. If the displacement vector falls in the xy, yz, or zx plane, or in a plane parallel to these planes, then you need to give only one angle for the direction, usually any of the four angles with respect to the four axis directions in one of these planes would work for this purpose.

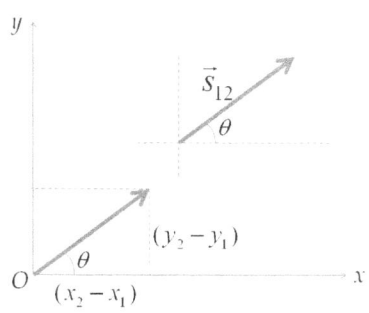

Figure 3.8: To determine the direction in the xy-plane, you can either draw coordinate axes with origin at the tail of the vector or drag the vector to the origin of the coordinate system keeping the physical orientation same.

Suppose, the vector falls in xy-plane, then you could give the angle the vector makes with either the positive x-axis, the positive y-axis, the negative x-axis, or the negative y-axis. Let θ be the angle with the positive x-axis, then you can use the following relation to calculate the angle.

$$\tan\theta = \frac{y_2 - y_1}{x_2 - x_1}.$$

Beware of the correction needed depending on the quadrant of the point $(x_2 - x_1, y_2 - y_1)$. A more detail discussion can be found in the chapter on vectors.

3.3. DISPLACEMENT VECTOR

3. Finally, if the displacement vector is neither along any axis or in any one of the planes mentioned, then we need the direction in the three-dimensional space. Now, you need two angles for which there are many choices. If you wish to assign the direction by the angles α, β, γ, the vector makes with the x, y and z-axes respectively, then you need only two of them since their cosines obey an identity.

$$\cos^2\alpha + \cos^2\beta + \cos^2\gamma = 1.$$

Alternately, you may use the polar and azimuthal angles of a spherical coordinate system to indicate the direction of a vector in three-dimensional space. These have been extensively discussed in the chapter on vectors (see Chapter 2).

Choice of Origin for Displacement Vector

We have already mentioned that the displacement vector does not require any reference point for its definition. We demonstrate here by analytic means that the displacement vector is independent of coordinate system as shown in Fig. 3.9. If you pick another place for the origin, say the point O' [read O-prime], and use a different orientation for the coordinate axes than the one for the origin O, then you will get different numerical values for the coordinates for points P_1 and P_2.

Let us denote the two coordinates systems by $Oxyz$ and $O'x'y'z'$ respectively. The coordinates for points P_1 and P_2 in the $O'x'y'z'$ system will be (x_1', y_1', z_1') and (x_2', y_2', z_2') respectively, which will be different from the coordinates (x_1, y_1, z_1) and (x_2, y_2, z_2) for the same points in the $Oxyz$ system. However, you will find that the difference in the coordinates in the two coordinate systems will be the same:

$$(x_1' - x_1',\ y_2' - y_1',\ z_2' - z_1') = (x_2 - x_1,\ y_2 - y_1,\ z_2 - z_1).$$

You can also see this directly in terms of the triangle of vectors in triangles $\triangle OP_1P_2$ and $\triangle O'P_1P_2$. The vector addition shows that, even though the pairs $\{\ \vec{r}_1,\ \vec{r}_2\ \}$ and $\{\ \vec{r}_1',\ \vec{r}_2'\ \}$ may be different, the displacement vector from P_1 to P_2 is the same in the two coordinate systems:

$$\vec{s}_{12} = \vec{r}_2 - \vec{r}_1 = \vec{r}_2' - \vec{r}_1'. \tag{3.4}$$

Example 3.3.1. Displacement vector for one-dimensional motion.
A box moves a distance of 2 m towards East. What is the displacement vector?

Solution. The statement of the problem already gives both the magnitude and direction of the displacement vector. So, no further work is necessary. The displacement is 2 m towards East as given.

Example 3.3.2. Displacement vector for one-dimensional motion.
A box moves a distance of 2 m towards East. A coordinate system is chosen

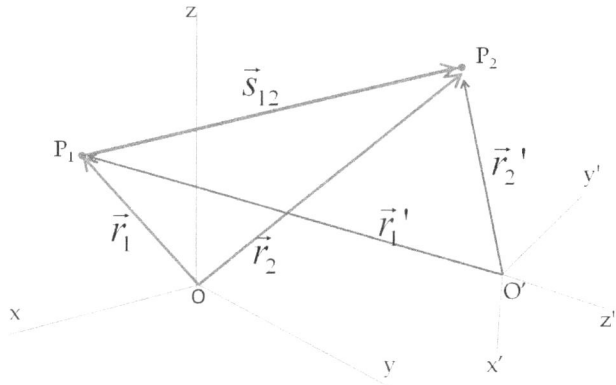

Figure 3.9: The displacement vector $\vec{s}_{12}$ from point P_1 to point P_2 from the perspectives of two different coordinate systems. The triangle of vectors in triangles $\triangle OP_1P_2$ and $\triangle O'P_1P_2$ show that, even though the pairs $\{\vec{r}_1, \vec{r}_2\}$ and $\{\vec{r}_1'\ \vec{r}_2'\}$ may be different, the displacement vector from P_1 to P_2 is the same in the two coordinate systems: $\vec{s}_{12} = \vec{r}_2 - \vec{r}_1 = \vec{r}_2' - \vec{r}_1'$.

which has the x-axis pointed towards East. What are the components of the displacement vector in the given coordinate system? Write the displacement vector using unit vectors along the Cartesian axes.

Solution. Suppose, the initial point is at the origin, then since the x-axis is in the direction of movement, the final coordinate would be $(2\text{ m}, 0, 0)$. Therefore, the x, y, and z-components of the displacement vector are 2 m, 0, and 0 respectively. This gives the following representation of the vector with respect to the axes: $\vec{s} = \{2\text{ m},\ East\} = (2\text{ m})\hat{u}_x$.

Example 3.3.3. Displacement vector for one-dimensional motion.
A box moves a distance of 2 m towards East. A coordinate system is chosen which has the y-axis pointed towards East. What are the components of the displacement vector in the given coordinate system? Write the displacement vector using unit vectors along the Cartesian axes.

Solution. This example together with Example 3.3.2 illustrates the arbitrariness of choice of coordinates. One may get different numerical values of components, but the choice does not affect the displacement vector. Suppose, the initial point is at the origin, then since the y-axis is now in the direction of movement, the final coordinate is $(0, 2\text{ m}, 0)$. Therefore, the x, y, and z-components of the displacement vector are 0, 2 m, and 0 respectively. Therefore, we now get the following representation of the displacement vector with respect to the new axes: $\vec{s} = \{2\text{ m},\ East\} = (2\text{ m})\hat{u}_y$. Note that the displacement is still 2 m towards East although the representation has changed.

Example 3.3.4. Displacement vector for one-dimensional motion.
A box moves a distance of 2 m towards East. A coordinate system is chosen which has the x-axis pointed towards 30° South of East and the y-axis is pointed 60° North of East. What are the components of the displacement vector in the given coordinate system? Write the displacement vector

3.3. DISPLACEMENT VECTOR

using unit vectors along the Cartesian axes.

Solution. This example together with Examples 3.3.2 and 3.3.3 illustrates the arbitrariness of the choice of coordinates. The choice of coordinates can make a one-dimensional motion problem appear two-dimensional as this example shows, but the physical vector is unaffected by the choice. Supposing the initial point to be at the origin, we can easily work out the coordinates of the final position to be $(2\ m\cos 30°, 2\ m\sin 30°, 0)$. Therefore, the x, y, and z-components of the displacement vector are ≈ 1.73 m, 1 m, and 0 respectively. Finally, we have the following representation of the displacement vector with respect to the new axes: $\vec{s} = \{2\ \text{m}, East\} = (1.73\ \text{m})\hat{u}_x + (1\ \text{m})\hat{u}_y$.

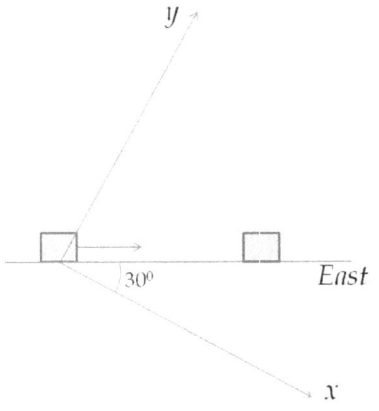

Figure 3.10: Example 3.3.4.

Once again, the displacement is $\{2\ \text{m}, East\}$, but the representation is different here than was in Examples 3.3.2 and 3.3.3. The differences are entirely due to the choice of the Cartesian axes. These examples illustrate that Cartesian axes are only calculational devices. Physics does not depend on your choice of coordinates, but simplicity or tediousness of the calculations may.

Example 3.3.5. Direction of Displacement Vector in a Plane from Components. You walk around in a room and after some time the **change in your coordinates** with respect to a particular Cartesian coordinate system is given as $(3\ \text{m}, 4\ \text{m}, 0)$. What is your displacement? Find both magnitude and direction.

Solution. Note that the given information allows you to write the displacement vector in the analytic form in terms of the unit vectors along the Cartesian axes.

$$\vec{s} = (3\ \text{m})\hat{u}_x + (4\ \text{m})\hat{u}_y.$$

The components can be used to determine the magnitude and direction.
Magnitude: $s = \sqrt{(3\ \text{m})^2 + (4\ \text{m})^2} = 5$ m.
Direction: Since the motion is in xy-plane, we can find one angle and refer the direction using that angle. Since, the vector is in the first quadrant of the xy-plane, we work out the angle the vector makes with respect to the positive x-axis. This is directly given by the arc tangent.

$$\theta = \arctan\left(\frac{4\ \text{m}}{3\ \text{m}}\right) \approx 53°.$$

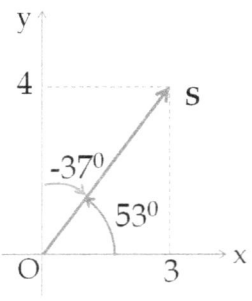

Figure 3.11: Example 3.3.5. The direction of a vector can be specified any angle with reference directions.

This says that the direction of the vector is in the xy-plane of the gives coordinate system at an angle counter-clockwise from the positive x-axis. Note again that the angle itself is not the direction - if you choose to give the angle with respect to the positive y-axis, the value will be $-37°$ for the same direction in the plane. So, the value of angle, in itself, is not the full information about the direction - you need to state what the angle means in the real space. Minus in $-37°$ refers to the clockwise direction as you look down from the positive z-axis.

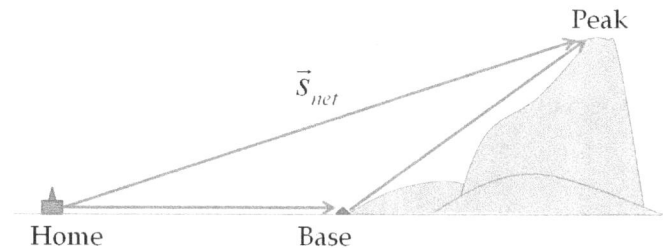

Figure 3.12: Example 3.3.6.

Example 3.3.6. Adding Displacement Vectors. You travel from your home to the base of a mountain, and then, climb the mountain. The coordinates of your home, base of the mountain and the top of the mountain in a particular coordinate system are: home (0,0,0), base (4000 m, 3000 m, 0), peak (4500 m, 2500 m, 800 m). Find the magnitude and direction of the net displacement from home to the peak (a) directly, and (b) by adding the two displacements, from home to the base and from base to the peak.

Solution. (a) To find the displacement directly, we use the coordinates of the home and the peak. This gives us the representation of the net displacement in the given Cartesian coordinate system, which can be used to determine the magnitude and direction of the net displacement vector. We will suppress writing units in calculations and put the units back at the end. The net displacement written in terms of the unit vectors along the Cartesian axes are

$$\vec{s}_{net} = (4500 - 0) \text{ m}\hat{u}_x + (2500 - 0) \text{ m}\hat{u}_y + (800 - 0) \text{ m}\hat{u}_z.$$

The magnitude of the net displacement is

$$s_{net} = \sqrt{(4500 \text{ m})^2 + (2500 \text{ m})^2 + (800 \text{ m})^2} \approx 5210 \ m. \tag{3.5}$$

The direction can be given by polar and azimuthal angles of a spherical coordinate system with the following values. For clarity in the calculations, we will not writs units in the intermediate steps.

$$\text{Polar angle: } \theta = \arctan\left(\frac{\sqrt{4500^2 + 2500^2}}{800}\right) = \arctan\left(\frac{5}{1}48800\right) \approx 81°.$$

$$\text{Azimuthal angle: } \phi = \arctan\left(\frac{2500}{4500}\right) \approx 29°.$$

Since, the polar angle and azimuthal angles are in the first octant, we do not need to make any adjustments. The angles can be read off to say that the direction is $\approx 29°$ counterclockwise from the positive x-axis and and $\approx 81°$ towards the xy-plane from the z-axis.

As mentioned above, one can also find the direction of a vector by working out the angles, α, β and γ, the vector makes with the x, y and z-axes respectively by using the direction cosines. We will not pause here

3.4. AVERAGE VELOCITY

to elaborate on this. An interested student should try to see how that can be done.

(b) The two displacements from home to base and from base to the peak are

$$\vec{s}_{12} = (4000 - 0) \text{ m}\hat{u}_x + (3000 - 0) \text{ m}\hat{u}_y + (0 - 0)\hat{u}_z$$
$$\vec{s}_{23} = (4500 - 4000) \hat{u}_x + (2500 - 3000) \text{ m}\hat{u}_y + (800 - 0) \text{ m}\hat{u}_z$$

Vector equation for the net displacement $\vec{s}_{net}$ is

$$\vec{s}_{net} = \vec{s}_{12} + \vec{s}_{23},$$

which gives the same expression for the $\vec{s}_{net}$ as in Eq. 3.3 since the components will add separately.

$$\begin{aligned}\vec{s}_{net} &= \vec{s}_{12} + \vec{s}_{23} \\ &= [(4000 - 0) \text{ m}\hat{u}_x + (3000 - 0) \text{ m}\hat{u}_y + (0 - 0)\hat{u}_z] \\ &\quad + [(4500 - 4000) \text{ m}\hat{u}_x + (2500 - 3000) \text{ m}\hat{u}_y + (800 - 0) \text{ m}\hat{u}_z] \\ &= 4500 \text{ m}\hat{u}_x + 2500 \text{ m}\hat{u}_y + 800 \text{ m}\hat{u}_z.\end{aligned}$$

Putting the units back the answer is

$$\vec{s}_{net} = (4500 \text{ m}) \hat{u}_x + (2500 \text{ m}) \hat{u}_y + (800 \text{ m}) \hat{u}_z.$$

3.4 AVERAGE VELOCITY

The rate at which position changes with time gives us information about the flow of motion. A simple measure of the rapidity of a motion, both how fast the movement occurs and in which direction the movement is taking place, is obtained by dividing the displacement vector by the interval since the displacement vector has both the magnitude and direction information. This quantity is called the **average velocity** of the object in that interval. Let there be a displacement $\vec{s}_{12}$ between time t_1 and t_2 then the average velocity $\vec{v}_{ave}$ during this interval will be

$$\boxed{\vec{v}_{ave} = \frac{\vec{s}_{12}}{t_2 - t_1}.} \quad (3.6)$$

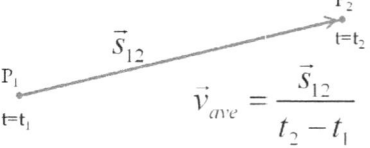

We saw above that the displacement vector can be written in a coordinate system using the position vectors for the starting and ending points of the interval. That is, if the position vectors at t_1 and t_2 are $\vec{r}_1$ and $\vec{r}_2$, respectively, then $\vec{s}_{12} = \vec{r}_2 - \vec{r}_1$. Therefore, we can write the average velocity in terms of the change in position vectors also.

$$\vec{v}_{ave} = \frac{\vec{r}_2 - \vec{r}_1}{t_2 - t_1}. \quad (3.7)$$

One often writes this relation in another notation. Let $\Delta t(= t_2 - t_1)$ denote the time interval and $\Delta \vec{r}(= \vec{r}_2 - \vec{r}_1)$ the displacement during that interval. Then, the average velocity is

$$\vec{v}_{ave} = \frac{\Delta \vec{r}}{\Delta t}, \tag{3.8}$$

which can be rearranged to get the change in position in the finite interval Δt

$$\Delta \vec{r} = \vec{v}_{ave} \Delta t. \tag{3.9}$$

Furthermore, using Eq. 3.3 we can express displacement in terms of the changes in coordinates, $(\Delta x, \Delta y, \Delta z)$ and write average velocity as

$$\vec{v}_{ave} = \left(\frac{\Delta x}{\Delta t}\right)\hat{u}_x + \left(\frac{\Delta y}{\Delta t}\right)\hat{u}_y + \left(\frac{\Delta z}{\Delta t}\right)\hat{u}_z. \tag{3.10}$$

The quantities multiplying the base vectors are called the x, y and z-**components of the average velocity**, and as you have studied in the last chapter, they are labeled with x, y and z subscripts similar to the labeling of components of other vectors.

Components of   Average Velocity:

$$\boxed{v_x^{ave} = \frac{\Delta x}{\Delta t}} \tag{3.11}$$

$$\boxed{v_y^{ave} = \frac{\Delta y}{\Delta t}} \tag{3.12}$$

$$\boxed{v_z^{ave} = \frac{\Delta z}{\Delta t}} \tag{3.13}$$

Note: The components v_x^{ave}, v_y^{ave}, and v_z^{ave} are not vectors!

Sometimes these components, v_x^{ave}, v_y^{ave}, and v_z^{ave} are called the average x, y and z-velocities respectively. Note that the names average x, y and z-velocities do not imply that they are vectors; they are the factors with which you must multiply the unit base vectors of the x, y, and z-axes and sum the resulting vectors along the axes to construct the actual vector, the average velocity. That is why it would be incorrect to put arrows over the symbols of the components since that would imply that they are vectors when they are not.

Example 3.4.1. Average Velocity of a Motion in a Straight Line. A train starts from Boston at 7 : 00 AM and goes towards New York City. In the first 2 minutes, the train moves on a straight track facing West and covers a direct distance of 3000 m from the starting point. What is the average velocity?

Solution. From the given description of the motion we need to deduce two parts of the average velocity vector: (1) magnitude and (2) direction. Since, there are two viewpoints of vectors, often there are two ways to address these problems: (a) Geometrically, and (b) Analytically.

(a) Geometric approach: For a geometric answer, we find the magnitude of average velocity vector by dividing the magnitude of displacement

3.4. AVERAGE VELOCITY

vector by the interval. The magnitude of the displacement is equal to the direct distance between the points of space at the end points of the interval, which is given to be 3000 m. The interval has a duration of 2 min. Therefore, the magnitude of the average velocity is 3000 m/2 min = 1500 m/min. It is usually customary to write numerical values in standard units, which would be kg, m, and (sec or s). Therefore, we rewrite the answer in m/s, which gives the magnitude of the velocity to be 25 m/s. The direction is towards West and can be written as such. Or, you could draw an arrow and name the direction "West".

(b) Analytic approach: Analytically, we would proceed by choosing a coordinate system and describe the direction using that coordinate system. Suppose, we choose a coordinate system such that its x-axis is pointed West with its origin at the starting point on the track. Then, the unit base vector $\hat{u}_x$ will be pointed towards the West direction. The coordinates of the beginning and end points are $(0,0,0)$ and $(3000\,\text{m}, 0, 0)$ respectively. Therefore, the change in x-coordinates, Δx, in the two-minute interval is 3000m. Dividing Δx by Δt we find the x-component of the average velocity vector to be 25 m/s. Therefore, using the chosen coordinate system, we will write the velocity vector as $\vec{v}_{ave} = (25\,\text{m/s})\hat{u}_x$. The answer makes sense when you refer to what the direction $\hat{u}_x$ means for the chosen coordinate system. Here, it means the direction towards West. Therefore, $(25\,\text{m/s})\hat{u}_x$ means that the velocity is { 25 m/s, West }.

Further Observations

Note that, in the analytic method, the form of the final answer has a reference to the coordinate system employed for calculations. Now, suppose, we had chosen the positive x-axis to point towards East, instead of West. What, then, will be our answer? Clearly, in the new coordinate system, the coordinates of the two points under study will be $(0,0,0)$ and $(-3000\,\text{m}, 0, 0)$.

Note the x-coordinate of the final point will be negative in the new system. The x-component of the average velocity, obtained by dividing the final x minus the initial x by the interval, now gives -25 m/s. We need to multiply the unit base vector by this number to obtain the average velocity vector. The unit base vector along the x-axis now points towards East, which is just the opposite of the direction the unit base vector was pointing in the previous choice of coordinate system. Therefore, let us use a prime to denote the unit base vectors in this system, which gives the average velocity vector in this system to be $\vec{v}_{ave} = (-25\,m/s)\hat{u}'_x$. [You should confirm that this is the right result.]

On the surface, the answer now does not appear to be same as the answer using the other coordinate system: We have a minus sign in the answer here and we didn't have the minus sign before. But, is the average velocity different? No, the average velocity is same, as an interpretation of the meaning of these numerical answers immediately reveals. Since $\hat{u}'_x$ points towards East, its negative will point towards West. Therefore, $(-25 \text{ m/s})\hat{u}'_x$ is a velocity of 25 m/s pointed towards West as before.

You do not have to use x-axis for the track. Try using the y-axis to point along the track. What do you get for the answer?

Example 3.4.2. Average Velocity for a Motion in a Straight Line.

For simplicity, consider an object moving in a straight line such as a train on a straight track. When we place a coordinate system such that its x-axis falls on the straight line of motion, the displacement will be a vector that can point either towards the positive x-axis or towards the negative x-axis only. Therefore, displacement in this coordinate system can be written entirely as a scalar multiple of the unit vector $\hat{u}_x$, which will mean that the average velocity has only the x-component v_x^{ave}

$$\vec{v}_{ave} = v_x^{ave}\hat{u}_x. \tag{3.14}$$

If $v_x^{ave} > 0$, then the average velocity will be in the same direction as the vector $\hat{u}_x$, i.e. towards the positive infinity of the x-axis, and if the $v_x^{ave} < 0$, then the average velocity will be in the direction opposite to that of the vector $\hat{u}_x$, i.e. towards the negative infinity of the x-axis. The direction towards the positive infinity of the x-axis is also referred to as moving to the right or moving toward the positive x-axis. Similarly we say moving to the left or towards the negative x-axis for the direction towards negative infinity of the axis.

We will use similar descriptions of motions on a straight line, if we happen to use a coordinate system oriented such that its y or z-axis coincides with the straight line where the motion takes place.

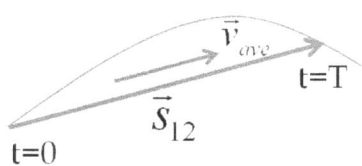

Example 3.4.3. Average velocity of a motion in a plane. A projectile is flying through the air in a plane. At $t = 0$, the projectile is at the origin of a coordinate system, and at $t = T$, the coordinates of the projectile are $(X, Y, 0)$. What is the average velocity of the projectile over the time interval 0 to T?

Solution. The analytic viewpoint of average velocity tells us that we can construct the average velocity vector from its components, which we can readily find here.

$$v_x^{ave} = \frac{X - 0}{T - 0} = \frac{X}{T}$$
$$v_y^{ave} = \frac{Y - 0}{T - 0} = \frac{Y}{T}$$
$$v_z^{ave} = \frac{0 - 0}{T - 0} = 0$$

Therefore, the average velocity has the following representation in the given coordinate system.

$$\vec{v}_{ave} = \left(\frac{X}{T}\right)\hat{u}_x + \left(\frac{Y}{T}\right)\hat{u}_y.$$

3.4. AVERAGE VELOCITY

The magnitude and direction of the average velocity can be determined as usual from the components. The magnitude of the average velocity is

$$v_{ave} = \frac{\sqrt{X^2 + Y^2}}{T}.$$

Since, the motion occurs in a plane, one angle will be sufficient to specify the direction of the vector. The angle θ from the positive x-axis in the xy-plane is given by

$$\theta = \arctan\left(\frac{Y}{X}\right).$$

This is the angle we usually specify when we want to give direction of a vector in the xy-plane.

3.4.1 Average Speed

Often we are interested in the total distance traveled over some time interval and not necessarily in the direction of the motion. For instance, if you travel by car between two cities, the road will not always point in the same direction. The distance you must travel will be more than the direct distance between the cities. Let D be the actual distance moved in total time T, then the ratio D/T, called the **average speed** is a measure of how fast the distance was covered. We will denote average speed by v_s.

$$\boxed{v_s = \frac{D}{T}.} \quad (3.15)$$

Average speed and magnitude of average velocity may be very different, especially in a motion where there are changes in direction. An extreme example will be a motion where an object returns to the original place after some time - the magnitude of average velocity in this case will be zero but average speed will not be zero.

Example 3.4.4. Average speed of a road trip. Google maps shows that a road trip from Boston to Los Angeles will cover 4800 km and take two days and zero hour. What is the average speed for the trip in km/h? From the longitudinal and latitude of the two cities the direct distance can be calculated assuming the Earth to be a sphere. The direct distance between the two cities is approximately 4100 km. What is the average velocity for the trip?

Solution. We divide the distance by time to obtain the average speed:

$$v_s = \frac{4800 \text{ km}}{48 \text{ hr}} = 100 \text{ km/hr}.$$

The magnitude of the average velocity will be equal to the direct distance divided by time, which gives $v_{ave} = 4100/48 = 84\ km/hr$. The direction is pointed from Boston towards Los Angeles.

Question to student: Will the magnitude of average velocity be always less than or equal to average speed? Why or why not? Can you prove your claim?

Figure 3.13: How would you calculate direct distance between cities since the Earth is not flat? Map credits: www.cia.gov

3.5 VELOCITY AND SPEED

Recall that average velocity gives us only the overall rate of change of position for an entire interval of time. If the time interval is long, we miss the details of the motion, such as the change in velocity during the interval. Clearly, average velocities over smaller intervals will give us a more detailed picture of the motion. It turns out that, if you examine average velocity for successively smaller intervals around each instant of time, you discover a way to deduce another quantity which can be identified as an instantaneous rate of change of the position vector. To gain an insight into the process that gives rise to the definition of **velocity at each instant** it is helpful to examine a simple one dimensional motion.

3.5.1 Velocity in One-dimensional Motion

Suppose we drop a ball from a 20-meter tower. How fast will the ball be moving at the $t = 1 \ sec$ mark? Recall that the average velocity is defined over an interval, not at an instant. So, how do we find how fast the ball is moving at a particular instant?

If we try to find the average velocity near the $t = 1$ sec mark for different time intervals near the 1 sec mark, what will we find?

We start by recording the location of the ball at various instants in time. The data for a hypothetical experiment is shown in Table 3.1, where we have displayed time to one microsecond (μs) precision and distance to 10 micrometer (μm) precision. From this table, we can find displacements for various size time intervals near $t = 1$ sec, and corresponding average velocities.

In Table 3.1 we have the data for the position of the ball as we approach the 1 sec mark as well as at instants immediately after that. We use the data in Table 3.1 to construct another table, Table 3.2, which contains average velocity for successively smaller intervals around the 1 sec instant. The details of constructing Table 3.2 is left as an important exercise for

Table 3.1: Data for a ball dropped from 20 m.

Time (s)	Height (m) Above Ground
0.900000	16.03100
0.990000	15.19751
0.999000	15.10980
0.999900	15.10100
0.999990	15.10010
0.999999	15.10001
1.000000	15.10000
1.000001	15.09999
1.000010	15.09990
1.000100	15.09902
1.001000	15.09020
1.010000	15.00151
1.100000	14.07100

3.5. VELOCITY AND SPEED

the student.

Table 3.2: Average velocity in successively smaller time intervals. As the interval nears $t = 1$ sec shrinks, the average velocity reaches closer and closer to a limit of { 4.9 m/s, down}.

Time interval (s)	Average velocity	
	Magnitude (m/s)	Direction
$1 \leq\leq 1.1$	5.414500	Down
$1 \leq\leq 1.01$	4.949245	Down
$1 \leq\leq 1.001$	4.904902	Down
$1 \leq\leq 1.0001$	4.900490	Down
$1 \leq\leq 1.00001$	4.900049	Down
$1 \leq\leq 1.000001$	4.900005	Down

Table 3.2 shows that, as we shorten the time interval, the average velocity gets closer and closer to { 4.9 m/s, down}. What would happen if we had a continuous record of the position in Table 3.1 instead of only the 10 μm resolution in distance and 1 μs resolution in time?

It is easier to discuss the continuous space and continuous time case in the analytic approach for vectors. To describe the motion analytically, we need to choose a coordinate system first. Here we will choose a coordinate system with z-axis pointed up with origin at the ground level as shown in Fig. 3.14. The choice of z is arbitrary, we could just as well have chosen x or y.

In this coordinate system, the z-coordinate of the ball equals its height from the ground. The negative z values are for points below the ground. The displacements for the vertically falling ball are all along the z-axis pointed towards negative z-axis.

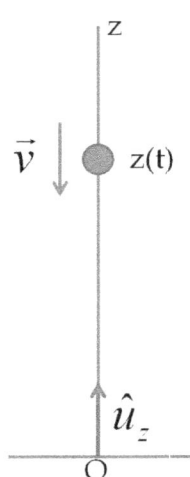

Figure 3.14: Coordinate system for analytic description of falling ball.

The position is denoted by the position coordinate z, and the position vector is, of course the vector from the origin to the position of the ball, that is, $\vec{r} = z\,\hat{u}_z$, where the unit vector $\hat{u}_z$ is pointed up as shown in Fig. 3.14. The average velocity is given along the z-axis also, and can be written in terms of change in the z-coordinate over a time interval and the unit base vector.

$$\vec{v}_{ave} = v_z^{ave}\hat{u}_z = \left[\frac{z(t_2) - z(t_1)}{t_2 - t_1}\right]\hat{u}_z \qquad (3.16)$$

Table 3.2 can now be interpreted as giving us the values of $v_z^{ave}\,\hat{u}_z$. Since, the unit vector is pointed up, the down direction will mean $v_z^{ave} < 0$. According to Table 3.2, v_z^{ave} has different values depending upon the size of the interval: $(-5.414500$ m/s$)$ when the interval has a size of 0.1 sec, $(-4.949245$ m/s$)$ for an interval size of 0.01 sec, ..., $(-4.900049$ m/s$)$ for an interval size of 0.00001 sec, and $(-4.900005$ m/s$)$ for an interval size of 0.000001 sec, etc.

Important observation: The sequence of the values for v_z^{ave} for successively smaller time intervals near $t = 1 \ sec$ appears to approach a unique limiting value of $(-4.90000 \ m/s)$.

If you repeat the process described above for intervals around other instants in time, we obtained different values of v_z. For instance, v_z would be zero for $t = 0$ and $-19.6 \ m/s$ at $t = 2$ sec, etc.

The limiting value obtained by the procedure outlined above is called the instantaneous rate of change of the z-coordinate of the object, which is the z-component, v_z, of the instantaneous velocity vector $\vec{v}$. It is customary to write the process in the following notation:

$$\text{Symmetric difference: } v_z = \lim_{\Delta t \to 0} \left[\frac{z(t + \Delta t/2) - z(t - \Delta t/2)}{\Delta t} \right], \quad (3.17)$$

It can be shown that, instead of taking a symmetric interval of time for the change in z-coordinates, one may take forward or backward coordinate differences without affecting the limit of the sequence.

$$\text{Forward difference: } v_z = \lim_{\Delta t \to 0} \left[\frac{z(t + \Delta t) - z(t)}{\Delta t} \right] \quad (3.18)$$

$$\text{Backward difference: } v_z = \lim_{\Delta t \to 0} \left[\frac{z(t) - z(t - \Delta t)}{\Delta t} \right] \quad (3.19)$$

It is instructive to write the change in coordinates as Δz for the interval of time Δt. Then, the symmetric, forward, and backward difference formulas can be combined into one form.

$$v_z = \lim_{\Delta t \to 0} \left[\frac{\Delta z}{\Delta t} \right], \quad (3.20)$$

The procedure described above for obtaining v_z defines v_z as the time derivative of the z-coordinate.

$$\boxed{v_z(t) = \frac{dz}{dt}.} \quad (3.21)$$

This says that the z-component of velocity is equal to the rate of change of the z-coordinate of the particle. The velocity at time t for the particle that moves only along the z-axis is given by multiplying the unit vector for the z-axis by v_z.

$$\vec{v}(t) = v_z(t) \ \hat{u}_z. \quad (3.22)$$

The change in z-coordinate of the particle can be obtained from the z-component of the velocity by integrating. Let the particle be at z_1 at time t_1 and at z_2 at t_2, then

$$\boxed{z_2 - z_1 = \int_{t_1}^{t_2} v_z(t) dt.} \quad (3.23)$$

Further Observations

3.5. VELOCITY AND SPEED

What would happen if we choose the x or y-axis to point up in place of the z-axis used above? Had we chosen the x-axis to point in the vertical direction, our analysis would have been in terms of the x-coordinate of the ball. This would have led us to the velocity as $\vec{v} = v_x \hat{u}_x$, where v_x would be the rate of change of the x-coordinate, called the x-component of the velocity or the x-velocity.

$$v_x = \frac{dx}{dt}. \tag{3.24}$$

Similarly, if the y-axis was chosen in the vertical direction, we would have written the velocity as $\vec{v} = v_y \hat{u}_y$, where v_y is the rate of change of the y-coordinate, i.e. the y velocity.

$$v_y = \frac{dy}{dt}. \tag{3.25}$$

For a general motion in three dimensions, all three coordinates of the object will change with time at their respective rates, and we will write velocity as the vector sum of velocities along the axes.

$$\vec{v} = v_x \hat{u}_x + v_y \hat{u}_y + v_z \hat{u}_z = \left(\frac{dx}{dt}\right)\hat{u}_x + \left(\frac{dy}{dt}\right)\hat{u}_y + \left(\frac{dz}{dt}\right)\hat{u}_z. \tag{3.26}$$

Components of velocity:

$$v_x = \frac{dx}{dt}$$
$$v_y = \frac{dy}{dt}$$
$$v_z = \frac{dz}{dt}$$

We will have more to say about the three-dimensional case below. Here we continue the discussion of one-dimensional motion.

Example 3.5.1. Instantaneous velocity as derivative of position
You are driving a race car towards East on a straight East-West road. Your position from an intersection is given in a particular reference in which origin is at the intersection and the x-axis is along the road with positive x-axis in the direction of East from the intersection. The x-coordinate of your car with time is given by the function, $x(t) = 20 + 15t + 4t^2$, where x is in meters and t in seconds. Find the velocity of the car at (a) $t = 0$, (b) $t = 5$ sec, and (c) $t = 10$ sec?

Solution. This problem demonstrates a direct calculation of the x-component of velocity, v_x from the calculation of derivative of $x(t)$. Since the car is moving only on the x-axis, the velocity of the car at an arbitrary instant t will be given by

$$\vec{v}(t) = v_x(t)\,\hat{u}_x.$$

First, let us calculate the derivative of $x(t)$ and then evaluate the resulting x-component of the instantaneous velocity at different instants.

$$\begin{aligned} v_x &= \frac{dx}{dt} \\ &= \frac{d}{dt}(20 + 15t + 4t^2) \\ &= 15 + 8t. \end{aligned} \tag{3.27}$$

(a) Now, we put $t = 0$ in Eq. 3.27 to find the x-component of velocity at $t = 0$.

$$v_x(0) = 15 + 8(0) = 15 \text{ m/s}.$$

Therefore, the velocity at $t = 0$ mark is

$$\vec{v}(0) = (15 \text{ m/s}) \, \hat{u}_x.$$

(b) The instant now is $t = 5$ sec. Therefore, we put $t = 5$ sec in Eq. 3.27 to find the instantaneous velocity at $t = 5$ sec.

$$v_x(5 \text{ sec}) = 15 + 8(5) = 55 \text{ m/s}.$$

Therefore, the velocity at 5 sec mark is

$$\vec{v}(5 \text{ sec}) = (55 \text{ m/s}) \, \hat{u}_x.$$

(c) Same process gives $v_x(10 \text{ sec}) = 95 \text{ m/s}$, and $\vec{v}(t) = (95 \text{ m/s}) \, \hat{u}_x$.

Example 3.5.2. Change in x-coordinate from constant v_x. A ball is rolling on the floor such that its x-coordinate in a particular Cartesian coordinate system changes at a constant rate u_0. If the x-coordinate is x_0 at $t = t_0$ what will be the x-coordinate at $t = T$? [NOTE: This example and the next use integration. Integrations appear naturally when dealing with varying rates as we will see in the next chapter.]

Solution. From our discussion we know that the rate of change in a coordinate is given by the derivative of the coordinate with time.

$$\frac{dx}{dt} = \text{Rate of change of } x \text{ coordinate}.$$

Here the rate is given to be u_0. Therefore, we have the following equation for the derivative of $x(t)$.

$$\frac{dx}{dt} = u_0.$$

This equation can also be written for differential elements, dx and dt by multiplying both sides by dt.

$$dx = u_0 dt,$$

which we can integrate on both sides. On the left side, the limit of integration is x_0 to $x(T)$ and the limit on the right side is the times for those x values, viz. from t_0 to T.

$$\int_{x_0}^{x(T)} dx = \int_{t_0}^{T} u_0 dt,$$

which gives the following result.

$$x(T) - x_0 = u_0(T - t_0).$$

Further Observations

An integration is actually not necessary here since the rate of change is constant in time. The constant rate implies that the average rate is the same as instantaneous rate of change. Therefore,

$$\frac{\Delta x}{\Delta t} = u_0,$$

To the student: If you cannot do integrations at this point in time, you can skip these two examples and continue in the text without much loss of continuity.

$$\frac{dx}{dt} = v_x(t) \implies$$

$$\int_{x_1}^{x_2} dx = \int_{t_1}^{t_2} v_x(t) dt.$$

3.5. VELOCITY AND SPEED

which can be multiplied both sided by Δt to solve for Δx. This immediately gives the change over the interval from t_0 to T as

$$x(T) - x_0 = u_0(T - t_0).$$

Example 3.5.3. Change in x-coordinate from varying v_x. A particle moves in the xy-plane of a coordinate system. The rate of change of its x-coordinate varies linearly with time such that the rate at any particular time is $v_x = u_0 + bt^2$, where u_0 and b are constants. If the x-coordinate is x_0 at $t = t_0$ what will be the x-coordinate at $t = T$?

Solution. The discussion of the last example takes us to the following equation for the differentials

$$dx = \left(u_0 + bt^2\right) dt,$$

which can be integrated to give

$$x(T) - x_0 = u_0(T - t_0) + \frac{b}{3}\left(T^3 - t_0^3\right). \qquad (3.28)$$

Further Observations - Computing the average x-velocity from changing v_x.

Here we must resort to integration since the rate is changing with time. The short-cut method applicable for constant rate does not give the correct answer. The average rate is also not equal to the average of the x-component of velocity at the end and the x-component of velocity at the beginning.

$$v_x^{ave} \neq \frac{v_x(T) + v_x(t_0)}{2}.$$

The x-component of the average velocity is equal to the change in the x-coordinate divided by the interval. Using Eq. 3.28 we find v_{ave} to be

$$v_x^{ave} = \frac{x(T) - x_0}{T - t_0} = u_0 + \frac{b}{3}\left(T^2 + t_0 T + t_0^2\right),$$

which can also be obtained by integrating the x-component of the instantaneous velocity over time and dividing by the interval.

$$v_x^{ave} = \frac{\int_{t_0}^{T} v_x(t) dt}{T - t_0} = \frac{\text{Integration of } x\text{-instantaneous velocity}}{\text{Duration}}.$$

This gives the same result.

Computing Derivatives from the Slopes of Tangents

In experiments we measure position at a finite number of instants and deduce the instantaneous velocity by making a plot of position versus time. You will now see that the components of velocity, i.e. $v_x(t)$, $v_y(t)$ and $v_z(t)$, can be determined from the plots of x vs t, y vs t and z vs t respectively, since the derivative of a function is also equal to the slope of the tangent of the curve when the function is plotted along the ordinate (the vertical axis) and time along the abscissa (the horizontal axis).

Let us look at an example of the x-component of velocity from the x vs t plot. Consider the graphs of a function $x(t)$ given in Fig. 3.15. Suppose we are interested in the x-component of velocity at $t = 1$ sec. The slopes of the secants joining a point of the curve before $t = 1$ sec and a point after $t = 1$ sec give the x-component of the average velocity for various time intervals. The figure shows that, as we examine shorter and shorter time intervals, the secants tend to become parallel to the tangent to the curve at $t = 1$ sec, the instant of interest. Therefore, the slope of the tangent will equal the instantaneous rate of change. This gives us another way of computing derivatives, which is particularly important for relating to experimental results.

$$\frac{dx}{dt} = \text{Slope of the tangent to the } x \text{ vs } t \text{ curve.} \qquad (3.29)$$

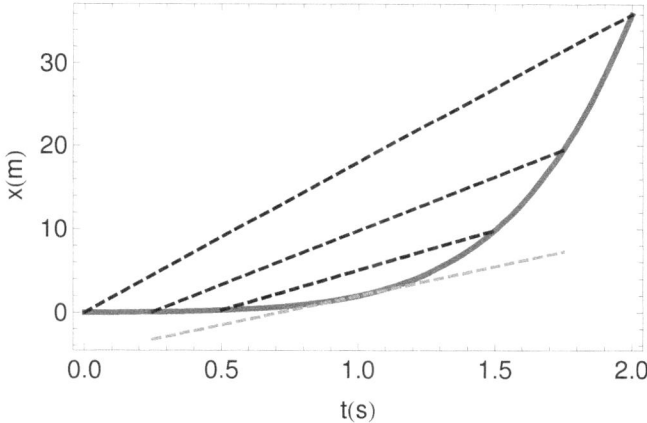

Figure 3.15: The x-component of the instantaneous velocity at the instant $t = 1$ sec is equal to the slope of x vs t curve at $t = 1$ sec. The slopes of the secants for various time intervals around $t = 1$ sec give average velocities in those time intervals. This figure shows visually that the slopes of secants change as we zero in near the instant of interest, eventually the secant becomes parallel to the tangent to the curve.

Example 3.5.4. Instantaneous velocity from slope. Figure 3.16 shows an example of calculation of the x-component of instantaneous velocity from a plot of x vs t. To obtain the x-velocity at an instant, say at $t = t_1$, we draw a tangent to the curve at the time of interest, i.e., $t = t_1$ as shown in the figure. The rise of the tangent line gives the change in position Δx and the run of the tangent gives the duration Δt for that change. Therefore, dividing Δx by Δt is equal to the rate of change of the

3.5. VELOCITY AND SPEED

x-coordinate of the object at the instant $t = t_1$, which is the x-component of instantaneous velocity at time $t = t_1$.

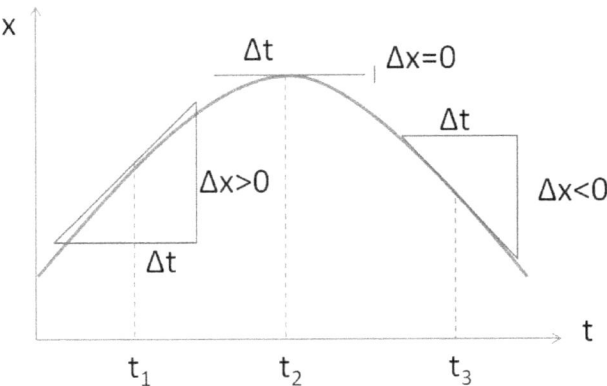

Figure 3.16: Example 3.5.4. The slope of tangents to the curve $x(t)$ gives the x-component of the instantaneous velocity. The slopes of the tangents obtained by dividing the rise Δx by the run Δt at instants $t = t_1$, $t = t_2$, and $t = t_3$ give the values of the x-component of velocity at those instants. Here, $v_x > 0$ at $t = t_1$, $v_x = 0$ at $t = t_2$ and $v_x < 0$ at t_2.

At time $t = t_1$, the slope of the tangent, i.e., v_x, is positive, meaning the velocity is pointed in the positive x direction for a one-dimensional motion on the x-axis. Just to remind you that velocity for one-dimensional motion on the x-axis is $\vec{v} = v_x \hat{u}_x$, where $\hat{u}_x$ is the unit vector pointed towards the positive x-axis. Note that at time $t = t_2$ in Fig. 3.16, the slope is zero, meaning that the x-velocity is zero there. At time $t = t_3$, the slope is negative, meaning the velocity will be pointed in the negative x-axis direction for a one-dimensional motion.

We wish to emphasize here that an x vs t plot gives you only v_x. In a three-dimensional motion, you will need x vs t, y vs t, and z vs t so that you can deduce x, y, and z-components of velocity, viz. v_x, v_y, and v_z, and construct the velocity vector from them.

Example 3.5.5. Velocity from slope. The position of a box in a one-dimensional motion is recorded by placing the y-axis on the line of motion. The data is plotted as a y vs t plot and shown in Fig. 3.17. Find the y-component of velocity at $t = 1$ sec, 3 sec, 5 sec, and 7 sec.

Solution. The plot of y vs t is a segment-wise straight line, which makes finding slopes easier. We do not need to draw any tangents since the tangent to a straight line is the straight line itself. The slopes at $t = 1$ sec and $t = 3$ sec are equal, $v_y = 1$ m/s. The slope at $t = 5$ sec is zero since the line is flat and rise is zero. Therefore at $t = 5$ sec, $v_y = 0$. Finally, the slope is negative at $t = 7$ sec. The y-coordinate changes by -2 m in 3 sec steadily. This gives $v_y = -2/3$ m/s.

Further remarks: Note that in a plot of x, y, or z vs t, there cannot be sharp corners since the velocity of an object cannot change abruptly. That

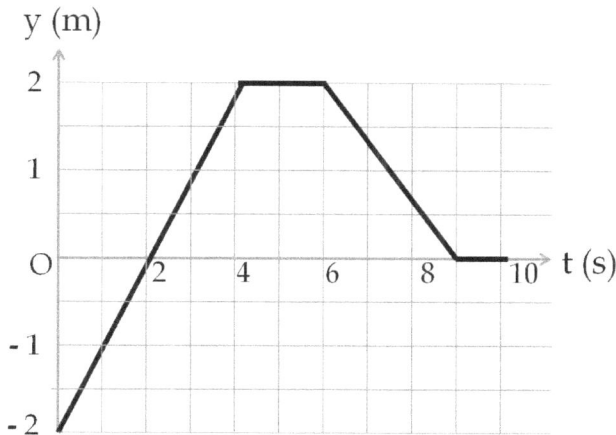

Figure 3.17: Example 3.5.5. Assume corners to be smooth.

is, the velocity at an instant before the corner must be arbitrarily close to the velocity an instant after the corner. In the limit of infinitesimal interval around the corner, the velocity obtained at an instant before a corner will not equal the one for an instant after the corner. Therefore, the plots of coordinates versus time must be smooth at all points, meaning that $x(t)$, $y(t)$ and $z(t)$ are smooth functions of t, not just continuous functions.

3.5.2 Velocity - General

We now generalize the discussions of the one-dimensional cases presented in the last subsection. As usual for vectors, it is helpful to look at the velocity vector from both geometric and analytic viewpoints. The geometric viewpoint gives a pictorial view of vectors and the analytic viewpoint based on a particular choice of coordinates is more suited for calculations. Note that, while in the geometric viewpoint one always works with the complete vector, in the analytic approach we work with components of the vector and in the end we must recover the magnitude and direction of the vectorial physical quantity from the components.

Geometric Viewpoint

Consider the trajectory of an arbitrary motion given in Fig. 3.18. What would be the velocity vector at time t when the object is at the point labeled P? We define the velocity vector in the same way here as we did for the one-dimensional motion: the displacements for increasingly smaller time intervals around point P are divided by their time intervals to evaluate average velocity vectors for those intervals. The process is shown schematically in Fig. 3.18. The limit of the sequence of average velocity vectors { $\vec{s}_{12}/(t_2 - t_1)$, $\vec{s}_{34}/(t_4 - t_3)$, $\cdots$ } for increasingly smaller time intervals is the instantaneous velocity at time t. Informally, we will

3.5. VELOCITY AND SPEED

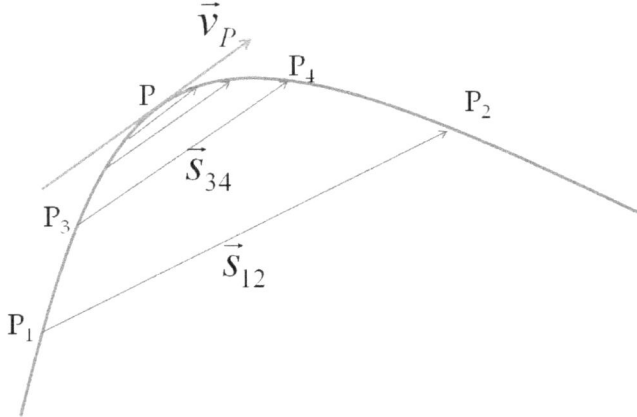

Figure 3.18: The instantaneous velocity $\vec{v}_P$ at the instant t when the object is at point P is pointed in the direction of tangent to the trajectory at point P. The object moves on the trajectory in the direction of P_1, P_3, P, P_4, P_2, $\cdots$. The limit of the sequence of average velocity vectors: $\{\ \vec{s}_{12}/(t_2 - t_1),\ \vec{s}_{34}/(t_4 - t_3),\ \cdots\ \}$, as time segments become increasingly smaller around point P defines the velocity at point P. Here t_1, t_2, t_3, t_4 correspond to time when the object is at P_1, P_2, P_3, P_4, respectively.

write velocity at time t as

$$\vec{v} = \lim_{\Delta t \to 0} \left[\frac{\Delta \vec{r}}{\Delta t}\right], \qquad (3.30)$$

where $\Delta \vec{r}$ is the displacement in the interval from $(t - \Delta t/2)$ to $(t + \Delta t/2)$. Here, $\Delta r = \vec{r}(t + \Delta t/2) - \vec{r}(t - \Delta t/2)$ is the change in the position vector over the interval. As the interval near t is made increasingly smaller, we obtain a limiting vector which represents the instantaneous rate of change of the position vector. The limit is called the **instantaneous velocity** and can be formally written as

$$\boxed{\vec{v} = \frac{d\vec{r}}{dt}.} \qquad (3.31)$$

Analytic Viewpoint

For the analytic viewpoint we make use of a Cartesian coordinate system and follow the changes in the Cartesian coordinates of the points on the trajectory of the motion. The x, y, and z-coordinates of the points on the trajectory change with time. The trajectory is said to be described by a triplet of functions $\{\ x(t), y(t), z(t)\ \}$ such that the position vector of the object at time t is given by the following vector:

$$\vec{r}(t) = x(t)\ \hat{u}_x + y(t)\ \hat{u}_y + z(t)\ \hat{u}_z, \qquad (3.32)$$

where $\hat{u}_x$, $\hat{u}_y$, and $\hat{u}_z$ are the usual unit vectors pointed in the directions of positive x, y, and z-axes respectively. The process of taking the limit of the sequence of average velocity vectors $\{\ \vec{s}_{12}/(t_2 - t_1),\ \vec{s}_{34}/(t_4 - t_3),\ \cdots$

} for increasingly smaller time intervals separates into similar processes for the three coordinate functions since the unit vectors have the same orientations for every point on the trajectory. Therefore, a separation of motion into motions along different axes occurs as illustrated in Fig. 3.19 for a two-dimensional case.

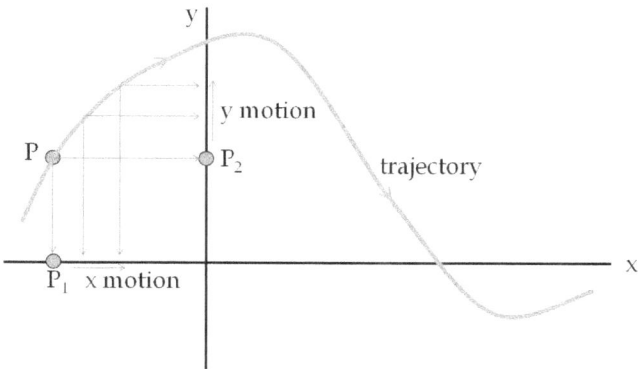

Figure 3.19: The coordinates of a particle P moving in the xy-plane change independently giving an equivalent picture of the motion. One particle P along the actual trajectory is analyzed as the motion of two particles P_1 and P_2, one moving along the x-axis and the other moving along the y-axis. The motions of P_1 and P_2 reproduce the changes in the x and y-coordinates of the original particle. The separate motion of the coordinates is a consequence of the vector nature of kinematic quantities.

By looking at the way the x-coordinates change with time, we determine the x-component of the velocity vector.

$$v_x = \frac{dx}{dt}.$$

Similarly for the y and z-components of the velocity vector. Therefore, the analytic expression for the velocity at time t is given by adding the velocities along the axes. Formally,

$$\vec{v}(t) = \frac{dx}{dt}\,\hat{u}_x + \frac{dy}{dt}\,\hat{u}_y + \frac{dz}{dt}\,\hat{u}_z, \qquad (3.33)$$

which can be written more compactly as the expression in Eq. 3.31, viz.

$$\vec{v} = \frac{d\vec{r}}{dt}.$$

The magnitude of the velocity vector can be obtained from its components as

$$\text{Magnitude: } |\vec{v}| = \sqrt{v_x^2 + v_y^2 + v_z^2}, \qquad (3.34)$$

and the information about the direction is obtained from the components by drawing a vector from $(0,0,0)$ to (v_x, v_y, v_z) in a three-dimensional coordinate system whose axes are marked with velocity component values.

3.5. VELOCITY AND SPEED

For a three-dimensional situation, you will need two angles for the specification of the direction of the vector, which are usually θ and ϕ of the spherical coordinate system. And, for a two dimensional situation, the direction can be specified by one angle only, which is usually the angle going counter-clockwise from the x-axis in the xy-plane. The way to figure out the directions of vectors have been treated earlier in this chapter and in the chapter on vectors (see Chapter 2).

Direction of $\vec{v}$:

> One-dimensional along (x) axis: $x > 0$ or $x < 0$.
>
> Two dimensional in (xy) plane: θ counter-clockwise from $+x$ axis.
>
> Three dimensional: θ and ϕ of spherical coordinate system

Example 3.5.6. The Components of Velocity in a Plane. Suppose a cannon ball is fired with a speed v_0 at an angle θ with the horizontal direction. A coordinate system is chosen such that the x-axis points horizontally and the y-axis points vertically up. What are the x and y-components to the velocity vector?

Solution. It is helpful to draw the vector and the axes as shown in Fig. 3.20. The hypotenuse of the right-angled triangle $\triangle OPQ$ has length equal to the magnitude of the velocity vector, and the sides OQ and PQ are the x and y-components, to be denoted as v_{0x} and v_{0y} respectively. Trigonometry yields the following expressions for the components immediately.

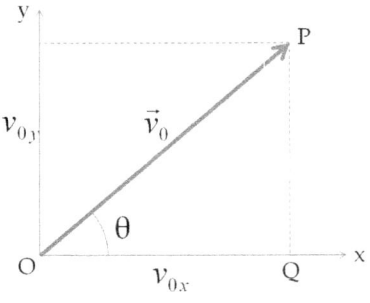

Figure 3.20: Example 3.5.6.

$$v_{0x} = v_0 \cos\theta$$
$$v_{0y} = v_0 \sin\theta$$

Example 3.5.7. Velocity of a Projectile at Different Points. The trajectory of a projectile is shown in Fig. 3.21 with the assumption that the initial velocity was 80 m/s at the angle of 72° from the horizontal direction. The position of the projectile at successive 1 sec instants are shown in the figure. You can see that the direction as well as the magnitude of the velocity of the projectile changes with time. As the projectile rises, the distance covered in each subsequent second decreases as a result of slowing down. At the top of the trajectory, the projectile only has horizontal velocity since the vertical component of the velocity is zero there. The velocity is never pointed vertically straight down since the projectile always has a non-zero horizontal component of the velocity.

Example 3.5.8. Instantaneous Velocity of a Projectile. The velocity of the projectile whose trajectory is shown in Fig. 3.21 at $t = 12$ sec has the following components: $v_x = 24.7$ m/s, $v_y = -41.6$ m/s, $v_z = 0$. Find the magnitude and direction of the velocity at that instant.

Solution. The magnitude of the velocity vector is easy to work out from its components.

$$v = \sqrt{v_x^2 + v_y^2 + v_z^2} = 48.4 \text{ m/s}.$$

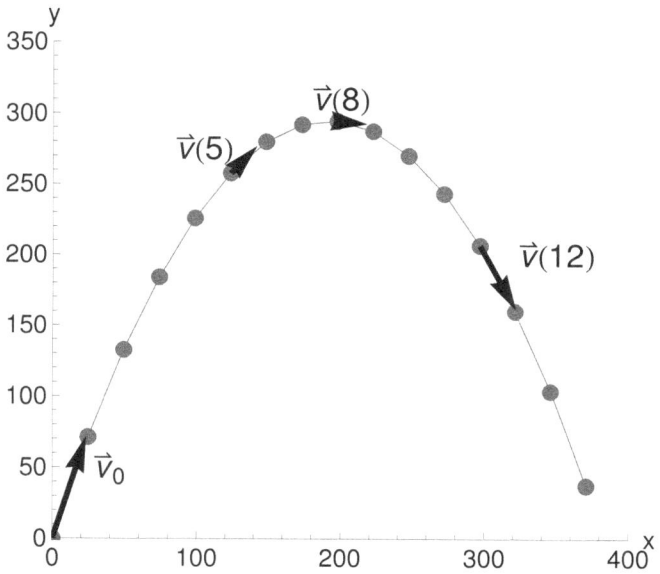

Figure 3.21: Example 3.5.7. The velocity vectors at $t = 0$, 5 sec, 8 sec, 12 sec are shown to demonstrate the changing magnitude and direction of velocity.

Since velocity vector is contained in a plane, here the xy-plane, we need only one angle. Normally we give angle from the positive x-axis. The x-component of the velocity is positive and y-component negative, which places the direction in the fourth quadrant. Therefore, the angle we use is either the clockwise angle from the positive x-axis which will be negative or the counterclockwise angle from the positive x-axis as shown in Fig. 3.22.

$$\text{Clockwise angle from positive } x \text{ axis} = \arctan\left(\frac{-41.6}{24.7}\right) = -59.3°.$$
$$\text{Counter-clockwise angle from positive } x \text{ axis} = 360° - 59.3° = 300.7°.$$

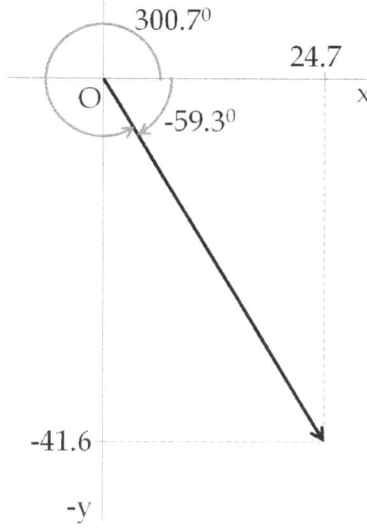

Figure 3.22: Example 3.5.8.

Example 3.5.9. Analyzing Two-dimensional Data. A ball is rolling on the floor. The position of the ball is recorded as (x, y) of a coordinate system. The z-coordinate of the ball is always zero in this coordinate system and therefore ignored. The plots of x vs t and y vs t are displayed here. Find the velocity of the ball at (a) $t = 3$ sec, (b) $t = 5$ sec, and (c) $t = 7$ sec.

Solution. From the given plots, we can read off slopes at the indicated times to obtain the x and y-components of the velocity. They are:

$$t = 3 \text{ sec}: \quad v_x = 0.5 \text{ m/s}, \ v_y = 1.0 \text{ m/s}.$$
$$t = 5 \text{ sec}: \quad v_x = -0.5 \text{ m/s}, \ v_y = 0.0 \text{ m/s}.$$
$$t = 7 \text{ sec}: \quad v_x = -0.5 \text{ m/s}, \ v_y = -1.0 \text{ m/s}.$$

From the components we determine the magnitude and directions of the

3.5. VELOCITY AND SPEED

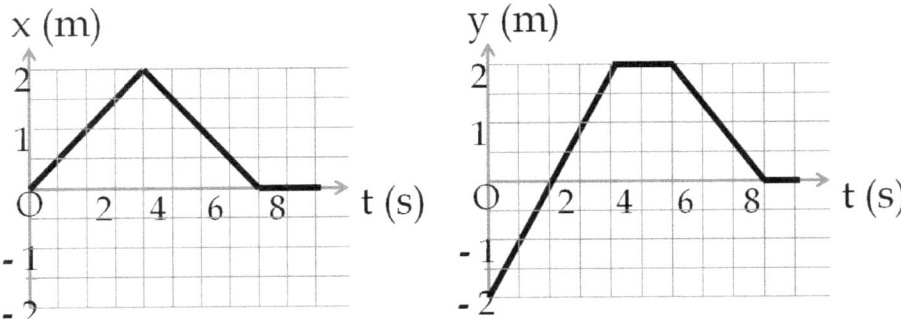

Figure 3.23: Example 3.5.9.

velocity the indicated instants as usual with the following results.

$t = 3$ sec : Magnitude: $v = \sqrt{v_x^2 + v_y^2} = 1.1$ m/s;

Direction: 63° counterclockwise from +x-axis.

$t = 5$ sec : Magnitude: $v = \sqrt{v_x^2 + v_y^2} = 0.5$ m/s;

Direction: towards negative x-axis.

$t = 7$ sec : Magnitude: $v = \sqrt{v_x^2 + v_y^2} = 1.1$ m/s;

Direction: 63° clockwise from the +x-axis.

3.5.3 Speed

The rate at which an object covers physical distance on the trajectory of its motion is called its **instantaneous speed** or simply speed. At any instant the infinitesimal distance ds covered on the actual trajectory, whether curved or not, will be identical to the magnitude of the infinitesimal displacement $d\vec{r}$ in that interval.

$$\boxed{ds = |d\vec{r}|}$$

Therefore, speed at an instant is equal to the magnitude of the velocity vector at that instant.

Instantaneous speed = Magnitude of instantaneous velocity.

Therefore, we denote speed by the same symbol as the velocity except for the arrow over the symbol.

$$\boxed{\text{Instantaneous speed, } v = |\vec{v}|.} \quad (3.35)$$

Since instantaneous speed is only the magnitude of instantaneous velocity vector, it does not have the information about the direction of motion. For example, a missile moving at 10 m/s towards East and another one at 10 m/s towards North, both have the same speed of 10 m/s, but their velocities are different, the first one being { 10 m/s, East} and the second

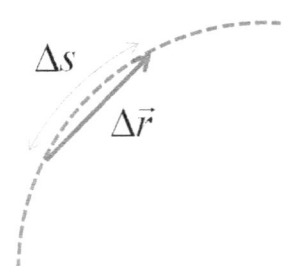

Figure 3.24: For finite interval of time Δt, the actual distance Δs traveled on the trajectory may differ from the magnitude of the displacement, $|\Delta \vec{r}|$. When the time interval becomes infinitesimal, the difference between Δs and $|\Delta \vec{r}|$ disappears for all motions.

one { 10 m/s, North }. While the velocity vector is represented by an arrow, speed is simply a non-negative real number.

If the velocity vector is examined analytically in a coordinate system, we will have its x, y and z-components v_x, v_y and v_z. In that case, we can write the speed in terms of the components of the velocity vector.

$$\boxed{\text{Instantaneous speed,}\ v = |\vec{v}| = \sqrt{v_x^2 + v_y^2 + v_z^2}.} \tag{3.36}$$

Note that, although average speed and magnitude of average velocity were not related since they correspond to any arbitrary size interval, the instantaneous speed is actually equal to the magnitude of the instantaneous velocity since the interval size is infinitesimal. In an infinitesimal interval the change in position is equal to the distance covered since there is not change in direction.

3.6 ACCELERATION

You know that objects can pick up speed after being at rest, or can slow down and come to rest. One example of speeding up happens at the beginning of a race as shown in Fig. 3.25.

Figure 3.25: Accelerating situation # 1: At the start of a race, the runner picks up speed due to acceleration. Photo Credits: Peter Griffin at www.publicdomainpictures.net.

Besides speeding up or slowing down, acceleration takes place when the direction of motion changes with time as shown for the cars as they round a bend in Fig. 3.26.

These examples illustrate the fact that the magnitude as well the direction of the velocity can change with time. The rate of change of velocity is called **acceleration**; the change in velocity due to a change in either magnitude or direction or both.

3.6. ACCELERATION

Figure 3.26: Accelerating situations #2: The direction of the velocities of the race cars change as a result of acceleration of the cars. Photo Credits: Michael Miloserdoff at www.publicdomainpictures.net.

We have already worked out a procedure for calculating the rate of change of a vector when we defined instantaneous velocity as the rate of change of the position vector. Briefly, to obtain the average rate of change of the position vector you subtract the position vector at time t from the position vector at time $(t+\Delta t)$ and then divide the result by the duration Δt. When the duration Δt was made arbitrarily small, the method led to an exact value of the rate of change at the instant t.

This procedure should work for the rate of change of any other vector. Therefore, we compute the rate of change of the velocity vector in an analogous way. Thus, average acceleration ($\vec{a}_{ave}$) is obtained by subtracting the velocity vector at time t from the velocity vector at time $(t+\Delta t)$, and then dividing the result by the duration Δt.

$$\boxed{\vec{a}_{ave} = \frac{\vec{v}(t+\Delta t) - \vec{v}(t)}{\Delta t}}, \tag{3.37}$$

which can be rearranged to obtain the change in velocity in a finite time interval as

$$\Delta \vec{v} = \vec{a}_{ave}\, \Delta t, \tag{3.38}$$

where $\Delta \vec{v}$ stands for the change in velocity vector over the interval from t to $(t+\Delta t)$.

$$\Delta \vec{v} = \vec{v}(t+\Delta t) - \vec{v}(t).$$

Making the time interval Δt arbitrarily small gives us the instantaneous rate of change, and consequently, we obtain the instantaneous acceleration, i.e. acceleration <u>at time</u> t. Formally,

$$\vec{a} = \lim_{\Delta t \to 0}\left[\frac{\Delta \vec{v}}{\Delta t}\right], \tag{3.39}$$

which is also written as

$$\boxed{\vec{a} = \frac{d\vec{v}}{dt}}, \tag{3.40}$$

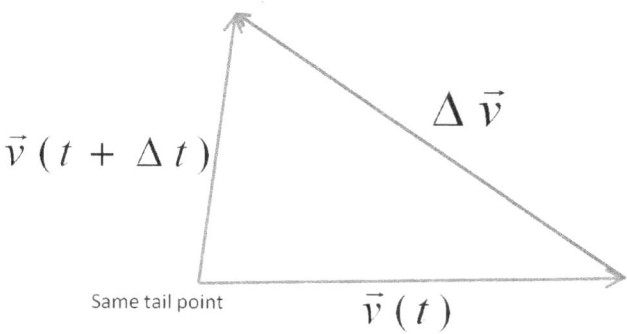

Figure 3.27: Change in velocity. When velocity vectors at instants $t + \Delta t$ and t are drawn with their tails at the same point, the vector from the tip of the velocity vector at t to the tip of the velocity vector at $t + \Delta t$ gives the change in velocity during the interval.

Similar to other vectors, there are two useful viewpoints for looking at the acceleration vector, namely geometric and analytic. We now study them in detail.

Geometric Viewpoint

The average acceleration is equal to the change in velocity divided by the time interval. To compute the change in velocity, we recall that in the geometric viewpoint, in order to subtract a vector $\vec{B}$ from another vector $\vec{A}$, we start with drawing them so that their tails are at the same point. Then, the vector from the tip of $\vec{B}$ to the tip of $\vec{A}$ is the vector $(\vec{A} - \vec{B})$.

Therefore, we draw the velocity vectors $\vec{v}(t)$ and $\vec{v}(t + \Delta t)$ from the same point, although on the trajectory of the motion the vector $\vec{v}(t)$ is tangent to the trajectory at time t and the vector $\vec{v}(t + \Delta t)$ is tangent to the trajectory at time $(t + \Delta t)$. Once we have drawn the velocity vectors corresponding to two nearby instants we obtain the vector for the change in velocity as shown in Fig. 3.27.

The change in velocity obtained by following the graphical procedure shown in Fig. 3.27 is divided by the time interval Δt to compute the average acceleration in the interval from t to $(t + \Delta t)$. One can compute average velocity for successively smaller intervals generating a sequence of average accelerations whose limit would give the instantaneous acceleration at time t.

Analytic Viewpoint

We now look at acceleration from analytic viewpoint by setting up a Cartesian coordinate system first. This makes computing the change in velocity simple. The change in velocity vector amounts to changes in the x, y and z-components separately. Therefore, average acceleration in the interval t

3.6. ACCELERATION

to $(t + \Delta t)$ can be written explicitly in terms of the components.

$$\vec{a}_{ave} = \frac{\Delta \vec{v}}{\Delta t} = \frac{\Delta v_x}{\Delta t}\hat{u}_x + \frac{\Delta v_y}{\Delta t}\hat{u}_y + \frac{\Delta v_z}{\Delta t}\hat{u}_z.$$

This shows that the components of the average acceleration vector $\vec{a}_{ave}$ are just the average rates of change of components of the velocity vector.

$$a_x^{ave} = \frac{\Delta v_x}{\Delta t}$$
$$a_y^{ave} = \frac{\Delta v_y}{\Delta t}$$
$$a_z^{ave} = \frac{\Delta v_z}{\Delta t}$$

The components of instantaneous acceleration at t is obtained by letting the interval near t become infinitesimally small, which says that the components of the instantaneous acceleration are simply time derivatives of the components of instantaneous velocity.

$$\boxed{a_x = \frac{dv_x}{dt}} \quad (3.41)$$

$$\boxed{a_y = \frac{dv_y}{dt}} \quad (3.42)$$

$$\boxed{a_z = \frac{dv_z}{dt}} \quad (3.43)$$

The instantaneous acceleration vector is the sum of vectors along the x, y, and z-axes obtained by multiplying the components with the unit vectors along the axes.

$$\boxed{\vec{a} = a_x\,\hat{u}_x + a_y\,\hat{u}_y + a_z\,\hat{u}_z = \frac{dv_x}{dt}\hat{u}_x + \frac{dv_y}{dt}\hat{u}_y + \frac{dv_z}{dt}\hat{u}_z.} \quad (3.44)$$

The magnitude of the acceleration vector is obtained in the usual way from the components.

$$\boxed{\text{Magnitude: } a \equiv |\vec{a}| = \sqrt{a_x^2 + a_y^2 + a_z^2}} \quad (3.45)$$

As usual, the direction of acceleration vector requires two angles if the motion is in three dimensional space, one angle if the motion is in a plane, and sign of the vector component if the motion is along one of the Cartesian axes.

Recall that the derivative of a function $f(t)$ evaluated for some instant $t = t_1$ is also equal to the slope of the tangent to f vs t curve at $t = t_1$. Therefore, the components of acceleration can be obtained from slopes of respective plots of velocity components versus time. That is, a plot of v_x versus t can be used to determine the x-component of acceleration, and similarly for other components.

1. a_x = slope of tangent to v_x vs t curve.

2. a_y = slope of tangent to v_y vs t curve.

3. a_z = slope of tangent to v_z vs t curve.

Finally, since the velocity components are rates of changes of position coordinates it is also possible to write acceleration components in terms of two derivatives of the position coordinates.

$$a_x = \frac{d^2x}{dt^2} \qquad (3.46)$$

$$a_y = \frac{d^2y}{dt^2} \qquad (3.47)$$

$$a_z = \frac{d^2z}{dt^2} \qquad (3.48)$$

Example 3.6.1. Average acceleration in one dimension. A runner runs on straight East-West road. At time $t = 0$, her velocity is 10 m/s towards East. After running for 1000 seconds, her velocity is still pointed towards East but she has slowed to a speed of 5 m/s. At a later time at $t = 5000$ sec, she is back to a speed of 10 m/s but moving in the opposite direction, i.e. towards West. What are her average accelerations between the following intervals of time: (a) $t = 0$ to $t = 1000$ sec, (b) $t = 1000$ sec to $t = 5000$ sec, and (c) $t = 0$ to $t = 5000$ sec?

Solution. Although the motion is in one straight line, the direction of motion reverses in the duration of interest. In such situations, it is very helpful to place the motion on one of the Cartesian axes and work with the corresponding components of vectors. To be concrete, we will work with the x-axis such that the positive x-axis points towards East. Now, we can translate the given information into x-component of velocity at different times.

$$v_x(0) = +10 \text{ m/s}.$$
$$v_x(1000 \text{ s}) = 5 \text{ m/s}.$$
$$v_x(5000 \text{ s}) = -10 \text{ m/s} \quad \text{(since } \vec{v} \text{ towards negative } x\text{-axis)}.$$

Now, we can use the definition of the x-component of average acceleration for each interval.

(a) Note that when doing numerical calculations, it is often helpful to do calculations on the units separately from the calculations on the numbers as illustrated here.

$$a_x^{ave} = \frac{v_{2x} - v_{1x}}{t_2 - t_1}$$
$$= \frac{5 - 10}{1000 - 0} \left[\frac{\text{m/s}}{\text{s}}\right] = -5.0 \times 10^{-3} \text{ m/s}^2$$

Therefore, average acceleration is $\vec{a}_{ave} = \left(-5.0 \times 10^{-3} \text{ m/s}^2\right) \hat{u}_x$, or 5.0×10^{-3} m/s^2 towards West.

(b)

$$a_x^{ave} = \frac{-10 - 5}{5000 - 1000} \left[\frac{\text{m/s}}{\text{s}}\right] = -3.8 \times 10^{-3} \text{ m/s}^2$$

Therefore, average acceleration is $\vec{a}_{ave} = \left(-3.8 \times 10^{-3} \text{ m/s}^2\right) \hat{u}_x$, or 3.8×10^{-3} m/s^2 towards West.

(c)

$$a_x^{ave} = \frac{-10 - 10}{5000 - 0} \left[\frac{\text{m/s}}{\text{s}}\right] = -4.0 \times 10^{-4} \text{ m/s}^2$$

Therefore, average acceleration is $\vec{a}_{ave} = \left(-4.0 \times 10^{-4} \text{ m/s}^2\right) \hat{u}_x$, or 4.0×10^{-4} m/s^2 towards West.

Example 3.6.2. Acceleration and velocity from a given position as a function of time. A missile is fired in a straight line such that its position with respect to time in a particular coordinate system is given by only its x-coordinates in meters as $x(t) = 3\,t + 2\,t^3$, where t is in seconds. Find its velocity and acceleration at (a) $t = 1$ sec, and (b) $t = 2$ sec.

Solution. First we find the x-components of velocity and acceleration for an arbitrary time t from the given $x(t)$, and then find the values for the particular instants by substituting the specific values of time given in (a) and (b). The velocity and acceleration will be obtained by multiplying the values of the x-components by the unit vector $\hat{u}_x$ along the x-axis, which will show whether the vector is in the direction of the positive x-axis or in the direction of the negative x-axis. Since the motion is in a straight line which coincides with the x-axis, the vector uses only the base vector for the x-axis.

$$v_x(t) = \frac{dx}{dt} = \frac{d}{dt}\left[3\,t + 2\,t^3\right] = 3 + 6t^2. \ [\text{m/s}]$$
$$a_x(t) = \frac{dv}{dt} = \frac{d}{dt}\left[3 + 6\,t^2\right] = 12t. \ [\text{m/s}^2]$$

Now we use $t = 1$ sec and $t = 2$ sec in these equations to find answers for parts (a) and (b) respectively.

(a) $t = 1$ sec:

$$v_x(1) = 3 + 6(1)^2 = 9 \text{ m/s}.$$
$$a_x(1) = 12(1) = 12 \text{ m/s}^2.$$
$$\vec{v}(1) = (9 \text{ m/s})\hat{u}_x; \quad \vec{a}(1) = (12 \text{ m/s}^2)\hat{u}_x;$$

(b) $t = 2$ sec:

$$v_x(2) = 3 + 6(2)^2 = 27 \text{ m/s}.$$
$$a_x(2) = 12(2) = 24 \text{ m/s}^2.$$
$$\vec{v}(2) = (27 \text{ m/s})\hat{u}_x; \quad \vec{a}(2) = (24 \text{ m/s}^2)\hat{u}_x;$$

Note: Since we do not know which way in space the x-axis is pointed, we cannot go further than the analytic representation of the final answer.

Example 3.6.3. Components of acceleration from plots of components of velocity. The x-component of velocity of an object changes with time according to the graph shown in Fig. 3.28. (a) From this graph find the x-component of acceleration at $t = 0$, $t = 1$ sec, $t = 2$ sec, $t = 3$ sec and $t = 4$ sec. (b) What can you say about the velocity and acceleration of the object?

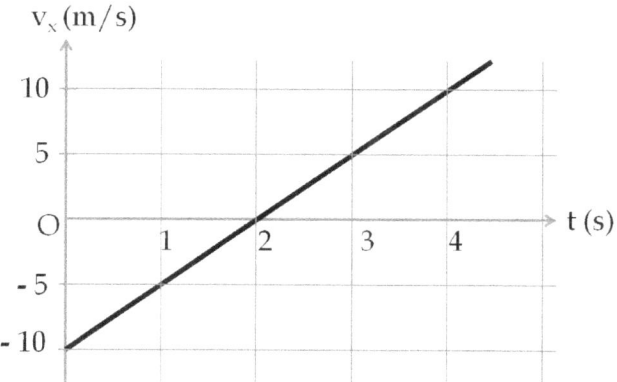

Figure 3.28: Example 3.6.3. Calculating the x-component of the acceleration from the rate of change of the x-component of velocity.

Solution. (a) From the definition of acceleration we know that each Cartesian component of acceleration is equal to the time derivative of corresponding component of velocity. Since the slope of the tangent to a curve when a quantity is plotted against the independent variable equals the derivative, we can use the given data to calculate the slope and find the derivative that way.

As the curve of the plot of v_x versus t is a straight line, the tangent to the curve is the line itself. Therefore, the slope of the tangent is the same regardless of the instant in time: there is only one tangent, and so there is only one slope to work out.

The slope of the line is

$$a_x = \frac{\text{Rise}}{\text{Run}} = \frac{20 \text{ m/s}}{4 \text{ s}}. \quad \text{Note: units in the rise and run!}$$

This gives the x-component of the acceleration at any point in time to be 5 m/s^2.

(b) In part (a) we found that the x-component of the acceleration was constant. We do not have data for y and z-components of velocity, so we cannot find the y and z-components of acceleration. Without all the three components of velocity and acceleration, we cannot find their magnitudes and directions of the velocity and acceleration.

Example 3.6.4. Acceleration from slope of velocity. An object moves in a straight line. A coordinate system is chosen so that the x-axis coincides with the line of motion. The x-component of velocity of an object varies as a function of time as shown in Fig. 3.29. Find the

3.6. ACCELERATION

x-component of the instantaneous acceleration at the following instants (a) $t = 1$ sec, (b) $t = 3$ sec, and (c) $t = 5$ sec.

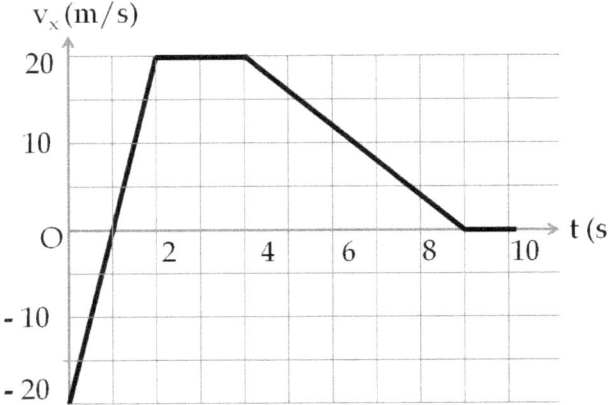

Figure 3.29: Plot of the x-component of velocity as a function of time.

Solution. Since the plot has only straight line segments, the tangents to the curves coincide with the segments. Therefore, here, it is relatively easy to find the slopes at different instants. At $t = 1$ sec, the slope can be calculated by noting that the velocity in this segment increases by 40 m/s in 2 sec interval at a constant rate. The slope is $40(\text{m/s})/2\text{s}$, or 20 m/s^2. The slope at $t = 3$ sec is clearly zero since velocity is not changing in this segment.

Finally, the slope at $t = 5$ sec is seen to be negative since the velocity goes down by 20 m/s in about 4 sec interval in this segment, from $t = 4$ sec to $t = 8$ sec. This yields a slope of -5 m/s^2. Summarizing the results:

(a) $a_x = 10 \text{ m/s}^2$ at $t = 1$ s.

(b) $a_x = 0$ at $t = 3$ s.

(c) $a_x = -5 \text{ m/s}^2$ at $t = 5$ s.

Example 3.6.5. Acceleration from slope of velocity - two dimensional case. An object moves in a plane. A coordinate system is chosen so that the xy-plane coincides with the line of motion. The x-component of velocity of an object varies as a function of time as shown in Fig. 3.30. Find acceleration at the following instants (a) $t = 2$ sec, and (b) $t = 6$ sec.

Solution. From the plots of v_x and v_y versus t we find the following values of the components of acceleration. Note the z-component of acceleration is zero and therefore ignored.

(a) $t = 2\ sec$:

a_x = Slope of v_x vs t = 4.0 m/s^2

a_y = Slope of v_y vs t = 7.5 m/s^2

Magnitude: $a = \sqrt{a_x^2 + a_y^2 + 0} = 8.5 \text{ m/s}^2$;

Direction: $\arctan(4.0/7.5) = 62°$ counterclockwise from the positive x-axis.

Further Remarks: Unlike velocity, acceleration can change abruptly. Therefore, the acceleration after an instant may not equal the acceleration before that instant. Thus, corners are allowed in v_x or v_y or v_z vs t plots. Of course, these plots are not allowed to have discontinuities since velocity cannot change abruptly.

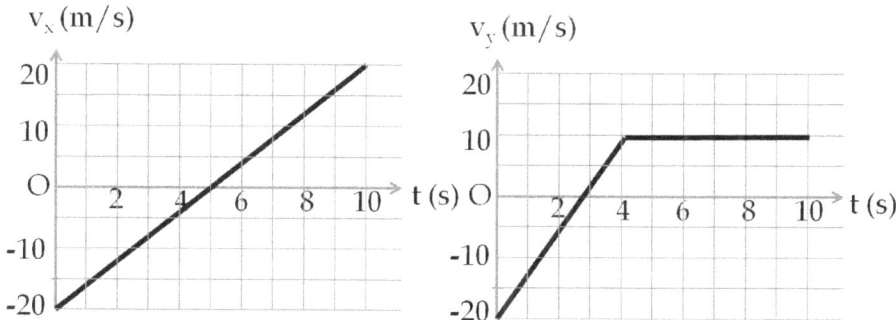

Figure 3.30: Plots of the x and y-components of velocity as a function of time.

(b) $t = 6\ sec$:

$$a_x = \text{Slope of } v_x \text{ vs } t = 4.0\ \text{m/s}^2$$
$$a_y = \text{Slope of } v_y \text{ vs } t = 0$$

Magnitude: $a = 4.0\ \text{m/s}^2$;

Direction: towards the positive x-axis.

Example 3.6.6. Acceleration and Deceleration in One Dimension.
In one-dimensional motion, acceleration can be either in the same direction as the velocity or in the opposite direction as illustrated in Fig. 3.31.

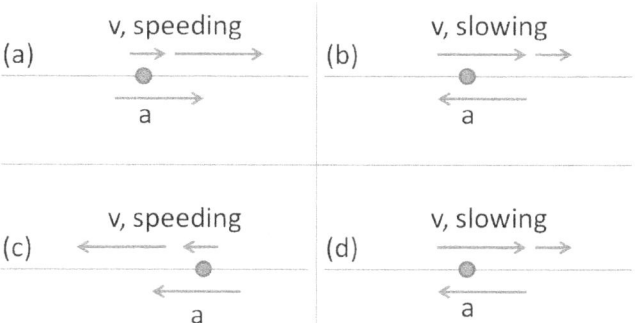

Figure 3.31: Example 3.6.6. Relative directions of velocity and acceleration determine if the motion will speed up or slow down. If acceleration is in the same direction as the velocity, as in (a) and (c), the magnitude of the velocity increases, and if the acceleration is in the opposite direction to the velocity as in (b) and (d), the magnitude of the velocity decreases.

If acceleration has the same direction as the velocity, it leads to an increase in the magnitude of the velocity vector. This situation is referred to as accelerating the motion (see Fig. 3.32).

If acceleration is in the opposite direction to the velocity, the immediate result is a slowing of motion, the object eventually coming to a rest, and finally, a reversal of the direction of motion occurs. The initial slowing effect of velocity and acceleration being in the opposite directions is also referred to as decelerating the motion.

3.6. ACCELERATION

We will see below that if acceleration is neither in the direction of the velocity nor opposite to it, the direction of the velocity changes.

Example 3.6.7. Acceleration for Vertically Falling or Rising Objects. If air resistance can be ignored, then a vertical motion of an object near Earth's surface has a constant acceleration of magnitude 9.81 m/s^2 and direction downward, called the acceleration due to gravity. The symbol g is used to denote this acceleration value.

$$g = 9.81 \text{ m/s}^2. \text{ (Acceleration due to gravity)} \qquad (3.49)$$

Note that the acceleration for a freely falling object is g pointed down regardless of whether the object is moving up or moving down or momentarily at rest. You should expect this to be the case since **acceleration is related to the change in velocity and not to the value of velocity at any instant.** To illustrate this case, we consider a ball thrown vertically up so that its initial velocity is pointed up as shown in Fig. 3.32.

Initially, as the ball goes up, its velocity and acceleration vectors are in opposite directions. As a result the ball slows down and comes to rest momentarily at the very top of the trajectory. After that, acceleration changes the velocity from zero at the resting point to non-zero value in the next instant, but now, the velocity is pointed down. On the way down, the ball picks up speed as it falls since velocity and acceleration are in the same direction.

This example shows that the same object may decelerate, meaning decreasing speed, in one time domain and accelerate, meaning increasing speed, in another time segment, all without a change in the acceleration. Therefore, you need to be careful when you see a decelerating motion since a decelerating motion does not mean the object will not pick up speed later after it has come to rest.

Example 3.6.8. Acceleration Perpendicular to Velocity If velocity and acceleration are collinear, the magnitude of the velocity either increases or decreases and the direction of the velocity either remains unchanged or reverses by 180°. What happens when the acceleration is not collinear with velocity? For instance, the acceleration of a moving charged particle caused by a magnetic force is perpendicular to the velocity of the particle.

In Fig. 3.33, the change in velocity as a result of an acceleration that is perpendicular to the velocity is shown. The velocity at time $t + \Delta t$ is equal to a vector sum of velocity at time t and the change in velocity $\vec{a}\Delta t$.

As we see from the figure, the direction of the velocity changes. The magnitude of the velocity at time $t + \Delta t$ is given by the hypotenuse of the triangle $\triangle OPQ$. From the triangle $\triangle OPQ$ you can deduce that the relative change in speed is of order $(a\Delta t/v)^2$ while the change in the direction of the velocity as given by the angle $\Delta\theta$ which goes as $(a\Delta t/v)$.

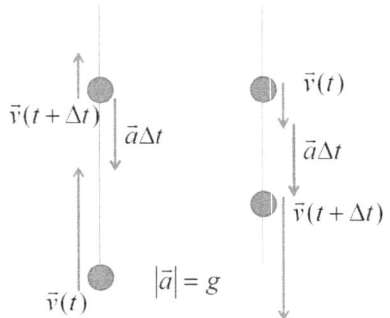

Figure 3.32: The acceleration of a freely falling object is same whether ball goes upward or downward. When the ball is going up, the acceleration is in the opposite direction of motion (i.e. of velocity). When the ball is going down, the acceleration is in the same direction as the direction of motion.

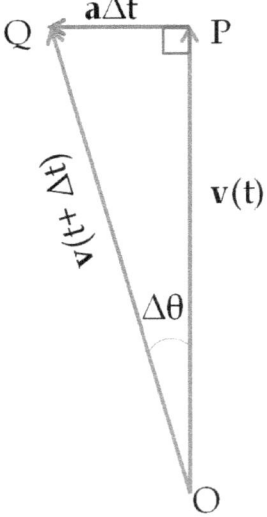

Figure 3.33: Example 3.6.8. Here $\vec{v}(t+\Delta t) = \vec{v}(t) + \vec{a}\Delta t$. When Δt becomes infinitesimal, dt, we get $|\vec{v}(t+dt)| = |\vec{v}(t)|$, but the direction of $\vec{v}(t+dt)$ is different from that of $\vec{v}(t)$. In this figure, vectors are indicated with bold letters.

When we reduce the time interval, we find that the magnitude of the velocity does not change as quickly as this angle, i.e. the direction. As a result, in the limit $\Delta t \to 0$, only the direction changes if $\vec{a}$ is always perpendicular to $\vec{v}$. Note that, in order for the acceleration to be perpendicular to the velocity at all times, the acceleration must also rotate along with the velocity vector.

3.7 MOTION USING POLAR COORDINATES

Some planar motions are more effectively analyzed in a different coordinate system than the Cartesian coordinates illustrated above. Many coordinate systems exist that help one make use of particular symmetries of the motion. For instance, polar coordinates are more natural for circular and elliptical trajectories. In this section, we will introduce polar coordinates and define new unit vectors for analysis of vectors.

3.7.1 Polar Coordinates

A point P with coordinates (x, y) in the xy-plane of a Cartesian coordinate system is at a distance r from the origin and at an angle θ counter-clockwise from the the positive x-axis. The distance r and angle θ are two coordinates of the **polar coordinate** (r, θ) as shown in Fig. 3.34. The polar coordinate r is called the radial coordinate and the polar coordinate θ the angular coordinate. You can see that the polar coordinates are simply

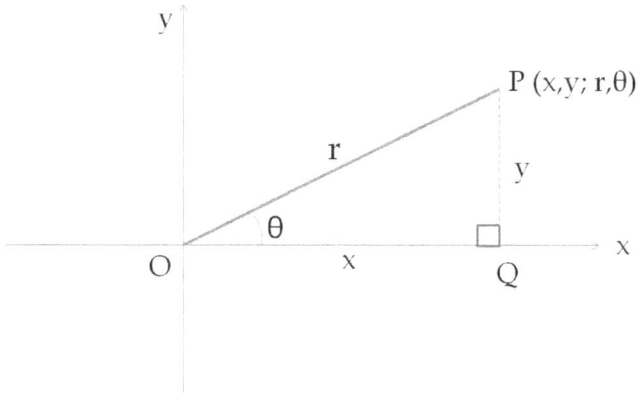

Figure 3.34: Polar coordinates r and θ. The right-angled triangle $\triangle OPQ$ shows the relation between the Cartesian coordinates (x, y) and polar coordinates (r, θ) for the same point P.

the magnitude (r) and direction (θ) of the position vector $(\vec{r} = x\hat{u}_x + y\hat{u}_y)$. The relations between the polar coordinates (r, θ) and the Cartesian coordinates (x, y) for the same point are

$$x = r\cos\theta \tag{3.50}$$
$$y = r\sin\theta \tag{3.51}$$
$$r = \sqrt{x^2 + y^2} \tag{3.52}$$
$$\theta = \arctan(y/x) \quad \text{Beware of quadrants!} \tag{3.53}$$

Polar coordinates in xy-plane together with the Cartesian z is called Cylindrical coordinates. Cylindrical coordinate system is useful for studying properties that have a symmetry about an axis, which is taken to be the z-axis.

3.7.2 Unit Vectors $\hat{u}_r$ and $\hat{u}_\theta$

Any vector in a plane can be written as a sum of two mutually independent vectors. Two vectors are mutually independent if their directions are different and one cannot be turned into the other by a multiplication with a scalar number. We have seen an application of this fact by constructing a vector in the xy-plane from the sum of a vector along the x-axis and another along the y-axis; the vector along the x-axis was obtained by multiplying the unit vector $\hat{u}_x$ pointed towards the positive x-axis with a real number, called the x-component of the vector, and the vector along the y-axis was obtained by multiplying the unit vector $\hat{u}_y$ pointed towards the positive y-axis with a real number, called the y-component of the vector.

Although, vectors along the Cartesian axes are very important for analysis of motion, they are not necessarily the only choice - any other mutually independent vectors would work also. In writing kinematic equations in polar coordinates, one often uses other sets of mutually independent vectors than vectors parallel to the x and y-axes.

As it turns out, there are infinitely many pairs of perpendicular vectors, one pair for each ray from the origin in the xy-plane, are defined for writing kinematics in polar coordinates. Consider a particular ray from origin to infinity passing through point P with Cartesian coordinates (x, y) and polar coordinates (r, θ) shown in Fig 3.35. The vector quantities, such as the position, velocity, and acceleration vectors, for a particle whose position falls on this ray can be expressed using the following pair of two mutually perpendicular unit vectors $\hat{u}_r$ and $\hat{u}_\theta$ as we will show below.

$$\hat{u}_r = \cos\theta\ \hat{u}_x + \sin\theta\ \hat{u}_y \tag{3.54}$$
$$\hat{u}_\theta = -\sin\theta\ \hat{u}_x + \cos\theta\ \hat{u}_y \tag{3.55}$$

It is easy to see that these vectors are mutually perpendicular by explicitly working out their dot product.

$$\hat{u}_r \cdot \hat{u}_\theta = 0. \tag{3.56}$$

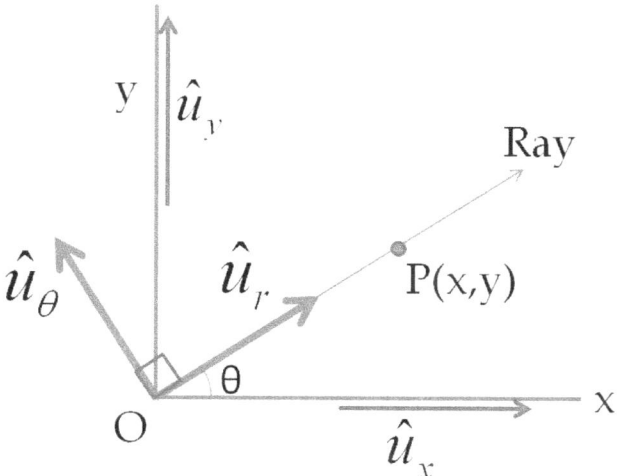

Figure 3.35: Polar unit vectors $\hat{u}_r$ and $\hat{u}_\theta$ in the direction of the ray from origin to infinity at an angle θ with respect to the x-axis. Unlike $\hat{u}_x$ and $\hat{u}_y$ for the Cartesian axes directions, there are infinitely many polar vectors $\hat{u}_r$ and $\hat{u}_\theta$, one pair for each direction in the plane given by the angle θ. For instance, if P is on the x-axis, $\hat{u}_r = \hat{u}_x$ and $\hat{u}_\theta = \hat{u}_y$, and if P is on the y-axis, $\hat{u}_r = \hat{u}_y$ and $\hat{u}_\theta = -\hat{u}_x$.

You may find in other books that the vectors $\hat{u}_r$ and $\hat{u}_\theta$ are denoted as simply $\hat{r}$ and $\hat{\theta}$. For pedagogical reasons we will continue to practice our notation, although at times it may appear too cumbersome to write u all the time and you may want to use the simpler notation when there is no confusion.

Equations 3.54 and 3.55 show that the polar unit vectors depend on the angle θ, which can vary from 0 to 2π radians, but the unit vectors on the x and y-axes do not depend on θ. Therefore, there will be infinitely many $\hat{u}_r$ and $\hat{u}_\theta$ vector pairs, one for each value of θ, while the Cartesian unit vectors $\hat{u}_x$ and $\hat{u}_y$ are fixed once you have chosen the direction of the Cartesian axes.

Note that Eqs. 3.54 and 3.55 can be inverted so that we can write the unit vectors $\hat{u}_x$ and $\hat{u}_y$ in terms of $\hat{u}_r$ and $\hat{u}_\theta$.

$$\hat{u}_x = \cos\theta\ \hat{u}_r - \sin\theta\ \hat{u}_\theta \tag{3.57}$$
$$\hat{u}_y = \sin\theta\ \hat{u}_r + \cos\theta\ \hat{u}_\theta \tag{3.58}$$

How can we write a vector in terms of polar unit vectors if the Cartesian components of the vector are known? Suppose we have an arbitrary vector $\vec{A}$ which has Cartesian components A_x and A_y. Then, by using Eqs. 3.57 and 3.58 we can replace $\hat{u}_x$ and $\hat{u}_y$ in terms of $\hat{u}_r$ and $\hat{u}_\theta$.

$$\begin{aligned}\vec{A} &= A_x \hat{u}_x + A_y \hat{u}_y \\ &= A_x(\cos\theta\ \hat{u}_r - \sin\theta\ \hat{u}_\theta) + A_y(\sin\theta\ \hat{u}_r + \cos\theta\ \hat{u}_\theta) \\ &= (A_x\cos\theta + A_y\sin\theta)\hat{u}_r + (-A_x\sin\theta + A_y\cos\theta)\hat{u}_\theta\end{aligned}$$

3.7. MOTION USING POLAR COORDINATES

This shows that every vector in the xy-plane can be written as sum of a real number times $\hat{u}_r$ and another real number times $\hat{u}_\theta$. The multiplying factor of these unit vectors are called r and θ components, although there will be infinitely many of them since there are infinitely many $(\hat{u}_r, \hat{u}_\theta)$ pairs. We can denote these components with a similar notation as used for the Cartesian components, namely by attaching a subscript to the symbol for the vector without the arrow.

$$\vec{A} = A_r \hat{u}_r + A_\theta \hat{u}_\theta. \tag{3.59}$$

3.7.3 Velocity and Acceleration in Polar Coordinates

In this section we will deduce the formula for velocity and acceleration vectors written in polar coordinates. We start with noting that the position vector $\vec{r}$ is readily written in polar coordinates by simply substituting x and y in its definition in the Cartesian representation.

$$\begin{aligned}
\vec{r} &= x\hat{u}_x + y\hat{u}_y \\
&= r\cos\theta\,(\cos\theta\,\hat{u}_r - \sin\theta\,\hat{u}_\theta) + r\sin\theta\,(\sin\theta\,\hat{u}_r + \cos\theta\,\hat{u}_\theta) \\
&= \left(r\cos^2\theta + r\sin^2\theta\right)\hat{u}_r
\end{aligned}$$

Therefore,

$$\boxed{\vec{r} = r\hat{u}_r.} \tag{3.60}$$

To deduce the polar form of the velocity vector we recall the following two facts: (1) velocity is equal to the derivative of the position vector and (2) the Cartesian directions are constant. Expressing the position vector in terms of Cartesian coordinates and taking derivatives leads to

$$\begin{aligned}
\vec{v} &= \frac{d\vec{r}}{dt} \\
&= \frac{dx}{dt}\hat{u}_x + \frac{dy}{dt}\hat{u}_y \quad (\text{since } d\hat{u}_x/dt = 0 = d\hat{u}_y/dt) \\
&= \left(\cos\theta\frac{dr}{dt} - r\sin\theta\frac{d\theta}{dt}\right)\hat{u}_x + \left(\sin\theta\frac{dr}{dt} + r\cos\theta\frac{d\theta}{dt}\right)\hat{u}_y \\
&= \frac{dr}{dt}\hat{u}_r + r\frac{d\theta}{dt}\hat{u}_\theta \quad (\text{replacing unit vectors and rearranging terms.})
\end{aligned}$$

Therefore, the velocity vector in polar coordinates is

$$\boxed{\vec{v} = \frac{dr}{dt}\hat{u}_r + r\frac{d\theta}{dt}\hat{u}_\theta.} \tag{3.61}$$

The rate of change of the radial coordinate, i.e. the quantity dr/dt is called radial velocity, and the rate of change of the angular coordinate, i.e., the quantity $d\theta/dt$ is called the angular velocity. Note that neither dr/dt nor $d\theta/dt$ is a vector since they do not have directions. More appropriate names would be radial velocity component and angular velocity component, but we will stick with the standard names for these quantities and make a mental note that, despite their names, the rates of change of radial and angular coordinates are not vectors. The angular velocity $d\theta/dt$ is often denoted by another symbol, ω.

$$\boxed{\text{Magnitude of the angular velocity: } \omega = \frac{d\theta}{dt}.} \tag{3.62}$$

When we study rotation, we will encounter the true angular velocity vector which would have both a magnitude and direction. We will find that the direction of angular velocity vector of a rotating particle is perpendicular to the position vector and velocity vector.

Making use of ω simplifies the writing of the formulas for velocity and acceleration as we will see below. Thus, the velocity of an object in the xy-plane can be written as

$$\vec{v} = \frac{dr}{dt}\hat{u}_r + \omega r \hat{u}_\theta. \tag{3.63}$$

Therefore, the radial and angular components of the velocity vector $\vec{v}$ are

$$\boxed{v_r = \frac{dr}{dt}.} \tag{3.64}$$

$$\boxed{v_\theta = r\frac{d\theta}{dt} = r\omega.} \tag{3.65}$$

Similarly, starting with $\vec{r} = x\hat{u}_x + y\hat{u}_y$ and taking two time derivatives gives the following expression for the acceleration vector.

$$\boxed{\vec{a} = \left[\frac{d^2r}{dt^2} - r\left(\frac{d\theta}{dt}\right)^2\right]\hat{u}_r + \left[2\left(\frac{dr}{dt}\right)\left(\frac{d\theta}{dt}\right) + r\left(\frac{d^2\theta}{dt^2}\right)\right]\hat{u}_\theta.} \tag{3.66}$$

This gives the radial and angular components of the acceleration vector $\vec{a}$ as

$$\boxed{a_r = \frac{d^2r}{dt^2} - r\left(\frac{d\theta}{dt}\right)^2.} \tag{3.67}$$

$$\boxed{a_\theta = 2\left(\frac{dr}{dt}\right)\left(\frac{d\theta}{dt}\right) + r\left(\frac{d^2\theta}{dt^2}\right).} \tag{3.68}$$

In the case of circular motions, these components, a_r and a_θ with their corresponding unit vectors are also called centripetal and tangential accelerations respectively.

3.7.4 Circular Motion

Speed and Velocity

Polar coordinates are natural for an object moving in a circle. Let us denote the radius of the circle by the capital letter R. With origin at the

3.7. MOTION USING POLAR COORDINATES

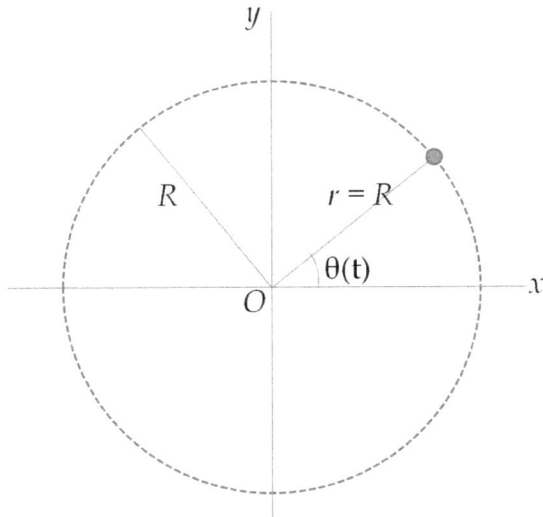

Figure 3.36: Radial coordinates in circular motion.

center of the circle, the radial polar coordinate r is equal to the radius of the circle and remains constant in time (see Fig. 3.36). Incidentally, note that if the origin is not at the center of the circle, then the radial coordinate will change with time even for a particle moving in a circle.

$$\text{Radial coordinate: } r = R, \text{ constant.} \tag{3.69}$$

Therefore, the only coordinate that changes with time is the angular coordinate of the object, $\theta(t)$. Recall that the rate of change of the angular coordinate is given by its own symbol, $\omega = d\theta/dt$. Hence, for a circular motion, the position and velocity vectors take the following simpler forms.

Circular Motion:

$$\boxed{\vec{r} = R\hat{u}_r.} \tag{3.70}$$

$$\boxed{\vec{v} = R\omega \hat{u}_\theta.} \tag{3.71}$$

The magnitude of the velocity is equal to the product of the radius of the circle and the angular velocity. This makes sense if you think about the distance on the circle being covered with time. If in time Δt, the angle coordinate changes by $\Delta \theta$, distance moved on the arc of the circle would be equal to $R\Delta\theta$. Therefore, your average speed will be

$$\text{Average speed} = \frac{R\Delta\theta}{\Delta t}.$$

In the limit of infinitesimal time, we find the formula for instantaneous speed as in terms of the derivative of the angular coordinate.

$$\boxed{\text{Circular motion instantaneous speed} = R\left|\frac{d\theta}{dt}\right| = R\,|\omega|,}$$

where I have used the absolute sign so that speed is positive regardless of which way the object moves in the circle - movement in one direction

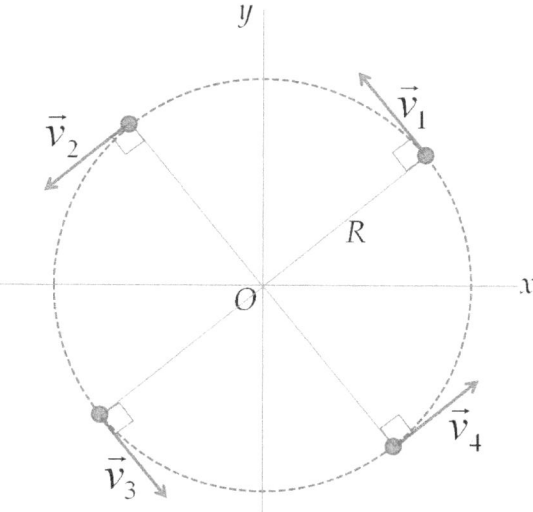

Figure 3.37: The direction of the velocity of a particle at a particular point in a circular motion is along the tangent to the circle at that point. Since tangent is changing, the direction of the velocity is always changing. Therefore, a particle in circular motion will always have non-zero acceleration.

on the circle leads to increasing angle values, giving a positive ω, and in the opposite direction on the circle corresponds to decreasing angle, and hence negative ω.

Eq. 3.71 for velocity tells us that the velocity of an object moving in a circle is tangential to the circle, i.e. in the direction of the unit vector $\hat{u}_\theta$ or in the opposite direction to $\hat{u}_\theta$. Since the tangential direction in space changes from place to place in the circle, the velocity of a particle moving in a circle always changes with time as shown in Fig. 3.37.

Tangential and Centripetal Accelerations

For circular motion we set $r = R$, a constant, in the definition of acceleration in Eq. 3.66 to find

$$\boxed{\textbf{Circular motion: } \vec{a} = -R\omega^2 \hat{u}_r + R\left(\frac{d\omega}{dt}\right)\hat{u}_\theta,} \quad (3.72)$$

where we find that the acceleration at any time is a sum of a radially pointed vector, the first term, and a tangentially pointed vector, the second term. These two contributions to acceleration are called **centripetal and tangential accelerations**, respectively.

$$\boxed{\text{Centripetal acceleration: } \vec{a}_c = -R\omega^2 \hat{u}_r.} \quad (3.73)$$

$$\boxed{\text{Tangential acceleration: } \vec{a}_T = R\left(\frac{d\omega}{dt}\right)\hat{u}_\theta.} \quad (3.74)$$

Figure 3.38 shows $\vec{a}_c$ and $\vec{a}_T$ at four points in a circle. Figure shows that the two parts of the acceleration of an object moving in circle are perpen-

3.7. MOTION USING POLAR COORDINATES

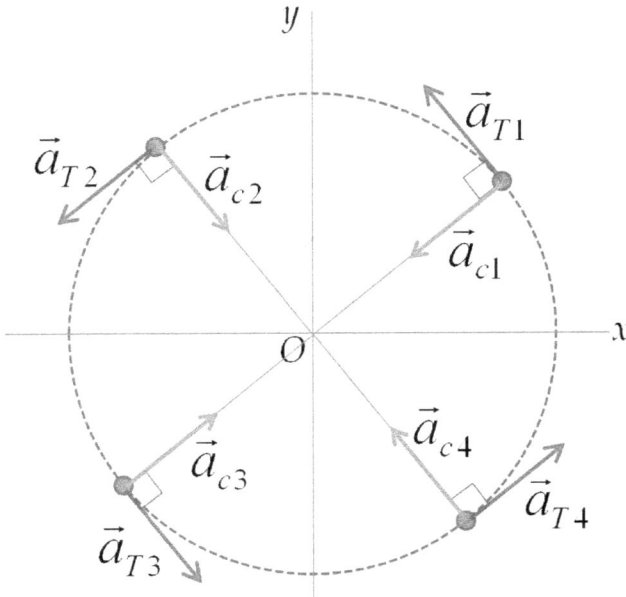

Figure 3.38: Tangential and centripetal accelerations of a particle in a circular motion are pointed in different directions at different points on the circle. The tangential acceleration is the acceleration along the tangent to the circle and the centripetal acceleration is the acceleration towards the center of the circle.

dicular to each other at all time and their directions change throughout the motion of the object. Often the magnitudes of $\vec{a}_c$ and $\vec{a}_T$ are themselves called centripetal and tangential accelerations with the tacit understanding the direction will be stated by drawing arrows on a circle at the instant of interest.

$$\text{Magnitude of Centripetal acceleration: } a_c = R\omega^2. \quad (3.75)$$

$$\text{Magnitude of Tangential acceleration: } a_T = R\left(\frac{d\omega}{dt}\right). \quad (3.76)$$

Uniform circular motion

Since the velocity of a circular motion is tangential, only a non-zero tangential acceleration can change the magnitude of the velocity. Therefore, if $a_T = 0$, then speed does not change. We see from the formula for the tangential acceleration, that if $d\omega/dt = 0$, then $a_T = 0$. The circular motion of constant speed or constant angular speed is called **uniform circular motion**.

The acceleration vector for the uniform circular motion points towards the center of the circle. Since ω for a uniform circular motion does not change with time, the magnitude of the centripetal acceleration is constant. However, that does not mean that the acceleration vector is constant. For an object moving in a circle the radially inward direction $(-\vec{u}_r)$ is pointed in different directions in the plane for different points on the

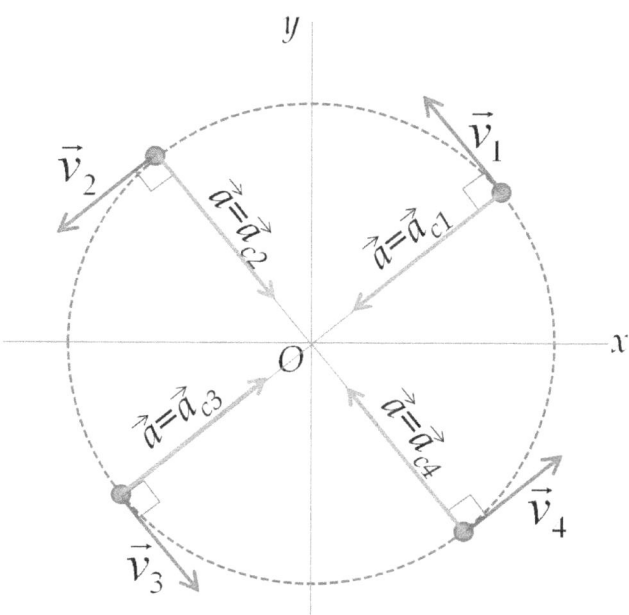

Figure 3.39: Uniform circular motion. The tangential acceleration is zero so that speed is constant although velocity is not constant, $|\vec{v}_1| = |\vec{v}_2| = |\vec{v}_3| = |\vec{v}_4|$ at points around the circle. The acceleration is entirely centripetal at each point.

circle (see Fig. 3.39).

For instance, when the object is at the x-axis, the acceleration vector in Eq. 3.73 is pointed towards the negative x-axis, and when the object is at the y-axis, the acceleration vector is pointed towards the negative y-axis. Therefore, the acceleration of an object in circular motion, always changes with time, even when the magnitude of the acceleration may be constant.

The magnitude of the centripetal acceleration can also be written in terms of the constant speed in the uniform circular motion by replacing ω^2 with $(v/R)^2$.

Centripetal acceleration for uniform circular motion:

$$\text{Magnitude: } |\vec{a}| = |\vec{a}_c| = \frac{v^2}{R} \text{ (since } a_T = 0\text{).} \quad (3.77)$$

Now, I will illustrate the use of polar coordinates with numerical and non-numerical problems.

Example 3.7.1. A rotating platform. Consider a rotating platform of radius 5 meters that is rotating at a constant angular velocity of 0.5 radians per second. (a) What will be your speed when you are standing at the edge? (b) What will be your acceleration? (c) Find your speed and acceleration when you are standing 2 meters from the center.

Solution. Note that parts (a) and (b) are about the circular motion when the radius of the circle is $R = 5$ m and part (c) is for a circular motion in a different radius, viz., $R = 2$ m. Using the formulas presented above for

3.7. MOTION USING POLAR COORDINATES

Figure 3.40: Example 3.7.1. A rotating platform.

uniform circular motions, we immediately find the following answer. (a) Speed, $v = R\omega = (5 \text{ m}) \times (0.5 \text{ rad/s}) = 2.5$ m/s. Note that m × rad = m, since the radian is dimensionless as it is the ratio of arc length to radius, both of dimension length.

(b) Now, since the question is about a vector, we must find the magnitude and direction.

Magnitude of acceleration = $R\omega^2 = (5 \text{ m}) \times (0.5 \text{ rad/s})^2 = 1.25$ m/s^2.

Direction of acceleration: It is pointed towards the center. The physical direction in space depends on where on the circle the object is at the time.

(c) Similar calculations as those of parts (a) and (b) give the following answer. Speed = 1.0 m/s and magnitude of acceleration 0.5 m/s^2. The direction of the acceleration is pointed towards the center whose orientation in plane changes with time.

Example 3.7.2. Pebble on a rotating tire. A pebble is stuck in the grooves of a rotating tire of radius 25 cm. If the car moves at a constant velocity of 20 m/s what is the angular velocity of the pebble?

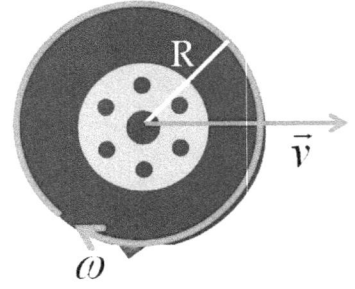

Figure 3.41: Example 3.7.2. Pebble stuck at the edge of a tire.

Solution. Notice that when the center of the tire moves a distance $2\pi R$, i.e. a full circle of the tire, the pebble rotates by an angle of 2π radians in a circle about the center of the tire whose radius is equal to the radius of the tire R. Since the center of the tire moves 20 m in 1 sec, the pebble rotates an angle of 20m/R radians in that time. Therefore, the angular speed of the pebble is 80 rad/s, since $R = 0.25$ m.

Example 3.7.3. Particle in uniform circular motion. A particle moves in a circle of radius a with a uniform circular motion of constant angular velocity $\omega = \omega_0$. The particle moves in the xy-plane in a counter-clockwise manner when observed from the positive z-axis with the center of the circle of motion at the origin. At $t = 0$ the particle is at the point whose Cartesian coordinates are $(a, 0, 0)$ with respect to origin at the center of the circle of motion. Where will the particle be at an arbitrary time T?

Solution. Since the particle is moving in a circle, it is advantageous to work in polar coordinates. We have the radial coordinate given to be $r = a$, which is independent of time. Therefore, we need to find the angular coordinate at time T. The angular coordinate changes with angular velocity ω, which is written in terms of derivative of the angular variable.

$$\frac{d\theta}{dt} = \omega_0.$$

This equation can be solved for general time t.

$$\theta(t) = \theta(0) + \omega_0 t.$$

The angle at zero time is zero since the particle was located on the x-axis at that instant. Therefore, the angle coordinate at the desired time is

$$\theta(T) = \omega_0 T.$$

We have found the polar coordinates at the desired time to be $(r, \theta) = (a, \omega_0 T)$. We can convert these into Cartesian coordinates if we so desire.

Example 3.7.4. Particle in a non-uniform circular motion A particle moves in a circle of radius a with a varying angular velocity $\omega = bt$ counterclockwise, where b is constant. At $t = 0$ the particle has Cartesian coordinates are $(a, 0, 0)$ with respect to the origin at the center of the circle of motion which is at the origin. Where will the particle be at an arbitrary time T?

Solution. In this example we have kept the data similar to the last example, except that the angular velocity is not constant any more. Since the particle is moving in a circle, it is again advantageous to work in polar coordinates. We have the radial coordinate given to be $r = a$, which is independent of time. Therefore, we need to find the angular coordinate at time T. Now, making use of the definition of angular velocity in terms of the derivative of the angular coordinate we find that the angular coordinate changes according to the following equation.

$$\frac{d\theta}{dt} = bt.$$

This equation can be solved for general time t.

$$\theta(t) = \theta(0) + \frac{1}{2}bt^2.$$

The angle at $t = 0$ is zero since the particle was located on the x-axis. Therefore, the angular coordinate at the desired time is

$$\theta(T) = \frac{1}{2}bT^2.$$

We have found that the polar coordinates at the desired time is $(r, \theta) = (a, 1/2\, bT^2)$. We can convert these into Cartesian coordinates by standard procedure with the result $x = a \cos\left(\frac{1}{2}bT^2\right)$ and $y = a \sin\left(\frac{1}{2}bT^2\right)$..

3.8 EXERCISES

Displacement, Average Velocity and Speed

Ex 3.8.1. Figure 3.42 shows three displacement vectors. (a) Read off the x and y-components of each displacement vector. (b) Use a ruler and a protractor to find the magnitude and directions of the displacement vectors. The scale is given in the axes. (c) Determine the magnitude and direction of each vector from its x and y-components. (d) Check if the magnitude and direction of the vectors found from the x and y-components agree with the magnitude and direction you found by graphical methods. Give a reason for any discrepancies you find.

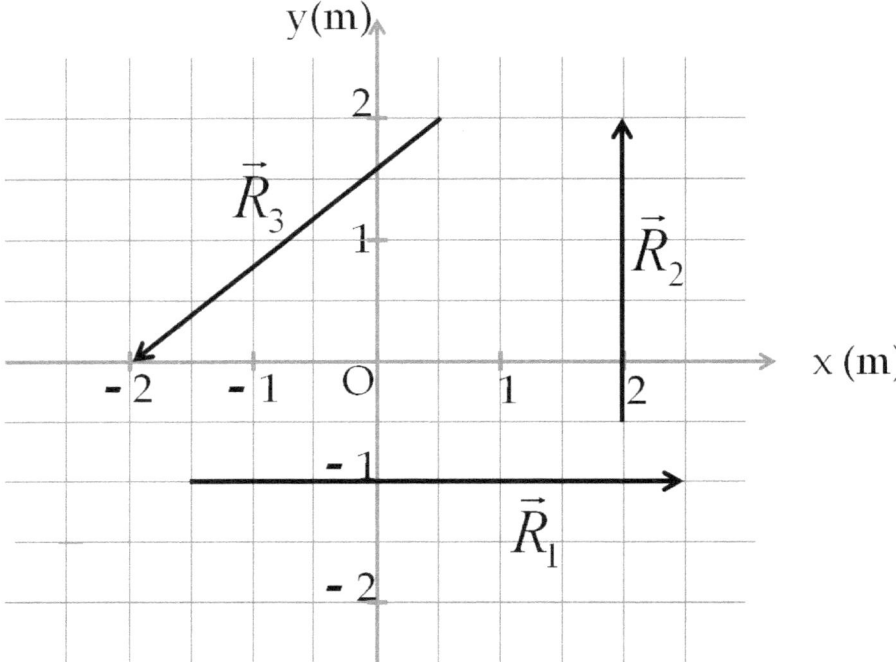

Figure 3.42: Exercise 3.8.1.

Ex 3.8.2. Figure 3.43 shows twelve displacement vectors which has only three unequal vectors. (a) Indicate which vectors are equal. (b) Using the scale given in the drawing, read the magnitude and directions of the displacement vectors. (c) Find the x and y-components of each displacement vector by either reading off from the figure or some other way. State your method. (d) Determine the magnitude and direction of each vector from its x and y-components. (e) Check if the magnitude and direction of the vectors found from the x and y-components agree with the magnitude and direction you found by graphical methods. Give a reason for any discrepancies.

Ex 3.8.3. Add the three displacement vectors in Figure 3.42 by two methods: (a) graphically, and (b) analytically. In each case give the magnitude and direction of the sum.

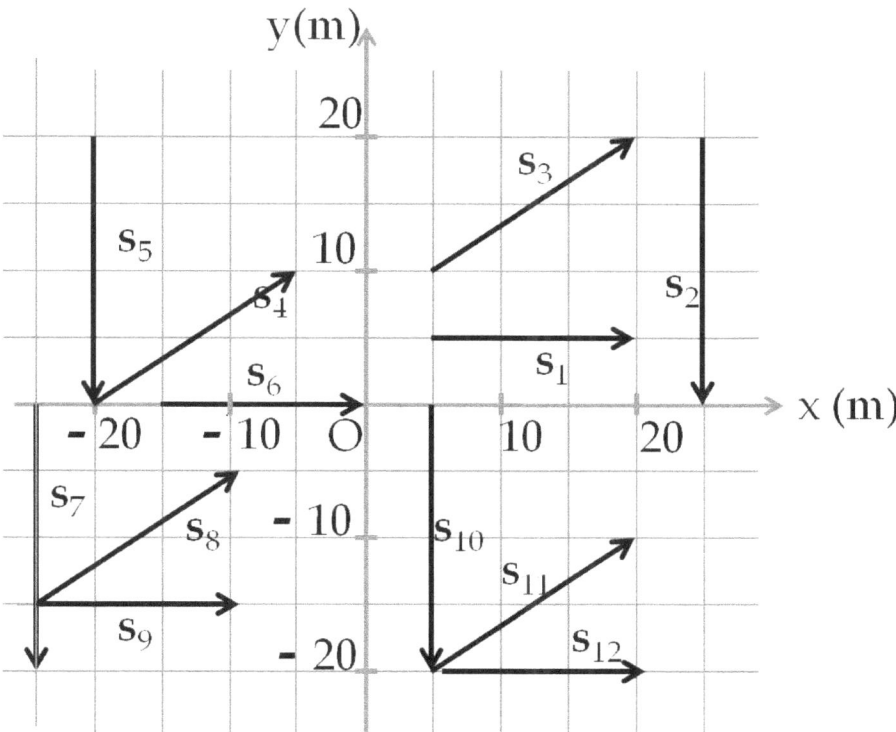

Figure 3.43: Exercise 3.8.2.

Ex 3.8.4. A steel ball rolls in straight track as shown in Fig. 3.44. Six positions on the track are marked $A - F$. The distance on the track are drawn according to the scale shown in the figure. (a) Draw the displacement vectors for displacements from A to B, from B to C, from C to D, from D to E, and from E to F. (b) Find the magnitudes and directions of the displacement vectors by using graphical or analytic method, whichever you find convenient. (c) If the ball takes 10 seconds between successively marked points in the figure, what are the values of the average speed in m/s between A and B, B and C, C and D, D and E, and E and F? (d) What would be the average velocity vectors between A and B, B and C, C and D, D and E, and E and F. Don't forget about the directions of the vectors.

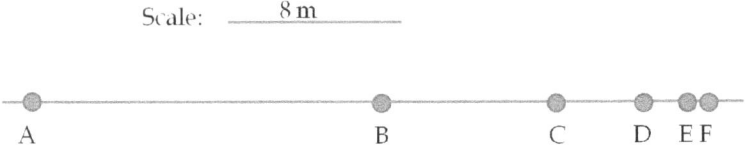

Figure 3.44: Exercise 3.8.4.

Ex 3.8.5. A truck travels north at a constant speed on a straight North-South road covering a distance of 35 km in 30 min. The driver realizes that he forget to pick up a package, and turns the truck around, and heads straight back to the original place. It took him 25 min on the return trip.

(a) Draw two displacement vectors, one for the Northward motion and the other for the return motion. (b) Find the displacement, average velocity and average speed for the first part of the motion to the north. (c) Find the displacement, average velocity and average speed for the second part of the motion. (d) Find the displacement, average speed and average velocity for the entire trip. Express your speed and velocity in km/h.

Ex 3.8.6. A sprinter runs 100 meters on a straight track in 10.1 seconds. Then he walks back to the starting place taking 5 minutes. After reaching the starting place he runs again on the same track, this time taking only 10.0 sec. (a) Find the displacement, the average velocity and the average speed of the athlete during the following time intervals given in seconds. (i) $[0, 10.1]$, (ii) $[10.1, 310.1]$, (iii) $[0, 310.1]$, (iv) $[310.1, 320.1]$, and (v) $[0, 320.1]$. (b) Draw the displacement vectors for these time intervals using the same scale. (c) Draw the average velocity vectors for these time intervals using the same scale.

Ex 3.8.7. A subway train travels on a straight rail between two stations. Taking one of the stations as the origin, placing the x-axis on the track, the x-coordinate of the train is measured at different times. The result is displayed in Fig 3.45. (a) Draw displacement vectors for the following intervals given in minutes: (i) $[0, 5]$, (ii) $[5, 15]$, (iii) $[15, 25]$, (iv) $[25, 45]$, and (v) $[45, 50]$. (b) Estimate the displacement, average velocity and average speed for these intervals and for the [0,50 min] interval.

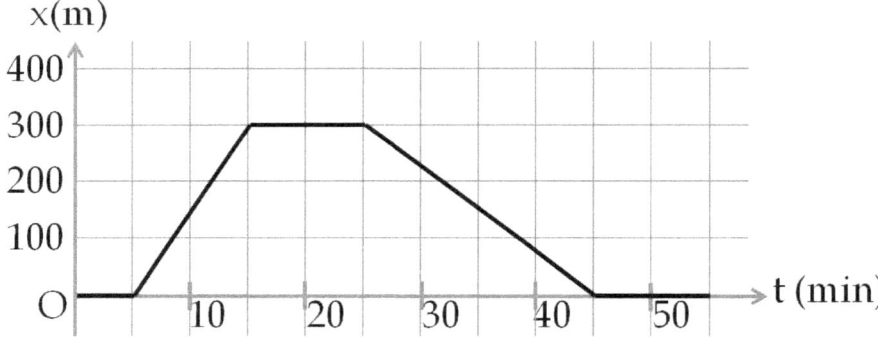

Figure 3.45: Exercise 3.8.7.

Ex 3.8.8. A train runs on a straight track all day repeating its route every 60 minutes, stopping at only the designated stations. Placing the origin at one of the stations and the x-axis on the track, the position of the train is recorded as its x-coordinate. Fig. 3.46 shows a plot of the x-coordinate of the train with time. (a) On a separate sheet of paper, draw displacement vectors for each successive 5 minute interval. (b) Assuming the train stops only at the stations, how many stations are there in the train's route? (c) For how long does the train stop at each station? (d) What are the average velocities of the train between different stations? Give the magnitude and velocity between pairs of the stations which are visited successively. (e) What is the average velocity over any 60-minute interval? (f) What is the average speed over any 60-minute interval?

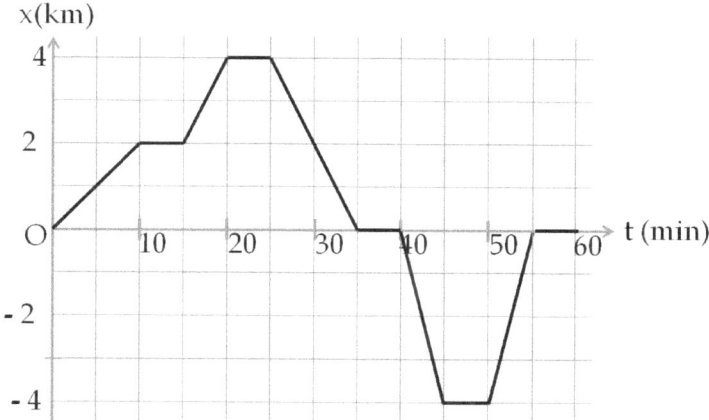

Figure 3.46: Exercise 3.8.8.

Ex 3.8.9. A steel ball rolls in a circular track as shown in Fig. 3.47. Six positions on the track are marked $A - F$. (a) In the figure, draw the displacement vectors for the displacements from A to B, from B to C, from C to D, from D to E, and from E to F. (b) Find the magnitudes and directions of the displacement vectors by using the graphical or the analytic method, whichever you find convenient. (c) If the ball takes 2 sec to complete one full revolution and if we assume that that ball rolls at constant speed, what is the value of the average speed in m/s? (d) What would be the average velocity vectors between A and B, B and C, C and D, D and E, and E and F? (e) Even though the speed is constant, why would you have five different average velocities for your answer?

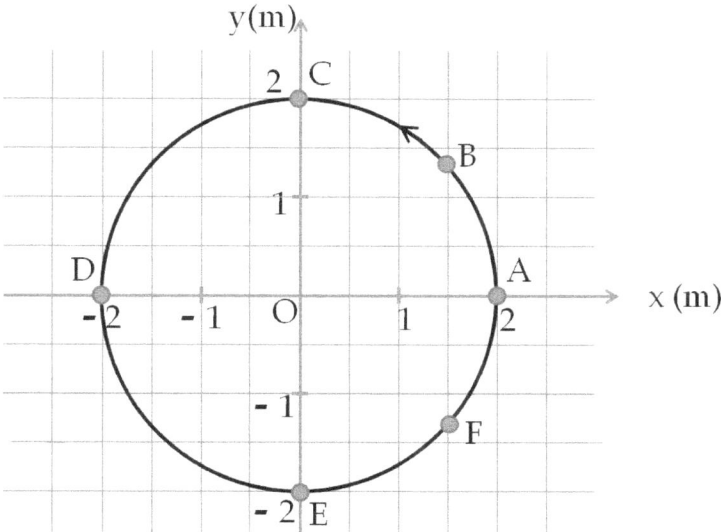

Figure 3.47: Exercise 3.8.9.

Ex 3.8.10. A race car is being driven on an elliptical track as shown in Fig. 3.48. Six positions on the track are marked $A - F$. (a) In the figure draw the displacement vectors for displacements from A to B, from B to

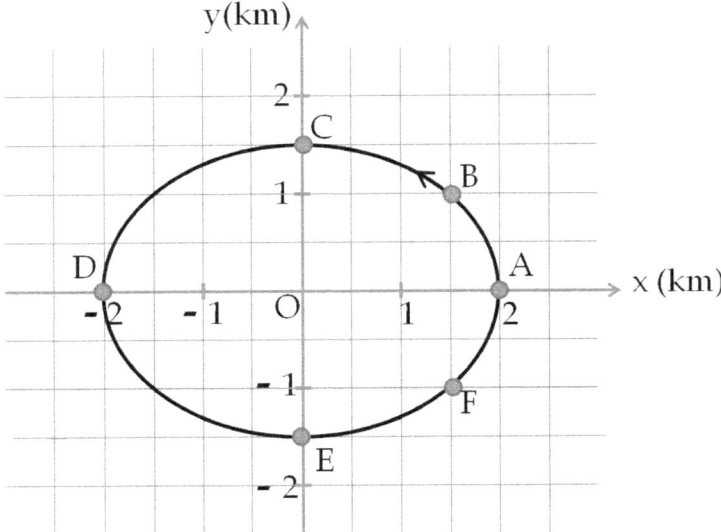

Figure 3.48: Exercise 3.8.10.

C, from C to D, from D to E, and from E to F. (b) Find the magnitudes and directions of the displacement vectors by using the graphical or the analytic method, whichever you find convenient. (c) If the car takes 3 min to complete one full revolution and if we assume that that ball rolls at constant speed, what is the value of average speed in km/hr and in m/s? (d) What would be the average velocities vectors between A and B, B and C, C and D, D and E, and E and F? (e) Even though the speed is constant, why would you have five different average velocities for your answer?

Instantaneous velocity and speed

Ex 3.8.11. Fig. 3.49 shows instantaneous speeds in m/s at various times when a ball is thrown vertically up. Decide on a scale for the velocity vectors and draw the velocity vectors at those instants. You may copy the figure on a larger piece of paper to show your drawings more clearly.

Ex 3.8.12. Refer to Fig. 3.47 for a ball rolling in a circular track. Suppose the ball rolls with a constant speed of 10 m/s. Decide on a scale for velocity and draw velocity vectors at points marked A, B, C, D, E, and F on the figure.

Ex 3.8.13. Refer to Fig. 3.48 for a car moving in an elliptical racetrack. Suppose the car moves with a constant speed of 250 km/h. Decide on a scale for the velocity and draw the velocity vectors at points marked A, B, C, D, E, and F on the figure.

Ex 3.8.14. Fig. 3.50 shows three positions of a car rounding a bend. The speedometer of the car has readings of 40 mph, 15 mph, and 30 mph when the car is at A, B, and C, respectively. Here, mph means miles per hour.

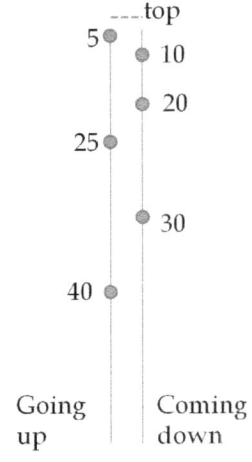

Figure 3.49: Exercise 3.8.11. Speeds in m/s.

Figure 3.50: Exercise 3.8.14.

Decide on a scale for the velocity vectors and draw the velocity vectors at those instants. Show your scale in the drawing.

Ex 3.8.15. Fig. 3.51 shows five points on the trajectory of motion of a pendulum bob. The pendulum bob goes from A to E and back to A. The speed of the bob is 0, 4.2 m/s, 6.4 m/s, 4.2 m/s, and 0 when it is at A, B, C, D, and E respectively. Decide on a scale for the velocity vectors and draw the velocity vectors at each location for all instants in one full swing. Not that the velocity vectors at some locations will be different for different instants. Show your scale in the drawing.

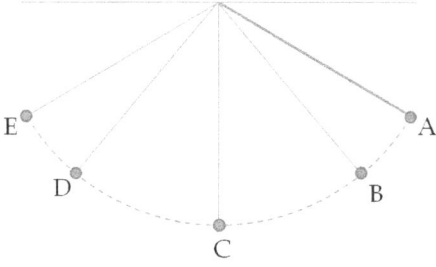

Figure 3.51: Exercise 3.8.15.

Ex 3.8.16. A boy throws a ball straight up. The vertical position of the ball is given by the y-coordinate in a coordinate system in which the positive y-axis is pointed up and the origin is at the point where the ball leaves the hand. The x and z-coordinates of the ball do not change with time. The y-coordinate of the ball as a function of time is shown in Fig. 3.52. (a) From the graph, estimate the instantaneous velocity of the ball at the following instants in time. (i) $t=0$, (ii) $t=0.4$ sec, (iii) $t=1$ sec, (iv) $t=1.2$ sec, and (v) $t=2$ sec. (b) Draw these velocity vectors on

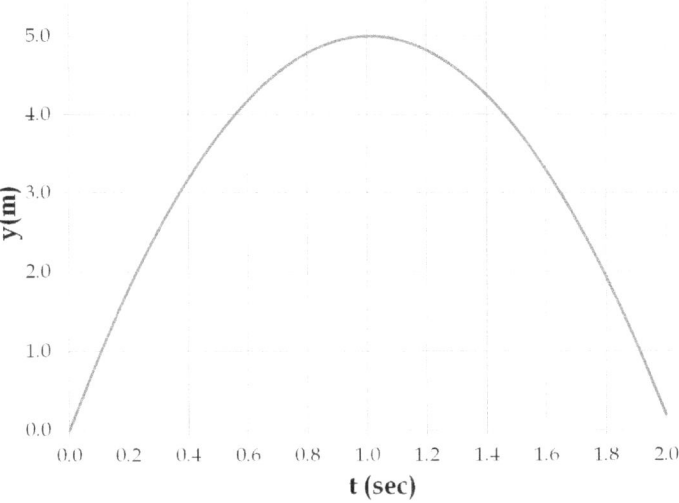

Figure 3.52: Exercise 3.8.16.

a sketch of the vertical motion of the ball. (c) From the graph, estimate the average velocity between the following intervals. (i) $[0, 0.4 \text{ sec}]$, (ii)

[0, 1 sec], (iii) [0.4 sec, 1 sec], (iv) [1 sec, 1.2 sec], and (v) [0.4 sec, 2 sec].
(b) How would your answer be affected if the x and z-coordinates also changed with time? Could you still determine the instantaneous velocities from the given data in Fig. 3.52?

Ex 3.8.17. The x-coordinate of a moving object varies in time according to the graph shown in Fig. 3.53, while the y and z-coordinates do not change with time. Assume the corners in the figure to be smooth. Determine the instantaneous velocity at the following instants (i) $t = 0$, (ii) $t = 5$ min, (iii) $t = 15$ min, (iv) $t = 35$ min, and (v) $t = 45$ min.

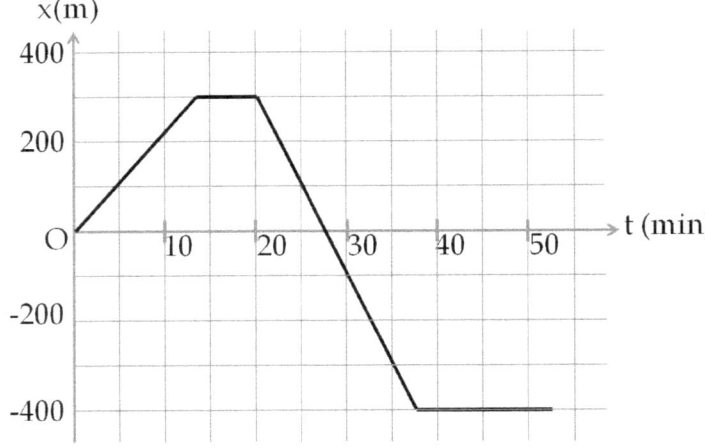

Figure 3.53: Exercise 3.8.17.

Ex 3.8.18. A golf ball is shot and its motion is recorded with respect to a Cartesian coordinate system. The z-coordinate of the ball does not change with time. The change in the x and y-coordinates of the ball with time is displayed in Fig. 3.54.

(a) From the graphs, estimate the instantaneous velocity of the ball at the following instants in time giving both the magnitude and direction: (i) $t = 0$, (ii) $t = 0.4$ sec, (iii) $t = 1$ sec, (iv) $t = 1.4$ sec, and (v) $t = 2$ sec.
(b) Draw these velocity vectors in the xy-plane using a convenient scale so that all vectors can be drawn on the same graph. Give the scale for your drawing.

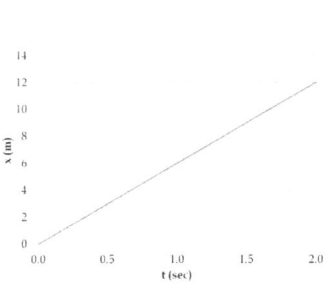
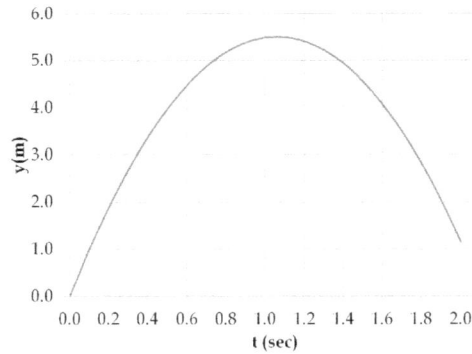

Figure 3.54: Exercise 3.8.18. Plot of x and y-coordinates of a projectile.

Ex 3.8.19. The motion of a steel ball rolling in a fixed track is recorded with respect to a Cartesian coordinate system. The z-coordinate of the ball does not change with time. The change in the x and y-coordinates of the ball with time is displayed in Fig. 3.55. (a) From the graphs, estimate the instantaneous velocity of the ball at the following instants in time giving both the magnitude and direction: (i) $t = 0$, (ii) $t = 0.5$ sec, (iii) $t = 1$ sec, (iv) $t = 1.5$ sec, and (v) $t = 2$ sec. (b) Draw these velocity vectors in the xy-plane using a convenient scale so that all vectors can be drawn on the same graph. Give the scale for your drawing. (c) Can you guess the shape of the track? Why do you think so?

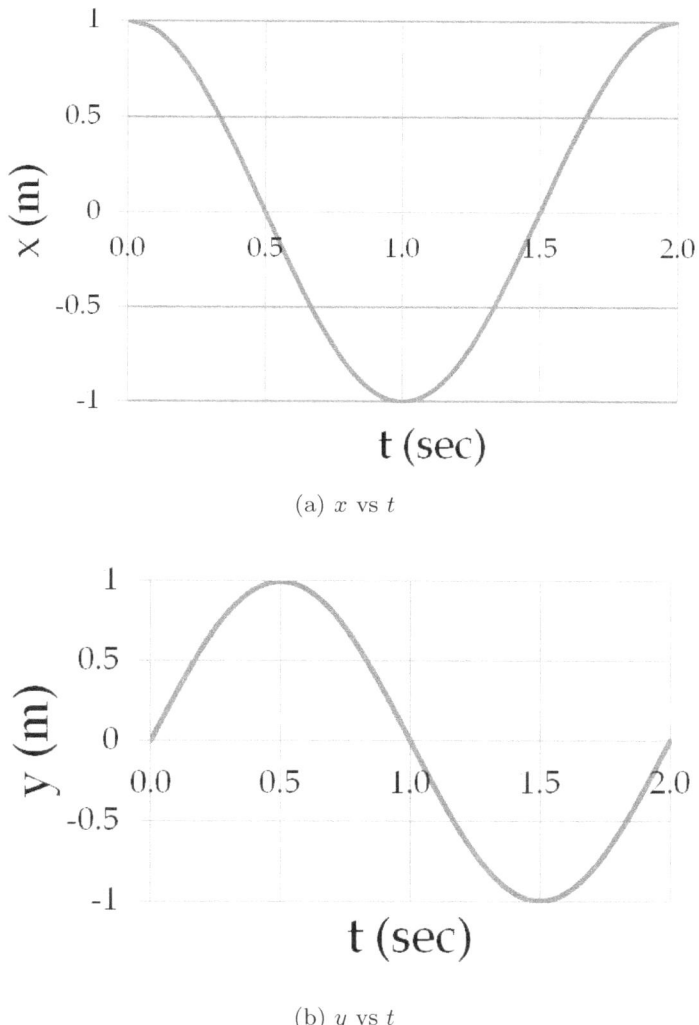

(a) x vs t

(b) y vs t

Figure 3.55: x and y coordinates of position with time t.

Ex 3.8.20. The position of a box sliding in a straight line is given by the x-coordinate of one of its corners. It is found that, while the y and z-coordinates do not change, the x-coordinate varies as a function of time as $x(t) = 4 + 18t - 5t^2$, where t is in seconds and x in meters. (a) Determine the instantaneous velocity and speed of the box at following instants in time. (i) $t = 0$, (ii) $t = 1$ sec, (iii) $t = 2$ sec, and (iv) $t = 3$ sec. (b) Find

3.8. EXERCISES

the average velocity of the box during the following intervals. (i) [0, 1 sec], (ii) [0, 2 sec], (iii) [1 sec, 2 sec], (iv) [0, 3 sec], and (v) [1 sec, 3 sec].

Ex 3.8.21. A block of copper is attached to a spring that executes a one-dimensional motion along the direction of the spring. A Cartesian coordinate system is chosen so that the motion of the block occurs along the x-axis only. In this coordinate system, the position of the center of the block varies with time according to the following functions, $x(t) = 5.0 \cos(2\pi t)$, $y(t) = 0$, $z(t) = 0$, where t is in sec and x in meters. (a) Plot x vs t for the time domain $(0, 3\ sec)$. (b) Determine the instantaneous velocity and speed at the following instants: (i) $t = 0$, (ii) $t = \frac{1}{4}$ sec, (iii) $t = \frac{1}{2}$ sec, (iv) $t = \frac{3}{4}$ sec, and (v) $t = 1$ sec. (c) Find the average velocity during the following intervals. (i) $[0, \frac{1}{4}$ sec], (ii) $[\frac{1}{4}$ sec, $\frac{1}{2}$ sec], (iii) $[0, \frac{1}{2}$ sec], (iv) $[0, \frac{3}{4}$ sec], and (v) $[0, 1$ sec].

Ex 3.8.22. After a football is thrown in air, the ball follows a parabolic trajectory that lies entirely in one plane. A coordinate system is chosen such that z-coordinate of the ball is always equal to zero. In this coordinate system, the x-axis is horizontal and the y-axis is pointed up. The changing x and y-coordinates of the ball are given by the following functions: $x(t) = 5t$ and $y(t) = 5t - 5t^2$, where t is in sec and x and y in meters. (a) Find the expressions for the x and y-components of the instantaneous velocity. (b) Determine the instantaneous velocity of the ball at the following instants in time. (i) $t = 0$, (ii) $t = 0.25$ sec, (iii) $t = 0.5$ sec, and (iv) $t = 1$ sec. (c) Draw the velocity vectors on a graph paper with an appropriate scale. Show the scale on your graph.

Ex 3.8.23. A bug is flying in a helical path. The position of the bug in a Cartesian coordinate is given by its coordinates (x, y, z). These coordinates vary with time according to: $x(t) = 2\cos\left(\frac{\pi}{4}t\right)$, $x(t) = 2\sin\left(\frac{\pi}{4}t\right)$ and $z = 0.1\ t$, where t is in sec and the coordinates are in meters. (a) Find the expressions for the x y, and z-components of the instantaneous velocity. (b) Determine the instantaneous velocity of the bug at following instants in time. (i) $t = 0$, (ii) $t = 10$ sec, (iii) $t = 20$ sec, and (iv) $t = 30$ sec. (c) Draw the velocity vectors at these instants with an appropriate scale.

Acceleration

Ex 3.8.24. The velocity vectors of a car at three instants are shown in Fig. 3.56. Use the arrow for the velocity at $t = 0$ for the scale for vectors in the figure. Use the geometric approach of vectors to find the average acceleration in the following intervals: (i) from $t = 0$ to $t = 50$ sec, (ii) from $t = 50$ sec to $t = 100$ sec, (iii) from $t = 0$ to $t = 100$ sec.

Ex 3.8.25. The velocity vectors of a rocket at two instants are shown in Fig. 3.57. (a) Use the geometric approach of vectors to find the average acceleration during an interval from $t = 0$ to $t = 2$ sec. (b) Choose a coordinate system so that the two velocity vectors fall in the xy-plane, and determine the x and y-components of the two velocity vectors. (c)

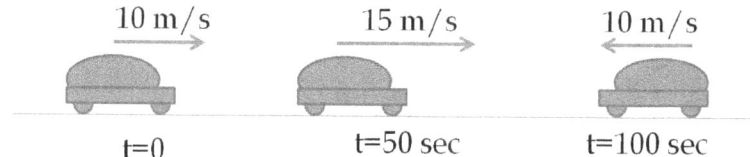

Figure 3.56: Exercise 3.8.24.

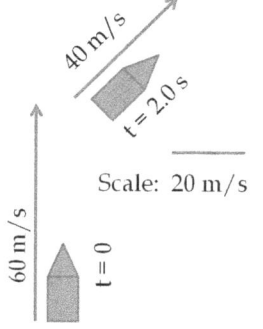

Figure 3.57: Exercise 3.8.25.

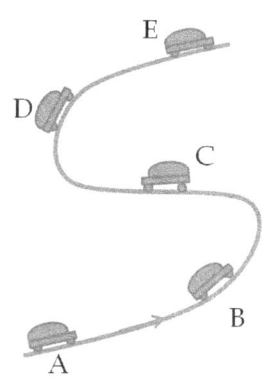

Figure 3.58: Exercise 3.8.26.

Find the average acceleration using the rate of change of the components of the velocity during the interval from $t = 0$ to $t = 2$ sec.

Ex 3.8.26. A car is moving at a constant speed of 5 m/s on a curved road as shown in Fig. 3.58. Suppose the time interval between successive points A, B, C, D, and E is 30 sec. (a) Draw the velocity vectors to scale at points A, B, C, D, and E. State the scale for your drawings. (b) Find the average acceleration between successive points marked on the trajectory. Give both the magnitude and the direction.

Table 3.3: Exercise 3.8.27

$t(s)$	v_x(m/s)	$t(s)$	v_x(m/s)
0	6.0	10	11.0
2	6.2	12	13.2
4	6.8	14	15.8
6	7.8	16	18.8
8	9.2		

Ex 3.8.27. A motion detector is used to measure the velocity of a cart that moves along a straight track. The data is recorded as the x-component of the velocity by placing the x-coordinate on the track. The y and z-components of the velocity are zero here. The recorded x-component of the velocity is shown the Table 3.3. Plot the data, and, from the graph, determine the acceleration at the following instants:(i) $t = 0$, (ii) $t = 5$ sec, and (iii) $t = 10$ sec.

Table 3.4: Exercise 3.8.28

$t(s)$	$x(m)$	$t(s)$	$x(m)$
0	0	9	−3
1	5	10	−35
2	10	11	−35
3	15	12	−32
4	20	13	−26
5	25	14	−17
6	25	15	−5
7	23	16	10
8	15		

3.8. EXERCISES

Ex 3.8.28. The position of a glider on a straight track is determined at various times. The x-coordinate of the glider at various times is shown in Table 3.4, and $y = z = 0$ at all times. (a) From the given table or a plot of x vs t from the data in the table, determine the instantaneous velocities at sufficient number of instants [10 may be enough] to draw a smooth curve for the v_x vs t plot. (b) Draw the v_x vs t plot from the data generated in part (a). (c) From the v_x vs t plot you generated, estimate the values of the instantaneous acceleration at the following instants: (i) $t = 2$ sec, (ii) $t = 8$ sec, and (iii) $t = 16$ sec.

Ex 3.8.29. The data for the velocity of a projectile in space is shown in Fig. 3.59. The axes are chosen so that the x-axis is horizontal and the y-axis vertically up. The z-coordinate of the projectile is not changing with time. (a) Find the instantaneous velocity of the projectile at the following instants: (i) $t = 0$, (ii) $t = 0.5$ sec, (iii) $t = 1$ sec, (iv) $t = 1.5$ sec, and (v) $t = 2$ sec. (b) Find the instantaneous acceleration of the projectile at the following instants: (i) $t = 0$, (ii) $t = 0.5$ sec, (iii) $t = 1$ sec, (iv) $t = 1.5$ sec, and (v) $t = 2$ sec.

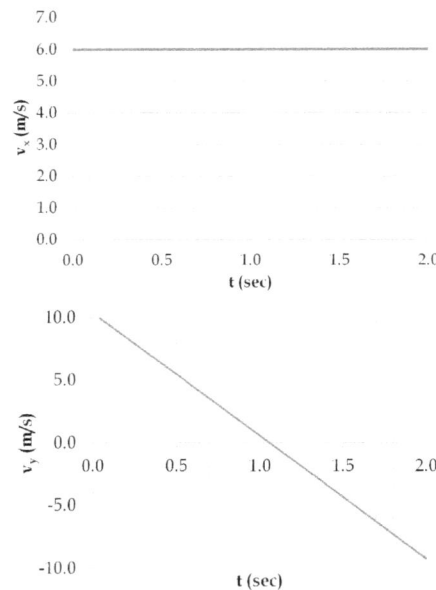

Figure 3.59: Exercise 3.8.29. Plot of v_x and v_y of projectile fired in space.

Ex 3.8.30. The data for the velocity of a soccer player in a field is shown in Fig. 3.60. The axes are chosen so that the x-axis is pointed from one goal post to the other and the y-axis from side to side. The z-coordinate of the player does not change during the time interval shown. (a) Find the instantaneous velocity of the soccer player at the following instants: (i) $t = 0$, (ii) $t = 5$ sec, (iii) $t = 25$ sec, (iv) $t = 35$ sec, and (v) $t = 45$ sec. (b) Find the instantaneous acceleration of the player at the following instants: (i) $t = 0$, (ii) $t = 5$ sec, (iii) $t = 12$ sec, (iv) $t = 25$ sec, (v) $t = 35$ sec, and $t = 45$ sec.

Ex 3.8.31. The x-coordinate of a particle moving in space is given by

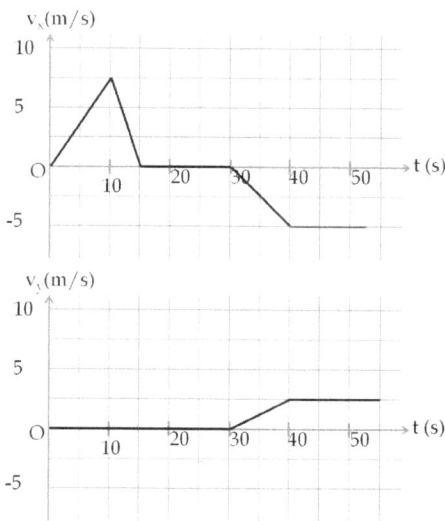

Figure 3.60: Exercise 3.8.30. Plot of v_x and v_y of a soccer player.

$x(t) = 3 + 5t + 9t^2$, where t is in sec and x in meter. (a) Determine the x-components of velocity and acceleration at (i) $t = 0$, (ii) $t = 1$ sec and (iii) $t = 2$ sec. (b) Determine the average velocity for the following intervals: (i) [0, 1 sec], (ii) [1 sec, 2 sec], (iii) [0, 2 sec]. (c) Determine the average acceleration for the following intervals: (i) [0, 1 sec], (ii) [1 sec, 2 sec], (iii) [0, 2 sec].

Ex 3.8.32. A car moves on a straight road such that its position from a reference point is given as $x(t) = 3 + 5t + 2t^2 + 0.4t^3$, $y = 0$, $z = 0$, where t is in sec and x in meter. (a) Determine the velocity and acceleration at (i) $t = 0$, (ii) $t = 0.5$ sec and (iii) $t = 1$ sec. (b) Determine the average velocity between $t = 0$ and $t = 1$ sec. (c) Determine the average acceleration between $t = 0$ and $t = 1$ sec.

Ex 3.8.33. The position of a projectile in the air is monitored with respect to a reference point in a Cartesian coordinate system. The x-axis of the Cartesian coordinate system is in the horizontal direction and the y-axis vertically up. The z-coordinate of the projectile does not change with time. The data for the x and y-coordinates are plotted and the following functions fit the x vs t and y vs t curves: $x(t) = 40t$ and $y(t) = 60t - 5t^2$. (a) Find the instantaneous velocity of the projectile at the following instants: (i) $t = 0$, (ii) $t = 4$ sec, (iii) $t = 8$ sec, and (iv) $t = 10$ *sec*. (b) Find the instantaneous acceleration of the projectile at the following instants: (i) $t = 0$, (ii) $t = 4$ sec, (iii) $t = 8$ sec, and (iv) $t = 10$ sec.

Ex 3.8.34. A car moves on a circular track. A coordinate system is chosen whose origin coincides with the center of the circle and the circle falls in the xy-plane. The position of the car with time varies according to the following functions of time: $x(t) = 2\cos\left(\frac{\pi}{100}t\right)$, $y(t) = 2\sin\left(\frac{\pi}{100}t\right)$, $z = 0$. Determine the velocity and acceleration of the car at (i) $t = 0$, (ii) $t = 50$ sec, (iii) $t = 100$ sec, (iii) $t = 150$ sec, and (iv) $t = 200$ sec.

3.8. EXERCISES

Polar Coordinates

Ex 3.8.35. Convert the following Cartesian coordinates (x, y) into polar coordinates. (i) $(0, 1)$; (ii) $(1, 0)$; (iii) $(-1, 0)$; (iv) $(0, -1)$; (v) $(3, 4)$; (vi) $(-3, 4)$; (vii) $(-3, -4)$; (vii) $(3, -4)$.

Ex 3.8.36. Convert the following polar coordinates (r, θ) into Cartesian coordinates. The angle is given in radians counterclockwise from the positive x-axis as seen from the side of the positive z-axis. (i) $(1, 0)$; (ii) $(1, \pi/3)$; (iii) $(1, 2\pi/3)$; (iv) $(1, 3\pi/3)$; (v) $(1, 4\pi/3)$; (vi) $(1, 5\pi/3)$; (vii) $(1, 6\pi/3)$; (viii) $(2, 29\pi/3)$.

Ex 3.8.37. (a) Draw the unit vectors $\hat{u}_r$ and $\hat{u}_\theta$ for the following rays. The rays are specified by the angles they make with the positive x-axis. (i) $\theta = 0$, (ii) $\theta = \pi/3$, (ii) $\theta = 2\pi/3$, (iv) $\theta = 3\pi/3$, (v) $\theta = 4\pi/3$, (vi) $\theta = 5\pi/3$, (vii) $\theta = 6\pi/3$. (b) Give the expression of these unit vectors in terms of the unit vectors $\{\hat{u}_x, \hat{u}_y\}$ along the Cartesian axes.

Ex 3.8.38. A displacement vector is given in the xy-plane by its magnitude 2 m and direction of $30°$ with respect to the x-axis. Write the displacement vector in terms of the unit vectors (i) $\{\hat{u}_x, \hat{u}_y\}$, (ii) $\{\hat{u}_r, \hat{u}_\theta\}$ for $\theta = 30°$, (iii) $\{\hat{u}_r, \hat{u}_\theta\}$ for $\theta = 45°$, and (iv) $\{\hat{u}_r, \hat{u}_\theta\}$ for $\theta = 90°$.

Ex 3.8.39. A velocity vector is given in the xy-plane by its magnitude 10 m/s and direction of $60°$ with respect to the x-axis. Write the velocity vector in terms of the unit vectors (i) $\{\hat{u}_x, \hat{u}_y\}$, (ii) $\{\hat{u}_r, \hat{u}_\theta\}$ for $\theta = 60°$, (iii) $\{\hat{u}_r, \hat{u}_\theta\}$ for $\theta = 120°$, (iv) $\{\hat{u}_r, \hat{u}_\theta\}$ for $\theta = 180°$, (v) $\{\hat{u}_r, \hat{u}_\theta\}$ for $\theta = 240°$, and (vi) $\{\hat{u}_r, \hat{u}_\theta\}$ for $\theta = 300°$.

Ex 3.8.40. The position of a particle moving in the xy-plane is given by $\vec{r}(t) = a\cos(\pi t)\hat{u}_x + a\sin(\pi t)\hat{u}_y$, where t is in sec and a in cm. (a) Show that the motion is a circular motion, and find the value of the radius of the circle, and find the Cartesian coordinates of the center of the circle. (b) Find an expression for the velocity of the particle at an arbitrary instant t. (b) Find an expression for the angular speed of the particle at an arbitrary time t. (c) Find an expression of the acceleration of the particle at an arbitrary instant t.

Ex 3.8.41. A ball is rolling in a circular track of radius R which is in the xy-plane but the center of the track is not at the origin of the Cartesian coordinate system in use. Instead, the center of the track is at (x_0, y_0). The ball has a constant speed of v_s. (a) Give an expression of the position vector as a function of time $\vec{r}(t)$ in terms of the unit vectors $\{\hat{u}_x, \hat{u}_y\}$ of the Cartesian coordinates. (b) Use your expression for $\vec{r}(t)$ to find an expression for the velocity of the particle at an arbitrary time t. (c) Use your expression for $\vec{r}(t)$ to find an expression of the acceleration of the particle at an arbitrary instant t.

Ex 3.8.42. A particle moves in a circle of radius 20 cm and centered at $(x = 1 \text{ m}, y = 0)$ in the xy-plane with a constant angular speed of 50 rad/sec. At $t = 0$, the particle is at $(x = 1.2 \text{ m}, y = 0)$. (a) For the particle moving in a counterclockwise sense, find the velocity at $t = 0$? (b) Find the position of the particle at an arbitrary time t?

Ex 3.8.43. A particle moves in a circular motion in the xy-plane centered at the origin. At $t = 0$ it is at $(x = 30 \text{ cm}, y = 0)$. Its angular speed is given by $\omega = 2 + 3t$, where ω is in rad/sec^2 and t in seconds. (a) What is the radius of the circle? (b) Find the change in the angular coordinate θ between $t = 0$ and $t = 1$ sec. (c) Find the change in the angular coordinate θ between $t = 1$ sec and $t = 2$ sec. (d) Find the velocity of the particle at $t = 0$. (e) Find the velocity of the particle at $t = 1$ sec. (f) Find the angular component a_θ of the acceleration at $t = 1$ sec. (g) Find the radial component of acceleration at $t = 1$ sec (h) Find the acceleration at $t = 1$ sec. (h) Find the angular component, radial component of the acceleration at $t = 2$ sec (i) Find the acceleration at $t = 2$ sec.

Ex 3.8.44. A bicycle with wheels of radius R is rolling without slipping. The center of the wheel moves with a constant speed v_s in a straight line towards the East. Write the position, velocity and acceleration vectors of a point on the outer edge of the wheel assuming the point is at $x = 0$ at $t = 0$ and the x-axis is pointed towards the East.

Ex 3.8.45. A platform is rotating about its center at constant angular speed ω_0. A man is standing at a distance R from the center of the platform and rotating with the platform. Find the position, velocity and acceleration vectors of the man.

3.9 PROBLEMS

The general problems for this chapter and the next are placed at the end of the next chapter.

Chapter 4

APPLICATIONS OF KINEMATICS

Contents

4.1	CONSTANT VELOCITY	124
4.2	CONSTANT SPEED	125
4.3	CONSTANT ACCELERATION	125
	4.3.1 Planar Motion	126
4.4	ONE DIMENSIONAL MOTION WITH CONSTANT ACCELERATION	128
	4.4.1 Free Fall: Application of Constant Acceleration	131
4.5	TWO DIMENSIONAL MOTION WITH CONSTANT ACCELERATION	137
	4.5.1 Projectile Motion	137
4.6	VARIABLE ACCELERATION	145
4.7	RELATIVE MOTION	149
	4.7.1 Observers Moving at Uniform Velocity	150
	4.7.2 Observers Moving at Uniform Acceleration . .	152
4.8	EXERCISES	154
4.9	PROBLEMS	162

In the last chapter we studied basic definitions of displacement, speed, velocity and acceleration. In this chapter we will apply these definitions to understand motion. We will take examples of motion of wide interest. We will look at increasingly complex cases, starting with constant velocity, followed by constant acceleration, and then finally, we will discuss variable acceleration.

4.1 CONSTANT VELOCITY

Constant velocity is perhaps the simplest motion. An object moving with a **constant velocity** with respect to a reference point has a constant speed and a constant direction. Let v_s be the constant speed. The distance s traveled in a duration t for a constant velocity object will be given by

$$\boxed{\text{Case: constant speed:} \quad s = v_s t.} \tag{4.1}$$

Since the direction of motion for a motion with constant velocity is fixed, we can choose one of the Cartesian axes to point in the direction of the constant velocity. This gives a simple one-component description of the velocity vector. Let $\vec{v}$ be the velocity vector and let the direction of motion be along the unit vector $\hat{u}_x$. The constant velocity vector can be written using the constant speed and the unit vector.

$$\boxed{\text{Case: constant velocity:} \quad \vec{v} = v_s \hat{u}_x.} \tag{4.2}$$

How would the position change with time? With this choice, the constant speed corresponds to the rate at which the x-coordinate changes with time.

$$\frac{dx}{dt} = v_s, \tag{4.3}$$

and the y and z-coordinates are fixed in time. The x-equation can be easily integrated to give the change in the x-coordinate during a duration t.

$$x - x_0 = v_s t, \tag{4.4}$$

where x is the x-coordinate at an arbitrary time t and x_0 is the x-coordinate of the particle at time $t = 0$.;

Example 4.1.1. Constant Velocity. A car is moving on an East-West road towards the East with a constant speed 30 m/s. If the car crosses a particular cross-section at $t = 0$, where will the car be 10 sec later?

Solution. Let the positive x-axis point in the direction of the constant velocity with the origin at the intersection. Then, the x-coordinate at an arbitrary time will be given by Eq. 4.4 with $x_0 = 0$.

$$x = v_s t = 30 \text{ m/s} \times 10 \text{ s} = 300 \text{ m}.$$

The car will be 300 m East of the intersection.

4.2 CONSTANT SPEED

The case of the motion with a constant speed covers many more situations than the constant velocity case. An object moving at a **constant speed** covers equal distance in equal time regardless of the direction of motion. A motion with a constant speed may be a complicated three-dimensional motion. However, as long as the object covers the same distance in equal time, no matter how complicated the trajectory, the motion would have a constant speed. For instance, when a charged particle enters a region of non-zero magnetic field, the particle's direction of motion changes based on the charge and the magnetic field, but the speed of the particle remains constant as displayed for a positively charged particle between the poles of a magnet in Fig. 4.1.

On the other hand, a motion with a constant velocity must always be a motion in a straight line since the direction for this type of motion is fixed in time.

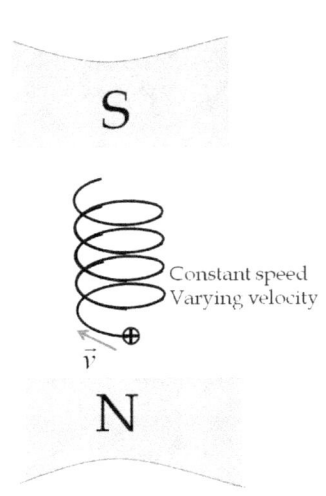

Figure 4.1: A positively charged particle moves in a helical path in a uniform magnetic field with a constant speed but varying velocity.

Example 4.2.1. A runner runs on an oval 400-m long track at constant speed of 8 m/s. The first 100 m of the track is straight, but after that the track rounds a corner. (a) How far does the runner go in 20 sec? (b) What is the velocity of the runner at any instant during the 20 sec period?

Solution. (a) Since speed does not depend upon the direction, we can use Eq. 4.1 to find the distance.

$$s = v_s t = 8 \text{ m/s} \times 20 \text{ s} = 160 \text{ m}.$$

(b) The velocity requires both magnitude and direction. We know that the magnitude of velocity is constant and equals 8 m/s. But, the direction of the motion of the runner changes; the runner has one direction of velocity when he is in the straight part of the track but his direction of motion changes as he round the corner. Therefore, the velocity of the runner is actually changing. Since we do not have information about the direction of the track after 100 m we cannot assign one velocity for all instants in the duration of 20 sec. In the first 100/8 sec the velocity has the magnitude of 8 m/s and is pointed in the forward direction on the straight track.

4.3 CONSTANT ACCELERATION

Motion with a constant acceleration is our most important example in this chapter. Throughout physics you will find many applications of the material presented in this section. An object moves with **constant acceleration** with respect to a reference point provided that both the magnitude and the direction of the acceleration vector do not change with time.

The analysis of the motion with constant acceleration is normally done in the analytic approach of vectors. The selection of axes for constant acceleration motion is aided by the constant direction of the acceleration. We point one of the Cartesian axes in the direction of the acceleration.

Note that direction of acceleration may or may not be same as direction of velocity: while velocity is in the direction of motion, acceleration is in the direction of force as we will learn in the next chapter.

Traditionally, different axes have served this role in different types of problems. For instance, for an object sliding down an incline, where the acceleration is pointed either down the incline or up the incline, one chooses the x-axis along the incline, but for a projectile motion whose acceleration is pointed vertically down, one chooses the y-axis to point vertically.

As an example, suppose, we choose the positive x-axis to be the direction of the constant acceleration of magnitude a. Then, the acceleration vector will have the following components.

$$a_x = a$$
$$a_y = 0$$
$$a_z = 0$$

The acceleration vector will have the following representation in this coordinate system.

$$\vec{a} = a\hat{u}_x,$$

where $\hat{u}_x$ is the unit vector pointed towards the positive x-axis. The velocity and position vectors may have all three components nonzero.

$$\vec{r} = x(t)\hat{u}_x + y(t)\hat{u}_y + z(t)\hat{u}_z$$
$$\vec{v} = v_x(t)\hat{u}_x + v_y(t)\hat{u}_y + v_z(t)\hat{u}_z$$

Since the velocity and position vectors can be in any direction, a motion with constant acceleration does not have to be a one-dimensional motion unlike the motion with a constant velocity. In the following we will see that by choosing our axes appropriately, one of the coordinates can always be made to be zero. This will prove that the most general type of motion of an object with a constant acceleration is a planar motion, i.e. the motion will occur in a flat plane. We call such motions **planar motions**.

4.3.1 Planar Motion

When an object has constant acceleration, the components of velocity in the plane perpendicular to the direction of the acceleration cannot change. Therefore, the motion is confined to a plane. This plane contains the vectors of the net velocity and the acceleration.

For instance, suppose the acceleration is pointed towards the positive x-axis. Then, the y and z-components of the velocity cannot change. That

4.3. CONSTANT ACCELERATION

means $v_y\hat{u}_y + v_z\hat{u}_z$ will be a constant vector, which allows us to choose new y- or z-axis in the direction of $v_y\hat{u}_y + v_z\hat{u}_z$ vector. Therefore, the original motion can be described completely in terms of the x-axis and the new y-axis.

Hence, in the case of a constant acceleration motion we can always choose axes such that the motion is completely confined to the xy-plane. That would simplify the position, velocity and acceleration vectors to the following:

$$\vec{r} = x(t)\hat{u}_x + y(t)\hat{u}_y$$
$$\vec{v} = v_x(t)\hat{u}_x + v_y(t)\hat{u}_y$$
$$\vec{a} = a_x\hat{u}_x \quad (a_x \text{ constant})$$

Therefore, we will have the following equations for the variation of the components of displacement and velocity.

$$x\text{-component:} \quad (a)\ dv_x/dt = a_x;$$
$$(b)\ dx/dt = v_x(t) \tag{4.5}$$
$$y\text{-component:} \quad (a)\ dv_y/dt = 0;$$
$$(b)\ dy/dt = v_y(t) \tag{4.6}$$

Now, we ask: what do these equations tell us about the velocity and position at an arbitrary time if they are given at some time $t = t_0$?

Change in velocity

Let us work out the change in velocity first. From Eqs. 4.5(a) and 4.6(a) we immediately see that if the velocity at time $t = t_0$ has components (v_{0x}, v_{0y}), then the components (v_x, v_y) at an arbitrary time t will be given by the following equations. (We will write a_x instead of a for the acceleration to emphasize the choice of x-axis).

$$x\text{-component:} \quad v_x = v_{0x} + a_x(t - t_0), \tag{4.7}$$
$$y\text{-component:} \quad v_y = v_{0y}, \tag{4.8}$$

Change in position

We use the expressions for velocity components in Eqs. 4.5(b) and 4.6(b) to obtain the following equations for the rate of change in coordinates.

$$x\text{-component:} \quad dx/dt = v_{0x} + a_x(t - t_0)$$
$$y\text{-component:} \quad dy/dt = v_{0y}$$

Multiplying these equations with dt and integrating with appropriate limits we obtain

$$x\text{-component:} \quad \int_{x_0}^{x(t)} dx = \int_{t_0}^{t} [v_{0x} + a_x(t - t_0)]\, dt$$
$$y\text{-component:} \quad \int_{y_0}^{y(t)} dy = \int_{t_0}^{t} v_{0y}\, dt$$

Integrating these equations and rearranging terms you can show that the result is

$$x\text{-component:} \quad x(t) - x_0 = v_{0x}(t - t_0) + \frac{1}{2}a_x(t - t_0)^2 \quad (4.9)$$

$$y\text{-component:} \quad y(t) - y_0 = v_{0y}(t - t_0) \quad (4.10)$$

Equations 4.7, 4.8, 4.9, and 4.10 give us the change in components of velocity and position for a motion which has a constant acceleration pointed along x-axis. We can simplify these equations further by choosing initial time to be $t_0 = 0$ and initial position to be at the origin.

$$\text{Choose}: \quad t_0 = 0 \quad x_0 = 0; \quad y_0 = 0; \quad z_0 = 0.$$

With these choices, the x and y-components of the position and velocity vectors at an arbitrary time when the constant acceleration is pointed towards the positive x-axis are

$$x\text{-component:} \quad \boxed{v_x(t) = v_{0x} + at} \quad (4.11)$$

$$\boxed{x(t) = v_{0x}t + \frac{1}{2}a_x t^2} \quad (4.12)$$

$$y\text{-component:} \quad \boxed{v_y(t) = v_{0y}} \quad (4.13)$$

$$\boxed{y(t) = v_{0y}t} \quad (4.14)$$

Table 4.1: Constant Accel. Relations

$(x_0 = y_0 = z_0 = 0)$
$(t_0 = 0; a_y = 0; v_z = 0)$

1. $v_x(t) = v_{0x} + a_x t$
2. $x(t) = v_{0x}t + \frac{1}{2}a_x t^2$
3. $v_y(t) = v_{0y}$
4. $y(t) = v_{0y}t$

The time variable t can be eliminated from the two equations of the x-components of the position and velocity vectors to obtain the following useful formula.

$$x\text{-component:} \quad \boxed{v_x^2 - v_{0x}^2 = 2a_x x} \quad (4.15)$$

One can also eliminate t from x and y- equations to obtain the equation of the trajectory in the xy-plane.

$$x = \left(\frac{v_{0x}}{v_{0y}}\right) y + \left(\frac{a_x}{2v_{0y}^2}\right) y^2. \quad (4.16)$$

Equations 4.11 to 4.16 are the basic equations that describe the motion of an object that has a constant acceleration pointed towards x-axis. We see that the direction of the initial velocity with respect to the direction of acceleration vector decides if both the x and y-components are needed.

4.4 ONE DIMENSIONAL MOTION WITH CONSTANT ACCELERATION

Suppose the constant acceleration is towards the positive x-axis, then, if the initial velocity has zero y and z-components, the motion will occur only along the x-axis. Even if you start with a zero x-component of velocity, the acceleration in the positive x-direction will change the x-component

4.4. ONE DIMENSIONAL MOTION

of the velocity. Since, there is no y- or z-component of acceleration, which happened as a result of the choice of coordinates, there is no way of changing the y- or z-component of velocity. That is, if v_y and v_z were zero at the beginning, they will remain zero throughout, keeping the motion entirely along the x-axis. In this case, we say that the motion is one-dimensional. In **one-dimensional motion**, only one component is needed.

1-D motion x-components only:
$$x(t) = v_{0x}t + \frac{1}{2}a_x t^2$$
$$v_x(t) = v_{0x} + a_x t$$

We now illustrate the use of these results with several examples.

Example 4.4.1. A Basic One-Dimensional Constant Acceleration Problem. A car moves on a horizontal road with a constant acceleration of 2 m/s² pointed towards the East. The velocity of the car at $t = 0$ is 15 m/s pointed towards the West. (a) Find the magnitude and direction of the velocity of the car at $t = 10$ sec. (b) Find the position of the car at $t = 10$ sec.

Solution. This problem exemplifies a typical scenario of a one-dimensional constant-acceleration kinematics problem. Note that the description of the problem does not tell us about the axes - we need to place one axis on the line of motion and choose the location of origin and also zero time reference. It is convenient to place x-axis on the line of motion with the positive x direction towards East so that the acceleration is towards the positive x-axis.

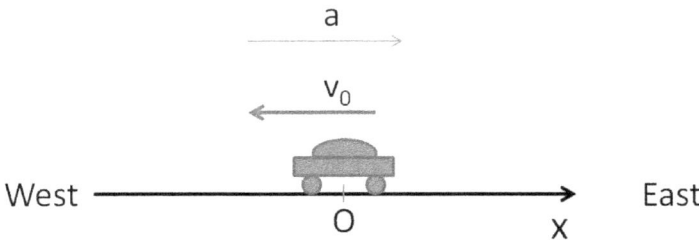

The origin is chosen to be where the car is at time $t = 0$. Since the direction of acceleration is towards the positive x-axis, therefore the value of x-component of acceleration will be positive. The direction of initial velocity is towards the negative x-axis, therefore the value of x-component of initial velocity will be negative. After we have settled on the axis and the positive and negative values of the given data, we summarize the known values.

$$t_0 = 0$$
$$x_0 = 0$$
$$v_{0x} = -15 \text{ m/s}$$
$$a_x = +2 \text{ m/s}^2$$
$$t = 10 \text{ s}$$

The list of unknowns are:

$$x$$

$$v_x$$

Now, it is simply a matter of choosing the appropriate equation from the list of the constant acceleration kinematics equations given in Table 4.1.

(a) In this part we need to find the velocity at $t = 10$ sec. Therefore, we see that Eq. 1 in Table 4.1 will immediately give the answer. All the lengths are in meters and times in seconds, so we will suppress the units while doing the calculation. It is often a good idea to take care of unit conversions before you input numerical values in your calculations. Then you can simply put the units at the end of the calculations. Here v_x will turn out in m/s.

$$v_x = v_{0x} + a_x t = -15 + 2 \times 10 = 5 \text{ m/s}$$

Since we have solved for the x-component of the velocity here (the y and z-components being zero), the sign of the answer does correspond to the direction of the actual velocity. The positive value of the x-component of the velocity at $t = 10$ sec means that the velocity is pointed towards the East which is the positive direction of the x-axis we have chosen here.

(b) In this part we need to find the x-coordinate at $t = 10$ sec. We have already found v_x in part (a). So, we will use Equation 2 in the Table 4.1.

$$x = v_{0x} t + \frac{1}{2} a_x t^2 = -15 \times 10 + \frac{1}{2} \times 2 \times 10^2 = -50 \text{ m}.$$

Since the value of x is negative here, the car is 50 m from the starting place towards the West - direction of the negative x-axis.

Example 4.4.2. More Questions about the accelerating car. Let us analyze the accelerating car given in the last example, Example 4.4.1. (a) Note that velocity at initial time is in the opposite direction to acceleration. Therefore, the car will slow down and eventually come to rest. When does the car come to rest momentarily? (b) How far away from the starting place does the car come to rest?

Solution. In the question, we have a different set of knowns and unknowns than was the case in the last example. We are told that the car comes to rest and we are asked about that instant in time. We will denote the time segment as $0 \le t \le T$. Let us list the new set of knowns, which will include $v_x = 0$.

$$x_0 = 0$$
$$v_{0x} = -15 \text{ m/s}$$
$$a_x = +2 \text{ m/s}^2$$
$$v_x = 0 \text{ (since at } t = T, \text{ the car is at rest.)}$$

4.4. ONE DIMENSIONAL MOTION

The list of unknowns are:

$$x$$
$$T$$

The tools are the same set of equations given in Table 4.1.

(a) For this part we need to find T. By looking at the equations in Table 4.1, it is clear that the first equation will readily give us the time.

$$v_x = v_{0x} + a_x T \implies 0 = -15 + 2T$$

solving for T we find $T = 7.5$ sec. Thus, it will take 7.5 sec for the car to come to rest.

(b) For this part we need to find x. We know one more data now from the solution of part (a): $T = 7.5$ sec. Once again, the equations in Table 4.1 show that the second equation in Table will immediately give us the value of x.

$$x = v_{0x} T + \frac{1}{2} a_x T^2 = 0 - 15 \times 7.5 + \frac{1}{2} \times 2 \times 7.5^2 = -56 \text{ m}.$$

Thus, the car will come to rest at the coordinate $x = -56$ m, which is 56 m from the starting point towards the West.

4.4.1 Free Fall: Application of Constant Acceleration

Free fall refers to an idealized vertical motion near earth when the air friction is negligible and the only force on the object is the gravitational force of earth. Although the unaided motion near the Earth is called free fall, the object itself may be either going up or coming down. For instance, when you throw a ball vertically up, it will be rising initially, and then, later it will be coming down.

The entire motion of the ball, after leaving hand and before getting caught or hitting the ground, is called free fall. It is confusing that the term "free fall" refers to motion in both directions but that is how we will call the motion of an object upon which the only force is the force of gravity.

Falling objects have been studied for a long time. Aristotle believed that when an object is released from rest, it acquires speed instantaneously in proportion to its weight (or on some other quality, such as its "fiery" or "earthy" character). Galileo rejected this idea based on simple arguments such as the following.

According to Aristotle, a two-pound rock will fall faster than a one-pound rock since the two-pound rock has more "Earthiness". Now, if you split the two-pound rock and join the halves by a light string, then we have an inconsistency in the prediction - on the one hand, each half

Figure 4.2: Examples of free fall. The motions of arrow, cannonball and soccer can be approximated as free fall when we ignore air resistance. Picture credits: archer: APoincot, cannon: Alfo23 from Svizzera, and soccer: Rdikeman, all at www.creativecommons.org.

should fall at one speed, and on the other, the two together should fall at a higher speed. But, an object must have a unique speed. Hence, Galileo concluded that Aristotle's thinking must be wrong.

Galileo then set out to find a quantitative way to study the motion of falling bodies and in the process revolutionized science. Galileo is considered to be the father of modern science since he was the first person to successfully make a break with the Aristotelian way of thinking about the fundamental nature of motion, which had dominated the thinking of scientists for nearly two millennia. Galileo used quantitative methods and interplay between observations, theoretical predictions and experimental testing, much like modern physicists.

To cut down the enormous acceleration due to gravity and make the speeds in the experiments more manageable, Galileo conducted his observations of motion on inclined planes. Through experiments, Galileo established that the distance d travelled by an object during a time interval t varies as the square of time t when released from rest.

$$d = \frac{1}{2}at^2, \qquad (4.17)$$

where the value of acceleration a depends on the angle of the incline. As you increase the angle of inclination, the acceleration changes, having the largest value when the ball is dropped vertically. The acceleration in the vertical free fall is called the **acceleration due to gravity**, whose magnitude is commonly denoted by the symbol g, which has a standard value of 9.81m/s^2 although it varies from place to place on Earth.

$$\boxed{g = 9.81 \ m/s^2 \ \text{(near the surface of the Earth)}} \qquad (4.18)$$

The value of g varies with the latitude on earth and the altitude from earth. To keep calculations simple, we will use the standard value for g in this book as long as the distance of the object under study from the earth is much smaller than the radius of the Earth. Since radius of Earth is approximately 6.37×10^6 m or $6,370$ km and the thickness of the atmosphere above ground is approximately only 65 km, you could use the standard value of g for anything moving in the atmosphere without incurring much error. For larger distances, we will have to use a more exact law for for the gravitational force which you will study in a later chapter.

In our notation, the vertical motion is a one-dimensional motion along a Cartesian axis. It is a common practice to use the y-axis for the vertical motion with the positive y-axis pointed up.

Although nothing will go wrong if we call this direction the x or z-axis, but we will follow the standard practice and point the positive y-axis up. With this choice the equations of constant acceleration will have non-zero y-components, and the value of the constant y-component of the acceleration will be equal to $-g$ (Table 4.2).

Table 4.2: Free fall with the positive y-axis pointed up - one-dimensional motion

($t_0 = 0$)

1. $v_y = v_{0y} - gt$
2. $y = y_0 + v_{0y}t - \frac{1}{2}gt^2$
3. $v_y^2 = v_{0y}^2 - 2g(y - y_0)$

Note that these equations are equally valid for free rise as well as for free fall. With the positive y-axis pointed up, the y-component of the velocity will be positive when the object is rising freely, and negative when the object is falling freely, but the y-component of the acceleration will always be negative, $a_y = -g$.

Example 4.4.3. A Freely Falling Ball. A ball is let go from rest (not thrown) from a tall building. (a) What is the velocity of the ball at $t = 5\ sec$? (b) How far has the ball fallen during this interval?

Solution. Assume that the ball has enough room to fall during the 2 sec interval so that ball does not hit the ground in that time. Then, the acceleration will have the magnitude of the free fall. We can use the equations given in Table 4.2 to solve this problem.

(a) The y-component of the initial velocity is zero, $v_{0y} = 0$. Therefore, the velocity at $t = 5$ sec is

$$v_y = a_y t = -9.81\ m/s^2 \times (5\ s) = -49.05\ m/s.$$

Therefore, the velocity at this instant is 49.05 m/s pointed down.

(b) The distance fallen can be found by evaluating the change in the y-coordinate.

$$y - y_0 = v_{0y}t - \frac{1}{2}gt^2 = 0 - \frac{1}{2} \times 9.81\ m/s^2 (5\ s)^2 = -122.7\ m.$$

Therefore, the ball would fall a distance of 122.7 m. The CN tower in Toronto is 446.5 m. The tallest building in the world in 2012, Burj Khalifa in Dubai, is 829.84 m. So, this experiment could be done from these tall skyscrapers. Can you calculate how fast a small pebble will be moving if dropped from the top of Burj Khalifa?

Example 4.4.4. A Typical Free Fall Problem. A baseball is thrown vertically up. Immediately after leaving the hand its speed was 20.0 m/s. The ball is caught on its way down 5 meters above the place it left the hand. (a) How long does it take to reach the maximum height? (b) To what height does the ball rise before turning back? (c) How long does it take before it is caught? (d) What was its velocity immediately before it was caught? Assume air resistance negligible.

Solution. Before I present the solution of the problem at hand, I want to discuss two fundamental considerations that goes into almost every problem that has a freely falling body.

Issue 1: The time domain of free fall

Note that you must be sure that you are looking at only that part of the motion which is truly freely falling. In this problem, only the motion of the ball after leaving the hand and before it is caught is freely falling motion. When the ball is touching the hand, the hand gives the ball a push up so as for the ball to gain the initial y-component of the velocity to

Figure 4.3: The CN tower in Toronto is 446.5 m. A pebble falling from the peak of the CN tower will hit the street at approximately 94 m/s or 337 km/h. Credits: publicdomainpictures.net, Bobby Mikul.

go up. The y-component of the acceleration in that part of the motion of the ball is different than g downward as assumed for the free fall motion.

The same type of problem occurs when the ball is caught. Just before the ball is caught, the ball is moving downward with y-component of the acceleration $-g$ and a negative value of y-component of the velocity. When the ball touches the hand, the hand pushes the ball upward and provides an upward acceleration, which reduces the downward velocity to zero. When the ball is in touch with the hand, y-component of the acceleration of the ball is no longer $-g$.

Therefore, the only part of motion we can call freely falling is between the time <u>after</u> the ball has left the hand initially and <u>before</u> it touches the hand again. <u>Note that in every problem concerning the one-dimensional kinematics of constant acceleration, you will need to pay attention to these details to properly understand the problem.</u>

Issue 2: The time segment of interest

Now that you have determined the total time interval, say from $t = t_i$ to $t = t_f$ that the ball is freely falling, a particular part of the question may ask you to find things within a segment of time that falls within this time interval. Different parts of the same problem may refer to different time segments within the total time interval, after leaving the ground and before getting caught.

For instance, parts (a) and (b) refer to the motion after the ball leaves the hand and when it comes to rest at the top of the path, while parts (c) and (d) are concerned with the interval between the top of the path where it was momentarily at rest and immediately before the ball was caught, where it was still moving down. Transferring values of various quantities between parts must be done very carefully. <u>Use physics to decide if a particular quantity in one part has the same or related value to a similar quantity in another part. Using formulas for transferring quantities from one part to another will most likely result in stupid mistakes.</u>

Now, we are ready to discuss the solution.

(a) Let us list various known and unknown quantities for the segment of motion under consideration for this part of the problem. Basically, you must decide on four quantities: (1) the time interval in which acceleration is constant, (2) the positions at the two ends of the time interval, (3) the velocities at the two ends of the time interval, and (4) the value of acceleration.

- **Interval:**
    - After leaving the hand and when it comes to rest at the top.
    - Set $t = 0$ at the instant after the ball leaves the hand. This makes the use of equations in the Table 4.2 possible without

4.4. ONE DIMENSIONAL MOTION

any modifications. Note the time t in equations in Table 4.2 is interval from the starting time.
- Set $t = T$ at the instant the ball arrives at the top. This time is not known.

- **Positions:**
  - Set $y_0 = 0$ at the instant after the ball leaves the hand. This is the choice for the origin.
  - Set $y = H$ at the instant the ball arrives at the top. This is not known.

- **Velocities:**
  - Note $v_{0y} = +20.0$ m/s at the instant after the ball leaves the hand. The y-component of the velocity is positive since the ball's velocity is pointed in the positive y direction.
  - Note $v_y = 0$ since the ball is at rest at the instant the ball arrives at the top.

- **Acceleration:**
  - The y-component of the acceleration is $a_y = -g$, where the value of $g = 9.81$ m/s^2, the standard value for free fall.

Now, in this part of the problem, we need to find T. Looking at the set of equations in Table 4.2, it is readily apparent that the Equation 1 in the Table would give the value of T in one step.

$$v_y = v_{0y} - gt \implies 0 = 20 - 9.81T, \text{ or, } T = 2.04 \text{ sec.}$$

When I used the calculator, I got $20/9.81 = 2.03873598369$. Should I present all the digits from the calculator? No. I used the rules of rounding off at the least significant digit. The data used in the calculation has three significant digits in g and three in v_{0y}. Therefore, I rounded off the answer to keep up to three digits. that is why I didn't write 2.0 or 2.039.

(b) This part also deals with the same interval as part (a). Therefore, we do not need to write out the table of values once again. We use the same table as in part (a). Furthermore, now we can also use the result $T = 2.04$ sec from part (a). We need to find $y = H$ in this part. It is clear that with the calculated T, equation number 2 in Table 4.2 will yield the value of the height H.

$$y = y_0 + v_{0y}t - \frac{1}{2}gt^2 \implies H = 0 + 20 \times 2.04 - \frac{1}{2} \times 9.81 \times 2.04^2 = 20.4 \text{ m.}$$

Once again, in the calculator I got a number with many digits, which I have rounded to three significant digits.

(c) This part and the next are about a <u>different time interval</u>. Therefore, we need to build another table. The table in part (a) does not apply anymore.

- **Interval:**
  - From the time the ball comes to rest at the top to the instant immediately before it is caught. Note: the instant the ball is caught is not a part of the interval.
  - Set $t = 0$ at the instant the ball comes to rest at the top. This makes the use of equations in the Table 4.2 possible without any modifications. Note the time t in equations in Table 4.2 is interval from starting time of the interval.
  - Set $t = T$ at the instant immediately before the ball is caught. This is not known.

- **Positions:**
  - Set $y_0 = 20.4$ m at the instant the ball leaves the top. We can leave the origin at the same place as in part (a). Although, leaving the origin alone is not required, it is often helpful to leave the origin alone. Sometimes, you may find it helpful to choose different origins for different parts.
  - Leaving the origin in the same location as above gives $y = 5$ m at the instant immediately before the ball is caught.

- **Velocities:**
  - Note $v_{0y} = 0$ at the instant the ball is at the top; this is the beginning of this interval. The velocity is zero since the ball is not moving at this instant.
  - Set $v_y = V$ at the instant immediately before the ball is caught. It is unknown. Note the ball is not caught yet. So V is not zero. This is one of the most common mistakes. We know that the value for V will be negative since the positive y-axis is pointed up and the velocity is pointed down.

- **Acceleration:**
  - The y-component of acceleration is $a_y = -g$, where $g = 9.81$ m/s^2.

In this part we need to find the value of T. Of course, this value of T is not the same as what we found in part (a). This T is for the present interval. So, using the same symbol should not confuse us as long as we know that each interval is a separate problem. Equation 2 of Table 4.2 appears to be the equation of choice since it will have only one unknown once we have substituted all the knowns into the equation.

$$y = y_0 + v_{0y}t - \frac{1}{2}gt^2 \implies 5 = 20.4 + 0 - \frac{1}{2} \times 9.81 T^2 \implies T = 1.77 \text{ sec.}$$

(d) This part has the same interval as part (c). Therefore, we can use the results of part (c) and we do not need to make another table of values. Here we seek the value of the velocity just before the ball is caught.

Equation 1 of Table 4.2 with the T value found in part (c) can help us find the required velocity.

$$v_y = v_{0y} - gt \implies V = 0 - 9.81 \times 1.77 \implies V = -17.4 \text{ m/s}.$$

The negative value of the velocity indicates the ball is moving towards the negative y-axis, which is pointed down, as expected since the ball is falling down at the moment.

4.5 TWO DIMENSIONAL MOTION WITH CONSTANT ACCELERATION

We have established above that the most general constant acceleration will occur in a plane. Suppose the acceleration is pointed towards the positive x-axis, then the y-axis can be chosen so that the motion happens entirely in the xy-plane. If $v_y = 0$ at the beginning in this set-up, then the motion will be one-dimensional, only along x-axis.

On the other hand, if v_y is non-zero at the beginning, the motion will be two-dimensional. You will need both the x and y-components for constructing the position and velocity vectors in the xy-plane. Table 4.1 summarizes the full set of equations for the planar constant acceleration motion.

Note that, if you choose axes such that the constant acceleration happens along the y-axis and not along the x-axis as assumed for the equations in Table 4.1, then you will need to switch x and y in the table. Now, you will have a_y non-zero and a_x zero.

$$\begin{aligned} x\text{-components:} \quad & v_x(t) = v_{0x} \\ & x(t) = v_{0x} t \\ y\text{-components:} \quad & v_y(t) = v_{0y} + a_y t \\ & y(t) = v_{0y} t + \frac{1}{2} a_y t^2 \end{aligned}$$

4.5.1 Projectile Motion

The projectile motion of an object thrown sideways or at an angle from the vertical illustrates the general principles of the motion with constant acceleration well. Note again that, in order for the motion under study to have a constant acceleration, you must restrict the time domain to a time segment <u>after</u> the projectile has been launched and <u>before</u> the projectile lands or hits any other object. This is necessary because when the projectile is launched, it has a different acceleration than the acceleration due to gravity, and similarly on the other end, the acceleration is different after the projectile hits the ground.

With a proper choice of time domain and the assumption of negligible air resistance, the vertical motion of a projectile is subject to a constant acceleration pointed down, but the horizontal motion has constant velocity. Although the direction of the constant acceleration is down, the traditional choice of coordinates is to point the positive y-axis up and to point the x-axis in the horizontal direction. With these choices for the coordinates the equations for the components of velocity and acceleration become

x-components	y-components
$v_x = v_{0x}$	$v_y = v_{0y} - gt$
$x - x_0 = v_{0x}t$	$y - y_0 = v_{0y}t - \frac{1}{2}gt^2 = \frac{v_y^2 - v_{0y}^2}{2g}$

When applying these equations to problems you will notice that the following steps are common to most problems.

- Step 1: Choose a Cartesian coordinate system.

- Step 2: List knowns and unknowns.

- Step 3: Work out components (this step and step 2 are often done together).

- Step 4: Set up equations and solve (what to solve for depends on particular question asked).

- Step 5: Present answer (often the numerical values need to be interpreted or translated into a descriptive language).

Example 4.5.1. Separating a planar motion into components

A cannonball is launched at a speed of 40 m/s speed at an angle of 30° from the horizontal. What will be the velocity of the cannonball after 3 sec?

Solution. I will use the present problem to illustrate the use of the systematic approach outlined above. Note that, although the statement of the problem does not make a reference to a coordinate system, we choose to work in the analytic approach for vectors and choose a convenient coordinate system first. We choose a coordinate system to perform the analysis. Here, as explained above, the acceleration is equal to (9.81 m/s², *down*). Therefore, one of the Cartesian axes will be vertical. It is a time-honored tradition to point the positive y-axis up. This makes the y-component of the acceleration negative. The horizontal direction is chosen to be the x-axis and the positive x direction is usually chosen in the direction of the increasing x-coordinate of the projectile under study. That is, if the projectile is going towards the North-east, we point the positive x-axis towards the Northeast, and if the projectile is going towards the North,

Step 1: Choose a Coordinate System

4.5. TWO DIMENSIONAL MOTION

then we point the positive x-axis towards the North, etc. These choices are shown in Fig. 4.4.

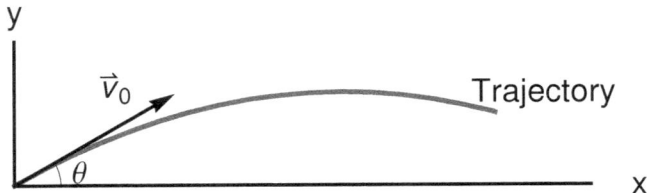

Figure 4.4: Example 4.5.1. The arrow labeled $\vec{v}_0$ is the velocity vector at time $t = 0$.

Now, we list the known and unknown quantities, usually in a table form, such as shown in Table 4.3. Unlike the one-dimensional problems, now we need to keep track of the x and y-components of all five vectors, namely, the initial position, the final position, the initial velocity, the final velocity, and the acceleration. The time domain of interest in this problem is from $t = 0$ to $t = 3$ sec.

Step 2: List Knowns And Unknowns

Table 4.3: Known and unknown components of vectors

Vector Quantity	x-component	y-component
Knowns:		
$\vec{r}(0)$	$x(0) \equiv x_0 = 0$	$y(0) \equiv y_0 = 0$
$\vec{v}(0)$	$v_x(0) \equiv v_{0x}$	$v_y(0) \equiv v_{0y}$
$\vec{a}$	$a_x = 0$	$a_y = -9.81 \text{ m/s}^2$
Unknowns:		
$\vec{v}(t)$	$v_x(t)$	$v_y(t)$
$\vec{r}(t)$	$x(t)$	$y(t)$

Step 3: Work Out Components

The components of the velocity vector at $t = 0$ are found by drawing the projections of the vector on the corresponding axes. In a two-dimensional situation, you need to draw the projection onto only one of the axes and use trigonometric relations for the resulting right-angled triangle as shown in Fig. 4.5.

From Fig. 4.5 we find the components of the initial velocity vector to be

$$v_x(0) = 40 \, \cos 30° = 34.6 \text{ m/s} \equiv v_{0x}$$
$$v_y(0) = 40 \, \sin 30° = 20.0 \text{ m/s} \equiv v_{0y}$$

Now, we are ready to answer the question about the velocity vector at $t = 3$ sec by working out its x and y-components. The data on the x-component given in Table 4.3 tells us that the x-component of the velocity does not change. Therefore,

Step 4: Set Up Equations and Solve

$$v_x = v_{0x} = 34.6 \text{ m/s}.$$

The data on the y-components given in Table 4.3 can be used to find v_y

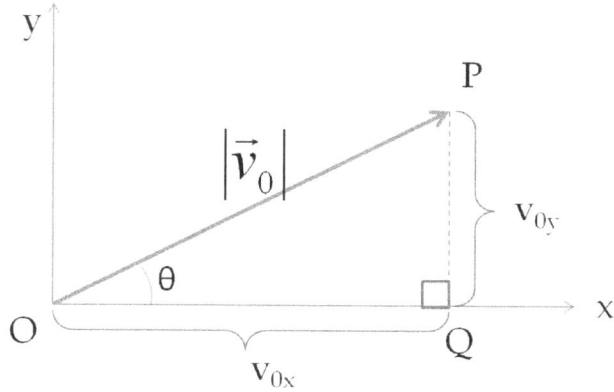

Figure 4.5: Components of initial velocity vector. The initial velocity has a magnitude $|\vec{v}_0|$ and direction O-to-P, which is at an angle θ counterclockwise from the positive x-axis. The right-angled triangle $\triangle OPQ$ is used to determine the components v_{0x} and v_{0y} of the initial velocity vector $\vec{v}_0$. Writing v_0 for the magnitude $|\vec{v}_0|$, we get $v_{0x} = v_0 \cos\theta$ and $v_{0y} = v_0 \sin\theta$.

as
$$v_y = v_{0y} + a_y t = 20.0 - 9.81 \times 3 = -9.43 \text{ m/s}.$$

We find that $v_y < 0$, which means that the y-axis vector, $v_y \hat{u}_y$, obtained by multiplying the y-component with the unit vector of the y-axis is pointed towards the negative infinity on the axis. In this problem, since the positive y-axis pointed up, $v_y \hat{u}_y$ is pointed down. Therefore, at $t = 3$ sec the ball is coming down. However, since $v_x > 0$, the ball is not coming straight down, but at an angle to the vertical, with the direction being actually in the fourth quadrant of the Cartesian axes chosen above.

Step 5: Present Answer

How do we construct the velocity vector at $t = 3$ sec? We have found its components, but the actual velocity is a vector. We use the components to obtain the magnitude and direction of the velocity vector. The magnitude is easy to work out from the components. [Note the z-component here is zero.]

$$\text{Magnitude of velocity vector:} \quad |\vec{v}| = \sqrt{v_x^2 + v_y^2} = 35.9 \text{ m/s}.$$

Work Out Magnitudes And Directions If Needed

How do we specify the direction of the velocity vector. There are two ways to do this here.

1. Draw the velocity vector in space.

2. Provide an angle with respect to one of the coordinate directions. Since we have a two-dimensional situation, we need to specify only one angle. The angle can be with respect to any of the four directions: (a) the positive x-axis, (b) the negative x-axis, (c) the positive y-axis, or (d) the negative y-axis. We also need to specify whether the angle is in the clockwise or counter-clockwise from the axis when we look down from the side of the positive z-axis.

4.5. TWO DIMENSIONAL MOTION

Drawing the vector is straight forward. Mark the scale on the axes according to the quantity being drawn. Here, we mark the axes by the scale of velocity for drawing approximately 35 m/s along x-axis and -10 m/s along the y-axis. We represent the velocity vector by drawing an arrow from the origin to the point (v_x, v_y) as shown in Fig. 4.6.

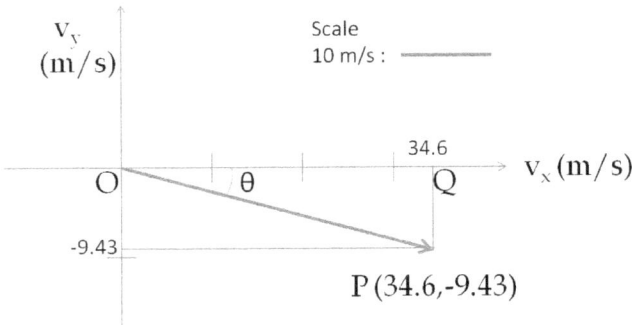

Figure 4.6: Graphically displaying direction of velocity vector. The x-component of the vector is plotted along x-axis and the y-component along the y-axis using the same scale, and the vector is then drawn from the origin to the point (v_x, v_y). Either angle θ clockwise from the positive x-axis or its complement angle counterclockwise from the negative y-axis are also used to indicate the direction.

In Fig. 4.6 the angle θ clockwise from the positive x-axis can be read off by a protractor or calculated from the components of the vector. If you use trigonometry on the right-angled triangle $\triangle OPQ$, you find that you get a negative value.

$$\theta = \arctan\left(\frac{v_y}{v_x}\right) = \arctan\left(\frac{-9.43}{34.6}\right) = -15.2°.$$

The usual practice is to state the angle as a positive number and say whether you have to go clockwise or counter-clockwise from the axis. Here, we will say that the direction is 15.2° clockwise from the positive x-axis. Sometimes, one also says 15.2° below the positive x-axis.

Beware of the value from your calculator when evaluating Arc-tan. Be mindful of the quadrant!

Example 4.5.2. Projectile motion problem with unknown time
A projectile is fired with speed 60 m/s at an angle of 40° with respect to the ground. Assume that the projectile has a constant acceleration of g (9.81 m/s^2) pointed down and lands on the ground that is at same height as the place of launch. (a) Find the horizontal distance the projectile flies before landing. (b) What is the hang time, i.e., the time between launching and landing? (c) What is the velocity with which the projectile will strike the ground?

Solution. Constant acceleration problems in two dimensional motion in which the time interval is not given often require passing the time variable between the x and y-components. In some problems, it may be helpful to eliminate the variable t in the equations for the x and y-components. In some other problems, you may be able to find t in x or y alone and you will need to pass the value of t to the other component.

All these problems start with first sketching the anticipated trajectory and a choice of the coordinate system as shown in Fig. 4.7. Then, we

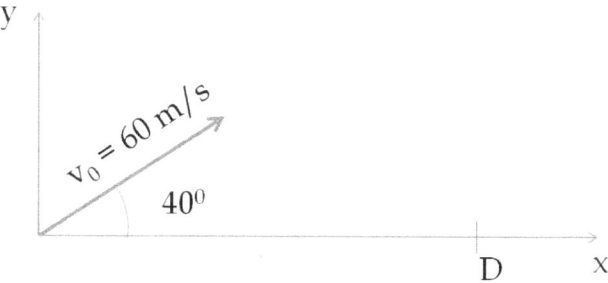

Figure 4.7: Physical setting and choice of coordinates for Example 4.5.2.

set up the table of knowns and unknowns and the two equations for each axis as shown in detail in the last example. We have once again placed the origin at the place of launch, and point the positive y-axis vertically up and the x-axis horizontally as in the previous problem. Let us denote the final time by capital letter T. Let D denote the horizontal distance traveled before landing.

Table 4.4: Known and unknown components of vectors

Vector Quantity	x-component	y-component
Knowns		
$\vec{r}(0)$	$x(0) = 0$	$y(0) = 0$
$\vec{v}(0)$	$v_x(0) \equiv v_{0x}$	$v_y(0) \equiv v_{0y}$
$\vec{a}$	$a_x = 0$	$a_y = -9.81$ m/s^2
Unknowns		
$\vec{r}(T)$	$x(T) = D$	$y(T) = 0$
$\vec{v}(T)$	$v_x(T) \equiv v_x$	$v_y(T) \equiv v_y$

Note that $\vec{v}(T)$ is not zero, since we are interested in the velocity an instant before the projectile lands. Once the projectile hits the ground its acceleration is no longer equal to the assumed constant value of (9.81 m/s^2, down).

Now, let us write down two independent constant acceleration kinematic equations for each axis.

x-component equations	y-component equations
$v_x = v_{0x}$	$v_y = v_{0y} + a_y T$
$D = v_{0x} T$	$0 = v_{0y} T + \frac{1}{2} a_y T^2$

So, we have four equations among four unknowns, D, T, v_x, and v_y. Since, we already have numerical values of some of the symbols in these equations, we will rewrite them with their specific values. First, note that we have the following for the x and y-components of the initial velocity

4.5. TWO DIMENSIONAL MOTION

vector.

$$v_{0x} = 60 \text{ m/s} \cos 40° = 46 \text{ m/s}.$$
$$v_{0y} = 60 \text{ m/s} \sin 40° = 39 \text{ m/s}.$$

The kinematic equations with numerical values include units, but they are cumbersome to carry around in calculations, so we will leave out the units. We will put the units back in at the end.

$$v_x = v_{0x} = 46 \tag{4.19}$$
$$D = v_{0x} T = 46\ T \tag{4.20}$$
$$v_y = 39 - 9.81\ T \tag{4.21}$$
$$0 = 39\ T - 4.91\ T^2 \tag{4.22}$$

The last equation, Eq. 4.22, can be solved for T which gives two values for T since T obeys a quadratic equation.

$$T = 0,\ 8 \text{ sec}.$$

The value of $T = 0$ refers to the initial time. Since, in our problem, the projectile has the y-coordinate equal to zero at the starting place and at the landing place, the equation includes the solution $T = 0$. We need to interpret the mathematical solution often to select the solution we seek. Therefore, we keep $T = 8\ sec$ for the final time.

Now, putting the value of T in Eq. 4.20 we obtain the distance D.

$$D = 366 \text{ m}.$$

Substituting the value of T in Eq. 4.21 gives us the last unknown, v_y.

$$v_y = -39.4 \text{ m/s}.$$

Let us now summarize our results.

$$D = 366 \text{ m}.$$
$$T = 8 \text{ sec}$$
$$v_x = 46 \text{ m/s}$$
$$v_y = -39.4 \text{ m/s}.$$

The values of D and T give the answers to parts (a) and (b). To obtain the answer to part (c), we need to construct the velocity vector from the components v_x and v_y. The procedure is already explained in the last example. Here we give the answer.

Magnitude of the velocity: $\sqrt{46^2 + (-39.4)^2} = 60$ m/s.

Direction of the velocity: As shown in the Fig. 4.8, $\theta = 40.3°$,

clockwise from the positive x axis.

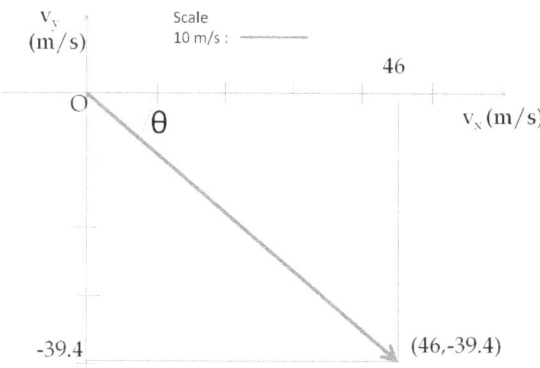

Figure 4.8: Graphically displaying direction of velocity vector. The x-component of the vector is plotted along x-axis and the y-component along the y-axis using the same scale, and the vector is then drawn from the origin to the point (v_x, v_y). Either angle θ clockwise from the positive x-axis or its complement angle counterclockwise from the negative y-axis are used to indicate the direction.

Example 4.5.3. Projectile motion problem with different heights.
An air plane flying horizontally at a speed of 400 km/h drops off a package form an altitude of 200 m from the ground. How far does the package travel horizontally from the point of its release?

Solution. Note that in the time interval after the package is let go and the instant before the package lands on the ground, the package is falling freely with the constant acceleration of 9.81 m/s^2 pointed down if we neglect the air resistance. We follow the tradition of choosing a coordinate system

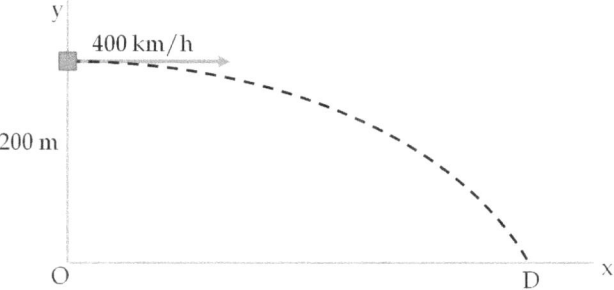

Figure 4.9: Example 4.5.3. Coordinate system used for the problem. The initial position of the package is at coordinate $(0, 200\ m)$ and the final position at $(D, 0)$.

with the positive y-axis pointed up as shown in Fig. 4.9. This will make a_y negative. The x-component of acceleration is, of course, zero. Let the value of time be $t = 0$ when the package is at point $(0, 200\ m)$ and $t = T$ when the package is at $(D, 0)$. We now collect the known and unknown quantities for the time segment $0 < t < T$ in Table 4.5, where we also include the kinematic equations for the corresponding axes.

With the values from the table, we obtain the following set of four

Table 4.5: Known and unknown components of vectors

Quantity	x-component	y-component
Known:		
$\vec{r}(0)$	$x(0) = 0$	$y(0) = 200$ m
$\vec{v}(0)$	$v_{0x} = 400$ km/h $= 111$ m/s	$v_{0y} = 0$
$\vec{a}$	$a_x = 0$	$a_y = -9.81$ m/s^2
Unknown:		
$\vec{r}(T)$	$x(T) = D$	$y(T) = 0$
$\vec{v}(T)$	$v_x(T) \equiv v_x$	$v_y(T) \equiv v_y$

equations, two from each axis (leaving the units out):

$$v_x = v_{0x} = 111 \tag{4.23}$$
$$D = v_{0x}T = 111\, T \tag{4.24}$$
$$v_y = -9.81\, T \tag{4.25}$$
$$-200 = -4.91\, T^2 \tag{4.26}$$

In this question, we need to find the value of D only. Therefore, we focus on those equations that would give us the value of D with least additional algebra. We see that, if we get T from Eq. 4.26 and plug into Eq. 4.24, we will get the value of D. Let us implement this strategy.

Solving Eq.4.26 for T we obtain two values of T.

$$T^2 = \frac{200}{4.91} = 40.7 \quad \Longrightarrow \quad T = \pm\sqrt{40.7} = \pm 6.38 \text{ sec.}$$

We get two values of T since the equation is quadratic in T. What are the meanings of the two values and which one would I need for the point $(D, 0)$ in the question? The negative value of T refers to a time before the package was released at $t = 0$. So, it is clearly not meaningful for the given physical setting. The negative T says that if the package was launched with appropriate initial velocity so that it has a horizontal velocity of magnitude 111 m/s at the top of the trajectory, then it would have taken 6.28 sec to reach the top of the trajectory. Our physical setting verifies that $T > 0$ at $(D, 0)$, the point where the package lands. Therefore, we use the positive value of T in Eq. 4.24 to obtain the required value of D.

$$D = 111 \times 6.38 \quad \Longrightarrow \quad D = 708 \text{ m.}$$

4.6 VARIABLE ACCELERATION

Although most of our examples in this book will have constant acceleration segments, it is worthwhile to look at a case of arbitrarily varying acceleration as an important extension of the methods for one-dimensional constant acceleration presented above.

The basic idea for handling an arbitrarily varying acceleration is to replace the original acceleration by an approximation obtained by dividing up the time interval into smaller time segments. If the time segments are small enough, we can approximate the original arbitrarily varying acceleration by a step-wise varying acceleration, where each step is an average acceleration in the corresponding time segment.

Clearly the original situation of a continuously varying acceleration is not the same as its replacement by the constant acceleration steps. However, Sir Isaac Newton showed that if the intervals were allowed to be arbitrarily small, then the predictions of the final position and velocity, based on step-wise approximate acceleration, can be made arbitrarily close to the exact answer (see Fig. 4.10).

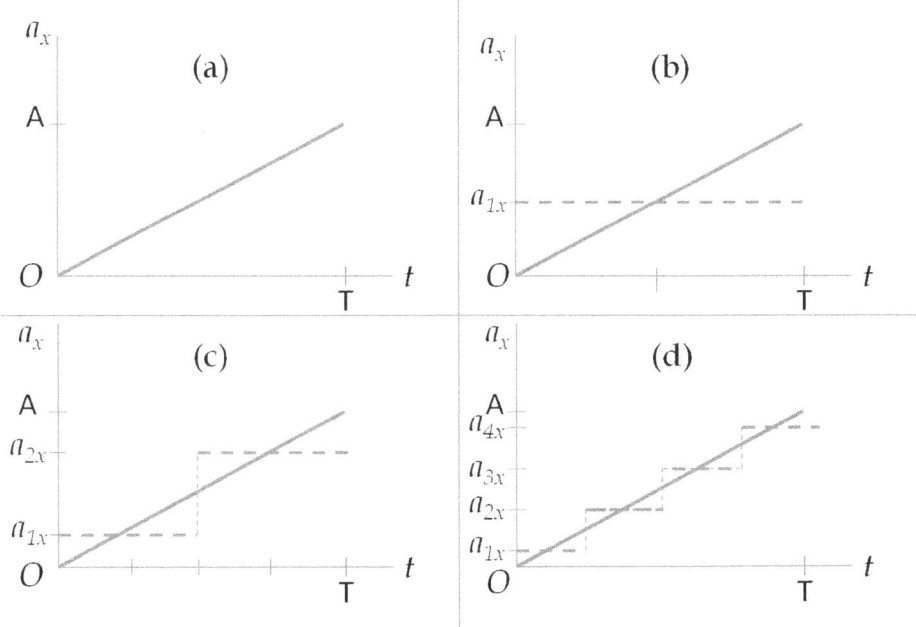

Figure 4.10: Successive approximations of a varying acceleration. The plots in this figure show the varying x-component of the acceleration. Fig. (a) contains the continuous function $a_x(t)$. In Fig. (b) the variable $a_x(t)$ is replaced by one constant step with the average x-component of the acceleration denoted as a_{1x}. In Fig. (c) $a_x(t)$ is replaced by two segments of constant acceleration steps of a_{1x} and a_{2x}, and in Fig. (d) by four steps. The process can be continued ad-infinitum. As you decrease the step size, or equivalently increase the number of steps, the original acceleration is covered more accurately by the approximation.

Recall that acceleration, velocity and position are vectors whose magnitudes and components can be written as functions and plotted in a graph. Since the kinematic equations separate into x-, y and z-components, it is sufficient to discuss one of the Cartesian components. To be specific, we will consider x-components of acceleration, velocity and position of an

4.6. VARIABLE ACCELERATION

object in which x-component of the acceleration changes with time and we wish to find the change in x-components of velocity and position with time.

As shown in Fig. 4.10, we divide the full interval $[0, T]$ into N subintervals, which we will denote by $[0, \Delta t]$, $[\Delta t, 2\Delta t]$, $[2\Delta t, 3\Delta t]$, $\cdots$ $[(N-1)\Delta t, N\Delta t]$, with $T = N\Delta t$.

Let the x-components of the average acceleration be a_{1x}, a_{2x}, a_{3x}, $\cdots$, a_{Nx} in the N time segments respectively. Let v_{1x}, v_{2x}, v_{3x}, $\cdots$, v_{Nx} be the x-components of velocity at the end of the corresponding intervals. Of course v_{Nx} is same as $v_x(T)$ or simply v_x at the end of the final time of the net interval.

Now, we show that our procedure leads us to the prediction of the x-component of the final velocity from the x-component of the initial velocity v_{0x}, the velocity at time $t = 0$.

Interval	Velocity at the end of the interval
0 to Δt	$v_{1x} = v_{0x} + a_{1x}\Delta t$
Δt to $2\Delta t$	$v_{2x} = v_{1x} + a_{2x}\Delta t$
$2\Delta t$ to $3\Delta t$	$v_{3x} = v_{2x} + a_{3x}\Delta t$
$\vdots$	$\vdots$
$(N-2)\Delta t$ to $(N-1)\Delta t$	$v_{N-1,x} = v_{N-2,x} + a_{N-1,x}\Delta t$
$(N-1)\Delta t$ to $N\Delta t = T$	$v = v_N = v_{N-1,x} + a_{Nx}\Delta t$
	Summing, $v_x = v_{0x} + \sum_{i=1}^{N} a_{ix}\Delta t$.

Summing the velocity change equation in each interval gives us the following equation for the change in velocity from v_{0x} to v over the entire time $[0, T]$.

$$v_x = v_{0x} + \sum_{i=1}^{N} a_{ix}\Delta t. \tag{4.27}$$

The approximation becomes better as Δt is made smaller. We denote the limit of the summation as Δt approaches zero by another symbol, called the definite integral of $a_x(t)$ from $t = 0$ to $t = T$.

Area under a_x vs t gives the change in velocity

$$v_x = v_{0x} + \int_0^T a_x(t)dt. \tag{4.28}$$

If the interval is other than $[0, T]$, e.g. from $t = t_1$ to $t = t_2$, then the integration will have to be done over the corresponding interval.

$$v_{2x} = v_{1x} + \int_{t_1}^{t_2} a_x(t)dt, \tag{4.29}$$

where v_{1x} is the velocity at t_1 and v_{2x} the velocity at t_2.

Another extremely useful interpretation of the result obtained in Eq. 4.27 is obtained by noting that $a_{1x}\Delta t$, $a_{2x}\Delta t$, $a_{3x}\Delta t$, $\cdots$, $a_{Nx}\Delta t$ are areas in the rectangle under the approximately constant x-component of the

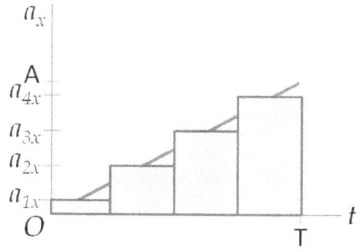

Figure 4.11: The change in velocity is equal to the area under the curve, approximated by the rectangles.

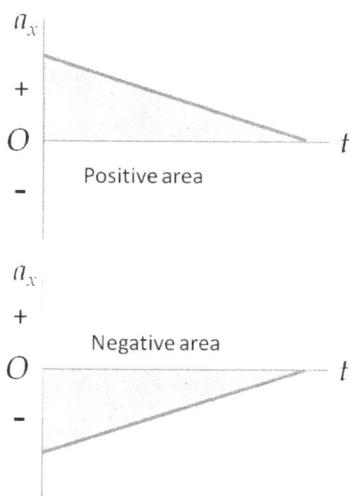

Figure 4.12: Positive and negative areas.

Figure 4.13: A US Marine Corp Paratrooper. Credits: Wikicommon.

acceleration steps in the graph of a_x vs t plot as shown in Fig. 4.11 for four segments. The sum in Eq. 4.27 is the net area under the curve. Therefore, we can read Eq. 4.27 as follows.

$$v_x = v_{0x} + (\text{Area under the curve of } a_x \text{ vs } t). \qquad (4.30)$$

The definite integral in Eq. 4.28 gives us an exact value of the area under the a_x vs t curve. Unlike the area of a polygon in space, though, the area under the curve can also be negative if the curve is below the horizontal since we would be multiplying a negative acceleration by a positive time segment resulting in a negative area (see Fig. 4.12.

Similar arguments will lead you to analogous relation between the change in the x-coordinate and the x-component of the velocity. We give the result for future reference and leave the derivation as an exercise for the student. Denoting the average velocities in segments $1, 2, 3, \cdots, N$ as $\bar{v}_{1x}, \bar{v}_{2x}, \bar{v}_{3x}, \cdots, \bar{v}_{Nx}$ we will obtain the following change in position $x - x_0$ for the total time $T = N\Delta t$. [We put a bar over the symbol of the average x-component of the velocity to distinguish the notation for the x-component of the velocity at the end of the segments.]

$$x - x_0 = \sum_{i=1}^{N} \bar{v}_{ix} \Delta t$$
$$\Longrightarrow x - x_0 = \int_0^T v_x(t) dt, \qquad (4.31)$$

(or, area under the curve of v_x vs t).

For an arbitrary interval $t = t_1$ to $t = t_2$, the integration will be done accordingly.

$$x_2 - x_1 = \int_{t_1}^{t_2} v_x(t) dt. \qquad (4.32)$$

Example 4.6.1. Area under the curve of acceleration as change in velocity. A skydiver drops off an air plane at zero speed and opens her parachute after 10 seconds. We will assume that parachute opens instantaneously at 10 second mark. For the first 10 seconds the acceleration of the paratrooper is 7m/s^2 pointed down. After the parachute opens at the 10-sec mark, the magnitude of the acceleration drops steadily to zero in another 20 seconds as shown in Fig. 4.14 in a coordinate system in which the positive y-axis is pointed up. Find the velocity of the paratrooper at (a) $t = 10$ sec, (b) $t = 30$ sec?

Solution. The area under the curve method for the changing acceleration can be applied to determine the change in velocity over any period. Part (a) is actually over an interval where the acceleration is constant. Therefore, for part (a) we could use either the constant acceleration equations or the area under the a_y vs t method. For part (b) we cannot use the constant acceleration formulas and must resort to the variable acceleration method.

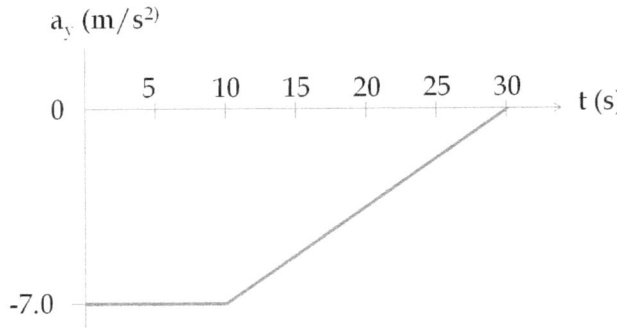

Figure 4.14: Example 4.6.1. Variable acceleration of a parachuter.

(a) The area under the curve method can be applied to the segment of curve from $t = 0$ to $t = 10$ sec. The area is that of a rectangle with one side equal to -7 m/s^2 and the other side equal to 10 sec. Therefore, the area is -70 m/s, which is the change in the y-component of the velocity of the paratrooper. Since the initial velocity of the paratrooper is given to be zero, the velocity at 10-sec mark is -70 m/s. The negative sign correctly gives the direction as pointed down since pointed up has been taken to be the positive direction of the y-axis. We can verify that we get the same result from using constant acceleration equations. Here $v_{0y} = 0$, $t = 10$ sec, $a_y = -7$ m/s^2. Therefore $v_y = v_{0y} + a_y t$ gives $v_y = -70$ m/s, same as the area under the curve method.

(b) This part has a variable acceleration, so it would be a mistake to use the constant acceleration formulas. The area under the y-component of the acceleration versus time curve is that of a triangle of height -7 m/s^2 and the base equal to 20 sec. This gives the change in the y-component of the velocity to be

$$\Delta v_y = \frac{1}{2}\left(-7 \ m/s^2\right)(20 \ sec) = -70 \ \text{m/s}.$$

Since the y-component of the velocity at the beginning of this interval was also -70 m/s, the y-component of the velocity at the end of the interval will be $(-70$ m/s$) + (-70$ m/s$)$, or -140 m/s. Thus, the velocity at time $t = 30$ sec is 140 m/s pointed down.

4.7 RELATIVE MOTION

Have you ever wondered whether physics is same from the perspectives of two observers who are in a relative motion with respect to each other, for instance one inside a moving train and another on a fixed railway platform? In this section you will study the relation between the kinematics quantities the two observers assign to the same object. To simplify our discussion we will orient axes for the two observers such that their axes have the same directions in space and one observer moves along the x-axis of the other observer. This movement will be followed by the displacement

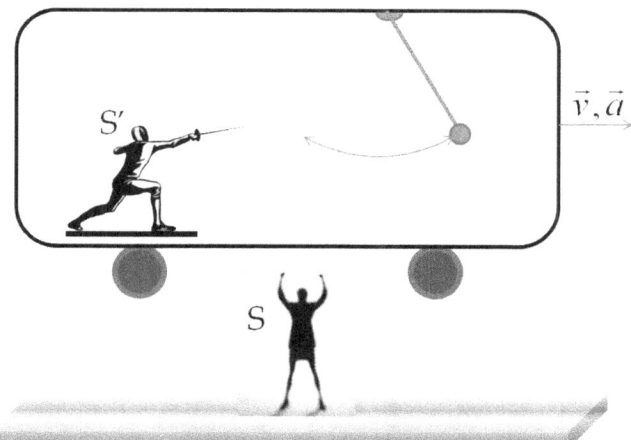

Figure 4.15: Observers S and S' give different values to the position, velocity and acceleration of the pendulum bob. What are their relations?

of the origin of one coordinate system with respect to the origin of the other coordinate system. The two coordinate systems are called frames.

We will also synchronize their clocks so that the two frames set their clocks to read $t = 0$ when they are on top of each other and their origins and axes coincide. We will also assume that the clocks run at the same rate in the two frames. This assumption turns out to lead to trouble in physics as shown by Albert Einstein in 1905 in his theory of special relativity.

The mistake caused by the assumption of equal rate of time elapse in the two moving frames, however, is very small for ordinary velocities, and becomes significant only when the relative speed of the two frames approaches the speed of light which is approximately 3×10^8 m/s. Most situations encountered in everyday life have speeds that are much smaller than the speed of light, and hence we will not bother about the difference in the rates of clocks in the two frames until we come to study motions with speeds near the speed of light. The study of relative motion of frames in this limit is called **Galilean relativity**.

4.7.1 Observers Moving at Uniform Velocity

In Fig. 4.16, I have drawn two frames S and S' in which S is at rest and S' moves towards the positive x-axis with speed V. The relative velocity $\vec{v}_{rel}$ has only x-component in the two frames. In frame S, the relative velocity is $\vec{v}_{rel} = V\hat{u}_x$ and in frame S', the relative velocity is $\vec{v}'_{rel} = -V\hat{u}'_x$. We wish to follow the motion of a point particle P with respect to the two frames.

With $t = 0$ when the origins O and O' are on top of each other, the x-coordinate of the particle P at time t in the frames S and S' will be x and $x - Vt$ respectively. I will denote position, velocity and acceleration in the two frames by the same symbols but add a prime for quantities in

4.7. RELATIVE MOTION

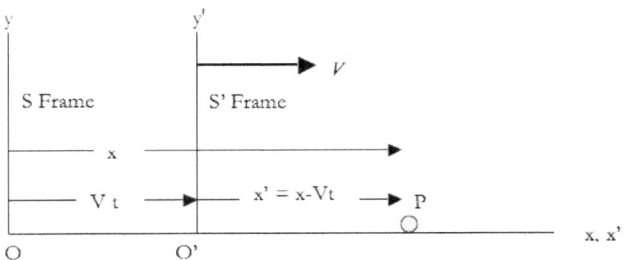

Figure 4.16: The position of an object P with respect to two frames S and S' which are in a uniform relative motion with respect to each other.

S' frame.

$$x' = x - Vt \tag{4.33}$$

The y and z-coordinates will be same in the two frames since there is no relative motion along y- or z-axis.

$$y' = y; \quad z' = z. \tag{4.34}$$

The relation between velocities in the two frames can be obtained by taking derivatives of position with time. Therefore, the components of the velocity of the particle P in the two frames are related as follows.

$$\begin{aligned} v'_x &= \frac{dx'}{dt} = \frac{d(x-Vt)}{dt} = \frac{dx}{dt} - \frac{d(Vt)}{dt} = v_x - V \\ v'_y &= \frac{dy'}{dt} = \frac{dy}{dt} = v_y \\ v'_z &= \frac{dz'}{dt} = \frac{dz}{dt} = v_z \end{aligned} \tag{4.35}$$

We find that the x-component of the velocity of the particle is different in the two frames, but the y and z-velocities are the same. Since the relative speed V is constant, the x-acceleration, which is equal to the derivative of the x-component of the velocity with time, will turn out to be same in the two frames. The y and z-accelerations are, of course, same in the two frames because the y and z-motion of the particle is unaffected by the relative motion along x-axis of the train and the platform.

$$\begin{aligned} a'_x &= \frac{dv'_x}{dt} = \frac{d(v_x - V)}{dt} = \frac{dv_x}{dt} = a_x \\ a'_y &= a_y \\ a'_z &= a_z \end{aligned} \tag{4.36}$$

Example 4.7.1. Relative speed of a runner. Three runners A, B, and C are running on a straight road that run East to West. The runners A and B have velocities {5 m/s, East} and {2 m/s, West}, respectively, with respect to the runner C. What is the relative velocity of the runner A with respect to the runner B?

Solution. It is best to perform the calculation analytically in a Cartesian coordinate system. Let the positive x-axis of the runner C be pointed towards East. Then, we can write the velocities of the runners A and B in this frame.

$$\vec{v}_A = (5 \text{ m/s})\hat{u}_x$$
$$\vec{v}_B = (-2 \text{ m/s})\hat{u}_x$$

The relative velocity of A with respect to B will be $\vec{v}_A - \vec{v}_B$ will be

$$\vec{v}_A - \vec{v}_B = (5 \text{ m/s})\hat{u}_x - (-2 \text{ m/s})\hat{u}_x = (7 \text{ m/s})\hat{u}_x.$$

Therefore, the velocity of A is 7 m/s towards the East with respect to B.

4.7.2 Observers Moving at Uniform Acceleration

Now, consider observing the motion of a particle P from two frames S and S' which have a relative acceleration. Let the frame S' have a constant acceleration of magnitude A directed towards the positive x-axis with respect to the frame S as shown in Fig. 4.17. Once again, we set $t = 0$ at the instant when the two origins coincide. Let V_0 be the relative speed at time $t = 0$.

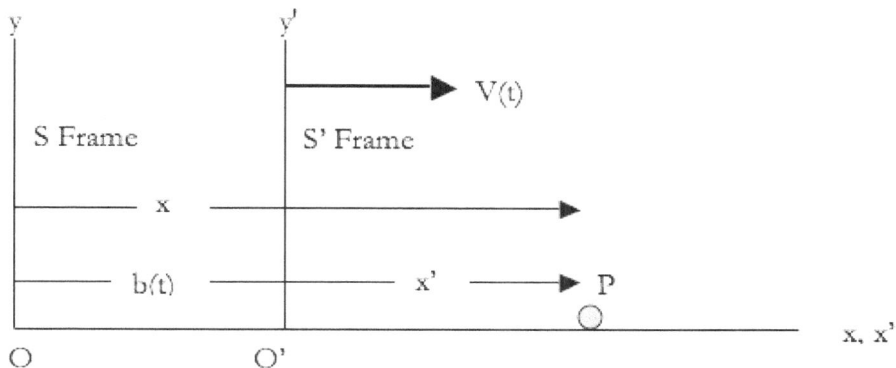

Figure 4.17: Observations on a particle P from a stationary frame S and an accelerating frame S'.

The position of the origin O' of frame S' at time t is obtained from the relative acceleration $A\hat{u}_x$ and the initial velocity $V_0\hat{u}_x$. We note that O' moves at a constant acceleration, hence $b(t)$ is obtained by using the kinematics equations for constant acceleration in one-dimension along x-axis.

$$b(t) = V_0 t + \frac{1}{2}At^2$$

Therefore, the coordinates of the particle P in the two frames are related as follows.

$$x' = x - b(t) = x - V_0 t - \frac{1}{2}At^2$$
$$y' = y$$
$$z' = z \tag{4.37}$$

4.7. RELATIVE MOTION

Now we find the relation between the velocities of the particle P with respect to the two frames.

$$v'_x = \frac{dx'}{dt} = \frac{d(x - b(t))}{dt} = \frac{dx}{dt} - \frac{db}{dt} = v_x - V_0 - At$$
$$v'_y = \frac{dy'}{dt} = \frac{dy}{dt} = v_y$$
$$v'_z = \frac{dz'}{dt} = \frac{dz}{dt} = v_z \qquad (4.38)$$

Acceleration of particle P is now also different in the two frames.

$$a'_x = \frac{dv'_x}{dt} = \frac{d(v_x - V_0 - At)}{dt} = \frac{dv_x}{dt} \frac{d(At)}{dt} = a_x - A$$
$$a'_y = a_y$$
$$a'_z = a_z \qquad (4.39)$$

Example 4.7.2. Freely falling marble in an accelerating elevator. A marble is dropped in an elevator that is accelerating at 3 m/s² upward with respect to the ground. Find the acceleration of the marble with respect to the ground and with respect to an observer in the elevator.

Solution. We draw a picture of the situation given in the problem in Fig. 4.18 where we denote the ground-based observer as O and the observer in the elevator as O'. The accelerations of the marble with respect to the observers O, and O' are related as follows.

$$\vec{a}_0 = \vec{A} + \vec{a}_0{}'$$

The vector relation here has only vertical component. Therefore, we obtain the following for the magnitude of $\vec{a}_0{}'$.

$$-9.81 = 3 + a'_0,$$

which give the following value for the acceleration of the marble for the observer in the elevator.

$$a'_0 = -12.81 \ m/s^2.$$

Note that the acceleration observed in the elevator frame is independent of the velocity of the accelerator. Therefore, the acceleration of the marble will be -12.8 m/s² in the elevator frame, regardless of whether the elevator is moving up or down and the value of the speed of the elevator as long as the acceleration of the elevator with respect to ground is 3 m/s² pointed up.

We see that the "effective g" for an observer in an elevator with acceleration pointed up is higher than 9.81 m/s². You can also deduce that if the acceleration of the elevator were pointed down then the value of the acceleration would be less than 9.81 m/s². For instance, if the elevator itself has an acceleration of 9.81 m/s², pointed down, the acceleration of a freely falling object as seen by the a person moving with the elevator will be zero, i.e., the freely falling objects with respect to a freely falling elevator is weightless!

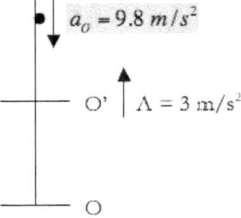

Figure 4.18: Example 4.7.2.

4.8 EXERCISES

Constant Speed and Constant Velocity

Ex 4.8.1. A ball is rolling in a straight groove with a constant speed of 1.5 m/s. (a) How much distance does the ball go in 20 sec? (b) There is a stop at the end of the track 100 m from the place it was let go at $t = 0$, what is the velocity of the ball at times (i) $t = 20$ sec, (ii) $t = 70$ sec?

Ex 4.8.2. A car is moving on a straight road in a fixed direction at a constant speed of 50 km/h with respect to the road. You wish to state the kinematic vectors of the motion of the car by using a Cartesian coordinate system whose positive x-axis is pointed in the direction of the motion of the car and the origin is fixed at some point on the road. (a) What is the expression for the velocity of the car? (b) What is the position vector at time $t = 0.01$ hr?

Ex 4.8.3. The rate of change of the x-coordinate of a moving object is found to be constant 10 m/s. If the x-coordinate at $t = 2$ sec was 100 m, what is the x-coordinate at $t = 10$ sec?

Ex 4.8.4. A function $f(t)$ of time t has a constant rate of change u given by

$$\frac{df}{dt} = u.$$

Find the change in the value of f in the interval between $t = t_1$ and $t = t_2$.

Ex 4.8.5. A function $h(t)$ of time t changes linearly with time, i.e. as one power of t, as given by

$$h(t) = b\,t + c,$$

where b and c are constants. Find the rate of change of h.

Ex 4.8.6. A function $q(t)$ of time t changes linearly with time, i.e. as one power of t. It has a value q_1 when $t = t_1$ and a value q_2 when $t = t_2$. Find an analytic expression for the function $q(t)$.

Ex 4.8.7. A function $s(t)$ of time t has a constant rate of change u. At $t = 0$ the value of s is 1. Find an analytic expression for the function $s(t)$.

Ex 4.8.8. The y-coordinate of a particle varies at a constant rate of 4 m/s. At $t = 0$, the y-coordinate was found to be 3 m. Find an analytic expression for the function $y(t)$.

Constant Acceleration - 1 Dimension

Ex 4.8.9. The x-coordinate of an object varies with time according to $x(t) = 3 + 5t + 9t^2$ where t is in seconds and x is in meters. (a) Find the x-component of the average velocity between $t = 0$ and $t = 1$ sec. (b) Find the x-component of acceleration at $t = 0.5$ sec and $t = 1$ sec.

4.8. EXERCISES

Ex 4.8.10. A car moves on a straight road such that its position from a reference point is given as $\{x(t) = 3 + 5t + 2t^2 + 0.4t^3, y = 0, z = 0\}$, where t is in seconds and x in meters. (a) Is the acceleration of the car constant? (b) Find the acceleration at $t = 0.5$ sec and $t = 1$ sec. (c) Find the average velocity between $t = 0$ and $t = 1$ sec. (d) Find the average acceleration between $t = 0$ and $t = 1$ sec.

Ex 4.8.11. Starting from rest a box slides down an incline at a constant acceleration of 2 m/s² pointed down the incline (not vertically down). (a) Find the speed of the box at $t = 1.5$ sec. (b) Find the average velocity of the box between $t = 0$ and $t = 1.5$ sec (beware of direction!). (c) Determine the distance traveled on the incline in 1.5 sec. (d) Verify that the displacement during the 1.5 sec interval from $t = 0$ to $t = 1.5$ sec is equal to the product of the average velocity for the interval and the time interval.

Ex 4.8.12. A hockey puck is hit up an incline. The puck slows down at a constant acceleration of magnitude 5 m/s² and stops after 1.2 sec. Let x-axis be pointed up the incline. (a) What are the directions of velocity and acceleration at $t = 0.5$ sec? (b) Find the initial velocity, i.e. the velocity of the puck immediately after leaving the stick. (c) How far up the incline does the puck go before coming to rest? (d) Find the position of the puck at $t = 1$ sec.

Ex 4.8.13. A bucket of water is hung with a rope which goes around a pulley. The other side of the rope is wound on a wheel that is rotated with a handle so that the bucket goes up with a constant acceleration of 0.5 m/s² starting from rest. (a) How much time will it take to raise the bucket by 1.5 m? (b) What will the bucket's velocity be when it reaches the 1.5 meter mark from the starting place?

Ex 4.8.14. A drum is rolled down an incline. It is found that the speed of the center of the drum increases steadily with time (i.e. with a constant acceleration) from 2 m/s to 4 m/s over a distance of 1.5 m on the incline. (a) Find the acceleration of the center of the drum. (b) Find the time interval for the given data. (c) How much time will it take to cover the next 1.5 m on the incline?

Ex 4.8.15. A basketball is rolled on a rough horizontal road. The velocity of the center of the ball is found to decrease steadily with time. In a particular roll, the ball stops 20 m from the starting place in 10 sec. (a) What is the speed of the ball at the initial time? (b) What is the acceleration (i.e. deceleration) of the ball? (c) How much distance does the ball cover in the first 5 sec? (d) How much distance does the ball cover in the last 5 sec?

Ex 4.8.16. A skater after an initial push glides with a constant deceleration. He comes to a stop after reaching 120 m in 25 sec. (a) Find his initial speed. (b) Find the deceleration. (c) How much distance did he cover in the first 12.5 sec? (d) How much distance did he cover in the last 12.5 sec?

Ex 4.8.17. A train starts from rest and accelerates on a straight track at a constant acceleration of 3 m/s² for 10 sec. It then coasts at a constant speed for 20 sec. (a) Find the speed with which the train is coasting in the last 20 sec. (b) Find the total distance traveled in 30 sec from the start. (c) Find the distance traveled in the first 5 sec. (d) Find the distance traveled in the next 5 sec.

Ex 4.8.18. A car starts from rest and accelerates on a straight road at a constant acceleration of 2 m/s² for 20 sec. It then coasts at a constant speed v for another 20 sec. Finally it decelerates and comes to a stop in 10 sec. (a) Find the speed v with which the car is coasting in the second 2 sec constant speed stage. (b) Find the total distance traveled in first 40 sec. (c) Find the deceleration in the last 10 sec. (d) Find the distance traveled in the last 10 sec. (e) Find the total distance traveled over the 50 sec interval.

Free Fall - 1 Dimension

Ex 4.8.19. A steel ball is dropped from rest from a 100-m tower. Assume the effect of the air resistance to be negligible on its motion. (a) Find the time the ball will take to fall to the ground. (b) Find the speed with which the ball will strike the ground. (c) Find the time the ball will take to reach the 50-m mark. (d) Find the time the ball will take to cover the last 50 meters. (e) Find the speed of the ball when it is at the 50-m mark.

Ex 4.8.20. A rocket it shot straight up. It reaches the top of its flight in 3 sec. Assume the effect of the air resistance to be negligible on its motion. (a) Find the velocity of the rocket immediately after it leaves the launch pad. (b) How high does the rocket go?

Ex 4.8.21. A brass ball is launched vertically up with an initial speed of 30 m/s. Assume the effect of the air resistance to be negligible on its motion. (a) Find the location of the ball after 2 sec. (b) How fast is the ball moving at that time? (c) How high does the ball rise before coming to rest?

Ex 4.8.22. A boy throws a rock from the roof of a 10-story building straight down with an initial speed of 10 m/s. Assume 3 meter per story for the building. Assume the effect of the air resistance to be negligible on its motion. (a) How long will it take the rock to strike the ground? (b) With what speed will the rock strike the ground?

Ex 4.8.23. Two balls are launched straight up at different times from a tennis ball launching machine. The second ball is launched at the time the first ball is at the top of its flight. If the launching speed of the balls are 20 m/s, where would the two balls hit each other? Assume the effect of the air resistance to be negligible on the motion.

Ex 4.8.24. Two rocks are thrown vertically down from a 100 m tall tower at different times. While the first rock is let go from rest, the second rock,

4.8. EXERCISES

one second later, is thrown at some speed v_0 such that the second rock strikes the first rock 2.5 seconds later, i.e. at $t = 3.5$ sec from when the first rock was thrown. (a) Find the initial speed of the second rock. (b) Find the speeds with which the two rocks are moving when they strike.

Projectile Motion

Ex 4.8.25. A ball is launched from the end of a table horizontally with a speed of 10 m/s and lands on the floor 2 m below. (a) Where does the ball land on the floor? (b) How much time was the ball in air? (c) What is the velocity of the ball immediately before it strikes the floor? Ignore the air resistance.

Ex 4.8.26. A stone is thrown horizontally from a 30-meter cliff with an initial speed of 20 m/s. (a) How far does it travel before hitting the ground? (b) At what time does it strike the ground, if it left the cliff at $2:01:00\ PM$? (c) At what time does it reach a spot that is 5 m from the ground? (d) What is the horizontal distance of the point where the stone is at 5 m height from the ground? Ignore air resistance.

Ex 4.8.27. A basketball is thrown from a height of 1.6 m from the ground. Immediately after leaving the hand the ball had an initial velocity of 5 m/s and in the direction of $40°$ above the horizontal direction. (a) Where on the floor does the ball land? (b) With what velocity does the ball strike the floor? Ignore air resistance.

Ex 4.8.28. A stone is thrown from a 30-m cliff with initial speed of 20 m/s and at an angle of $60°$ above the horizontal direction. The stone lands on a plateau that is 10-m above the bottom of the cliff. How far away from the cliff, horizontally, is the place where the stone will land?

Ex 4.8.29. A cannonball must travel 300 m horizontally and 20 m vertically above from the launch site to land on the enemy position. With what speed should the ball be launched at $40°$ from the horizontal to make the hit? Assume no air resistance.

Ex 4.8.30. A cannonball must travel 400 m horizontally and 50 m vertically below from the launch site to land on the enemy position. With what speed should the ball be launched at $30°$ from the horizontal to make the hit? Assume no air resistance.

Ex 4.8.31. A rocket launcher launches missiles at a speed of 50 m/s. At what angle with the horizon should a missile be fired so that it strikes a target at 100 m horizontal distance and 30 m vertical distance from the launch site?

Ex 4.8.32. A rocket launcher launches missiles at a speed of 40 m/s. At what angle with the horizon should the missile be fired so that the missile strikes a target at 80 m horizontal distance and 25 m vertical distance from the launch site?

Ex 4.8.33. A football leaves the kicker's foot at 50° from the horizontal direction. The football is required to clear a horizontal bar at a height of 10 m above the ground at a horizontal distance of 40 m. Find the speed with which the ball must leave the kicker's foot so that it would barely go over the bar.

Variable Acceleration

Ex 4.8.34. A particle starts out at rest at $t = 0$ and accelerates in a straight line with an acceleration of 2 m/s^2 from $t = 0$ to $t = 3$ sec, and then with an acceleration of 4 m/s^2 pointed in the same direction from $t = 3$ sec to $t = 5$ sec. (a) Find the position and the velocity of the particle at $t = 3$ sec. (b) Find the position and the velocity of the particle at $t = 5$ sec.

Ex 4.8.35. At $t = 0$ a particle has a velocity of 20 m/s in the direction of the positive x-axis, and accelerates on the x-axis as shown in the Fig 4.19. The y and z-components of the acceleration are zero. (a) Find the position and the velocity of the particle at $t = 2$ sec. (b) Find the position and the velocity of the particle at $t = 5$ sec.

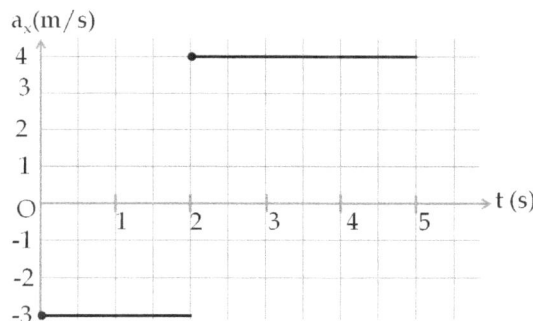

Figure 4.19: Exercise 4.8.35.

Ex 4.8.36. A particle starts out with a velocity of 20 m/s in the direction of the negative x-axis, and accelerates on the x-axis as shown in the Fig 4.20. (a) Find the position and the velocity of the particle at $t = 2$ sec. (b) Find the position and the velocity of the particle at $t = 3$ sec. (c) Find the position and the velocity of the particle at $t = 5$ sec.

Ex 4.8.37. A box slides on a floor in a straight path such that the magnitude of the acceleration is not constant in time. The data is collected with respect to a Cartesian axis system in which only the a_x is non-zero. The x-component of the acceleration is shown in Fig 4.21. As the figure shows, the direction of the acceleration is always pointed towards the positive x-axis, but the magnitude varies with time. (a) Find the velocity of the box at $t = 4$ sec if it starts out at rest at $t = 0$. (b) Find the velocity of the box at $t = 10$ sec.

4.8. EXERCISES

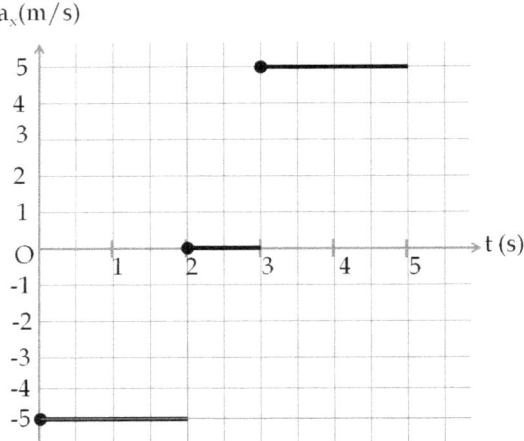

Figure 4.20: Exercise 4.8.36.

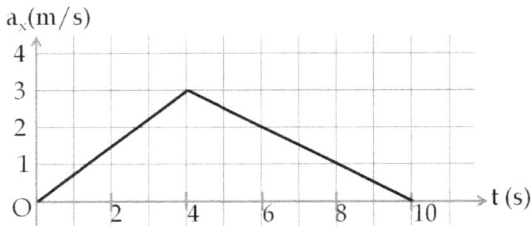

Figure 4.21: Exercise 4.8.37.

Ex 4.8.38. A hockey puck is shot on a surface that has different roughness at different places. As a result its acceleration varies from place to place. By placing the x-axis on the line of motion of the puck, we cast the acceleration vector in terms of its x-component which can be plotted with time. Note the acceleration vector cannot be plotted, since they are not ordinary functions; only the components can be plotted. The resulting x-component of the acceleration is shown in Fig. 4.22. The y and z-components of the acceleration are zero. The puck has a velocity of 2 m/s at $t = 5$ sec. Find the initial velocity of the puck at $t = 0$.

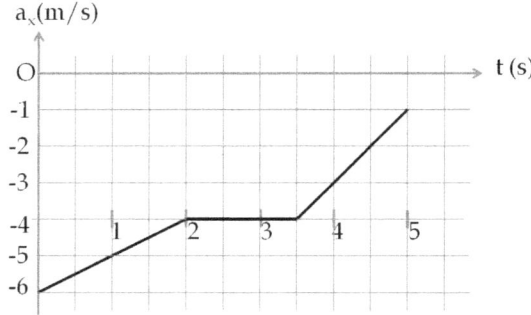

Figure 4.22: Exercise 4.8.38.

Ex 4.8.39. The x-component of the acceleration of a particle changes according to the following analytic expression, $a_x(t) = 4 + t^2/2$, where t is in sec and the acceleration in m/s². Find the x-components of the velocity and position vectors.

Ex 4.8.40. The x-component of the acceleration of a particle changes according to the following analytic expression, $a_x(t) = 1 + 3t$, where t is in sec and the acceleration in m/s^2. (a) Find the x-component of the velocity at $t = 3$ sec if the particle was at rest at $t = 0$. (b) Find the change in the x-coordinate of the particle in the interval $t = 1$ sec to $t = 3$ sec.

Ex 4.8.41. The x-component of the velocity of a ball is given as $v_x(t) = 20 + 3t^3$, where t is in sec and the velocity is in m/s. (a) Find the change in the x-coordinate between $t = 0$ and $t = 5$ sec. (b) Find the x-component of the acceleration at (i) $t = 0$, and (ii) $t = 5$ sec.

Integration of Equations

Ex 4.8.42. Find function $f(t)$ from the given conditions on the function. (a) $df/dt = 0$ and $f(0) = 1$; (b) $df/dt = 2$ and $f(0) = 5$; (c) $df/dt = 4t$ and $f(1) = 1$; (d) $df/dt = 5 + 3t$ and $f(0) = 2$; (e) $df/dt = (1/2)t^2$ and $f(0) = 7$; (f) $df/dt = \sin(t)$ and $f(0) = 1/2$; (g) $df/dt = \pi - t$ and $f(1) = 10$; (h) $df/dt = e - 3t$ and $f(0) = 2$.

Ex 4.8.43. A particle moves at a constant speed of 20 m/s along the positive x-axis of a Cartesian coordinate system. (a) Write its velocity? (b) If the particle was at $x = 1$ m at $t = 0$, where will it be at $t = 3$ sec? (c) If the particle was at $x = 1$ m at $t = 1$ sec, where will it be at $t = 3$ sec? (d) What is its acceleration?

Ex 4.8.44. In a particular Cartesian coordinate system, the y and z-components of the velocity are zero and the x-component varies as given by the function, $v_x(t) = 2 + 5t$, where t is in sec and v_x is in m/s. (a) Find the displacement vector in the following time intervals (i) $[0, 1\text{ sec}]$, (ii) $[0.5\text{ sec}, 1.5\text{ sec}]$, and (iii) $[2\text{ sec}, 3\text{ sec}]$. (b) Find the average velocity in the same intervals. (c) Find the instantaneous velocity at the following instants in time, (i) 1 sec, (ii) 2 sec, and (iii) 3 sec. (d) Find the instantaneous accelerations at the same instants in time. (e) If the particle was at $x = 0$ at $t = 0$, find the position of the particle at (i) 1 sec, (ii) 2 sec, and (iii) 3 sec.

Ex 4.8.45. In a particular Cartesian coordinate system, the y and z-components of the velocity are zero and the x-component varies as given by the function, $v_x(t) = 2 + 5t - 3t^2$, where t is in sec and v_x is in m/s. (a) Find the displacement in the following time intervals ((i) $[0, 1\text{ sec}]$, (ii) $[0.5\text{ sec}, 1.5\text{ sec}]$, and (iii) $[2\text{ sec}, 3\text{ sec}]$. (b) Find the average velocity in the same intervals. (c) Find the instantaneous velocity at the following instants in time, (i) 1 sec, (ii) 2 sec, and (iii) 3 sec. (d) Find the instantaneous accelerations at the same instants in time. (e) If the particle was at $x = 0$ at $t = 0$ and moved on the x-axis, find the position of the particle at (i) 1 sec, (ii) 2 sec, and (iii) 3 sec.

4.8. EXERCISES

Ex 4.8.46. In a particular Cartesian coordinate system, the y and z-components of the acceleration are zero and the x-component varies as given by the function, $a_x(t) = -20 + 10\,t$, where t is in sec and a_x is in m/s^2. The particle's velocity at $t = 0$ was pointed towards the positive x-axis and had a magnitude of 10 m/s. (a) Find the change in the velocity in the following time intervals (i) $[0, 1\text{ sec}]$, (ii) $[0.5\text{ sec}, 1.5\text{ sec}]$, and (iii) $[2\text{ sec}, 3\text{ sec}]$. (b) Find the average acceleration in the same intervals. (c) Find the instantaneous accelerations at the following instants in time, (i) 1 sec, (ii) 2 sec, and (iii) 3 sec. (d) Find the instantaneous velocity at the same instants in time. (e) If the particle was at $x = 0$ at $t = 0$, find the position of the particle at (i) 1 sec, (ii) 2 sec, and (iii) 3 sec.

Ex 4.8.47. In a particular Cartesian coordinate system, the y and z-components of the acceleration are zero and the x-component varies as given by the function, $a_x(t) = 5t - 3t^2 + 20\exp(-t)$, where t is in sec and a_x is in m/s^2. The particle's velocity at $t = 0$ was pointed towards the positive x-axis and had a magnitude of 10 m/s. (a) Find the change in velocity in the following time intervals (i) $[0, 1\text{ sec}]$, (ii) $[0.5\text{ sec}, 1.5\text{ sec}]$, and (iii) $[2\text{ sec}, 3\text{ sec}]$. (b) Find the average acceleration in the same intervals. (c) Find the instantaneous accelerations at the following instants in time, (i) 1 sec, (ii) 2 sec, and (iii) 3 sec. (d) Find the instantaneous velocity at the same instants in time. (e) If the particle was at $x = 0$ at $t = 0$, find the position of the particle at (i) 1 sec, (ii) 2 sec, and (iii) 3 sec.

Ex 4.8.48. In a particular Cartesian coordinate system, $y = 0$ and $z = 0$, and the x-coordinate of a particle varies with time as given by $x(t) = 2\sin(3t) + C$, where t is in sec and x in m, and C is a constant to be determined by the data. At $t = 0$ the particle was at $x = 1$ m and the velocity of the particle was pointed towards the positive x-axis and had a magnitude of 10 m/s. (a) Find the value of constant C. (b) Find the displacement in the following intervals, (i) $[0, 1\text{ sec}]$, (ii) $[0.5\text{ sec}, 1.5\text{ sec}]$, and (iii) $[2\text{ sec}, 3\text{ sec}]$. (c) Find the instantaneous velocity at the following instants in time, (i) 1 sec, (ii) 2 sec, and (iii) 3 sec. (d) Find the instantaneous accelerations at the following instants in time, (i) 1 sec, (ii) 2 sec, and (iii) 3 sec.

Ex 4.8.49. In a particular Cartesian coordinate system, $v_y = 0, v_z = 0$, and v_x varies with time as given by, $v_x(t) = 20\cos(5t) + C$, where t is in sec and v_x is in m/s, and C is a constant to be determined by the data. The particle's acceleration at $t = 0$ was pointed towards the negative x-axis and had a magnitude of 3 m/s^2. (a) Find the value of constant C. (b) Find the displacement in the following intervals, (i) $[0, 1\text{ sec}]$, (ii) $[0.5\text{ sec}, 1.5\text{ sec}]$, and (iii) $[2\text{ sec}, 3\text{ sec}]$. (c) Find the average velocity in the same intervals. (d) Find the instantaneous velocity at the following instants in time, (i) 1 sec, (ii) 2 sec, and (iii) 3 sec. (e) Find the instantaneous accelerations at the same instants in time. (f) If the particle was at $x = 0$ at $t = 0$, find the position of the particle at (i) 1 sec, (ii) 2 sec, and (iii) 3 sec.

Relative Motion

Ex 4.8.50. John and Betsy are separated by 20 m on a platform and watch a train approach the station on a straight track at a constant speed of 5 m/s. Find the position, velocity, and acceleration of the train at $t = 10$ s with respect to John and Betsy if the front of the train was 200 m from Betsy and 220 m from John at $t = 0$.

Ex 4.8.51. On a straight East-West road, two cars A and B are moving towards East with respect to a person P on the ground. According to P, the car A is moving with a constant velocity of 30 m/s and the car B with a constant acceleration of 5 m/s^2. At $t = 0$, car B was at rest with respect to P. (a) Find the velocity of the car B with respect to the car A at $t = 20$ s. (b) Find the velocity of the person P with respect to the car A at $t = 20$ s. (c) Find the acceleration of the car B with respect to the car A at an arbitrary time t.

Ex 4.8.52. Two cars A and B start from a junction as observed by a person P on the ground. Car A goes north at a constant acceleration of 5 m/s^2 having started out at rest at the junction. Car B goes East at a constant velocity of 25 m/s. (a) Find the velocity of person P with respect to car B at $t = 10$ s. (b) Find the velocity of car B with respect to car A at $t = 10$ s. (c) Find the acceleration of car A with respect to car B at an arbitrary time.

Ex 4.8.53. On a straight East-West road, two cars A and B are moving towards East with respect to a person P on the ground. According to person P, car A is moving with a constant velocity of 30 m/s and car B with a constant acceleration of 4 m/s^2. At $t = 0$, the car B was at rest with respect to P. (a) Find the velocity of person P with respect to the car B at $t = 30$ s. (b) Find the velocity of the car A with respect to the car B at $t = 30$ s. (c) Find the acceleration of the car A with respect to the car B at an arbitrary time.

4.9 PROBLEMS

Problem 4.9.1. Suppose you live on the 5th floor of a tall building. Your friend goes to the roof and drops a steel ball from rest while you time the ball as it falls. You find that the ball takes 0.110 sec to fall from the top of the window to the bottom of the window, a distance of 0.8 meter. How high above the top of the window of the 5th floor is the roof?

Problem 4.9.2. A rocket is rising at a constant velocity of 200 m/s. A piece of the rocket comes loose and falls freely to the ground. It takes 50 sec for the piece to fall to the ground after dislodging from the rocket. How far above the ground the piece came loose?

Problem 4.9.3. A ball is tossed with a speed v_0 and caught at a height H above the launch point on its way down. Find the time the ball was in flight. Note: the velocity of the ball when caught is not zero.

Problem 4.9.4. A person P_1 at rest on the sidewalk observes that the rain drops are falling vertically with speed 10 m/s. What is the velocity of raindrops in the frame of another person P_2 who is running towards the person P_1 in the rain in a straight line with speed 5 m/s?

Problem 4.9.5. A person standing on the sidewalk at rest observes that rain drops are falling at 30° with respect to the vertical direction with speed 10 m/s. Assume the velocities of all rain drops fall in one plane. What is the velocity of the raindrops in the frame of a person that is running in the rain in a straight line with speed 5 m/s in a direction that is perpendicular to the plane containing the velocity vectors of the rain drops?

Problem 4.9.6. Two runners are running in the opposite directions on a 400-m closed track with different speeds, which can be assumed to be constant. Let $t = 0$ be the time when they first pass each other, and let the distances be measured from that point O on the path. They are next seen to pass each other 180 m from O as measured on the track. (a) How far from O will they pass each other next? (b) If the speed of the faster runner is 8 m/s, what is the speed of the slower runner? (c) Let t_n be the nth time they will pass each other. Find a formula for t_n.

Problem 4.9.7. The shadow of a pendulum cast on a flat board moves on a straight line. By placing the x-axis on the straight line with the origin at the middle of the total path, the x-coordinate of the shadow is given by $x(t) = 50\cos(\pi t)$, where t is in seconds and x is in cm. (a) Find the speed of the shadow at (i) $t = \frac{1}{4}$ sec and (ii)$t = \frac{1}{2}$ sec. (b) Find the acceleration of the shadow's motion at $t = \frac{1}{3}$ sec. (c) How much distance does the shadow travel in 20 sec?

Problem 4.9.8. An ant is moving in a straight line with a variable speed. It covers the first meter in 5 sec, the next $\frac{1}{2}$ m in 5 *sec*, the next $\frac{1}{4}$ m also in 5 sec, and so on. That is, the ant covers $\frac{1}{2}$ of the previous step in the same amount of time as the last step. (a) How much distance would the ant travel in all before finally coming to rest? (b) How much time does the ant take to reach its final destination? (c) What is the average speed of the ant?

Problem 4.9.9. One way to measure the value of g is to launch a projectile vertically upward and record the return times at two different heights. Let T_1 and T_2 be the two return times, i.e.,the time for 1-3-1' and 2-3-2', at heights H_1 and H_2 respectively as shown in Fig. 4.23. Deduce the following formula for g from the heights and times.

$$g = \frac{8(H_2 - H_1)}{T_1^2 - T_2^2}.$$

Problem 4.9.10. An insect flies at a constant speed between two moving walls that are initially a distance D apart with the insect near one wall.

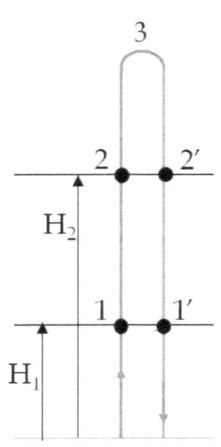

Figure 4.23: Problem 4.9.9.

The insect goes from one wall to the other and back without losing anytime in the turn around, and does this continuously. If both the walls move right with a constant speed u with respect to the floor and the insect's speed be v, find the distance the insect has traveled by the Nth turn around. Assume that at $t = 0$ the insect is moving to the right from its position just to the right of the left wall. Think of other variations of this problem, and work them out for fun.

Problem 4.9.11. A projectile is fired with speed v_0 at an angle θ from the horizontal. (a) Show that the horizontal distance D traveled by the projectile before landing at the same height from the ground is given by

$$D = \frac{v_0^2 \sin(2\theta)}{g}.$$

(b) From this formula, prove that the range is maximum when $\theta = 45°$.

Problem 4.9.12. A cannon is on an incline with an angle of inclination φ as shown in Fig. 4.24. A cannonball is fired with a speed v_0 at an angle θ with respect to the incline. The ball comes out of the cannon at a height h. How far up on the incline will the ball land if the air resistance is ignored?

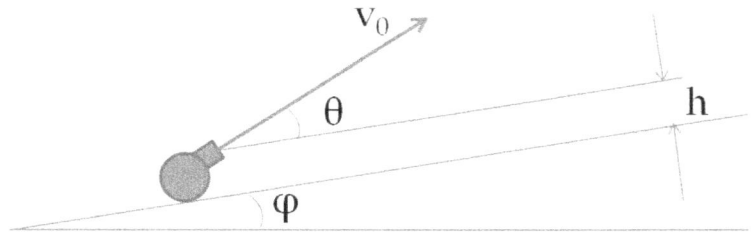

Figure 4.24: Exercise 4.9.12.

Chapter 5

FORCES AND STATIC EQUILIBRIUM

Contents

5.1	MEASURING FORCES		166
	5.1.1	Distinction Between Mass and Weight	166
	5.1.2	A Method of Measuring Forces	167
5.2	WHAT IS A FORCE?		168
	5.2.1	Forces and Newton's Third Law of Motion	168
5.3	FORCES AS VECTORS		171
	5.3.1	Resultant force	171
	5.3.2	Component of a Force in a Given Direction	172
5.4	SOME COMMON FORCES		174
	5.4.1	Weight and Gravitational Force	175
	5.4.2	Tension Force and Hooke's Law	177
	5.4.3	Spring force and Hooke's Law	178
	5.4.4	Normal Forces at a Contact Surface	179
	5.4.5	Static Frictional Force	183
	5.4.6	Kinetic or Sliding Frictional Force	186
	5.4.7	Rolling Friction	187
	5.4.8	Viscous drag	188
5.5	TORQUE OR MOMENT OF A FORCE		189
5.6	STATIC EQUILIBRIUM		195
	5.6.1	Static Equilibrium and Newton's First Law of Motion	195
	5.6.2	Problems of Static Equilibrium	197
5.7	EXERCISES		207
5.8	PROBLEMS		215

Figure 5.1: In power lifting the force by the muscles of the lifter must balance the force of weight of the lifted object. Photo credit: International Powerlifting Federation, at Wikicommons.

5.1 MEASURING FORCES

5.1.1 Distinction Between Mass and Weight

When you hold a heavy object in your hand, you feel the sensation of a downward force on your hand. The source of the downward force on the hand is the attractive pull of Earth on the object. The force of Earth on an object is called its weight. A power lifter must balance the weight of the load on the barbell. The more you put on the barbell more difficult it gets to lift the barbell.

The pull of Earth on an object is proportional to the amount of matter, but they are not the same quantities and we make a distinction between the mass (m) and the pull of Earth, which we call **weight** (W).

$$\boxed{W = m\ g,} \tag{5.1}$$

where the proportionality constant g is the acceleration due to gravity, the same quantity we have used for the acceleration of an object in the free fall. Although, the value of g is different at different places on Earth, we will use a standard value of 9.81 m/s² in our calculations. The SI unit of weight is obtained from multiplying the units of mass and acceleration, the result being kg.m/s², which is also called Newton and denoted by the symbol N.

$$1\ N = 1\ kg.m/s^2.$$

Therefore, the weight of 1-kg is equal to a force of magnitude 9.81 N, and a 1-lb weight is equal to 4.54 N.

Weight is a vector quantity while mass is a scalar quantity. Note that Eq. 5.1 gives only the magnitude of the weight vector. The

5.1. MEASURING FORCES

direction of weight vector is towards the center of Earth. It is normally too much to keep writing the phrase "the magnitude of weight" when we refer to mg. Therefore, we will simply drop the qualifier "the magnitude of" and call mg weight, although, we know that we also need the direction for weight.

Because it is relatively easy to set up physical systems that will give a reliable indication of magnitude of weight at a location, such as a spring balance to be described below, it is tempting to use Eq. 5.1 to gain the value of the mass of an object. However, since g varies from place to place, an absolute value of mass cannot be truly obtained by instruments that measure weight. We can only obtain mass of an unknown relative to an arbitrarily chosen standard mass by comparing their weights. This is what we do when we compare two masses on a balance. We compare (magnitudes of) their weights at the same location, and if the (magnitudes of) weights are equal, then masses are equal also.

$$\text{If } W_1 = W_2 \implies m_1 g = m_2 g \implies m_1 = m_2. \tag{5.2}$$

5.1.2 A Method of Measuring Forces

According to Eq. 5.1, if you hold a 2-kg object, then you should feel twice the force as you would experience for a 1-kg object. But, in practice, although a 2-kg object will feel much heavier than a 1-kg object, you can't really say if the 2-kg object is exactly twice as heavy as the 1-kg object reliably and consistently. Therefore, we need a better method for measuring forces than the sensation of a human hand.

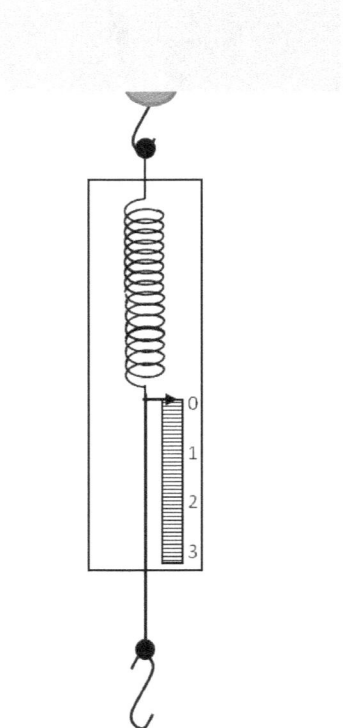

Figure 5.2: A spring balance measures weight. The balance is often calibrated to indicate mass value based on the weight at one location.

The stretching of a simple spiral spring provides a more reliable tool for comparing different pulls. A **spring balance** is constructed by fixing one end of a spring and attaching a pointer to the free end as shown in Fig. 5.2. Once a spring balance is calibrated at a particular location, we can use the spring balance not only for measuring weights but also for measuring other forces as well by comparing them to the force it takes to stretch the spring by a set amount. For instance, if you pull the spring so that the pointer points to 1-kg mark on the scale, then you have applied a force equal to the weight of 1-kg mass, or 9.81 N if g at that site is 9.81 m/s^2. In this way you can measure all other forces with the spring balance after it has been calibrated with reference weights.

Note that a spring balance measures force and not mass. You should be careful in deducing mass from a reading on the spring balance. Suppose you attach a mass to a spring balance and the reading says 1 kg. Would the reading be same if you took the same spring and same mass at another location on Earth? The answer is no. Since spring balance shows force of pull and Earth's pull changes from place to place on Earth, which is reflected in changing value of g, therefore, the spring will show less pull at a place where g is less. It should also be noted that the reading for the weight of an object on the spring balance will go down as you go away from the center of the Earth as the pull of the Earth will decrease then. In a zero gravity situation, the spring balance will read zero no matter the amount of mass attached to the bottom of the spring.

5.2 WHAT IS A FORCE?

5.2.1 Forces and Newton's Third Law of Motion

Two bodies interact and influence each other's motion through a force between them. Intuitively, a force is a push or a pull a body exerts on another body. When you push on a wall, you are aware of the fact that you are applying a force on the wall. What is not obvious is that in generating the force, you have helped create a force by wall also.

Fundamentally, every force requires two participating objects. When the force acts on one of the bodies, it has one direction, and when it acts on the other body, it has the opposite direction, as codified in the statement of the third law in Philosophiae naturalis principia mathematica, also known as Principia by Sir Isaac Newton published in 1686.

"To every action there is always opposed an equal reaction; or the mutual actions of two bodies upon each other are always equal, and directed to the contrary parts."

In Principia Newton further elaborated the meaning of the **third law**, "Whatever draws or presses another is as much drawn or pressed by that other. If you press a stone with your finger, the finger is also pressed by the stone. If a horse draws a stone tied to a rope, the horse (if I may so say) will be equally drawn back towards the stone: for the distended rope, by the same endeavor to relax or unbend itself, will draw the horse as much towards the stone, as it does the stone towards the horse, and will obstruct the progress of the one as much as it advances that of the other."

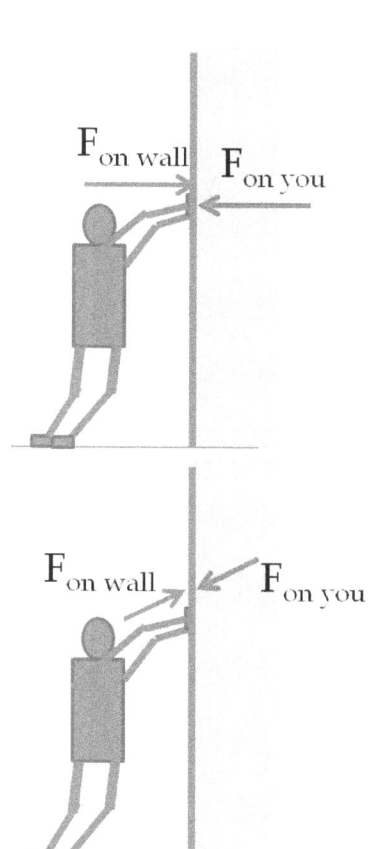

Figure 5.3: Forces pairs. You push on the wall, the wall pushes you back!

Example 5.2.1. A plough pulled by horses. In Fig. 5.4, two horses pull on a plough and a man is also pushing on the plough. Focusing on the forces between the horses and the plough, we note that there is a force between the horses and the plough since the they are linked by a belt under tension. When this force acts on the horses, we call the force $\vec{F}_{\text{on horses}}$, and when it acts on the plough, we call the force $\vec{F}_{\text{on plough}}$. Note the language here: the force is between the horses and the plough, but we do not have a force vector until we ask the question - on which object the force is acting? The same force generates two different forces, one on each object of the link.

Since, the two forces, $\vec{F}_{\text{on horses}}$ and $\vec{F}_{\text{on plough}}$, are not acting on the same object, they are not used simultaneously in any one problem con-

5.2. WHAT IS A FORCE?

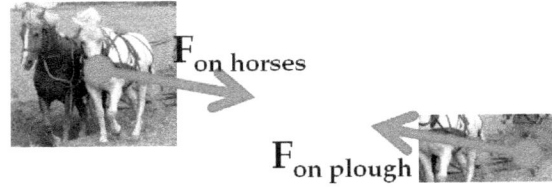

Figure 5.4: Forces act between two objects connected by a visible or an invisible link. The pictures here show the force between the horses and the plough that act through the linking belt which is under tension. The force between the horses and the plough can act on the horses or the plough. When the force acts on the horses, it generates $\vec{F}_{\text{on horses}}$, and when it acts on the plough, it generates $\vec{F}_{\text{on plough}}$. The force $\vec{F}_{\text{on horses}}$ influences the motion of horses only and does not affect the motion of plough, and the force $\vec{F}_{\text{on plough}}$ influences the motion of the plough and not that of the horses.

cerning either the motion of the horse or that of the plough. When we study the motion of the horses, the force of use is the force on the horses and not the force on the plough, although it is there. On the other hand, when we study the motion of the plough, the force on the plough is used.

Example 5.2.2. The forces between train cars. The cars of a train are pulled by forces between cars that act along the links between the cars as illustrated in Fig. 5.5. A shown in the figure, there is a force between car A and car B that acts through the link between them, and there is force between car B and car C that acts through the link between B and C. These two forces create two forces on car B as shown in the figure. Can you find how the force between A and B acts on A? Can you find how the force between B and C acts on C?

Figure 5.5: The forces between cars A, B and C of a train take place due to bolts and hooks linking A and B, and those linking B and C. The force between A and B and that between B and C acting on B generate two forces on B shown in the figure. The motion of B is governed by the combined effect of these two forces on B.

Example 5.2.3. Forces between Earth and Moon. The previous two examples show forces that act through a visible link between objects - the belt in the horses/plough and the bolts/hooks between the cars. Fig. 5.6 shows a different type of force - the force between the Earth and the Moon that appear to act without a visible link between them. Such forces are also called action-at-a-distance forces.

Figure 5.6: The gravitational force between the Earth and the Moon act through an invisible gravitational link between them, called the gravitational field. The gravitational force between Earth and Moon can act on the Earth or the Moon, generating a force on Earth and one on Moon. To study the motion of the Moon, you need the force on Moon and not the force on Earth.

The action-at-a-distance force between the Earth and the Moon is called gravitational force and the invisible force field linking the two objects is called the gravitational field. The force of gravitation acts between all objects in nature. The force between the Earth and the Moon can act on either the Earth or the Moon: when it acts on the Earth, it pointed towards the Moon, and when it acts on the Moon, it is pointed towards Earth. The force on Moon affects the motion of the Moon and the force on Earth affects the motion of the Earth.

Example 5.2.4. Electric force between electric charges. Figure 5.7 shows another action-at-distance force, the electric force between a positively charged object, such a proton, and a negatively charged object, such as an electron. The invisible force field linking the two objects is called the electric field. The electric force between the charges generates a force on the positive charge and one on the negative charge. To study the motion of the negative charge you need the force on the negative charge and not the one on the positive charge. Similarly for the motion of the positive charge, you will need only the force on the positive charge.

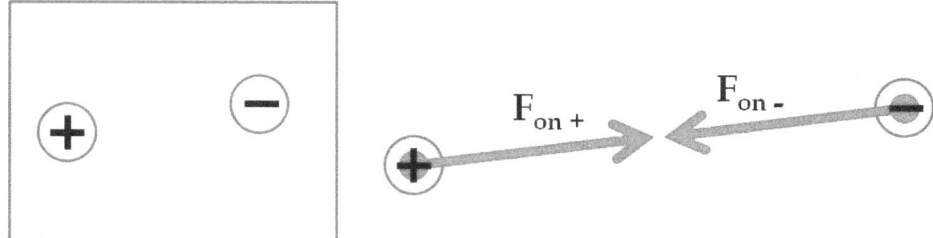

Figure 5.7: The electric forces between charges act through an invisible link between the charges, called the electric field. The electric force between a positive charge and a negative charge is shown. The electric force between the charges generates a force on the positive charge and one on the negative charge. To study the motion of the negative charge you need the force on the negative charge and not the one on the positive charge. Similarly for the motion of the positive charge, you will need only the force on the positive charge.

5.3 FORCES AS VECTORS

5.3.1 Resultant force

Force is a vector quantity since it has both magnitude and direction. The effect of a force on the motion of an object is independent of which body applies the force. For instance, an identical motion results from a force by someone pushing on a box or an equal force applied by a machine pulling on the box (Fig. 5.8). Therefore, when several bodies exert forces on one object, the net effect of all forces is identical to one force that is a vector sum of all forces on the body. The **net force** is also called the **resultant force**.

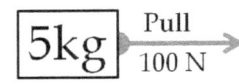

Figure 5.8: Push or pull cause same motion.

If two forces acting on a body are parallel to each other, then the resultant force has a magnitude equal to the sum of the magnitudes of the two forces. If two anti-parallel forces act on an object, then the resultant force has a magnitude equal to the difference of the magnitudes of the two forces, and the direction of the larger of the two forces. If two forces act at an angle, then you need to use the parallelogram law of addition of vectors to figure out the resultant force. Often, it is much easier to work analytically with components of forces as illustrated in Example 5.3.1.

Example 5.3.1. Resultant force

A box is pulled by a force of 40 N in the horizontal direction and a force of 30 N also horizontally but at an angle of 90° to the direction of the other force. Find the magnitude and direction of the resultant force.

Solution. This problem can be done in a number of ways. Here we will illustrate two most commonly used methods.

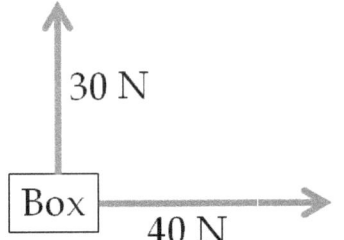

(a) **Geometrical method.** Place the second force at the tip of the first force (Fig. 5.9). Then the force from the tail of the first force to the tip of the second force is the resultant force. In the graphical method of addition, we use deduce the magnitude of the force by a scale for the drawing and use protractor to read off the angle. The scale on the drawing gives the magnitude of the force to be 50 N. A protractor gives the approximate angle of 35° vector $\vec{F}$ makes with the horizontal direction. Note that it is impossible to read the angle with 100% precision and we will obtain only an approximate value for the angle when using the graphical method.

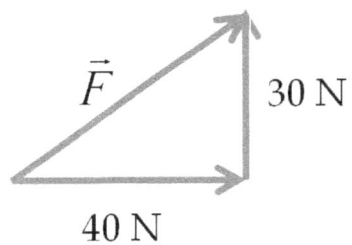

Figure 5.9: Example 5.3.1. Graphical addition of forces.

(b) **Analytic method:** First, we choose Cartesian axes and figure out the Cartesian components of the given forces. The components are easy to figure out if you place origin of the coordinate system at the tails of the forces as shown in Fig. 5.10. Then, we add the x components of the forces separately from their y and z-components to obtain the x, y and z-components of the resultant force. The magnitude and direction of the resultant force are then calculated from its Cartesian components. In the present situation, we have the z-components all zero since the given forces fall in one plane. Therefore, we have only the x and y-components

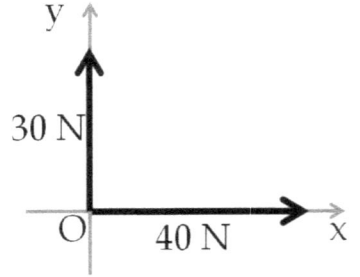

Figure 5.10: Example 5.3.1. Forces with tails at the origin.

to work with. A table usually helps in organizing these calculations as shown below.

Force	x-comp (N)	y-comp (N)	z-comp (N)
$\vec{F}_1$	40	0	0
$\vec{F}_2$	0	30	0
Resultant, $\vec{F}$	40	30	0

Therefore, the magnitude of the force is

$$F = \sqrt{F_x^2 + F_y^2} = \sqrt{40^2 + 30^2} = 50 \text{ N},$$

and the direction given as the angle from the positive x-axis is

$$\theta = \arctan\left(\frac{F_y}{F_x}\right) = \arctan\left(\frac{30}{40}\right) = 37°,$$

which is in general agreement with the result from the geometrical method given in part (a). It is clear that the geometrical approach gives the sum vector directly, while if you follow the analytic approach, you will get components from which you need to construct the magnitude and direction. Most of the time we will work using the analytic approach since it is usually easier to do and does not require us to draw vectors to scale.

5.3.2 Component of a Force in a Given Direction

The magnitude of the component of a force in a given direction tells us the effect of force on the motion in that direction.

Geometrically: To obtain geometrically the component of a force in a given direction, we draw a projection of the force on a line in that direction. The length of the projection is equal to the component of the force in the given direction.

Analytically: Let direction of interest have a unit vector $\hat{u}$ in that direction. The unit vector $\hat{u}$ may be $\hat{u}_x$, $\hat{u}_y$, $\hat{u}_z$, or some other unit vector. The component of force $\vec{F}$ in the direction of $\hat{u}$ is given by the dot product of $\vec{F}$ and the unit vector $\hat{u}$.

$$\text{The component of } \vec{F} \text{ in the direction of } \hat{u} = \vec{F} \cdot \hat{u}. \quad (5.3)$$

Example 5.3.2. Component of weight along an incline. The weight of an object of mass m has magnitude $W = mg$ and direction towards the center of Earth. The direction locally can be drawn vertically down. A cart of mass m is placed on an incline as shown in Fig. 5.11. What is the component of weight vector along the incline?

5.3. FORCES AS VECTORS

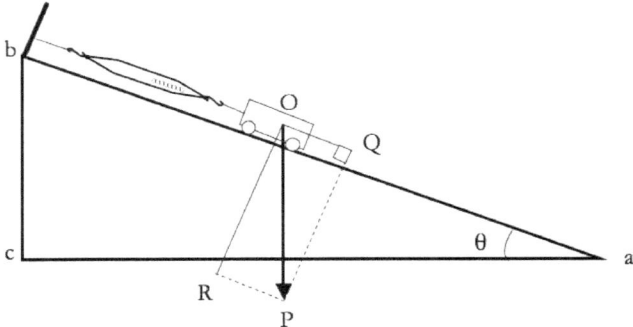

Figure 5.11: A cart on an inclined plane. The component of weight parallel to the inclined plane ab is equal to OQ with $OQ/OP = bc/ab$.

Solution. We will show the calculation both geometrically and analytically.

Geometrical approach: We seek the component of weight parallel to the inclined plane (Fig. 5.11). The weight W of the cart is pointed downward and is represented by arrow $\overrightarrow{OP}$ with length $OP = W$. The projection of $\overrightarrow{OP}$ on line ab is the component of weight parallel to the inclined plane. By drawing the projection, we see that the length OQ is equal to the component of weight we seek. How do we find the length of OQ in terms of the known length OP? We can find the ratio OQ/OP by looking at similar triangles $\triangle POQ$ and $\triangle abc$. Alternately we can use trigonometry. First, let us work out the similar triangle method. The triangles $\triangle POQ$ and $\triangle abc$ are similar triangles since $\angle POQ = \angle abc$, $\angle PQO = \angle acb$, and $\angle OPQ = \angle bac$. Therefore their sides are proportional also.

$$\frac{OQ}{OP} = \frac{bc}{ab} = \sin\theta. \tag{5.4}$$

The effect of the weight on the motion of the cart is reduced by a factor of $\sin\theta$, where θ is the inclination angle. To test this conclusion, we attach a spring balance to prevent the cart from sliding. The reading on the spring balance will be equal to the force given by OQ, i.e. $W\sin\theta$.

We can use trigonometry also. The angle $\angle QOP = 90° - \theta$. Therefore, in the right-angled triangle $\triangle OPQ$, the angle $\angle OPQ = \theta$. Using trigonometry of right-angle triangle we immediately find that

$$\sin\theta = \frac{OQ}{OP} \implies OQ = OP\,\sin\theta,$$

as found in Eq. 5.4.

Analytic approach: We introduce a Cartesian coordinate system so that we can work out the components. Since we seek component along the incline, it makes sense to place one of the Cartesian axes along the incline as shown in Fig. 5.12. Let the positive x-axis point down the incline, the positive y-axis normal to the incline, and the z-axis coming out of page. The weight vector is now not pointed along any of the axes. The vector

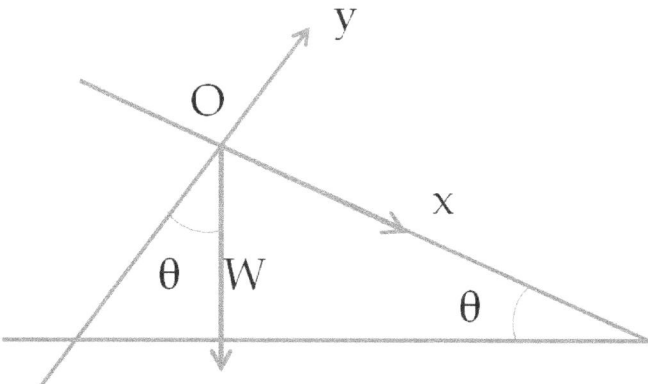

Figure 5.12: Example 5.3.2. Axes for computing component of weight along incline.

is actually pointed in the fourth quadrant of the Oxy plane with angle θ with respect to the negative y-axis, or $90° - \theta$ with respect to the positive x-axis. The x-component, W_x of the vector $\vec{W}$ is therefore easily worked out to give

$$W_x = \vec{W} \cdot \hat{u}_x = W \cos(90° - \theta) = W \sin\theta,$$

which is same as what we found by the geometrical method.

5.4 SOME COMMON FORCES

Experiments show that there are four **fundamental forces** in nature that can explain all other forces.

1. Gravitational force
2. Electromagnetic force
3. Weak nuclear force
4. Strong nuclear force

The weak and strong nuclear forces are short-range forces, becoming significant only at nuclear distances. We will not study the nuclear forces in this course.

The electromagnetic and gravitational forces are long-range forces. The electromagnetic force lumps together electric force and magnetic force since both are related to the same property of matter, i.e. the electric charge.

Besides these fundamental forces, we encounter many other forces in everyday life which are fundamentally caused by gravitational and/or electromagnetic force. For instance, the force of friction between two solid

5.4. SOME COMMON FORCES

surfaces is an average effect of the electric force between the electrons and protons of the two surfaces in contact. Similarly, the force applied by a spring, the force of tension in a string, the force of air resistance, and the force due to surface tension in a fluid, just to name a few of common application, all have their basis in the electric force.

5.4.1 Weight and Gravitational Force

We have already studied the basic aspects of the force called weight. The weight of an object is due to the **gravitation force** between the object and the Earth. In general, gravitational force between any two objects of masses m_1 and m_2 separated by a distance r is given by Newton's law of gravitation, which says that magnitude and direction of the gravitational force are:

Magnitude:

$$F = G_N \frac{m_1 \, m_2}{r^2} \qquad (5.5)$$

where G_N is a universal constant, called the Newton's gravitational constant, and has the following approximate value.

$$G_N = 6.67 \times 10^{-11} \, \frac{\text{N.m}^2}{\text{kg}^2}. \qquad (5.6)$$

Note that G_N is not g, the acceleration due to gravity. I have placed a subscript to G_N to avoid this common confusion. If you are not confused, you may drop the subscript and denote Newton's gravitational constant by G to save time.

For spherical objects, we will show in a later chapter that the distance in Eq. 5.5 would be center-to-center distance. It can be also shown that the second object is inside a spherical shell, then there would be no gravitational force. If the second object is inside a spherical object such as at a point inside the Earth, the mass of the sphere that is within the distance from the center has a net gravitational force; the gravitational force from masses outside cancel out.

Direction:

Force on m_1 is towards the other mass, m_2, and the pair force, i.e. the force on m_2 is towards m_1. The direction says that the force of gravitation between the masses is attractive.

Now, we have learned above that the magnitude of the gravitational force between Earth and an object of mass m, called the weight of the object, is approximately equal to mg, where $g = 9.81$ m/s^2. This turns out to be true only when the object is close to the surface of the Earth.

If an object is not near the surface of the Earth, for instance, if the distance of the object from the surface of the Earth is comparable to or

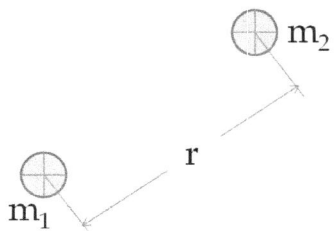

Figure 5.13: Two masses separated by a distance r have an attractive gravitational force between them.

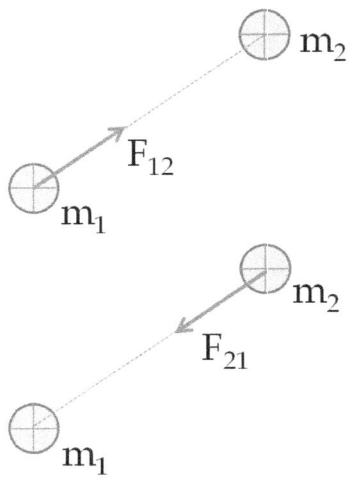

Figure 5.14: The direction of the gravitational force between two masses depends on the body the force acts. The force on 1 by 2, indicated as F_{12} is towards 2. The force on 2 by 1, indicated as F_{21} is towards 1.

exceeds the value of the radius of the Earth, the difference between mg and the actual value of the gravitational force between the object and Earth becomes significant and you cannot say that the weight is mg.

You can address this discrepancy by saying the "value of g is different than 9.81 m/s^2 when you are not near the surface of the Earth" and using a different value of g that would be appropriate for the situation. Although, this viewpoint is acceptable, one rather abandons mg in these situations and uses Eq. 5.5 for the force between Earth and the object.

Demonstration that mg is due to gravitational force

We now demonstrate the gravitational force between Earth and an object at the surface of Earth is equal to weight by applying the formula given in Eq. 5.5 to Earth/Object pair. Let us denote the mass of Earth by M_E and the radius of Earth by R_E. Then, the distance from the center of Earth (since Earth is almost spherical) to an object at the surface is equal to the radius of Earth. Therefore, we set $r = R_E$ in Eq. 5.5. This gives us the following magnitude for the gravitational force between Earth and an object of mass m located at the surface of Earth.

$$F = G_N \frac{M_E\, m}{R_E^2},$$

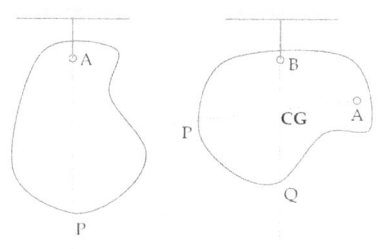

Figure 5.15: The force of gravity of Earth acts at all particles of a body. For the purposes of torque on the body, the force of gravity can be considered to act at a special point called **center of gravity (CG)**. The CG of a body can be located by hanging the body from different points (A and B in the figure) and locating the point where the vertical lines cross.

which can be rearranged so that we write it as m times something that is independent of m so that we can compare with mg formula,

$$W = m \left(G_N \frac{M_E}{R_E^2} \right).$$

The question before us is: Is this equal to mg with $g = 9.81$ m/s^2? Let us compare the quantity within the parenthesis to the standard value of g by plugging in the standard values of $M_E = 5.97 \times 10^{24}$ kg and $R_E = 6.37 \times 10^6$ on the right side.

$$G_N \frac{M_E}{R_E^2} = 6.67 \times 10^{-11} \frac{5.97 \times 10^{24}}{(6.37 \times 10^6)^2} = 9.81 \text{ m/s}^2,$$

which is identical to the value of g. This confirms that mg is equal to the magnitude of the gravitational force of Earth on an object at the surface of Earth.

Example 5.4.1. What is the gravitational force between Earth and a satellite of mass m at a distance $2R_E$ from the center of Earth?

Solution. At $r = R_E$, we found that the gravitational force was equal to mg. The gravitational force given in Eq. 5.5 decreases as distance squared. The distance here is 2 times R_E. Therefore, the force will be $\frac{1}{4}$ of the force at $r = R_E$. Hence, the gravitational force on the satellite will be equal to $\frac{1}{4}mg$, where $g = 9.81$ m/s^2.

5.4.2 Tension Force and Hooke's Law

A wire or string can be used to create forces between two objects. Suppose you tie one end of a wire to a post and pull at the other end (Fig. 5.16). When the wire is loose, there is no force between the post and the person. But, when the wire is taut, the two bodies get linked by the tension force in the wire. The tension force acts on the post in the direction away from the post and towards the person, and it acts on the person in the opposite direction.

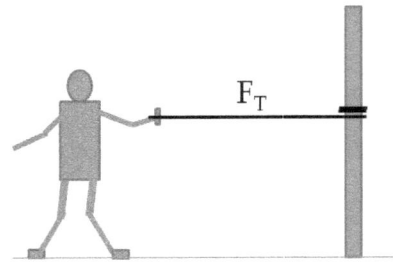

Figure 5.16: A tension force acts between the post and the person.

The magnitude of the tension force is proportional to the amount of stretch. One way to find the relation between stretching and the corresponding tension is to hang various masses at the bottom of the wire while fixing the top end of the wire to a ceiling as illustrated in Fig. 5.17. In Fig. 5.18 we show a schematic view of the way tension F_T varies with the stretching of wires.

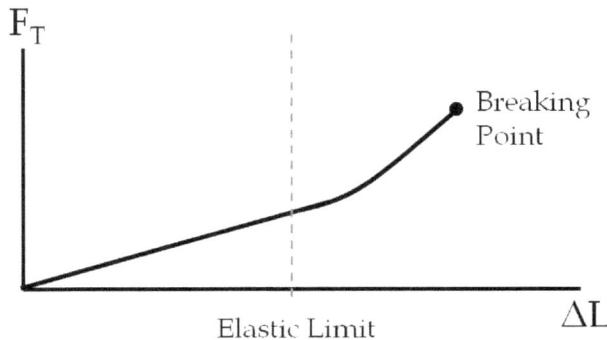

Figure 5.18: Tension in the wire as a function of stretching ΔL.

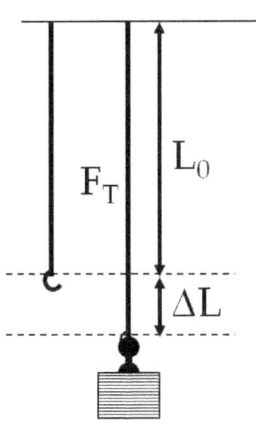

Figure 5.17: Stretching a wire by hanging masses.

If the stretching is less than the elastic limit of the wire, the wire goes back to its original length once you remove the tension. Beyond the elastic limit, the wire gets permanently changed. Within the elastic limit, the change in length and the tension force are directly proportional and we can write the tension force as

$$F_T = k\Delta L, \qquad (5.7)$$

where the proportionality constant k depends on the chemical makeup of the material, and the length and area of cross-section of the wire. This observation was first made by Robert Hooke (1635-1703), a contemporary of Sir Isaac Newton, and is often called **Hooke's law**, which states that "Ut tensio sic uis," in Latin, meaning "As the extension, so the force." Every elastic object obeys Hooke's law within its elastic limit.

The main point of Hooke's law is that if a system, whether a wire, a spring, or any other material body, is deformed from some stable structure, a restoring force develops which tends to bring the system back to the original stable point. In the next section, we will apply Hooke's law to spring force.

This is a commonly encountered behavior of most materials. The basic mechanism of the development of a tension force may be understood in terms of bonding of molecules. When the wire is not pulled, the molecules are positioned at certain distance from each other that balances the attractive and repulsive forces. As the wire is pulled, the distances between molecules increase, increasing both the attractive and repulsive forces making a new equilibrium between the forces on molecules. As long as the external pull is present, the new equilibrium is maintained. When the external pull is removed, the attractive force between the molecules is no longer balanced by the external pull and brings the wire to the original length.

In most problems concerning the tension force, we usually do not know k or ΔL. Therefore, we leave the tension force as F_T or T rather than express it in the corresponding $k\Delta L$ form.

Tension Force:
Magnitude: T, varies
Direction: Along the string, away from the body
Where: At the point(s) string tied or pressed to the body

The tension in a continuous wire is same throughout the wire if the wire is only pulling on the two objects connected at the ends and does not pressing on any other object. For instance, when a mass m is hung from the ceiling by using a wire, the tension in the entire wire is same. However, if the wire goes over a pulley, a ring, or a hook, then the tension now connects two sets of bodies on the two sides of the body in the middle. For instance, in Fig. 5.19, there are tension forces between the ring and the block m and the ring and the support S. The tensions in the two sides of the ring may be different: $T_1 \neq T_2$. Furthermore, while there is one tension force acting on mass m and another one acting on the support S, both tension forces act on the ring R. The net tension force on the ring is the vector sum of the two tension forces acting on it as shown in Fig. 5.19.

5.4.3 Spring force and Hooke's Law

When a force is applied on both ends of a spring the distance between coils can change causing a restoring force in the spring that acts to restore the original length of the spring. This force is called **spring force** which acts on the two objects linked by the spring. If the spring is extended, then spring forces on the two objects are such that the two objects are attracted to each other (Fig. 5.20). On the other hand, if spring is compressed, spring forces on the two objects are such that the two objects are repelled from each other.

Spring Force:
Magnitude: $k\Delta L$
Direction: Towards the equilibrium point
Where: At the point spring attached to the body

Spring force is similar to the tension force we have studied in the sense that the force is proportional to the change in length from a reference length as long as the extension or compression is not so large that the spring is permanently deformed. That is, Spring force obeys Hooke's law:

If a spring is either extended or compressed by an amount ΔL, the magnitude of the **spring force** on either object connected to the spring

5.4. SOME COMMON FORCES

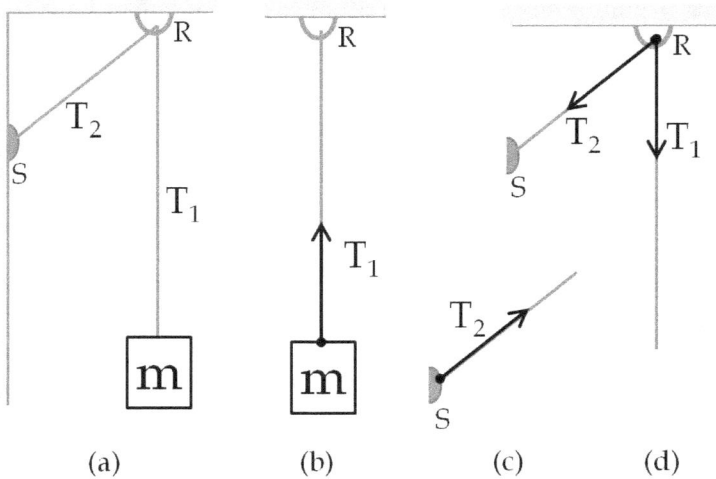

Figure 5.19: The tension in the string acts on three objects, the mass m, the ring R and the support S. The tension force between mass m and ring R may be different from the tension force between the ring and the support R although the same wire connects all three objects. In (b) the tension force on mass m is shown, in (c) the tension force on the support S is shown, and in (d) the tension forces on ring R is shown. The net tension force on ring R is the vector sum of the two tensions.

is

$$|\vec{F}_{spring}| = k\Delta L, \tag{5.8}$$

where the proportionality constant k is called the spring constant. Note that when the spring is neither extended or compressed, the spring force is zero.

The directions of the spring forces on the two objects on the two sides of the spring depends on whether the spring is extended or compressed as explained above. When the spring is extended compared to its natural length, the block and the support attract each other, i.e. the force on each is pointed towards the other as shown in Fig. 5.20. When the spring is compressed compared to the natural length, then the block and the support repel each other, i.e. the forces on them are pointed away from each other.

5.4.4 Normal Forces at a Contact Surface

When you press a solid body against another solid body, the bonds among molecules at the interface of the two bodies are compressed. These bonds act as small springs. The compressed "molecular springs" in the two bodies generate a repulsive force on each other. This repulsive force is called the normal force.

Normal force acts on the two bodies in contact through the surface of contact of the two bodies. The magnitude of the normal force depends in a complicated way on the degree of deformation of the bonds near the

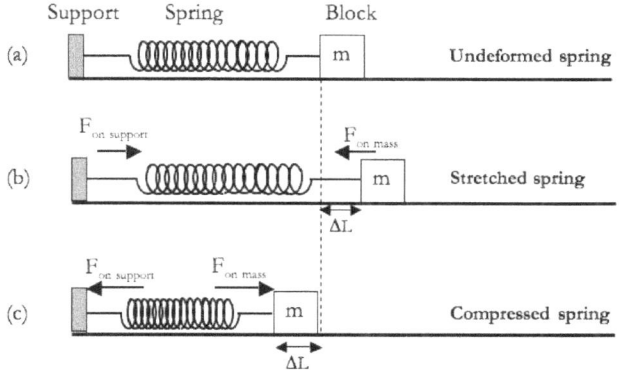

Figure 5.20: Spring force between block and support. The spring force has the magnitude $F = k\Delta L$ and acts on the block and the support. The directions of the forces on the block and support are such that the spring forces tend to restore the original length of the spring.

surface each body and the total area of contact. The force will be greater in the area where the surfaces are pressed together more and less where the surfaces are pressed less. If the surfaces are in contact but not pressing against each other, then the normal force will be zero.

Normal Force:

Magnitude: varies; symbol F_N or N

Direction: Towards the body, perpendicular to the surface of contact

Where: At points of the contact surface

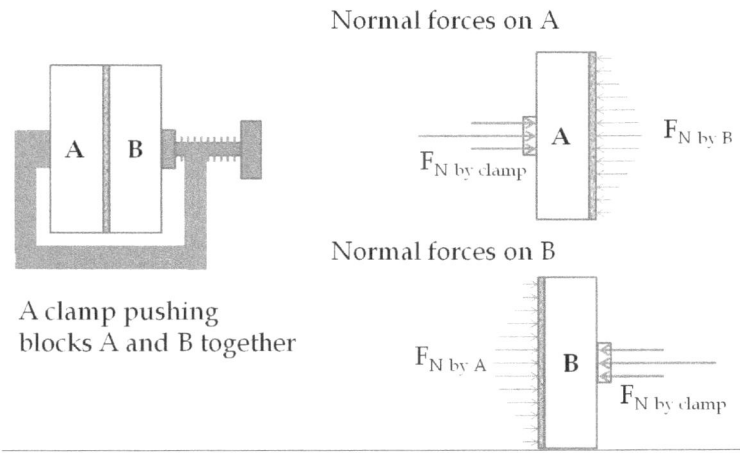

Figure 5.21: The normal forces between blocks A and B is created by pushing the blocks together. The forces by the clamp itself is transmitted through the surfaces between the clamp and the blocks. The normal force is distributed along the surface of contact. Normal force is higher where the surfaces are more pressed together and smaller where they are less pressed together.

Normal force is distributive in nature similar to the force of gravity since it does not act any one point. Despite this fact we will place a net normal force between two bodies to act at one point on the body for the purposes of calculation of torque from the normal force. Just as net torque by gravity of Earth on a body is same as if the net force of gravity acts on the center of gravity, we can envision a net normal force acting at some point of the body giving us the same net torque as would be obtained by calculating the net torque from the distributed actual normal forces

5.4. SOME COMMON FORCES

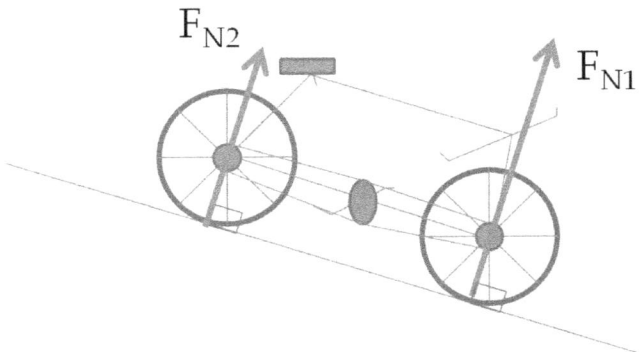

Figure 5.22: The normal forces on the front and back wheels on a bicycle on an incline are not equal.

between the two bodies. For this purpose, the net normal force may or may not pass through the center of gravity of the body. The net normal force on the body will act on that point of the body that is more pressed than other points. For instance, if a bicycle is on an incline with the front wheel down the incline and back wheel up the incline, then the normal force on the front wheel will be larger than the normal force on the back wheel.

Normal force is a reaction force. For a normal force to develop between two bodies, the two bodies need to be subjected to outside forces that are tending to push the bodies into each other.

Unlike weight, which has a formula mg near Earth, and spring force, which has a formula $k\Delta L$, there is no particular formula for a normal force. The magnitude of the normal force is determined by the external forces acting on the body that generate the normal force and the conditions of the motion of the two solid bodies.

Many formulas of normal force are found in elementary textbooks which are specific to the particular problem at hand and should not be memorized. In particular, one often encounters $F_N = mg$ and $F_N = mg\cos\theta$, neither of which are always applicable. The formula for a normal force in a particular situation must be "discovered" by setting up Newton's laws of motion in that situation. In this chapter and the next, we will work out problems where the process of setting up Newton's laws will be explained further.

The direction of the normal force on each body can be deduced from its repulsive nature. Thus, if body 1 is pressed against body 2, the normal force on body 1 will be pointed away from body 2. Similarly, the normal force on body 2 will be pointed away from body 1.

As an illustration of the normal force consider a book on a table (Fig. 5.23). We know that the weight of the book is supported by the table. But what is really going on at the interface between the book and the

WARNING:
$F_N \neq mg$
$F_N \neq mg\ \cos\theta$
F_N is determined by other forces and acceleration of the body.

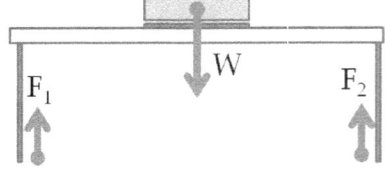

Figure 5.23: The downward weight (W) of the book and upward support forces $\vec{F}_1$ and $\vec{F}_2$ on the legs of the table compress the table and book together.

table?

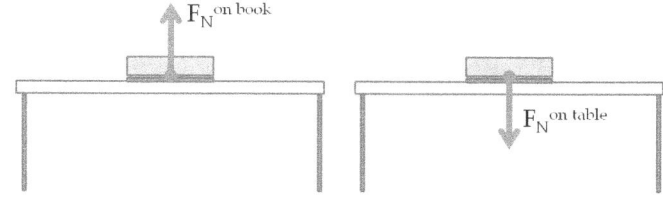

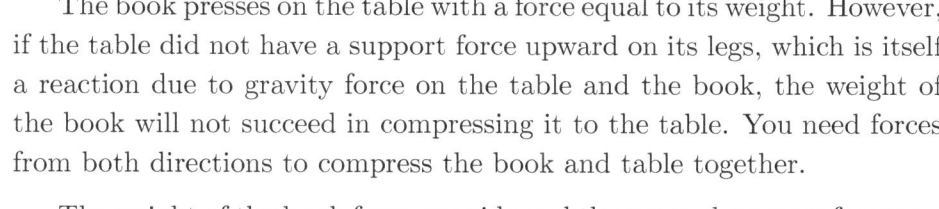

Figure 5.24: Normal force between book and table. When acting on the book, the force is pointed towards the book and away from the table, and when acting on the table it is pointed towards the table and away from the book.

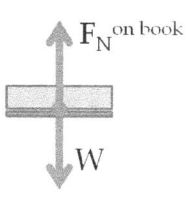

Figure 5.25: Forces on the book.

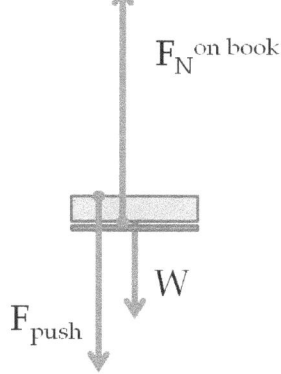

Figure 5.26: Your push on the book acts with the weight to generate a larger normal force. In order for torque to be also balanced, net normal force now acts at a point to the left of the center of gravity.

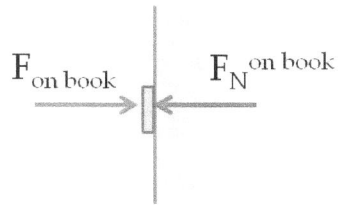

Figure 5.27: Your push on the book against the wall generates a normal force between the book and the wall. Weight and friction on the book are not shown in this figure to keep it simple.

The book presses on the table with a force equal to its weight. However, if the table did not have a support force upward on its legs, which is itself a reaction due to gravity force on the table and the book, the weight of the book will not succeed in compressing it to the table. You need forces from both directions to compress the book and table together.

The weight of the book from one side and the upward support forces on the legs of the table push the book and table into each other, which compresses the "molecular springs" of both the book and the table. Therefore, a normal force develops that acts on the book and the table (Fig. 5.24).

Now, if you look at the forces on the book, you will discover two forces - the force of gravity from Earth pointed down and a normal force from the table pointed up (Fig. 5.25). The two forces may or may not be equal in magnitude depending upon whether or not the book is in equilibrium as we will see later in this chapter.

If the book is in equilibrium, meaning forces on it are balanced, then the magnitude of the normal force will be equal to the weight of the book. Otherwise, if the book is not in equilibrium, the magnitude of the normal force may be greater that or less than weight.

Book in equilibrium (balanced forces): $F_N = mg$

Book not in equilibrium (unbalanced forces):
$$F_N > mg \quad \text{or} \quad F_N < mg.$$

Now, suppose you apply a force F_{push} on top of the book in the vertical direction (Fig. 5.26). How would the normal force between the book and the table change? The book will press on the table with a force equal to its weight plus F_{push}. The book/table interface will be even more compressed, giving rise to a normal force that would be greater than the normal force when you hadn't applied the force.

As another example, suppose you press the book against a wall at right angle to the wall (Fig. 5.27). What is going on at the interface of the book and the wall? Your force on the book compresses the molecular springs of the book and the wall, which generates a normal force at the interface

5.4. SOME COMMON FORCES

which acts on the book pushing the book away from the wall and also acts on the wall pushing the wall away from the book.

Let $\vec{F}$ be your push force on the book. Now, the normal force between the book and the wall will not be related to the weight of the book but to the push force.

Book in equilibrium (balanced forces): $F_N = F$

Book not in equilibrium (unbalanced forces):

$$F_N > F \quad \text{or} \quad F_N < F.$$

What happens if you press on the book at an angle other than 90° to the wall? The normal force now will be smaller since only a projection of your force is pushing the book's surface into the wall (Fig 5.28). Now, there will also be a force from the book on the wall parallel to the wall. This force is called frictional force which we will study next.

5.4.5 Static Frictional Force

Suppose you try to slide a book resting on a table by applying a horizontal force $\vec{F}$ on the book (Fig. 5.29). What would happen? Recall that due to gravity on the book the book is pressed into the table. As a result, a bonding also develops between the molecules of the two bodies at the interface.

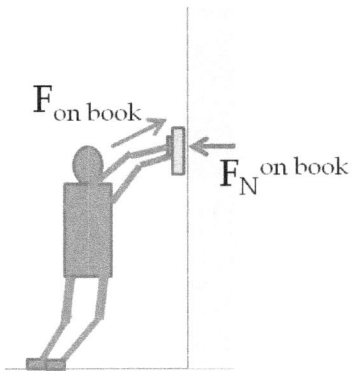

Figure 5.28: Your push on the book at an angle generates normal force and friction between the book and the wall. Weight and friction on the book are not shown in this figure to keep it simple. Note normal force is 90° to the surface of contact.

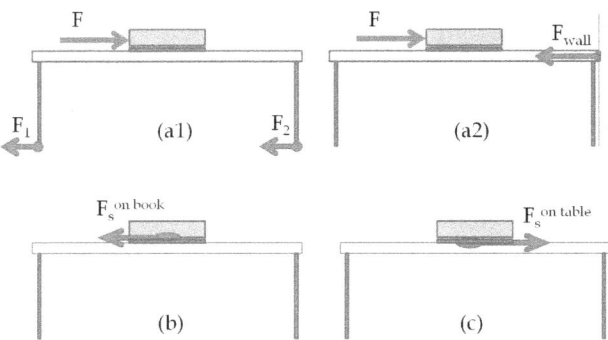

Figure 5.29: Static friction generated as a response to external forces horizontal to the surface of contact. (a1) Force $\vec{F}$ applied on the book is balanced by constraining forces at the floor of the legs of the table. (a2) Force $\vec{F}$ applied on the book is balanced by normal force by the wall. (b) Static friction force acting on the book. (c) Static friction force acting on the table. Other forces on the book not shown here: weight and normal.

Due to this bonding of the book to the table, when we apply a horizontal force on the book, the book attempts to pull the table along with it. That is, the force applied on the book is transmitted to the interface between the book and the table such that the book now applies a horizontal force on the table that is equal to the applied horizontal force.

If the table is free to move, then the horizontal force from the book at the interface will cause the table to accelerate. However, if the table cannot move, either because it is held against something else, or the legs are bonded to the floor by friction or by some other means, the table will react to the applied force with a generation of a frictional force at the surface between the book and the table. Thus, if an external force is applied parallel to the interface, a static frictional force will develop if there is also a normal force pressing the surface together or the two surfaces are bonded together.

Static frictional force acts on the two bodies in the tangential plane of the contact surface such that the force on frictional force on the body with the applied force on it has the direction of the frictional force opposite to that of the direction of the applied force. The direction of the frictional force on the other body is in the direction of the applied force. In our example above, the direction of the friction force on the table is in the direction of the applied force on the book, and the direction of the frictional force on the book is opposite of the direction of the applied force on it.

The magnitude of the frictional force is determined similarly to the way the magnitude of the normal force is determined. Since friction force is a reaction force, the magnitude depends on the applied force that helps generate frictional force. Therefore, the magnitude of the static friction force depends on the magnitude of the external applied force and whether or not the two bodies are in equilibrium.

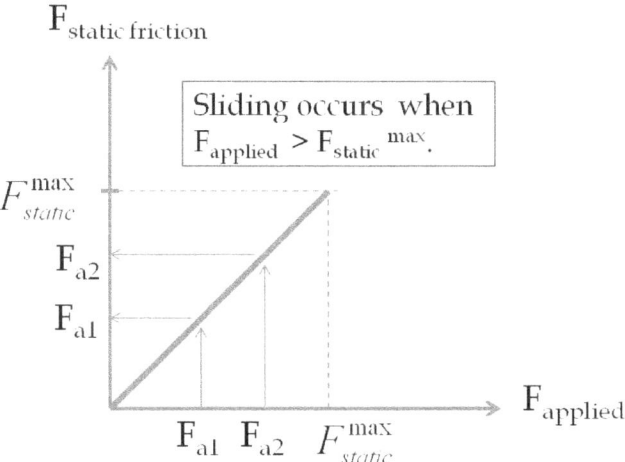

Figure 5.30: The magnitude of static friction depends on the applied force since the frictional force matches the external applied force as long as the later is less than F_{static}^{max}.

As a general rule, the magnitude of static frictional force will increase with the increase in applied force up to a limit when the static bonds between the two bodies break up and the two bodies start to slide past each other. Thus, static friction force can have any magnitude up to

5.4. SOME COMMON FORCES

this limit, which is called the **maximum static friction force**, $F_s^{\max}$, as shown schematically in Fig. 5.30. We are, of course, free to apply any amount of force on the bodies, but the static friction force cannot be more than $F_s^{\max}$. When the applied force is less than the maximum static friction force, i.e. when $F_{\text{appl}} < F_s^{\max}$, such as the values F_{a1} and F_{a2} shown in Fig. 5.30, the static friction force matches the applied force and there is no sliding.

The magnitude of the force required to overcome the static bonds depends on the nature of the two surfaces and the degree to which the two surfaces are pushed into each other. This says that $F_s^{\max}$ can be stated in terms of magnitude of the normal force, which is a measure of the extent to which the two surfaces are pushed into each other.

$$\boxed{F_s^{\max} = \mu_s\, F_N,} \qquad (5.9)$$

where μ_s is called the coefficient of static friction. The constant μ_s depends on the area of contact and the physical nature of the two surfaces in contact. Note that Eq. 5.9 is not a vector relation since friction and normal forces act perpendicularly to each other. Rather, Eq. 5.9 relates only the magnitudes of the maximum static friction and the normal force.

Static Friction Force:
Magnitude: varies up to a maximum magnitude; $F_s \leq F_s^{\max}$; Inequality law.
Direction: In the tangent plane of the surface of contact to be determined from the dynamical conditions of the body.

WARNING:
$F_s \neq \mu_s F_N$
$F_s \neq \mu_s mg$

Example 5.4.2. Maximum static friction force - 1

A 1.4-kg book rests on a horizontal wooden table. If the coefficient of static friction force is 0.8, what would be the horizontal force needed to overcome the maximum static friction between the book and the table?

Solution. From the discussion above, we know that we need to overcome static friction force to make the book slide. So, the minimum force we need must be equal to the maximum static friction on the book by the table. According to Eq. 5.9, the magnitude of the maximum static friction force is equal to the product of the coefficient of static friction force and the magnitude of the normal force. The normal force F_N is equal to the force with which the book is pressing on the table vertically. Here, the book is pushing on the table due to its weight. Therefore, the magnitude of the normal force in the present situation must be equal to the weight of the book.

$$F_N = 1.4 \text{ kg} \times 9.81 \text{ m/s}^2 = 13.7 \text{ N}.$$

Hence, the magnitude of the minimum horizontal force needed to slide the book

$$F_{\min} = F_{\text{static}}^{\text{maximum}} = \mu_s\, F_N = 0.8 \times 13.7 = 11.0 \text{ N}.$$

Example 5.4.3. Maximum static friction force - 2

A 1.4-kg book rests on a horizontal wooden table. A 2-kg steel block is placed upon the book. If the coefficient of static friction force between the table and the book is 0.8, what would be the minimum horizontal force needed to overcome the static friction between the table and the book?

Solution. Note that the force with which the book is pressing vertically on the table is larger in this example than was the case in the last example. The net force pressing the book into the table here is equal to the total weight of 3.4 kg, which is equal to 33.4 N. Therefore, the magnitude of the minimum horizontal force needed to slide the book now is

$$F_{\min} = 0.8 \times 33.4 = 26.7 \text{ N}.$$

Example 5.4.4. Maximum static friction force - 3

A 1.4-kg book rests on a horizontal wooden table. You push on it with a force of 20-N in the direction 30° below the horizontal direction. If the coefficient of static friction force is 0.8, will the book slide?

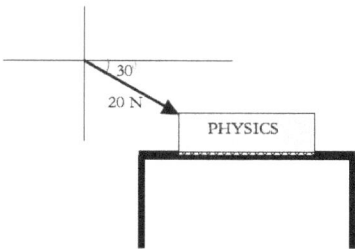

Solution. In the present example, some of the force with which the book is pressing comes from the vertical component of the applied force, and the rest from the weight of the book, which acts vertically. Therefore, the magnitude of the normal force F_N by the table on the book will be

$$F_N = W + F\ sin\theta = 13.7 + 20\ \sin\ 30° = 23.7 \text{ N}.$$

The minimum horizontal force required to slide the book must equal the maximum static friction force:

$$\text{Minimum horizontal force needed } = 0.8 \times 23.7 = 18.96 \text{ N}.$$

Now, the question is: Does the horizontal component of the applied force exceed this value?

$$\text{Horizontal component of force} = F\ \cos\theta = 20\cos\ 30° = 17.32 \text{ N},$$

which is less than 18.96 N required for overcoming the maximum static frictional force. Therefore, the book will not slide.

5.4.6 Kinetic or Sliding Frictional Force

What happens when you apply a horizontal force larger than the maximum static friction force? Once you overcome the maximum static friction force, the two bodies in contact will start to slide relative to each other, and you will find that it takes a smaller force than the maximum static friction force to maintain a steady constant velocity motion. There is still frictional forces on the sliding objects that oppose the sliding motion, called **kinetic**

5.4. SOME COMMON FORCES

or **sliding friction force**. The magnitude F of the force needed to slide an object in a uniform motion of constant velocity is usually proportional to the magnitude F_N of the normal force.

$$F_k = \mu_k \, F_N, \quad \text{(directed opposite to the sliding motion)}, \qquad (5.10)$$

where the proportionality constant μ_k is called the **coefficient of kinetic friction**. The constant μ_k depends on the area of contact between the surfaces and the physical nature of the surface of the two bodies such the roughness of the surface.

5.4.7 Rolling Friction

The static and kinetic frictions are caused by the interlocking of microscopically uneven surfaces of two solid bodies in contact which vary directly with the area of contact between the two bodies. When a circular hoop or a spherical ball rolls on a flat surface, the surface of contact is much reduced (Fig. 5.31). We say that a body is rolling over the other body without sliding if the point of contact of the bodies is momentarily at rest. In a sense, the rolling motion is due to a static friction between the bodies since the point of contact between the two bodies does not move in a rolling motion. Consequently, the rolling friction is considerably less

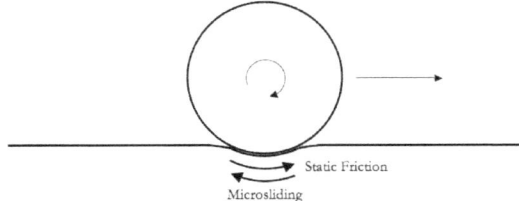

Figure 5.31: Schematic drawing of the two major forces involved in the rolling friction. If the static friction is not enough then the ball will slide instead of rolling. Both the rolling ball and the rolling surface are deformed at the point of contact.

than the sliding friction.

The contributing factors to rolling resistance are the deformation of the surface of contact due to weight of the wheel, adhesion of the two surfaces and relative micro-sliding between the surfaces. For instance, a surface that is easily deformable under pressure, such as rubber or sand will give a larger rolling resistance than a surface which is more resistant to deformation, such as steel and concrete. Thus, a car rolling on sand will slow down faster than a car on a concrete surface.

Since, a rolling motion is just a reduced static friction, we expect the rolling friction to depend upon the degree to which the bodies are dug into each other at the point of contact. Therefore, we also write the magnitude of the rolling friction F_r in terms of the magnitude of the normal force between the two as was done for the sliding friction.

$$F_r = k_r \ N, \tag{5.11}$$

where k_r is the coefficient of rolling friction. Rolling friction coefficients are typically $\frac{1}{100}$th of the sliding friction between the same two surfaces. Thus, between an iron block and an iron plate, the sliding friction coefficient is approximately 0.02, and the rolling friction of a wheel of same mass rolling on an iron plate is approximately 0.0002.

A wheel will roll as long as the horizontal force is greater than the rolling friction force needed to get the wheel rolling but less than the maximum static friction allowed between the two surfaces. If the force exceeds the maximum static friction force, then the wheel will start to slide. In winter one has to be careful in driving over frozen roads since static friction coefficient between rubber tires and ice is considerably less than that between tires and concrete, and it is easily exceeded if wheels are turning rapidly. The coefficient of static friction between rubber tires and concrete road is approximately $\mu_s = 1.7$ and the rolling friction coefficient is approximately $\mu_r = 0.01$. Therefore, the minimum force needed to get a 3000-kg car rolling on a horizontal concrete road is $\mu_r \ N = \mu_r \ m \ g = 294$ N, and the maximum force that can be exerted on the car before it starts to slide is $\mu_s \ N = \mu_s \ m \ g = 50,031$ N. At the maximum force on the car, it will have an acceleration of $1.7 \ g = 16.7$ m/s^2. If you try to accelerate the car at more acceleration than $1.7 \ g$, the car will slide.

Although rolling of a wheel avoids the sliding friction at the surface, but at the axle, sliding friction is present. Ball bearings between the axle and the wheel reduce friction by making use of the rolling friction of the ball bearings rather than sliding of the wheel against the axle.

5.4.8 Viscous drag

When a solid moves through a fluid as when a baseball moves through the air or car travels in air or a fish swims in water, it encounters a friction due to the viscosity of the fluid. The **fluid resistance** or the **viscous drag**, as it is commonly called, increases with the speed of the moving body. For slowly moving objects viscous drag ($\vec{F}_d$) is proportional to speed and acts in the opposite direction to the velocity.

$$\boxed{\vec{F}_d = -b \ \vec{v}, \quad (b > 0)} \tag{5.12}$$

where b is a constant that depends on the viscosity of the fluid and the area of cross-section of the object perpendicular to the velocity. For objects moving fast the drag force can vary as square of the speed, or even the cube of the speed. Due to the increasing drag force, the cost of incremental speed boost to an object is much higher at higher speeds. It is much more costly to increase the speed of a car from 150 km/h to 160 km/h than

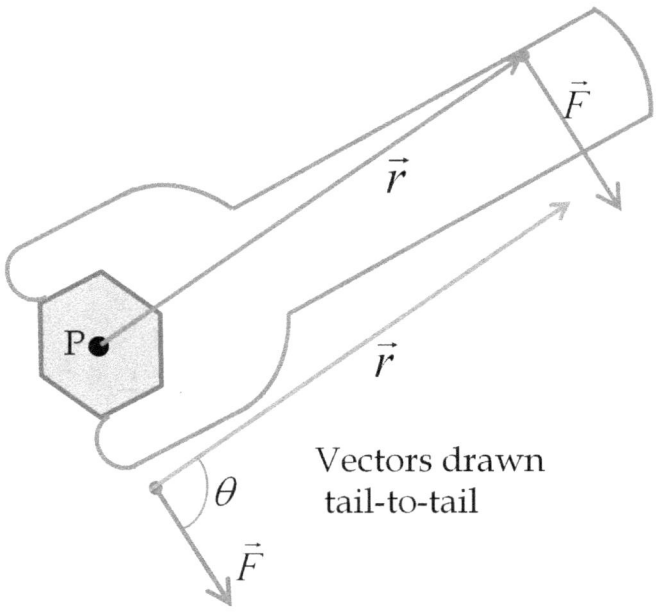

Figure 5.33: Point P about which torque of force $\vec{F}$ requires the displacement vector $\vec{r}$ from P to the point where the force acts. The tail-to-tail diagram helps define the angle between the vectors.

from 15 km/h to 25 km/h for the same additional 10 km/h rise. Drag is responsible for the existence of a safe limit of a terminal speed when you sky dive. When the parachute opens it increases the drag force by increasing the cross-section upon which air resistance can act. At some critical speed, called the terminal speed, v_t, the weight of the person plus the parachute is balanced by the drag force by air.

$$b\, v_t = m\, g \quad \Longrightarrow \quad v_t = \frac{mg}{b}. \tag{5.13}$$

5.5 TORQUE OR MOMENT OF A FORCE

The moment of a force, also called the torque, is related to the ability of a force to rotate a body about some axis. Suppose you want to rotate a bolt by fitting a wrench on the head of the bolt and applying a force on the handle of the wrench as in Fig. 5.32. The line through the center of the bolt is the axis about which the bolt and the wrench together will rotate. We find that the same force applied on the handle at a larger distance from the axis is more effective than when it is applied nearer to the axis. That is why a longer wrench is more effective than a shorter wrench.

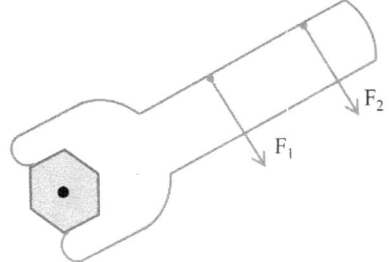

Figure 5.32: Force F_2 is more effective than F_1 in rotating the bolt since it is applied at a greater distance from the axis of rotation.

For rotation, we find that the location of the force on the body is as important as the direction and magnitude of the force. This is taken care of by introducing a quantity called **moment or torque of force** which has the following precise definition.

Let $\vec{r}$ be the displacement vector from some point P in space to the point in the body where we apply a force $\vec{F}$ as shown in Fig. 5.33. The moment or torque of the force $\vec{\tau}$ about point P is defined as the cross product of the displacement vector and the force. Note that the names moment and torque mean the same thing and are calculated identically, but the usage usually differs - moments are more commonly used in static situations and torque for both static and dynamic settings.

$$\vec{\tau} = \vec{r} \times \vec{F}. \tag{5.14}$$

The unit of torque in the SI system of units is N.m. If the point P, about which the torque is calculated, is fixed during the rotation, then, we say that P is the pivot point of rotation. The definition of the cross product gives us the rules for the magnitude and the direction of torque vector $\vec{\tau}$. We will recall those rules and apply them to the cross product for torque below.

The geometric picture:

To write the magnitude and direction using geometric picture of vectors, we need the angle θ between vectors $\vec{r}$ and $\vec{F}$ when they are drawn tail-to-tail. Then,

Magnitude of torque:

$$|\vec{\tau}| = |\vec{r}|\,|\vec{F}|\sin\theta. \tag{5.15}$$

Direction of torque:

Use the right-hand-rule as given in the chapter on vectors. Briefly, hold your right hand such that the thumb points in the direction of $\vec{r}$ and any of the other fingers points in the direction of $\vec{F}$, then the torque will be perpendicular to the two vectors and will be coming out of the palm. There are other ways of doing the right-hand rule also. For instance, if you point $\vec{r}$ in the direction of the positive x-axis, and $\vec{F}$ in the xy-plane, then $\vec{\tau}$ will be in the direction of the positive z-axis of a right-handed Cartesian coordinate system.

The lever arm picture:

The magnitude of a torque given in Eq. 5.15 gives rise to popularly used term called the **lever arm**. Suppose you draw the displacement and force vectors in their real settings and not move them for tail-to-tail drawing as illustrated in Fig. 5.34. Now, we drop a perpendicular from point P about which we want torque to the line of action of the force. The perpendicular distance $r_\perp$ is equal to $|\vec{r}|\sin\theta$ and is called the lever arm of the force about P. The larger the lever arm the larger the torque for the same force.

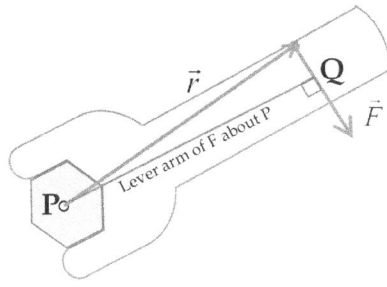

Figure 5.34: The lever arm of force $\vec{F}$ about point P is the distance PQ of the perpendicular from P to the line of the force.

$$\text{Lever arm,}\ r_\perp = |\vec{r}|\sin\theta. \tag{5.16}$$

Magnitude of torque in terms of lever arm:

$$|\vec{\tau}| = r_\perp |\vec{F}|. \tag{5.17}$$

5.5. TORQUE OR MOMENT OF A FORCE

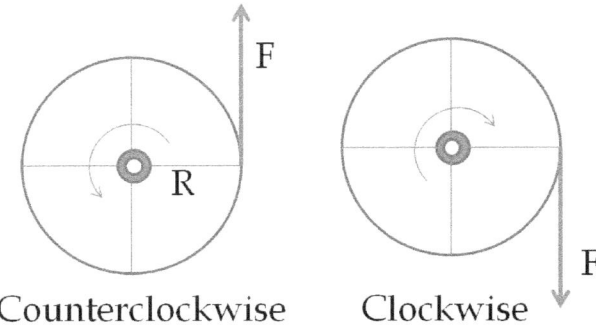

Figure 5.35: The "sense of rotation" for a torque. The torque on the left figure will tend to rotate the body counter-clockwise about an axis through the center if observed from above as in this figure. The body on the right has a clock-wise sense of rotation. Torque vector for counter-clockwise would be in the direction coming-out-of-page and that of the clockwise sense will be in-the-page.

The analytic picture:

The cross product in Eq. 5.14 can also be calculated analytically. First, we choose a Cartesian coordinate system and express the two vectors $\vec{r}$ and $\vec{F}$ into components. The cross product is usually done by organizing the components in a determinant. We summarize these steps now.

1. Choose a Cartesian coordinate system.

2. Write out the vectors in component forms

$$\vec{r} = x\hat{u}_x + y\hat{u}_y + z\hat{u}_z$$
$$\vec{F} = F_x\hat{u}_x + F_y\hat{u}_y + F_z\hat{u}_z$$

3. Expand the determinant form of the cross product.

$$\vec{r} \times \vec{F} = \begin{vmatrix} \hat{u}_x & \hat{u}_y & \hat{u}_z \\ x & y & z \\ F_x & F_y & F_z \end{vmatrix}$$

The sense of rotation and torque

We can assign a **sense of rotation** with a torque whether of not the object rotates. Suppose we place the displacement vector of where the force acts from the reference point, $\vec{r}$, and the force, $\vec{F}$, in the xy-plane. Then, the torque will be pointed towards either the positive z-axis or the negative z-axis, which are also referred to as "out-of-page" and "into-the-page" as shown in Fig. 5.35. If the torque is pointed out-of-page, i.e., towards the positive z-axis, then the torque will tend to rotate the object in the counter-clockwise fashion as observed from the positive z-axis. We say that the torque has a counter-clockwise "sense of rotation". On the other hand, if torque is pointed towards the negative z-axis, the sense of rotation will be clockwise.

Example 5.5.1. A peg goes through a hole in a metal bar at one end so that the bar can rotate about the peg. A force of 10 N is applied on the bar perpendicular to the bar a distance 20 cm from the peg as shown in the figure. What is the torque about the peg?

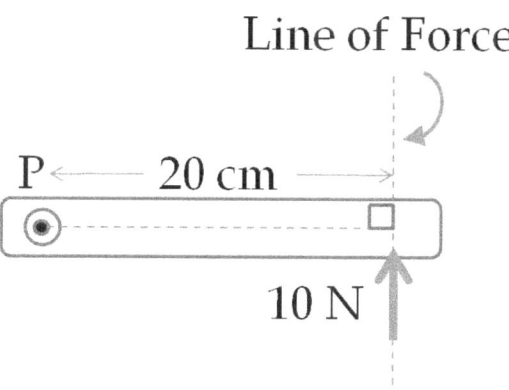

Solution. The lever arm is easy to work out here since the force is perpendicular to the bar. The projection from peg to the line of the force gives the lever arm to be 20 cm, which we convert to meter before using in the calculations. Therefore, the torque has the following magnitude

$$\text{Magnitude:} \quad \tau = (0.2)(10) = 2 \text{ N.m.}$$

The direction of the torque is obtained by the right-had rule. In the given figure, the direction is coming out of the page.

Example 5.5.2. A peg goes through a hole in a metal bar at one end so that the bar can rotate about the peg. A force of 10 N pulls the bar away from the peg as shown in the figure. What is the torque about the peg?

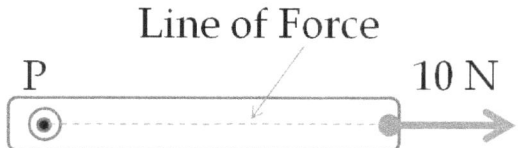

Solution. In this configuration, the line of force goes through point P located at the peg. Therefore, the lever arm is zero and so is the torque.

Example 5.5.3. The two examples presented above can be written more generally. From the definition of the torque, it is clear that torque will be zero if the angle between $\vec{r}$ and $\vec{F}$ is zero or 180°. This says that if we think of the force as a sum of two forces, one parallel to $\vec{r}$ and another perpendicular to this direction, then only the perpendicular part of the force will give a non-zero contribution to the torque. That is, if we write the force vector as

$$\vec{F} = \vec{F}_{\parallel} + \vec{F}_{\perp},$$

5.5. TORQUE OR MOMENT OF A FORCE

then
$$\vec{r} \times \vec{F} = r\, F_\perp,$$

where $F_\perp$ and r are magnitudes of $\vec{F}_\perp$ and $\vec{r}$ respectively.

Example 5.5.4. A peg goes through a hole in a metal bar at one end so that the bar can rotate about the peg. A force of 10 N is applied on the bar at an angle 30° to the bar a distance 20 cm from the peg as shown in the figure. What is the torque about the peg?

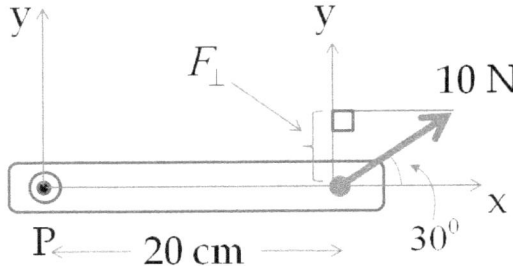

Solution. As illustrated in Example 5.5.3 the magnitude of the torque will be simply product of the perpendicular force and the distance from P. Let us place a coordinate system so that $\vec{r}$ and $\vec{F}$ are in xy-plane. Let the x-axis be along the bar, which is the direction of $vecr$. Then, to find the magnitude of the torque, we just multiply the y-component of the force and the distance from P to the point where force acts.

$$F_\perp = F_y = 10\text{ N}\sin(30°) = 5\text{ N} \quad \text{and} \quad r = 20\text{ cm} = 0.2\text{ m}.$$

Therefore, the magnitude of the torque is

$$\text{Magnitude:} \quad \tau = (0.2)(5) = 1 \text{ N.m.}$$

The direction of the torque is obtained by the right-had rule. In the given figure, the direction is coming out of the page.

Example 5.5.5. How to do the previous example by lever arm method?

Solution. To find the lever arm of the force about point P, we extend the line of force so that a perpendicular line from P on the line of force can be drawn. The extended line and the perpendicular are shown in Fig. 5.36.

The lever arm is equal to the length $r_\perp$, the side PQ in the right-angle triangle $\triangle PRQ$. We immediately find the lever arm to be 10 cm. The magnitude of the torque is equal to the product of the lever arm and the magnitude of the force, which gives the result 1 N.m. The direction of the torque is obtained as before by using the right-hand rule - the direction is coming out of page in the given figure.

Example 5.5.6. How to do the previous example by analytic method?

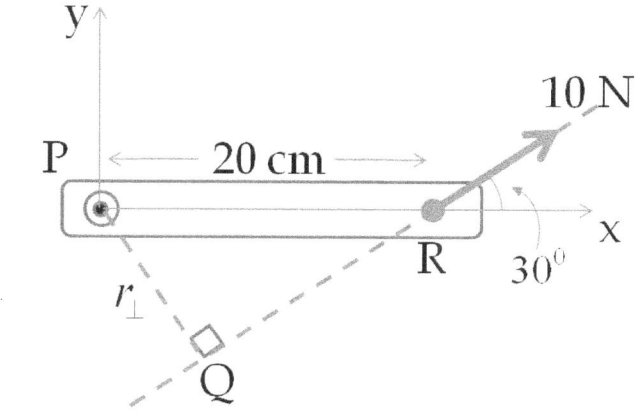

Figure 5.36: Example 5.5.5

Solution. We use the coordinate system given in Fig. ?? to write out the vectors in their component form so that we can compute the cross product. We wish to leave out units in the expressions, so we convert all units in meter-kilogram-second system of units before we use in equations.

$$\vec{r} = 0.2\hat{u}_x,$$
$$\vec{F} = 10\ \cos(30°)\hat{u}_x + 10\ \sin(30°)\hat{u}_y.$$

The torque is computed by directly working out the cross product.

$$\begin{aligned} \vec{\tau} &= \vec{r} \times \vec{F} \\ &= 0.2\hat{u}_x \times \left(5\sqrt{3}\hat{u}_x + 5\hat{u}_y\right) \\ &= (1.0\ \text{N.m})\ \hat{u}_z. \end{aligned}$$

Putting the units back, the magnitude of the torque is 1 N.m and the direction is towards the positive z-axis, which is pointed out of paper in the figure.

Further Remarks

Do not confuse torque with force. They are entirely different quantities. Some major difference between torque and force to keep in mind are

1. Torque depends on the choice of the point P about which torque is to be evaluated. Force has no such dependence.

2. Torque and force are perpendicular to each other since torque is defined as cross product of a displacement vector and force.

3. There can torque on a system even when net force is zero. For instance, if you apply equal magnitude forces in the opposite directions on opposite ends of a wheel, the net force will be zero, but the net torque about the center will not be zero (see Fig. 5.37).

4. Net force an a system may not be zero even when torque is zero. For instance, if you apply equal forces on opposite ends of a wheel, the net torque about the center will be zero, but the net force will not be zero (see Fig. 5.37).

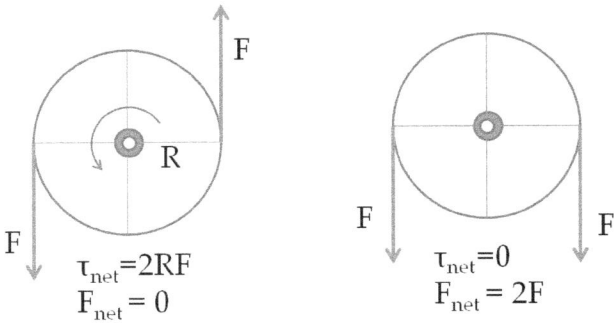

Figure 5.37: Torque versus force.

5.6 STATIC EQUILIBRIUM

5.6.1 Static Equilibrium and Newton's First Law of Motion

The basis for the first law of motion was discovered by Galileo from his observations on the motion of bodies along inclined planes. Newton stated his first law of motion as,

Every body perseveres in its state of being at rest or of moving uniformly straight forward, except insofar as it is compelled to change its state by force impressed.

You can see this law at play in many ordinary situations. For instance, your body in a moving car lurches forward when you suddenly stop the car, and similarly you tend to slide back when a car is suddenly started from rest with high acceleration. In both instances, the body is tending to resist the change in its state of motion, whether originally moving or at rest.

According to the first law of motion, a body at rest will remain at rest until a force acts on it. This principle underlies the stability of many physical structures. Questions like, how trusses can support large weights above them, how far above a ladder is it safe to climb, how much weight can be placed before a beam will give way, and many more, can be understood based on the balance of forces on and within physical structures.

To maintain a stable structure, no part of a body may have a left-over force or imbalance of forces. We say that a system is in static equilibrium if the net forces on each particle of the system is zero since if any particle of the system has a left over force, then, according to Newton's first law, a state of rest for that particle cannot be maintained.

Consider a system consisting of N parts. Let there be a net force $\vec{F}_1$ on part 1, $\vec{F}_2$ on part 2, etc. Each part may have more than one force on it, and the forces, $\vec{F}_1$, $\vec{F}_2$, etc, represent the vector sums of all forces on each part. Then, for the whole system to be in a static equilibrium we require the following condition.

$$\left.\begin{array}{l} \vec{F}_1 = 0 \\ \vec{F}_2 = 0 \\ \vdots \\ \vec{F}_N = 0 \end{array}\right\} \quad (5.18)$$

External and internal forces

The forces on each part of the system can be separated into two categories, internal and external. For instance, the net force $\vec{F}_1$ on part 1 is a vector sum of forces on part 1 by other parts, $2, 3, \cdots, N$ of the N-part system, which we call **internal forces**, and forces on part 1 from object that are outside the system which we call **external forces**. Suppose we sum all the internal forces on part 1 and call it $\vec{F}_1^{int}$ and sum all the external forces on part 1, call it $\vec{F}_1^{ext}$, then we can write $\vec{F}_1$ as

$$\vec{F}_1 = \vec{F}_1^{int} + \vec{F}_1^{ext}.$$

Similarly for forces on other parts of the system. The static equilibrium condition in Eq. 5.18 can now be expressed in terms of the internal and external forces on each part.

$$\left.\begin{array}{l} \vec{F}_1^{int} + \vec{F}_1^{ext} = 0 \\ \vec{F}_2^{int} + \vec{F}_2^{ext} = 0 \\ \vdots \\ \vec{F}_N^{int} + \vec{F}_N^{ext} = 0 \end{array}\right\} \quad (5.19)$$

From the third law of motion, we know that every force comes in pairs. Therefore, if there is force on 1 by 2, then there is also an oppositely directed force of the same magnitude on 2 by 1. This tells us that if we add all the equations in Eq. 5.19, the internal forces will all cancel each other, and we will obtain an equation that has only external forces on various parts.

$$\boxed{\vec{F}_1^{ext} + \vec{F}_2^{ext} + \cdots + \vec{F}_N^{ext} = 0} \quad (5.20)$$

We will prove below that if internal forces between parts are directed along the lines joining them, i.e. if force between parts 1 and 2 has the direction that is either parallel or anti-parallel to the displacement vector from part 1 to part 2, and the same for other internal forces, then the following identity for torques of external forces on different parts of the system can be proven from Eq. 5.19. Let $\vec{r}_1, \vec{r}_2, \cdots, \vec{r}_N$ be displacement vectors from an arbitrary point P to parts $1, 2, \cdots, N$ respectively. Then

$$\vec{r}_1 \times \vec{F}_1^{ext} + \vec{r}_2 \times \vec{F}_2^{ext} + \cdots + \vec{r}_N \times \vec{F}_N^{ext} = 0, \quad (5.21)$$

5.6. STATIC EQUILIBRIUM

which is the same as the sum of torques of external forces about an arbitrary point P.

$$\boxed{\vec{\tau}_1^{\text{ext}} + \vec{\tau}_2^{\text{ext}} + \cdots + \vec{\tau}_N^{\text{ext}} = 0} \tag{5.22}$$

Equations 5.20 and 5.22 give us an effective way to determine if a body consisting of N parts will be in static equilibrium.

Proof of Eq. 5.21

The force on i by j, denoted by $\vec{F}_{ij}$, and the force on j by i, denoted by $\vec{F}_{ji}$ are related by Newton's third law of motion.

$$\vec{F}_{ij} = -\vec{F}_{ji}.$$

Now, the sum of the torques of force $\vec{F}_{ij}$ on particle i and of force $\vec{F}_{ji}$ on particle j becomes

$$\vec{r}_i \times \vec{F}_{ij} + \vec{r}_j \times \vec{F}_{ji} = (\vec{r}_i - \vec{r}_j) \times \vec{F}_{ij},$$

where the vector $(\vec{r}_i - \vec{r}_j)$ is the displacement vector from particle i to particle j. If force between the particles is pointed along the line joining the two particles, then the internal torque of the forces $\vec{F}_{ij}$ and $\vec{F}_{ji}$ will be zero, leaving only net external torque.

Now, let us take cross product of the first equation in Eq. 5.19 by $\vec{r}_1$, the position of particle 1, of the second equation by $\vec{r}_2$, the position of particle 2, etc., and sum all resulting equations. The sum leads to vanishing of internal torque, leaving the result given in Eq. 5.21.

Deformable objects and static equilibrium

For an arbitrary structure to be in a static equilibrium, the vanishing of the net external force as given in Eq. 5.20 does not ensure static equilibrium. For instance, if you push on a piece of foam from two opposite sides, the net force and net torque on the foam are zero, but the foam gets deformed because of the imbalance between external and internal forces at the points of application of the forces. Therefore, you need to apply the complete condition for the static equilibrium given in Eq. 5.18, viz., the net force at each point of the system must vanish independently.

5.6.2 Problems of Static Equilibrium

We now consider some examples that will help us understand the use of static equilibrium in undeformable objects. The method for solving static problems follow a few basic steps outlined below. Make note of many decisions you will need to make when solving problems involving forces in equilibrium.

1. Pick the system. It can be either the entire structure, a part of the structure, or some small mass element inside the structure. The choice depends on the physical question that need addressing. Once,

you have decided on the objects that belong into the system, other objects that the system interacts with supply the external forces. This process helps us identify the external forces on the system.

2. Identify forces on the system by the external agents, making particular note of their directions and where they act. These usually fall into two categories: (a) long-distance, usually gravity in our case which can be taken to act at the center of gravity, and (b) contact forces at every surface of contact and every point of contact.

   Often, it is helpful to ask if any part of your system is pushing or pushing against something outside of the system. Then, you can use the third law of motion to "discover" the force that the other body would be applying on the system.

   Make a list of forces that you have identified. Remove any force that is by your system since we will be using only the forces on the system.

3. Calculations are usually mistake-free if done in analytic approach of vectors, although sometimes, it may mean more work than is necessary to solve a problem. The analytic approach begins with a judicious choice of Cartesian coordinates.

4. First, we work out components of the force vector. Often it is helpful to draw a separate diagram where all forces come out of the origin of the coordinate system. This diagram is called a free-body diagram and it is just a tool for calculations. Beware that forces may be acting at different points on the body, but they are redrawn in the tail-to-tail fashion in this diagram keeping their directions in space.

5. For the torque equation, we also need to choose a point P about which we will calculate torques. In a static situation, the choice of point P is arbitrary. Therefore, we pick point P such that torques are easy to calculate. A good location for P is where forces cross each other if you extend their vector lines in either or both directions. This choice will make the torques of the forces whose lines pass through P zero, and therefore, we will need to calculate the torques of other forces only.

6. Once point P has been chosen, write out all the displacement vectors $\vec{r}_i$ for $i = 1, 2, \cdots, N$. At this point you should have components of all forces and all displacement vectors if you are planning on working with the analytic approach. If you are working with the geometric approach, you would have these vectors with arrows drawn out now.

7. Next order of business is to calculate all the non-zero torque vectors, which you can do by any of the methods illustrated in the section on torques above. Most problems in this book involve vectors in

5.6. STATIC EQUILIBRIUM

one plane and the lever arm method for torques turns out to be convenient.

8. Finally, since Equations 5.20 and 5.22 are vector equations. they give rise to three equations per vector equation.

   Depending on how much is known about the vectors and displacements, and the particular question you are trying to answer, you may need just Eq. 5.20, or just Eq. 5.22, or both. Examples below will illustrate the decision making process that goes into selecting particular set of equations we work with in different situations.

Example 5.6.1. A simple problem. Lets begin with a really simple problem. A 10-kg block is hung from the ceiling using a light string. What is the force between the ceiling and the block?

Solution. Let us first understand the problem and get a sense of what we need for the answer. The force between the ceiling and the block is the tension in the string. The directions of the tension force on the block and ceiling are known here, so we only need to find the magnitude. Let T denote the magnitude of the tension force, although we will sometimes also write $\vec{T}_{\text{on block}}$ and $\vec{T}_{\text{on ceiling}}$ when we want to be specific about the individual force of the force pair acting on a particular body.

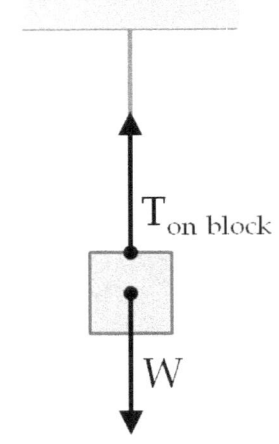

Figure 5.38: Example: 5.6.1.

What should be our system? Could block alone be enough to solve the problem? Or, do we have to include the ceiling as well? Let us look at this choice from the simplest to more complicated. It would be great if forces on the block alone could solve the problem. In this case, the two forces on the block, namely, the weight of the block and the tension in the string, are external forces (see Fig. 5.38). Although these two forces act at different points on the block, they act along the same line. These two forces must be balanced since we have an equilibrium. The balancing of forces on the block will generate one equation for one unknown T. Therefore, we have sufficient information if we choose the block alone to be the system.

Balancing the forces on the block can be done analytically, by placing one of the axes, say the y-axis along the common line of action of the forces in the free-body diagram as shown in Fig. 5.39. Only the y-components of forces are non-zero. Therefore, the y-component of the force-balancing equation, namely Eq. 5.20, results in the following relation.

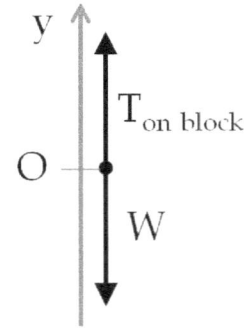

Figure 5.39: Example: 5.6.1. Tail-to-tail force vectors or a free-body diagram and axes for calculating components.

$$\text{y-component:} \quad T - 98.1 \text{ N} = 0.$$
$$\text{x-component:} \quad 0 = 0.$$
$$\text{z-component:} \quad 0 = 0.$$

Therefore, $T = 98.1$ N, where we now put the units back. Note that I have include the information about the x and z-components for pedagogical reasons. You do not need to include them in your answer since it is quite clear from the figure itself that the x and z-components are zero.

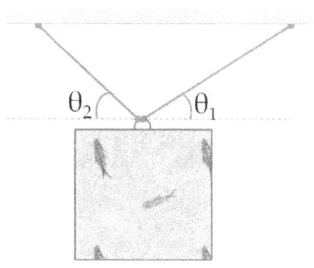

Figure 5.40: Example: 5.6.2. A picture frame hung from ceiling using two strings.

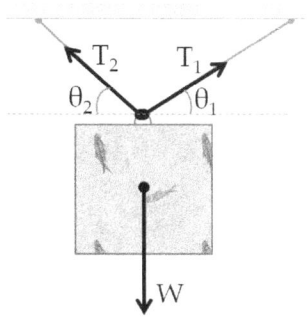

Figure 5.41: Example: 5.6.2. Forces on the picture frame.

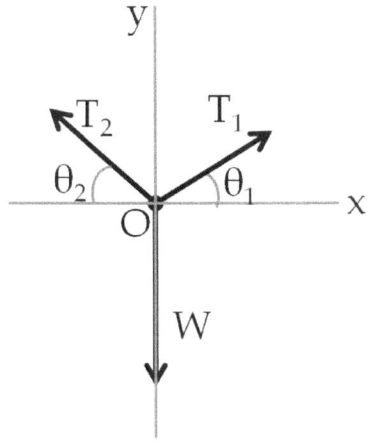

Figure 5.42: Example: 5.6.2. Tail-to-tail forces or free-body diagram and axes.

We write the answer as follows: The force between the ceiling and the block has a magnitude of $mg = 98.1$ N. The tension force on the block is pointed up towards the ceiling, and the tension force on the ceiling is pointed down towards the block. Note that, since the question is about force, we must state both the direction and magnitude of the force in the answer. The force between the ceiling and the block refers to both the force on the ceiling and the force on the block.

Example 5.6.2. Hanging a picture frame. A picture frame of mass m is hung by using two light strings. The two strings are tied to two different hooks in the ceiling such that the two strings make different angles with the horizontal direction as shown in Fig. 5.40. The picture hangs in equilibrium. What are the tensions in the two strings?

Solution. Once again, we start with first getting a feel for the problem. A rough drawing of the situation helps here. The situation is shown in Fig. 5.41. Since the strings are taut, there will be tension forces in the strings that would act on the frame as well as on the hooks in the ceiling. Now, since we have two strings, and there is no obvious symmetry arguments that can relate the magnitudes of the tensions in the two strings, we will use two different symbols, T_1 and T_2, for the magnitudes of the tensions in the two strings as shown in the figure.

Now, we need to decide about the system. Here, we have the choices: picture frame only, picture frame and the hooks, hooks only, etc. We again start with the simplest first. Fig. 5.41 shows three external forces on the picture frame. Here we know the directions of all three forces and the magnitude of one (W). So, we have two unknowns - the magnitudes T_1 and T_2. Since all forces fall in one plane, the force balancing equation will give us two relations. Therefore, the force balancing equation has sufficient information to solve the problem. Before we proceed to the solution, we point out that the torque equation will not give us any useful information since all three forces cross at one point when their lines of action are extended, which will give zero torque for each force with respect to the crossing point, which can be chosen for the arbitrary point P.

To implement the force balancing equation, we can work in the analytic picture of vectors as shown in Fig. 5.42. We draw the forces in the tail-to-tail fashion, called a free-body diagram, and then choose axes. With the choice of axes shown in Fig. 5.42, the z-components are all zero and the x and y-components of force balancing equations are:

$$\text{x-component:} \quad T_1 \cos\theta_1 - T_2 \cos\theta_2 = 0. \quad (5.23)$$
$$\text{y-component:} \quad T_1 \sin\theta_1 + T_2 \sin\theta_2 - mg = 0. \quad (5.24)$$
$$\text{z-component:} \quad 0 = 0.$$

Note the minus signs in the x-component of T_2 and the y-component of weight. The minus signs arise for the projection of vectors that fall on the corresponding negative axes. I have also listed the zero equal zero for the

5.6. STATIC EQUILIBRIUM

z-component for pedagogical reasons. When you are solving a problem and if it is obvious that a particular direction will give no information, you do not need to bother with listing that direction. I will also stop doing this pretty soon.

Now, we can solve Eqs. 5.23 and 5.24 for T_1 and T_2. Although, it isn't difficult to solve these equations, the answer looks complicated and we will give it below. As an example of a simpler calculation, let us solve these equations for particular numerical values of mass of the frame and the angles.

Case: $m = 0.5\ kg$, $\theta_1 = 30°$, $\theta_2 = 60°$

Putting these values in Eqs. 5.23 and 5.24 we find the following two equations in two unknowns.

$$\sqrt{3}T_1 - T_2 = 0.$$
$$T_1 + \sqrt{3}T_2 = 9.81.$$

Multiplying the first equation by $\sqrt{3}$ and adding the two equations gets rid of T_2 and we find

$$4T_1 = 9.81 \implies T_1 = 2.45\ \text{N},$$

which we put in the first equation to obtain $T_2 = 4.24$ N. A student may prove that for general m, θ_1, and θ_2, the solution of Eqs. 5.23 and 5.24 give the following for T_1 and T_2.

$$T_1 = \frac{mg}{\sin\theta_1 + \cos\theta_1 \tan\theta_2}. \quad (5.25)$$
$$T_2 = \frac{mg}{\sin\theta_2 + \cos\theta_2 \tan\theta_1}. \quad (5.26)$$

Although Eqs. 5.25 and 5.26 appear complicated, they give more insight into the physical content than just numerical values. These general solutions let you answer many other questions. For instance, we can ask, what would happen if the strings were symmetric in their directions? It is a simple matter to set $\theta_1 = \theta_2 = \theta$, where we now write θ for equal angle of the two, to obtain the following conclusion.

$$T_1 = T_2 = \frac{mg}{2\sin\theta} \quad \text{(Symmetric situation.)}$$

Figure 5.43: Seesaws in a Montreal park.

Example 5.6.3. A Seesaw A seesaw of mass M has a support in the middle and can be balanced if it is perfectly horizontal. A child of mass m_1 sits tight at a distance d_1 from the middle. At what distance from the middle should another child of mass m_2 sit so that the seesaw is in static equilibrium?

Solution. Once again, we face the decision about what should be the system that would help answer the question. Should it be the seesaw alone, or the seesaw and the two children? Here we need to find the distance of the second child from the pivot point of the seesaw. This information will be contained in the torque equation.

Let us see what happens if we think of only the seesaw as the system. Then the external forces on the seesaw will come from Earth, the support at the pivot, child 1 and child 2. These forces are: weight of the seesaw as Mg, the force from support as F_N, force by child 1 as F_1 and force by child 2 as F_2.

WARNING!! If the seesaw is not in equilibrium, which is the point of fun in the seesaw, then often-made assumption that the force by child is equal to his/her weight would be false. We will discuss non-equilibrium situations when we introduce Newton's second law of motion in the next chapter. Presently, we leave the forces by children undetermined, and use equilibrium of each child to deduce the magnitude of the force each child applies on the seesaw.

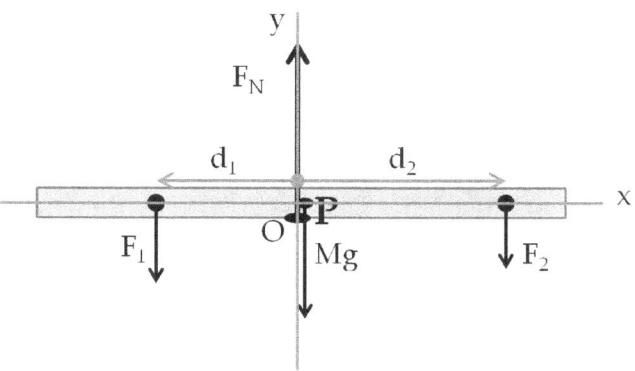

Figure 5.44: Example 5.6.3. External forces and their locations on the system consisting of seesaw only. The base of each force arrow shows the location where the force acts.

Fig. 5.44 shows all the forces on the seesaw. The base of each force arrow shows the location where the force acts. Figure also shows a choice of coordinate system. The force from the support is the normal force here, and is labeled as $\vec{F}_N$. The line of force for the force from the support goes through the center of the seesaw where the force of gravity also acts.

Now, let us figure out the force each child applies on the seesaw. To find the magnitude of the force by each child, we will need to set up a problem with that child as the system (Fig. 5.45). For instance, let us look at child 1 as the system. We find that there are two external forces on child 1 - his/her weight and the force by seesaw. The force by seesaw is equal to negative of force $\vec{F}_1$ by third law of motion. Since, the child 1 is in equilibrium, the forces must be balanced. This gives the following condition.

Figure 5.45: In equilibrium, the two forces on the child are balanced.

Child 1 in equilibrium: $\quad F_1 - m_1 g = 0 \implies F_1 = m_1 g.$

Similar arguments for child 2 gives

Child 2 in equilibrium: $\quad F_1 - m_2 g = 0 \implies F_2 = m_2 g.$

5.6. STATIC EQUILIBRIUM

Therefore, we find that if a child is in equilibrium, the child presses on the seesaw by a force equal to his/her weight.

Now, we know the directions of all forces, and the magnitudes of all but the force from the support. Force balancing equation will clearly give us the value of the magnitude of the force from the support, but we are not after that information here. If the question had asked about the force from support, we would write the force balancing equation at this point and solve for the force. Our question is about the distance of child 2 from the support. That means, we will need to work with the torque equation for sure.

Choosing point P for torque calculations: Recall that torque equation requires specification of a point P, whose choice is up to us, for the calculations of torques about that point. Now, if we place point P at the support, then we don't need to know the support force since it will have zero contribution in the torque equation. This way, the torques equation for the system with only the seesaw will produce an equation in the unknown distance.

Torque calculations: With the choice of point P at the center of the seesaw, the contribution of $\vec{W}$ and $\vec{F}_N$ to the torques equation are zero and we need to calculate only the torques of pushes by each child on the seesaw. A calculation of torque can be done using the geometric method or lever arm or analytic approach of vector calculations using components. For the sake of comparison, I will work out the geometric and analytic methods.

Geometric approach:
In the geometric picture for torque as across product we need to know the directions of each torque so that we can decide if they will add or subtract from each other. Since the force and displacement vectors are in one plane, the torques will be perpendicular to this plane, pointing either above the plane or below the plane.

In Fig. 5.44 we use the right-hand rule to find the direction of the torques. The torque from F_1 is pointed towards the positive z-axis (coming out of page), and the torque from F_2 is towards the negative z-axis (going into the page). Therefore, they are oppositely directed and their vector sum will actually be subtraction of their magnitudes.

The magnitudes of the two torques are obtained from the magnitudes of the forces and displacements and the angle between them. Since the angle between the displacement vectors from point P to the point of action of the forces are both $90°$, the magnitudes of the torques are simply multiplications of the magnitudes of the forces and the displacements.

$$\text{Torque of } \vec{F}_1 \text{ about P:} \quad \vec{\tau}_1 = \{m_1 g d_1, \text{ pointed out of the page}\}$$
$$\text{Torque of } \vec{F}_2 \text{ about P:} \quad \vec{\tau}_2 = \{m_2 g d_2, \text{ pointed into the page}\}$$

Let out-of-page direction be taken as positive, then the sum of the torques

becomes
$$m_1 g d_1 - m_2 g d_2 = 0,$$
which can be solved for the unknown d_2 to yield
$$d_2 = \left(\frac{m_1}{m_2}\right) d_1$$

Analytic approach:
The analytic approach for this problem is really not needed. But, for the sake of demonstration of the calculation, I will present the calculation of the torques in this problem using analytic method.

For the analytic method, we start with the components of the forces. The axes are given in Fig. 5.44. It is easy to show that the forces and displacements have the following representations in the given coordinate system.
$$\vec{F}_1 = -m_1 g \hat{u}_y$$
$$\vec{F}_2 = -m_2 g \hat{u}_y$$
$$\vec{d}_1 = -d_1 \hat{u}_x$$
$$\vec{d}_2 = d_2 \hat{u}_x$$

It is not difficult to work out the cross product here. You do not need to use the determinant. Just do it directly using the cross product results of the base vectors themselves.
$$\vec{\tau}_1 = \vec{d}_1 \times \vec{F}_1 = m_1 g d_1 \hat{u}_z$$
$$\vec{\tau}_2 = \vec{d}_2 \times \vec{F}_2 = -m_2 g d_2 \hat{u}_z$$

Therefore, the sum of the two is
$$\vec{\tau}_{net} = (m_1 g d_1 - m_2 g d_2)\, \hat{u}_z.$$

Setting the net torque to zero gives the same result as obtained above.

Example 5.6.4. A cantilever A beam that extends beyond its support

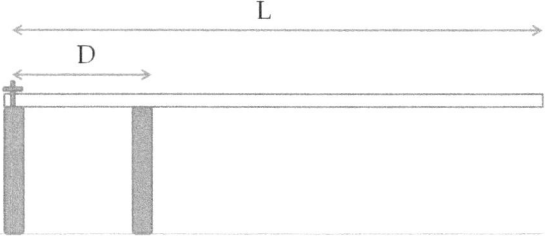

Figure 5.46: Example 5.6.4.

is called a **cantilever**. Consider a cantilever of length L and mass M resting and bolted to two supports, one at the end and the other is at a distance D from it. (a) Find the conditions on forces so that the cantilever is in static equilibrium. (b) For $L = 30$ m, $M = 1000$ kg, and $D = 12$ m meters, find the magnitude and directions of the forces on the beam from the supports.

5.6. STATIC EQUILIBRIUM

Solution. We will consider beam of the cantilever as the system. The beam has force from the support and bolt at the support at the end, an upward normal force from support in the middle, and the weight of the beam. As shown in Fig. 5.47, we will combine the forces of the bolt and the support at the end into a horizontal force $\vec{F}_{11}$ and a vertical force $\vec{F}_{12}$, the sum of the two forces at the end give us the net force $\vec{F}_1$ at that point.

$$\vec{F}_1 = \vec{F}_{11} + \vec{F}_{12}.$$

Normally, one writes the force $\vec{F}_{11}$ as $\vec{F}_{1x}$ and $\vec{F}_{12}$ as $\vec{F}_{1y}$. This notation is confusing since symbols F_{1x} and F_{1y} stand for the x and y-components of the vector $\vec{F}_1$ and are not vectors themselves. Therefore, we will use $\vec{F}_{11}$ and $\vec{F}_{12}$ for the the horizontally and vertically pointed vectors whose sum is the force at the end.

From physical grounds we expect that all four forces $\vec{F}_{11}$, F_{12}, $\vec{F}_2$ and the weight $\vec{W}$ will be in one plane, which we take to be the xy-plane of the coordinate system shown.

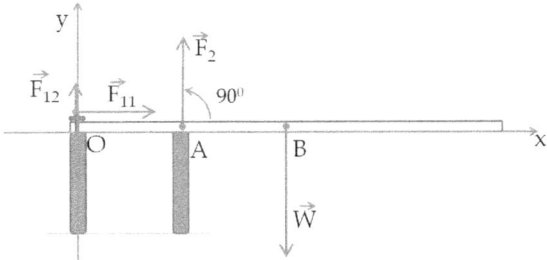

Figure 5.47: Example 5.6.4. Forces on the cantilever beam.

I leave for the student to do this problem using lever arm and geometric methods. Here, I will use the analytic approach. For the torque equation, we need to choose a suitable point P about which we will work out the torques. Since we do not know either $\vec{F}_1$ (or equivalently $\vec{F}_{11}$ and $\vec{F}_{12}$) or $\vec{F}_2$, it does not save us more work whether we place point P at O or A. Let us us pick O because it is nicely on one end so that we can measure displacements of the forces from the end point. We will also name the displacements as $\vec{r}_1$, $\vec{r}_2$ and $\vec{r}_w$ for the four forces. Let us organize the information regarding forces and displacements in a table.

Force	Force in components	Displacement	Torque
$\vec{F}_{11}$	$F_{11}\hat{u}_x$	$\vec{r}_1 = 0$	0
$\vec{F}_{12}$	$F_{12}\hat{u}_y$	$\vec{r}_1 = 0$	0
$\vec{F}_2$	$F_2\hat{u}_y$	$\vec{r}_2 = D\hat{u}_x$	$DF_2\hat{u}_z$
$\vec{W}$	$-Mg\hat{u}_y$	$\vec{r}_2 = (L/2)\hat{u}_x$	$-(L/2)Mg\hat{u}_z$

Balancing the forces and torques means setting the sum of the x, y, and z-components zero independently. Therefore, we obtain the following three equations in the present case from the force in component and torque columns in the table.

$$F_{11} = 0$$
$$F_{12} + F_2 = 0$$
$$DF_2 = \frac{1}{2}LMg$$

These equations can be solved to yield the following for the magnitudes of the forces.

$$F_{11} = 0 \tag{5.27}$$
$$F_{12} = Mg\left(1 - \frac{L}{2D}\right) \tag{5.28}$$
$$F_2 = \left(\frac{L}{2D}\right)Mg \tag{5.29}$$

Equation 5.27 says that there is no horizontal component of force $\vec{F}_1$ at the end, and Eq. 5.28 shows that if the second support is less that $L/2$ from the end, then the force in the vertical direction at the end support is downward, i.e. the force there is dominated by the push back from the bolt at the top of the beam.

(b) Plugging in the numerical values given we find the following.

$$F_{11} = 0,$$
$$F_{12} = -2500 \text{ N},$$
$$F_2 = 12000 \text{ N},$$

where I have rounded off to two significant digits.

5.7 EXERCISES

Balancing Forces

Ex 5.7.1. Find $\vec{F}_{net}$ in situation give in Fig. 5.48.

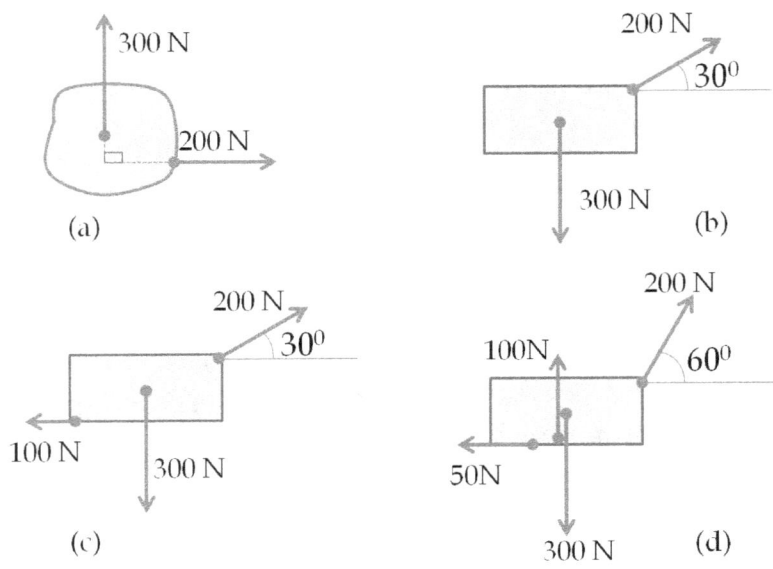

Figure 5.48: Exercise 5.7.1.

Ex 5.7.2. The net force in each situation in Fig. 5.49 is zero. Find the magnitude and direction of the force labeled F in each case.

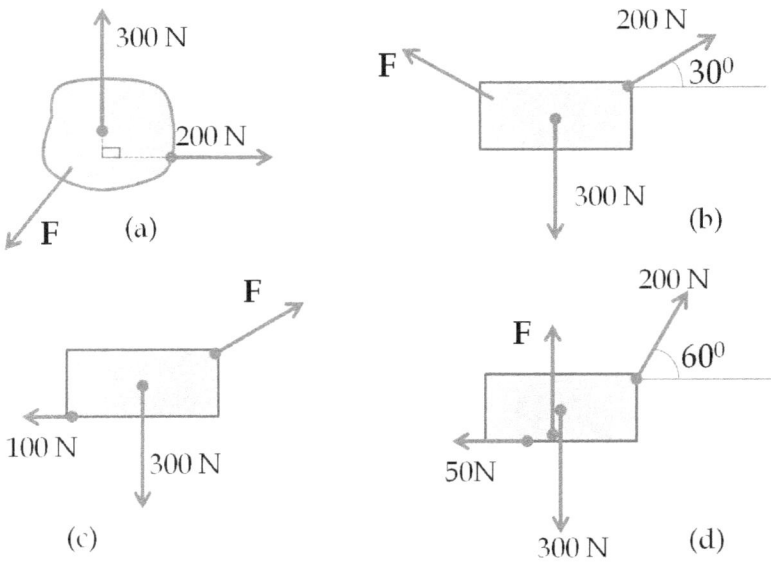

Figure 5.49: Exercise 5.7.2.

Gravitational Force

Ex 5.7.3. Find the gravitational force between two 10-kg lead spheres separated by 20 cm center-to-center distance. Provide both the magnitude of the force between the lead spheres and the directions of the forces on each sphere.

Ex 5.7.4. Look up the necessary masses and distances and find the force on the Earth by the Sun and the force on Sun by the Earth. Indicate the directions of the two forces by sketching a drawing.

Ex 5.7.5. Look up the necessary masses and distances, and find the magnitude of gravitational forces on the Earth by the Moon and by the Sun, and compare the two.

Ex 5.7.6. A 50-kg man is standing still on the surface of earth. (a) Find the force on the man exerted by the Earth. (b) Find the force on the Earth exerted by the man. Give directions of the two forces also.

Ex 5.7.7. Assuming the orbit of the Earth around the Sun to be circular and that of the Moon about the Earth be also circular, find the sum of the forces of the Moon and the Sun on the Earth in the two instances: (a) the solar eclipse, and (b) the lunar eclipse. Note: the solar eclipse happens when the Moon is in the line of sight between the Earth and the Sun, and the lunar eclipse when the Earth is between the Sun and the Moon.

Ex 5.7.8. Find the sum of the forces of the Earth and the Sun on the Moon when the Moon is in the position indicated in the Fig. 5.50.

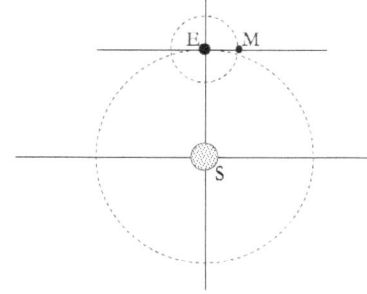

Figure 5.50: Exercise 5.7.8.

Weight

Ex 5.7.9. Using the mass and the radius of the Moon find the weight of a 1-kg brass block on the Moon.

Ex 5.7.10. What would be the weight of a 1-kg brass block on the Jupiter?

Ex 5.7.11. (a) Find the force of gravity of the Earth on a climber of mass 45-kg atop Mount Everest which is 8850 m above the sea level. Assume that the distance of the top of mount Everest from the center of the earth is 8850 m more than the average radius of the earth, which is approximately $6,370,000$ m. (b) Compare your answer to the weight of the climber at the sea level. You can compare the two by taking their ratios or by calculating the percentage difference.

Ex 5.7.12. A box of mass 3 kg is placed on a plane inclined at a 30° angle with the horizontal direction. (a) Find the weight of the box. (b) Find the component of the weight along the incline. (c) Find the component of the weight perpendicular to the incline.

Ex 5.7.13. A box of mass 10 kg is placed on a plane inclined at a 60° angle with the horizontal direction. (a) Find the weight of the box. (b) Find the component of the weight along the incline. (c) Find the component of the weight perpendicular to the incline.

Normal Force and Balancing Forces

Ex 5.7.14. A book of mass 1.5 kg is resting on a table which is fixed to the ground. (a) Draw a diagram showing with symbols and arrows all the forces acting on the book. (b) What is the weight of the book. Note: give both magnitude and direction. (c) What is the net force on the book? State how you can draw your conclusion about the net force on the book. (d) Find the magnitude and direction of the force on the book by the table. (e) Find the magnitude and direction of the force on the table by the book.

Ex 5.7.15. Two books A and B of masses m_1 and m_2 respectively are stacked on a table as shown in the Fig. 5.51. (a) Draw a diagram showing with symbols and arrows all the forces acting on the book A. (b) Assuming the net force on A to be zero what would be the vector equation relating all forces on A? (c) Find the magnitude of all forces on A in terms of the masses and g. (d) Repeat steps in (a)-(c) for the book B. (Note: the weight of A acts on A not on B. The force from A acting on B may or may not be the weight of A and depends on the situation. Think!)

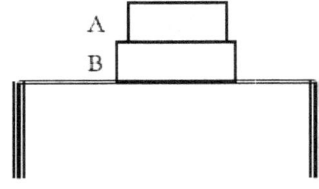

Figure 5.51: Exercise 5.7.15.

Ex 5.7.16. A 50-kg box is pushed against a wall by a force that acts horizontally and has a magnitude 200 N. Ignore any friction forces for this problem. (a) Draw a diagram showing with symbols and arrows all the forces acting on the book. (b) Choose a Cartesian coordinate system and compute the components of all forces on the box. (c) What are the magnitudes and directions of all forces on the box?

Ex 5.7.17. A book of unknown mass m can be supported against a vertical frictionless wall by applying a force 10 N at an angle 60° with the horizontal (Fig. 5.52). (a) Draw a diagram showing with symbols and arrows all the forces acting on the book. (b) Choose a Cartesian coordinate system and compute the components of all forces on the book. (c) What are the magnitudes and directions of all forces on the book?

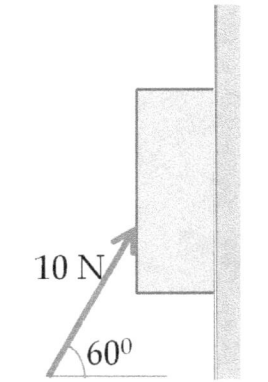

Figure 5.52: Exercise 5.7.17.

Static Frictional Force and Balancing Forces

Ex 5.7.18. A box of mass 1.5 kg is resting on a table which is fixed to the ground. Find the following forces: (a) the net force on the box, (b) the normal force on the box by the table, (c) frictional force on the box by the table (d) the normal force on the table by the box, and (e) the frictional force on the table by the box.

Ex 5.7.19. A book of mass 1.5 kg is at rest on a table which is fixed to the ground. The book is then pushed horizontally with a constant force of 10 N. Even with the 10 N horizontal force the book does not move. Draw a free-body diagram of forces on the book and find the following forces using the diagram. (a) the net force on the book, (b) the normal force on the book by the table, (c) the static frictional force on the book by the table,

(d) the normal force on the table by the book, (e) the static frictional force on the table by the book, (f) the minimum value of coefficient of static friction between the table and the book that is consistent with the given data, and (g) the total force applied by the book on the table. Just a reminder that you need to provide both the magnitude and direction of each force.

Ex 5.7.20. A book of mass 1.5 kg at rest on a table is pushed horizontally with a force of magnitude F_A, which can be varied on demand. (a) If the coefficient of static friction between the book's bottom surface and the table is 0.3, at what value of the applied force would the book start to move? (b) Find the magnitude of the force applied by the book on the table when the static friction has its maximum value.

Ex 5.7.21. A box of mass m is at rest on a table inclined at an angle θ as shown in Fig. 5.53. The static friction between the box and the table surface has a coefficient equal to μ_s. It turns out that there is not enough static friction and an additional force of magnitude F_A acting up the incline is required to stop the box from sliding. Find the magnitude of the force F_A in terms of m, θ, g and μ_s by balancing all forces acting on the box. Use a free-body diagram to help set-up the problem.

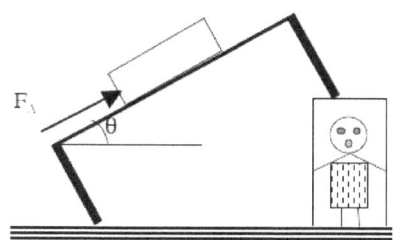

Figure 5.53: Exercise 5.7.21.

Kinetic Frictional Force and Balancing Forces

Note that forces on an object are balanced when the object moves with constant velocity. Therefore, if an object slides with a constant velocity, the net force will be zero. In the next chapter you will study problems where the net force in not zero; in that case, velocity would not be constant.

Ex 5.7.22. Consider a 1.5-kg book resting on a table. The coefficient of static friction between the book's bottom surface and the table is known to be 0.3. The book is pushed horizontally with a force of 5 N. (a) Show that there is not enough static friction to prevent the book from sliding. (b) When the book is sliding, a horizontal force of magnitude 2 N is enough to keep the book moving on the table at a constant velocity. Therefore, only 2 N force is now applied. By balancing forces on the book, using the help of a free-body diagram and a Cartesian axis system, determine the value of kinetic frictional coefficient.

Ex 5.7.23. A box of mass 5 kg is sliding at a constant velocity down an inclined plank with an angle of incline of 10°. Making use of a free-body diagram and a Cartesian axis system answer the following questions. (a) What must be the coefficient of kinetic friction between the bottom of the box and the surface of the plank? (b) What is the total force applied by the inclined plank on the box? (c) What is the total force applied by the box on the inclined plank? When providing answer for a force, give both the magnitude and direction of each force.

5.7. EXERCISES

Ex 5.7.24. A heavy equipment of mass 1000 kg in a lab needs to be moved from one side of the room to the other. You find that it takes 300 N horizontal force to get it to budge from its place, and then only 230 N is enough to keep it moving at a constant velocity. Making use of a free-body diagram and a Cartesian axis system determine the values of the coefficients of static and kinetic frictions. Assume flat horizontal floor.

Tension and Balancing Forces

Ex 5.7.25. A 250-gram copper ring is hung from the ceiling using a light string. Then a 500-gram steel ball is hung from the ring using another light string (Fig. 5.54). (a) Draw a diagram showing with symbols and arrows all the forces acting on the ring. Label forces in your diagram with standard notation. (b) Draw a diagram showing with symbols and arrows all the forces acting on the ball. Label forces in your diagram with standard notation. (c) Choose a Cartesian coordinate system, and determine the relations between components of forces on the ring. (d) Choose a Cartesian coordinate system, and determine the relations between components of forces on the ball. (e) Find the magnitude of all the forces on the ring and on the ball in terms of the masses of the ring and the ball and the acceleration due to gravity, g.

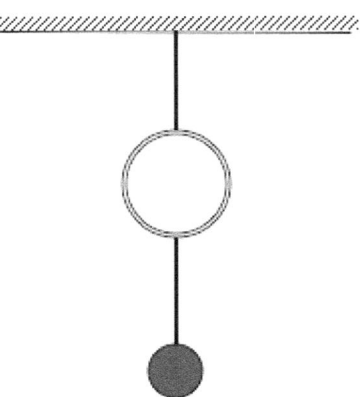

Figure 5.54: Exercise 5.7.25.

Ex 5.7.26. Two 1-kg blocks, A and B, are hung on the two sides of a pulley by using a light string. The pulley is then hung from a support in the ceiling. Once hung, the blocks remain motionless. (a) Draw a diagram showing with symbols and arrows all the forces acting on the block A. (b) Draw a diagram showing with symbols and arrows all the forces acting on the block B. (c) Draw a diagram showing with symbols and arrows all the forces acting on the pulley. (d) Find magnitudes and directions of all forces on the blocks and the pulley.

Assumptions: The pulley is considerably lighter than 1 kg and can be assumed to be massless. The pulley can also freely rotate about an axle so that the you can assume no friction when pulley rotates. This type of pulley is often referred to as a "massless and frictionless" or and an "ideal" pulley. The tensions in the strings on the two sides of a "massless and frictionless" pulley are equal.

Ex 5.7.27. A large crate of mass M is lifted by pulling with a force F on a string that goes over two "massless frictionless" pulleys as shown in Fig. 5.55. You can read more about "massless and frictionless" pulley in Exercise 5.7.26. Label the pulleys as 1 and 2, where 1 is the one with the crate attached to it. (a) Draw a diagram showing with symbols and arrows all the forces acting on the pulley # 1. (b) Draw a diagram showing with symbols and arrows all the forces acting on the pulley # 2. (c) Draw a diagram showing with symbols and arrows all the forces acting on the crate. (d) Find the magnitude of the force F in terms of M and g if the crate moves up with a constant velocity using the fact that forces on an

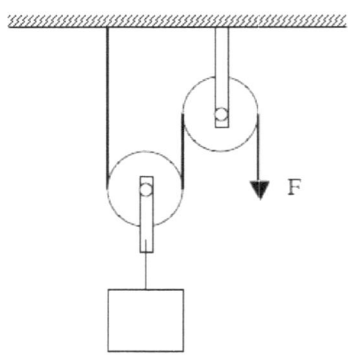

Figure 5.55: Exercise 5.7.27.

object are balanced if the velocity of the object is constant.

Explore further:

(e) Suppose, you were using only one pulley to pull the crate up. Show the arrangement, and prove that it would require more force than the two pulley system shown in this problem. (f) How many pulleys will you need and how would you arrange them so that you would need to apply only $\frac{1}{4}Mg$ for pulling the crate up at constant velocity? (g) Can you generalize the arrangement so that you would need only $\frac{1}{n}Mg$ for any n? Does your arrangement depend upon n being even or odd?

Ex 5.7.28. A porter is pulling a 50-kg box on a flat surface by tying the box with a rope (Fig. 5.56). The box is sliding on a relatively smooth floor at a constant velocity. Note that forces are balanced when an object is moving with a constant velocity. The coefficient of kinetic friction between the contact surface of the box and the floor is 0.1. (a) Draw a free-body diagram of the forces on the box, and choose a coordinate system to help with balancing forces on the box to answer the following questions. (b) What is the magnitude of the tension force in the rope? (c) What is the direction of the tension force when it acts on the box? (d) What is the direction of the tension force when it acts on the porter? (e) What should be the magnitude of the least static friction force on the porter from the floor so that he does not slide? What would be the direction of this static friction force?

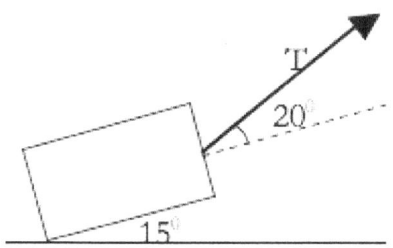

Figure 5.56: Exercise 5.7.28.

Spring Force and Balancing Forces

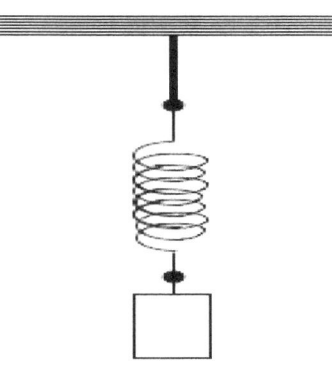

Figure 5.57: Exercise 5.7.29.

Ex 5.7.29. A 2-kg brass block is attached at the end of a spring of length 30 cm and spring constant 1000 N/m. The end of the spring is attached to an "unstretchable" string of length 25 cm. The string is then tied to a support in the ceiling so that the brass block hangs at rest (Fig, 5.57). (a) Draw a free-body diagram of forces on the brass block. (b) Draw a free-body diagram of the forces on the spring. (c) Draw a free-body diagram of forces on the string. (d) Find the magnitude and direction of all the forces you have included in (a), (b) and (c). (e) By how much does the spring stretch?

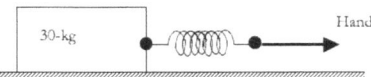

Figure 5.58: Exercise 5.7.30.

Ex 5.7.30. A 30-kg box on a smooth floor is attached to a spring and pulled horizontally (Fig. 5.58) by increasing force until the point when the box starts to slide. The length of the spring when unstretched is 20 cm. The spring constant has a value of 100 N/cm. The coefficient of static friction between the bottom of the box and the floor surface has been determined to be 0.2. (a) Draw a free-body diagram of the forces on the box. (b) Draw a free-body diagram of the forces on the spring. (c) Choose a coordinate system and work with components of forces to find out the magnitude of the maximum force that can be applied to the spring so that the box before the box starts to slide. (d) At the instant when the

5.7. EXERCISES

static friction is maximum, what is the length of the spring?

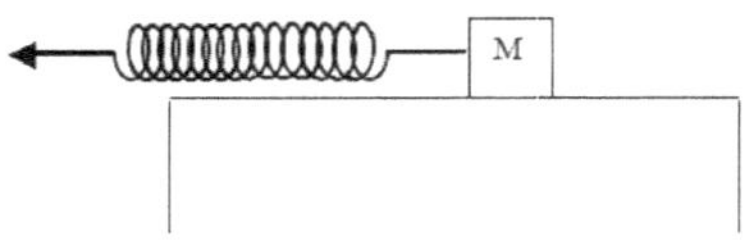

Ex 5.7.31. A block of mass m is attached to a stiff spring of spring constant k. The block is put on a rough surface so that the coefficient of static friction between the block and surface is μ_s (Fig. 5.59). The other end of the spring is then pulled with increasing force. Find the stretch of the spring in terms of m, k, μ_s, and g when the block just starts to slide.

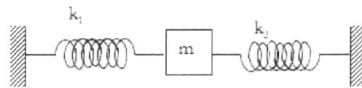

Figure 5.59: Exercise 5.7.31.

Ex 5.7.32. A block of mass m is attached to two springs of spring constants k_1 and k_2 (Fig. 5.60). The other ends of the springs are attached to fixed walls. The mass is in the middle when neither of the springs are stretched or compressed. The block is pulled to the right by a distance Δx and held there by a person. (a) Draw free-body diagram of the block. Do not forget to include the force by the person. (b) Use a coordinate system to write out the balanced force condition for the block. (c) Determine the magnitude and direction of the force on the block by the person.

Figure 5.60: Exercise 5.7.32.

Calculating Torque

Ex 5.7.33. Find the torque vectors on a rod from the forces shown in Fig. 5.61 about pivot points (a) O_1 and (b) O_2. All forces have magnitude 100 N and $L = 1$ m.

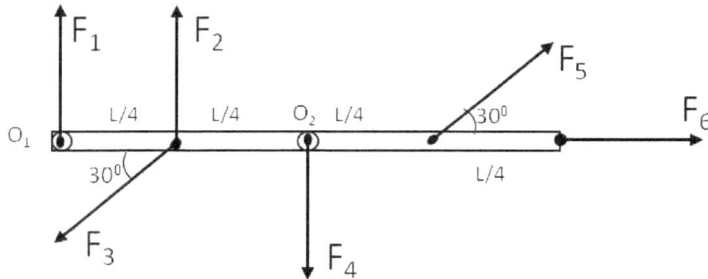

Figure 5.61: Exercise 5.7.33.

Ex 5.7.34. Find the torque vector on a wheel of radius R rolling down an incline about the pivot points (a) O_1 and (b) O_2 as shown in Fig. 5.62. Give your answer in terms of the forces shown, the radius of the wheel, and the angle of inclination. Note your answer will be a vector.

Ex 5.7.35. A wheel of radius R is rotated by applying two force that are applied at two different points at the rim of the wheel as shown in Fig. 5.63. The two forces are pointed in the opposite direction but have equal magnitude. The two forces are said to form a **couple**. Evaluate the net torque on the wheel about (a) the center, (b) A, and (c) B.

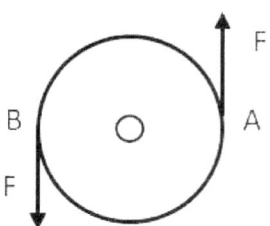

Figure 5.63: Exercise 5.7.35.

Static Equilibrium: Balancing Forces and Torques

Ex 5.7.36. A beam of mass M is placed on a support. When the support is in the middle of the beam, the beam remains in a horizontal position

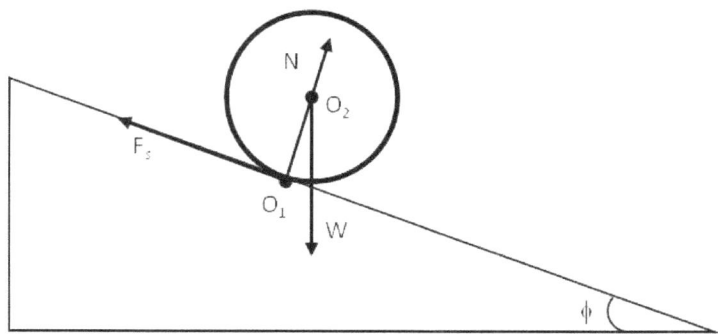

Figure 5.62: Exercise 5.7.34.

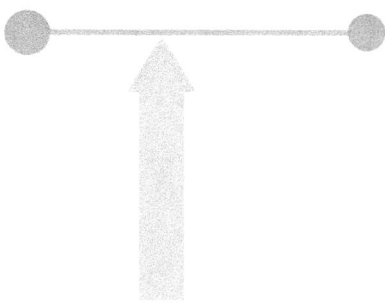

Figure 5.64: Exercise 5.7.37.

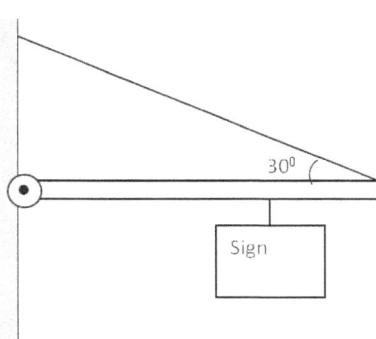

Figure 5.65: Exercise 5.7.39.

without any motion. What are the forces on the beam and where do the forces act on the beam?

Ex 5.7.37. A baton of length L has two different blocks attached on the two ends (Fig. 5.64). Assume the masses of the blocks to be m_1 and m_2. When the baton is placed on a pointed support, it rests horizontally without tipping over. (a) What are the forces on the beam and where do the forces act on the beam? (b) Where is the location of the special point relative to the end with mass m_1 where baton should be placed on the support?

Ex 5.7.38. A diving board of mass 100 kg and length 4 m is bolted at one end and resting on another support at a distance one meter from the bolted end. A diver of mass 50 kg is standing at the other end of the board. (a) Draw a diagram showing all the forces (in symbols) acting on the board, their directions and where they act on the board. For the force from the bolt, use two forces, a horizontal force and a vertical force from the bolt. (b) Use a coordinate system and generate the equations of equilibrium for forces and torques. (c) Determine the force on the board by the bolt. (d) Determine the force on the board from the other support.

Ex 5.7.39. A sign of mass 300 kg is hanging 1 m from the end of a horizontal beam of mass 50 kg and length 3 m held in place by a pin and a cable as shown in the Fig. 5.65. The cable makes an angle of 30° with the beam. (a) Draw a diagram showing all the forces on the beam (using symbols), their directions and where they act on the beam. (b) Choose a Cartesian coordinate system and find the relations among the components for a balanced force condition on the beam. (c) Find the relation between the lever arms and the forces for a balanced torque condition. (d) Use your results to determine the magnitude of the tension in the cable. (e) Find the magnitude and direction of the total force applied by the pin on the beam.

Ex 5.7.40. A board of mass M and length L is supported by two vertical cables on the two ends (Fig. 5.66). A painter of mass m is at the distance L_1 from the left end of the board. (a) Draw a diagram showing (in symbols) all the forces on the board, their directions and where on the board they act. Note: the tensions in the cables may be different. (b) Use the equations for the balanced forces and the balanced torques to determine the magnitudes of all forces in terms of M, m, L, L_1 and g.

Figure 5.66: Exercise 5.7.40.

Ex 5.7.41. A steel cable AB of of length 90 cm supports a beam CB of mass 2 kg and length 60 cm hinged at C as shown in the Fig. **??**. Assuming the length of the cable to be 65 cm, find the tension in the cable and the force on the pin if mass $m = 200$ kg.

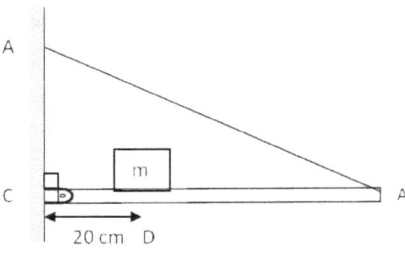

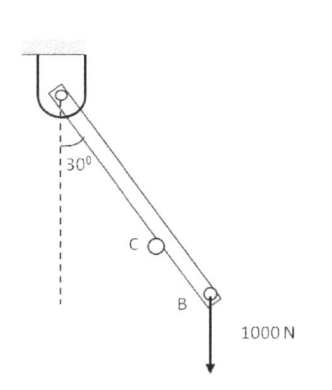

Ex 5.7.42. A rod AB of mass 30-kg and length 1 meter is supported by a pin at A and rests on a frictionless peg at C at a distance of 70 cm from A. A force of 1000 N is applied at B as shown in Fig. 5.67. Find the forces on the rod at A and C.

Figure 5.67: Problem 5.7.42.

5.8 PROBLEMS

Problem 5.8.1. A block of mass 5 kg is attached to two springs S_1 and S_2 of spring constants 300 N/m and 175 N/m respectively (Fig. 5.68). The block is then put on a rough surface so that the coefficient of static friction between the block and surface is 0.65. The spring S_2 is stretched

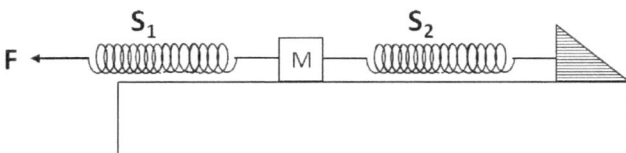

Figure 5.68: Problem 5.8.1.

by 2 cm and tied to a support without making the block move. The other end of the spring S_1 is then pulled with increasing force F. Find the stretch of the spring S_1 and the magnitude of the force F when the block just starts to slide.

Problem 5.8.2. A 100-kg person pulls a rope attached to a strong pole at an angle of 15° from horizontal (Fig. 5.69). The tension in the rope is found to be 800 N. Find the forces of friction and normal from the floor on the person.

Problem 5.8.3. A block is resting on a flat plane board whose angle of incline with respect to the horizontal direction can be varied by placing spacer at one end of the board. It is found that at a certain angle of

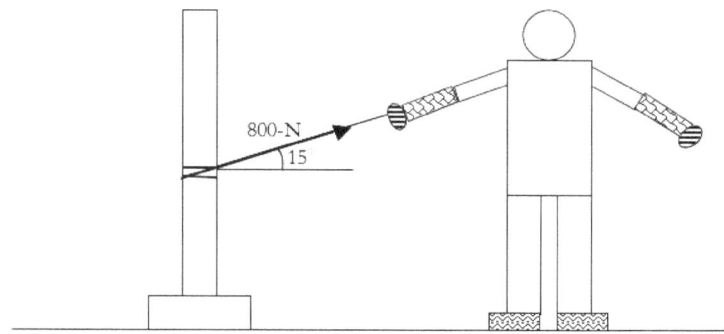

Figure 5.69: Problem 5.8.2.

inclination, say θ_0, the block starts to slide. Prove that the coefficient of static friction between the block and the board is given by $\mu_s = \tan\theta_0$.

Problem 5.8.4. A thin rod of mass 500 grams and length 50 cm rests on a cylinder of radius 20 cm (Fig. 5.70). The rod is attached to a collar of negligible mass, which is free to slide without friction over a vertical guide. Find angle at which the rod must rest so that the collar does not slide.

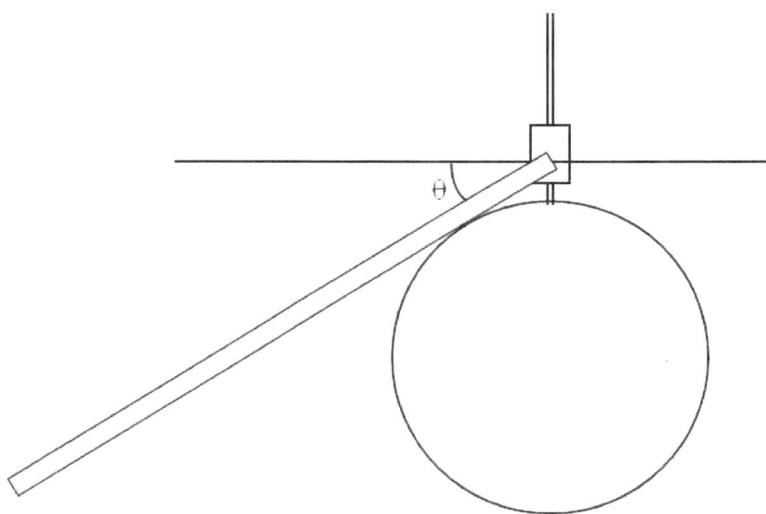

Figure 5.70: Problem 5.8.4.

Problem 5.8.5. A board of mass m and length L is resting against a vertical wall. The coefficient of static friction between the board and the wall and between the board and the floor are μ_w and μ_f, respectively. Find the minimum angle θ the board may be placed without slipping.

Problem 5.8.6. A spherical ball of mass M and radius R is resting against a step of height h with $h < R$ as shown in Fig. 5.71. A horizontal force F is applied to the ball so that its line of action always passes through the center of the ball. Find the minimum force needed to make the ball climb over the step.

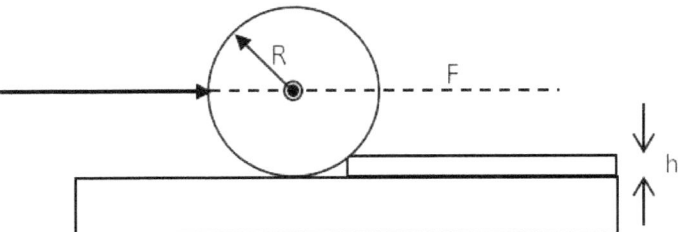

Figure 5.71: Problem 5.8.6.

Problem 5.8.7. A rectangular parallelepiped shaped wooden block of height b and square base of side a is placed on a smooth plane. There is a static frictional coefficient μ_s between the block and the plane. As the angle between the plane and the horizontal table is varied, the block may either tip first, or slide first depending upon the values of a, b and μ_s. Find the condition when the block tips first.

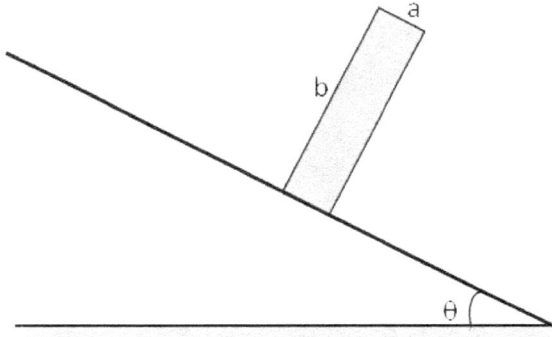

Figure 5.72: v.

Chapter 6

NEWTON'S LAWS OF MOTION

Contents

6.1	**NEWTON'S FIRST LAW OF MOTION** . . .	**220**
6.2	**NEWTON'S SECOND LAW OF MOTION** .	**223**
	6.2.1 Definition of Mass	223
	6.2.2 Definition of Momentum	223
	6.2.3 The Second Law	225
	6.2.4 Operational Definition of Force and Mass . . .	228
6.3	**NEWTON'S THIRD LAW OF MOTION** . . .	**230**
6.4	**PROBLEMS CONCERNING MOTION OF PARTICLES**	**232**
6.5	**PROBLEMS CONCERNING MOTION OF EXTENDED BODIES**	**238**
	6.5.1 General Considerations	238
	6.5.2 Examples of Translational Motion of one Body	240
	6.5.3 Coupled Systems	244
6.6	**EXERCISES**	**254**
6.7	**PROBLEMS**	**260**

INTRODUCTION

In the last chapter we studied various forces between objects. According to Newton's third law of motion, an interaction between two bodies produces forces on the two bodies. In the last chapter, we used Newton's first law of motion to study systems in equilibrium. We found that, if forces on a system are balanced, then the system remains at rest if it was at rest to begin with.

In this chapter, we will ask: what happens when forces are not balanced? The answer to this question in contained in Newton's second law of motion. The second law says that an unbalanced force will cause change in momentum of the system. For a system with unchanging mass, this means that force causes a system to accelerate, i.e a change in velocity. We will study several examples of systems out of equilibrium to get a sense of how to apply Newton's second law of motion.

Newton's second law of motion also presents us other ways of looking at Newton's first and third laws. Therefore, in this chapter we will take a second look at Newton's first and third laws. Newton's first law is not only about the objects at rest, but also about the objects in perpetual motion in a straight line if observed from a particular frame. Additionally, we will find that Newton's third law results in a powerful result concerning the conservation of momentum. To help us better appreciate the impact and the use of Newton on mechanics, we will present the three laws of motion in a coherent language in this chapter.

6.1 NEWTON'S FIRST LAW OF MOTION

Before Galileo Galilei (1564-1642), people generally believed that a continuous push was necessary to maintain the motion of an object, although some philosophers had suggested otherwise. From his experiments Galileo convincingly showed that

"A body moving on a level surface will continue in the same direction at constant speed unless disturbed".

It is easy to demonstrate this fact of nature by the simple experiment shown in Fig. 6.1. Suppose you take two inclined surfaces and roll a marble from some fixed place on one incline, and observe what happens on the second plane. If the surfaces are hard and the ball rolls without sliding, you will find that the ball reaches approximately the same height on the second plane. Now, notice what happens as you reduce the angle of the incline of the second inclined surface. The same height on the second incline now appears at a longer distance from the base. Galileo concluded that, if the second incline was a truly horizontal surface, then the ball will continue to travel at a constant speed indefinitely on the horizontal track.

6.1. NEWTON'S FIRST LAW OF MOTION

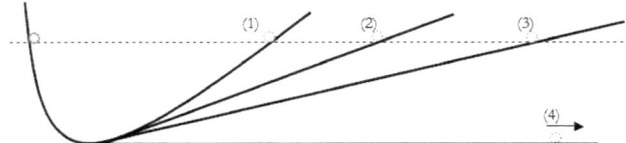

Figure 6.1: A ball is rolled on an inclined plane and allowed to continue on a second inclined plane. The ball reaches approximately the same height regardless of the angle of incline of the second surface. With less angle of incline, the ball goes farther, and in the limit of zero degree incline, i.e. a horizontal surface, the ball would continue for ever in a straight line with a constant speed.

Isaac Newton generalized Galileo's conclusions and enunciated his **first law** in *Principia* as follows.

"Every body preserves in its state of rest, or of uniform motion in a right line, unless it is compelled to change that state by forces impressed thereon".

The reason for continuation of uniform motion of a body is not known. The tendency of material bodies for preserving their state of rest or of motion is called **inertia**. In *Principia* Newton included the following definition of this innate tendency of all bodies and called it inertia.

"The *vis insita*, or innate force of matter, is a power of resisting, by which every body, as much as in it lies, endeavors to preserve in it present state, whether it be of rest, or of moving uniformly forward in a right line".

Newton went on to further explain the nature of this so-called innate force.

"This force is ever proportional to the body whose force it is; and differs nothing from the inactivity of the mass, but in our manner of conceiving it. A body, from the inactivity of matter, is not without difficulty put out of its state of rest or motion. Upon which account, this *vis insita*, may, by a most significant name, be called *vis inertia*, or force of inactivity. But a body exerts this force only, when another force, impressed upon it, endeavors to change its condition; and the exercise of this force may be considered both as resistance and impulse; it is resistance, in so far as the body, for maintaining its present state, withstands the force impressed; it is impulse, in so far as the body, by not easily giving way to the impressed force of another, endeavors to change the state of the other. Resistance is usually ascribed to bodies at rest, and impulse to those in motion; but motion and rest, as commonly conceived, are only relatively distinguished; nor are those bodies always truly at rest, which commonly are taken to be so. "

Unlike Newton, physicists now do not believe on any innate force as the reason for inertia. Instead, the first law of motion is understood to highlight the role of special frames of observation.

Role of the frame of observation

The tendency to maintain inertia is evident only if the motion is observed from particular types of reference frames, called **inertial frames**, which are themselves defined as frames of those observers who would observe this tendency. Thus, the definition of inertial frames only restates what is meant by inertia in a more operational sense.

For instance, if you were to observe a tree from the roadside, yourself at rest with respect to the road. Then, you would find that the tree is at rest with respect to you, which is no surprise at all.

However, if you were to look at the same tree from inside an accelerating car, the tree will appear to move with acceleration in the other direction to the car's acceleration, even when there is no net force on the tree. In this frame, the tree appear to acquire acceleration without any impressed force. Therefore, the observation from inside an accelerating car violates Newton's first law of motion.

How about an observation from inside a car that is moving at a constant velocity with respect to the road? This observer will assign a constant velocity to the tree in the other direction to the car's velocity with respect to the road. The observer will now say that the tree is moving with a constant velocity and has no net force on it. This observation agrees with the first law of motion.

Thus, we see that two observers, one at rest with respect to the road and the other at constant velocity with respect to the road, agree with the observation of Galileo that no force on the object would correspond to no change in velocity. But, the other observer, the one accelerating with respect to the road, does not agree with this conclusion. The observers with respect to whom the law of inertial is obeyed are called inertial observers. We state the first law of motion as a requirement of existence of these special frames in nature.

There exist reference frames in nature, called inertial frames, such that, when observed from these frames, an object will have no acceleration if there is no net force on the object.

Supposing that the center of the universe is at rest, then any frame that is at rest with respect to this center or moving at constant velocity with respect to this center will be an inertial frame. A frame fixed to earth is actually rotating and is not an inertial frame. But, for many practical purposes, we find that the non-inertiality of a frame fixed to the Earth does not make significant impact. Therefore, we will assume that an earth-based frame is an inertial frame until we discuss physics in non-inertial frames in a later chapter. There, you will study the phenomena in which the non-inertiality of earth-based frames plays a crucial role, and must be taken into account for properly making sense of the observations made by earth-based observers.

6.2 NEWTON'S SECOND LAW OF MOTION

Newton's second law addresses the central question of motion: what causes change in motion of an object? Before we can discuss the second law of motion, we need definitions of mass and momentum.

6.2.1 Definition of Mass

In *Principia* Newton gives a simple definition of mass based on the density and volume.

"The quantity of matter is the measure of the same, arising from its density and bulk conjunctly".

Newton went on to further clarify the definition, "Thus air of a double density, in a double space, is quadruple in quantity; in a triple space, sextuple in quantity. The same thing is to be understood of snow, and fine dust or powders, that are condensed by compression or liquefaction; and of all bodies that are by any causes whatever differently condensed".

This definition of mass was common in Newton's time and he did not feel the need to define mass in other ways. Now a days, it is more common to define mass operationally by the response of different materials to the same force.

The definition of mass based on response to a known force is independent of the nature of the force and is called **inertial mass**. The inertial mass is sometimes distinguished from the mass used for the gravitational force; the later is called **gravitational mass**.

Although the definition of mass given in *Principia* is quite adequate for our present purpose of introducing the basic aspects of the second law of motion, we will present below the definition of mass based on the second law of motion later in this section along with the operational definition of force.

6.2.2 Definition of Momentum

Newton defined a measure of the quantity of motion we call momentum now.

"The quantity of motion, is the measure of the same, arising from the velocity and quantity of matter conjunctly".

Let m be the mass of a body and $\vec{v}$ the velocity, then the **momentum** $\vec{p}$ is defined as

$$\vec{p} = m\,\vec{v}. \tag{6.1}$$

Momentum has the same direction as velocity since mass is always a pos-

itive real number. The dimensions of momentum are:
$$[p] = \frac{[M][L]}{[T]}.$$
Therefore, the metric unit for momentum is kg.m/s. A momentum of a shot put of mass 5 kg thrown with speed 10 m/s is 50 kg.m/s in the direction of the velocity.

If a system consists of more than one particle, then the momentum of the whole system will be equal to the vector sum of the momenta of all of its parts. Let $\vec{p}_1, \vec{p}_2, \cdots, \vec{p}_N$ be the momenta of N parts of a system, then the momentum of the whole, the net momentum is given as

$$\boxed{\vec{p}_{net} = \vec{p}_1 + \vec{p}_2 + \cdots + \vec{p}_N.} \qquad (6.2)$$

Note that the sum of momenta is a vector sum and one must be careful not to just add the magnitudes of the momenta since directions are equally important. For instance, the total momentum of two objects of equal mass and equal speed but moving towards each other in a straight line is equal to zero since the vector sum of two equal magnitude but oppositely directed vectors is zero.

Example 6.2.1. Momentum of one body. A car of mass 3000 kg is moving to the East at a speed of 30 m/s. How much momentum does the car have?

Solution. From the definition of the momentum, we can immediately state the magnitude and direction of the momentum of the car.

Magnitude $= mv = (3000 \text{ kg}) \times (30 \text{ m/s}) = 90,000$ kg.m/s.

Direction: Towards the East.

Example 6.2.2. Total momentum of two bodies moving along the same line. A car of mass 3000 kg is moving to the East at a speed of 30 m/s and a truck of mass 6000 kg is moving to the West at a speed of 20 m/s. What is the net momentum of the two?

Solution. By applying the definition of momentum to the car and the truck separately, we find that the car has a momentum of 90,000 kg.m/s pointed towards the East and the truck has a momentum of 120,000 kg.m/s pointed towards West. Now, we need to add these two vectors. Since their directions are opposite, they will subtract, giving a net momentum of 30,000 kg.m/s pointed towards the West.

You can also use the analytic method to add the two vectors by utilizing a Cartesian coordinate. To be concrete, let us point the x axis towards the East. Then, the two momentum vectors will have the following x-components (the y and z-components of both being zero).

$p_{car,x} = 90,000$ kg.m/s

$p_{truck,x} = -120,000$ kg.m/s

(negative since $\vec{p}_{truck}$ is pointed towards negative x-axis)

6.2. NEWTON'S SECOND LAW OF MOTION

Therefore, the net momentum has the following components.

$$p_{net,x} = -30,000 \text{ kg.m/s}$$
$$p_{net,y} = 0$$
$$p_{net,z} = 0$$

Therefore, the net momentum has the magnitude $30,000$ kg.m/s pointed towards the negative x-axis, which is towards the West.

Example 6.2.3. Total momentum of two bodies moving in a plane. A car of mass 3000 kg is moving to the East at a speed of 30 m/s and a truck of mass 6000 kg is moving to the North at a speed of 20 m/s. What is the net momentum of the two?

Solution. By applying the definition of momentum to the car and the truck separately, we find that the car has a momentum of $90,000$ kg.m/s pointed towards the East and the truck has a momentum of $120,000$ kg.m/s pointed towards the North. Now, we need to add these two vectors. Since the vectors are not in one line, we need to employ the parallelogram law of addition of vectors, either geometrically or analytically. We have seen that analytical method is often easier to implement.

To use the analytical method, as usual we start with a choice of coordinate system. Let us point the x-axis towards the East, the y-axis towards the North and the z-axis vertically up. Then, we see that the two momenta have the following representation in components.

$$\vec{p}_{car} = (90,000 \text{ kg.m/s}, 0, 0)$$
$$\vec{p}_{truck} = (0, 120,000 \text{ kg.m/s}, 0)$$

where x, y and z-components of each vector have been listed in order. Therefore, the net momentum has the following representation in the given coordinate system.

$$\vec{p}_{net} = (90,000 \text{ kg.m/s}, 120,000 \text{ kg.m/s}, 0)$$

From the components of the net momentum vector, we can easily determine the magnitude and direction.

$$\text{Magnitude} = \sqrt{(90,000)^2 + (120,000)^2} = 150,000 \text{ kg.m/s}$$
$$\text{Direction: } \arctan\left(\frac{120,000}{90,000}\right) = 53° \text{ from the East towards the North.}$$

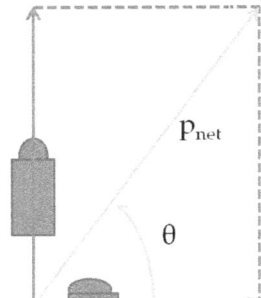

Figure 6.2: The net momentum of car and truck is the vector sum of their separate momenta.

6.2.3 The Second Law

The second law of motion addresses how the state of motion of an object changes when an external force acts on the object. In Newton's own words from *Principia*,

" The alteration of motion is ever proportional to the motive force impressed; and is made in the direction of the right line in which that force is impressed".

Newton explains the physical content of the law as follows.

"If any force generates a motion, a double force will generate double the motion, a triple force triple the motion, whether that force be impressed altogether and at once, or gradually and successively. And this motion (being always directed the same way with the generating force), if the body moved before, is added to or subdued from the former motion, accordingly as they directly conspire with or are directly contrary to each other; or obliquely joined, when they are oblique, so as to produce a new motion compounded from the determination of both".

The alteration of motion here is meant to be the rate of change of momentum. We have seen above the special role of the inertial frames. We will recast Newton's second law using modern terminology.

<u>Observed from an inertial reference frame, the rate of change of momentum of a body at an instant is proportional to the net force on the body acting at that instant.</u>

Therefore, if a force $\vec{F}$ acting on a body for a time Δt causes a change in momentum $\Delta \vec{p}$, then Newton's second law says that

$$\frac{\Delta \vec{p}}{\Delta t} \propto \vec{F}. \tag{6.3}$$

We can write this statement as equality as follows.

$$\frac{\Delta \vec{p}}{\Delta t} = k\vec{F}, \tag{6.4}$$

where k is a proportionality constant. We fix the value of the proportionality constant k to 1 by choosing the units of p, F and m. Therefore, we define 1 N to be a force that will change the velocity of a 1 kg object by 1 m/s if the force acts on the body for 1 s. With these units we have the following statement of Newton's second law.

$$\frac{\Delta \vec{p}}{\Delta t} = \vec{F}. \tag{6.5}$$

Now, taking the limit $\Delta t \to 0$ we obtain instantaneous relationship between the rate of change of momentum and the force at that instant.

$$\lim_{\Delta t \to 0} \frac{\Delta \vec{p}}{\Delta t} = \vec{F}, \tag{6.6}$$

which is written as

$$\boxed{\frac{d\vec{p}}{dt} = \vec{F}.} \tag{6.7}$$

Therefore, a more rapid change of momentum, whether direction or magnitude would require a larger force. Since, the momentum is a product of mass and velocity, you could change the momentum by either changing the mass of your system or the velocity. Rewriting Eq. 6.5 we obtain the change $\Delta \vec{p}$ in momentum in an interval Δt as

$$\Delta \vec{p} = \vec{F} \Delta t, \tag{6.8}$$

6.2. NEWTON'S SECOND LAW OF MOTION

which says that the longer a force acts on a body, the larger the change in momentum, and larger the force the larger the change in momentum.

The law makes sense when you think about your own experience with forces and motion. For instance, when you push a cart gently, it picks up a low velocity, and when you push the cart harder, the corresponding pick up in velocity is larger. Also, longer you push on a cart consistently the more speed it picks up. These qualitative observations let us verify the essential point of Eq. 6.7 that velocity changes as we exert a force on the system. controlled experiments in laboratory have demonstrated the correctness of the proportionality of force and change in momentum.

Other forms of second law

Newton wrote his second law of motion in the form we have presented above, namely in terms of the rate of change of momentum.

$$\frac{d\vec{p}}{dt} = \vec{F},$$

which is also written as

$$\vec{F} = \frac{d\vec{p}}{dt}. \tag{6.9}$$

In most elementary textbooks, we find another form of Newton's second law, which is obtained when mass of the system does not change with time. Let us deduce the form of second law for constant mass case.

We replace momentum in Eq. 6.9 by its definition, $\vec{p} = m\vec{v}$ to get

$$\vec{F} = \frac{d\vec{p}}{dt} = \frac{d}{dt}(m\vec{v}).$$

Now, the chain rule of differentiation gives two terms on the right side.

$$\vec{F} = \left(\frac{dm}{dt}\right)\vec{v} + m\left(\frac{d\vec{v}}{dt}\right),$$

where the first term is the rate of change of momentum due to changing mass, as in the case of a rocket when it is releasing burnt fuel from the exhaust, and the second term is the rate of change of momentum due to acceleration. For most systems, the mass does not change, and therefore the simpler equation results.

$$\vec{F} = m\left(\frac{d\vec{v}}{dt}\right),$$

which is the same thing as

$$\boxed{\vec{F} = m\vec{a}. \text{ (mass constant)}} \tag{6.10}$$

You should also note that Newton's second law, whether in the form given in Eq. 6.7 or in Eq. 6.10, are **instantaneous statements**. That is, the second law is a statement about the relation between the force acting on a body and the resulting rate of change of momentum or mass times acceleration at the same instant.

Another point to note is that Newton's second law gives **relation among vectors**, the force and the rate of change of momentum. Thus, equations 6.7 and 6.10 are **vector equations**. You must be very careful when making algebraic manipulations with them since the laws of vector addition and multiplication are very different from those for scalar numbers.

Algebraic manipulations are often easier in analytic picture. In the analytic picture, we work with components of vectors in a coordinate system of our choice. The components separate into separate equations as usual.

Component	Always correct	Correct when mass constant
x component	$F_x = dp_x/dt$	$F_x = m\, a_x$
y component	$F_y = dp_y/dt$	$F_y = m\, a_y$
z component	$F_z = dp_z/dt$	$F_z = m\, a_z$

If more than one force acts on a particle, each force contributes to the change in momentum. The net acceleration of the particle subject to more than one force is the vector sum of the contributions of each force.

Multiple forces on a single particle: $\left\{ \begin{array}{l} \vec{F}_1 = m\vec{a}_1 \\ \vec{F}_2 = m\vec{a}_2 \\ \ldots \\ \vec{F}_N = m\vec{a}_N \end{array} \right\} \implies \vec{F}_{\text{net}} = m\vec{a}_{\text{net}}.$

6.2.4 Operational Definition of Force and Mass

Newton's second law of motion transforms an intuitive notion of force into a quantifiable entity. The quantifiable definition of force also helps us develop an operational definition of mass. Let us first look at how mass is given a more fundamental meaning by second law of motion than just the notion of mass being the "quantity" of matter.

Operational definition of mass

Imagine pulling two objects on a frictionless table with equal forces. For example, you could attach identical springs to different blocks, and pull on them so that the stretch is the same in the two so as to make sure equal forces pull on them as shown in Fig. 6.3.

You will find the accelerations of the two objects will in general be different, but the ratio of the magnitudes of the accelerations for two

6.2. NEWTON'S SECOND LAW OF MOTION

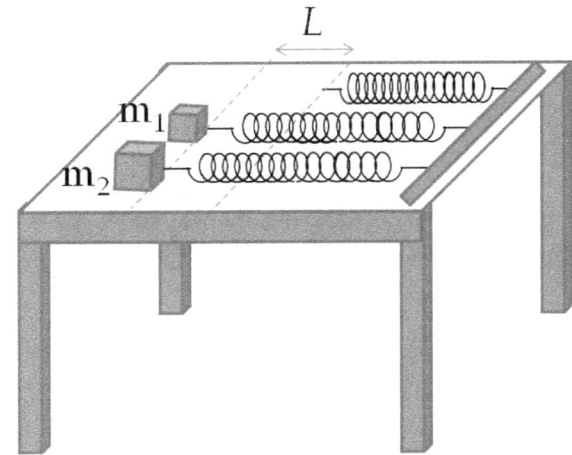

Figure 6.3: Two objects pulled with equal force on a frictionless surface have unequal accelerations.

masses subject to equal forces is a constant.

$$\frac{a_1}{a_2} = \text{constant} \quad \text{(same force on 1 and 2.)} \qquad (6.11)$$

The constant varies among different pairs of objects. The unequal accelerations for two different blocks corresponds well to the "quantity" of material in the two blocks and gives an operational way for defining mass. The acceleration for the objects with a higher "quantity" of matter gets less acceleration than the object with a lower "quantity" of matter. Therefore, mass can be defined by comparing the accelerations of two objects when subjected to the same force.

$$\frac{m_2}{m_1} = \frac{a_1}{a_2} \quad \text{(same force on 1 and 2.)} \qquad (6.12)$$

Now, if we choose one of the masses as a standard mass, the masses of all other objects can be determined by comparing their accelerations. The standard mass sanctioned by the second Conférence Générale des Poids et Mesures (CGPM) conference in 1889 is a 1-kilogram cylinder of platinum-iridium alloy kept at the Bureau International des Poids et Mesures (BIPM) in Paris in France. This is the only standard unit still being represented by a man-made sample rather than naturally occurring phenomenon or unit such as mass of a particular atom.

Operational definition of force

Our experience with push and pull give us an intuitive notion of force. But, how do we compare the magnitudes of two forces quantitatively? The second law of motion provides an operational way to define a force. Once again, it is easier to illustrate the idea of force using spring force, which is proportional to the stretch. Attach a spring to a mass m and pull it on a frictionless surface such that the spring is stretched by some length L.

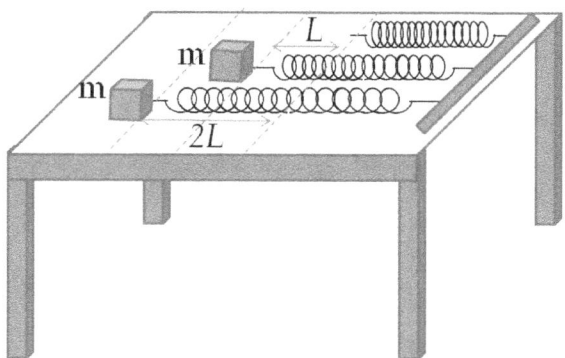

Figure 6.4: Twice the force on the same mass causes twice the acceleration.

Measure the acceleration of the mass as you let go of it. Now, if you pull twice as hard such that spring is stretched by twice as much, 2 L, and let go of the mass. This time, you will find that the resulting acceleration is twice as much with the stretch 2 L of the spring as when the stretch was only L. Thus, acceleration is directly proportional to the applied force.

Changing mass but keeping the force same gave us the conclusion that acceleration produced by the same force is inversely proportional to mass, while changing force but keeping the mass same shows that the magnitude of acceleration is proportional to the magnitude of force. Putting these two together, we find that acceleration a, force F and mass m are related as enunciated in the second law.

$$a \propto \frac{1}{m} \text{ and } a \propto F \implies a = k\frac{F}{m}$$

where k is a proportionality constant, which is absorbed in the units for force. For instance, if we call the magnitude of force that causes 1-kg mass to accelerate at 1 m/s^2 the unit force, we will be able to set $k = 1$. The unit so defined is called 1 Newton (N).

6.3 NEWTON'S THIRD LAW OF MOTION

The third law of motion addresses the nature of mutual interactions between two bodies as we have studied in the last chapter. Recall the statement of Newton's third law given in the last chapter.

<u>To every action there is always opposed an equal reaction; or the mutual actions of two bodies upon each other are always equal, and directed to the contrary parts.</u>

Let $\vec{F}_{AB}$ denote the force on body A by body B, and $\vec{F}_{BA}$ the force on B by A. Then, Newton's third law states that:

$$\vec{F}_{AB} = -\vec{F}_{BA}. \tag{6.13}$$

Here, the minus sign on right side of the equation tells us that the two forces are acting in the opposite directions. <u>Since, only one of these forces</u>

6.3. NEWTON'S THIRD LAW OF MOTION

acts on any body, you do not use both forces in any problem concerning only one of the two bodies.

What happens if we choose a system that includes both bodies? The consideration of this question leads to an important result of mechanics and gives us another view of Newton's third law.

Let us imagine a world where only two bodies A and B exist which may exert a force on each other. We say that the two bodies make up an isolated system. Any two bodies that interact with each other and not with anything else can be an example of such a world.

According to the second law of motion, the total momentum of the combined system cannot change with time since there is no external force on the isolated system. However, if you look at A or B, the momentum of each is changed by the force by the other. Since the total momentum does not change with time, the change in the momentum ($\Delta \vec{p}_A$) of A must be exactly equal in magnitude to the change in the momentum ($\Delta \vec{p}_B$) of B and must be pointed in exactly opposite direction.

$$\Delta \vec{p}_A = -\Delta \vec{p}_B \ (A \text{ and } B \text{ isolated from the rest of universe}). \quad (6.14)$$

This relation gives another view of forces: forces are vehicles for exchange of momentum between interacting objects - while one body gains momentum, the other body must lose momentum at the same time. Let us write Eq. 6.14 for an interval from t_1 to t_2. Let the momenta of A and B at t_1 be $\vec{p}_1^A$ and $\vec{p}_1^B$ and at t_2 be $\vec{p}_2^A$ and $\vec{p}_2^B$ respectively. Then Eq. 6.14 will be

$$\vec{p}_2^A - \vec{p}_1^A = -\left(\vec{p}_2^B - \vec{p}_1^B\right), \quad (6.15)$$

which can be rearranged so that the momenta at one instant are on one side of the equation to obtain

$$\vec{p}_1^A + \vec{p}_1^B = \vec{p}_2^A + \vec{p}_2^B, \quad (6.16)$$

Eq. 6.16 states a very important principle of physics - **the principle of conservation of momentum**:

If a system is isolated such that the net external force is zero then the total momentum is constant in time, even though momentum of its constituents may change with time.

Since the principle of conservation of momentum plays a major role in physics, we will take up the subject in a separate chapter.

Experimental test of the third law

One way of testing the third law experimentally involves two carts placed on a smooth horizontal track as shown in Fig. 6.5.

We attach a spring to one cart and use the other cart to press the two carts together to hold them still on the track. After we let go of the carts, each cart applies a force on the other through the spring. As a result the

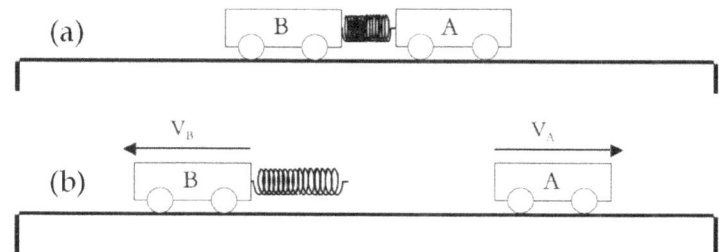

Figure 6.5: (a) Two carts with a compression spring attached to one cart and compressed between them are let go from rest so that their starting momenta are zero. (b) After release, the two carts move in opposite directions such that the magnitude of the change in momenta of the two carts are equal: $|M_A V_A| - 0 = |M_B V_B| - 0$.

momenta of the two carts change which can be determined by the speed of the carts once they are no longer in contact.

The measurement of speeds of the two carts shows that the amount of change in momentum of the two carts are equal. And since the carts are moving in opposite directions, the momentum change vectors of the carts are pointed in the opposite directions. Therefore, the total momentum of the two carts is still zero even when they are moving away from each other. Thus, momentum before release is same as the momentum afterwards, confirming the fact that the forces exerted by the two carts in each other are equal in magnitude and opposite in directions.

6.4 PROBLEMS CONCERNING MOTION OF PARTICLES

Example 6.4.1. Change in momentum due to a constant force. A particle of mass m interacts with another object A such that the force on the particle is constant with magnitude F and direction towards A. What is the change in momentum of the particle over the interval t_1 to t_2?

Solution. This problem is just an application of Newton's second law. Since, the rate of change of momentum is constant, we can write Newton's second law for finite interval of time rather than the instantaneous form.

$$\frac{\Delta \vec{p}}{\Delta t} = \vec{F},$$

which gives the change in momentum over the interval t_1 to t_2 as

$$\Delta \vec{p} = \vec{F} \Delta t = \vec{F}(t_2 - t_1).$$

Therefore, the change in momentum has a magnitude of $F \times (t_2 - t_1)$ and has the same direction as the direction of the force.

Example 6.4.2. Instantaneous acceleration due to a constant force. A particle of mass m interacts with another object A in the time interval

6.4. PROBLEMS CONCERNING MOTION OF PARTICLES

$t_1 \le t \le t_2$ such that the force on the particle is constant with magnitude F and direction towards A. What is the instantaneous acceleration at an arbitrary time t?

Solution. This problem is an application of Newton's second law for constant mass case. At any instant, the net force is equal to mass times acceleration. Therefore, acceleration is obtained by dividing the force on the particle by the mass of the particle.

$$\vec{a} = \frac{\vec{F}}{m}.$$

This says that the acceleration at as arbitrary instant t during t_1 to t_2 has a magnitude of F/m and has the same direction as the direction of the force.

Example 6.4.3. Change in velocity due to a constant force. A particle of mass m interacts with another object A such that the force on the particle is constant with magnitude F and direction towards A. (a) What is the change in velocity of the particle over the interval t_1 to t_2? (b) If the velocity of the particle at $t = 0$ was $\vec{v}_0$, what is its velocity $\vec{v}$ at an arbitrary instant t?

Solution. (a) We can make use of either Example 6.4.1 or 6.4.2. From Example 6.4.1, we know the change in momentum from t_1 to t_2 for the same situation, and since mass is constant here, we find the change in velocity by simply dividing by m.

$$\Delta \vec{v} = \frac{\vec{F}}{m}(t_2 - t_1).$$

(b) Now, setting $t_1 = 0$, $t_2 = t$, and $\Delta \vec{v} = \vec{v} - \vec{v}_0$ in the result of part (a) we find the velocity at an arbitrary time.

$$\vec{v} = \vec{v}_0 + \frac{\vec{F}}{m} t. \tag{6.17}$$

Example 6.4.4. Change in velocity due to a constant force. A particle of mass m falls near Earth under its weight. (a) What is the change in velocity of the particle over the interval t_1 to t_2? (b) If the velocity of the particle at $t = 0$ was $\vec{v}_0$, what is its velocity $\vec{v}$ at an arbitrary instant t?

Solution. (a) The constant force is equal to mg pointed down. Therefore, we find the change in velocity

$$\Delta \vec{v} = \{g(t_2 - t_1), \text{pointed down}\}.$$

(b) Now, setting $t_1 = 0$, $t_2 = t$, and $\Delta \vec{v} = \vec{v} - \vec{v}_0$ in the result of part (a) we find the velocity at an arbitrary time.

$$\vec{v} = \vec{v}_0 + \{gt, \text{pointed down}\}. \tag{6.18}$$

We can write these results analytically in a coordinate system. Let y-axis be pointed up, and the horizontal plane have x and z-axes. Then, we will get the following components for Eq. 6.18.

$$v_x = v_{0x},$$
$$v_y = v_{0y} - gt,$$
$$v_z = v_{0z},$$

which says that the vertical motion is a constant acceleration motion and the horizontal motion a constant velocity motion, as we expect in a situation of an object that is falling freely. This is the projectile motion problem we have studied before.

Example 6.4.5. Change in position due to a constant force. A particle of mass m interacts with another object A such that the force on the particle is constant with magnitude F and direction towards A. (a) What is the displacement of the particle over the interval t_1 to t_2? (b) If the position of the particle at $t = 0$ was $\vec{r}_0$, what is its position at an arbitrary instant t?

Solution. (a) Newton's second law relates force with the rate of change of momentum or acceleration if mass is constant. So, from the given information of a constant force on the particle, we obtain the constant acceleration. A constant acceleration gives velocity that changes linearly with time as given in Eq. 6.17. Let us write the velocity at an arbitrary time as $\vec{v}(t)$ to display the time-dependence of the velocity vector. We wish to determine the displacement vector, which can be obtained by making use of the definition of velocity.

$$\frac{d\vec{r}}{dt} = \vec{v}(t),$$

which can be written as a differential relation:

$$d\vec{r} = \vec{v}(t)dt.$$

Now we can integrate over the time interval. On the left side the limits are $\vec{r}_1$ and $\vec{r}_2$ corresponding to the limits on the right of t_1 and t_2.

$$\int_{\vec{r}_1}^{\vec{r}_2} d\vec{r} = \int_{t_1}^{t_2} \vec{v}(t)dt.$$

This integral is usually done by expressing the displacement vector in Cartesian components giving us three integrations, one each for x, y and z.

$$\int_{x_1}^{x_2} dx = \int_{t_1}^{t_2} v_x(t)dt.$$
$$\int_{y_1}^{y_2} dy = \int_{t_1}^{t_2} v_y(t)dt.$$
$$\int_{z_1}^{z_2} dz = \int_{t_1}^{t_2} v_z(t)dt.$$

The velocity for this problem given in Eq. 6.17 can be integrated to find the change in position. After performing the integrations separately, we combine the results and write it in the vector form.

$$\vec{r}_2 - \vec{r}_1 = \vec{v}_0\,(t_2 - t_1) + \frac{1}{2}\frac{\vec{F}}{m}(t_2 - t_1)^2. \tag{6.19}$$

(b) We put $t_1 = 0$, $t_2 = t$, $\vec{r}_1 = \vec{r}_0$ and $\vec{r}_2 = \vec{r}$ in Eq. 6.19 to obtain the standard constant acceleration kinematics equation in the vector form.

$$\vec{r} = \vec{r}_0 + \vec{v}_0 t + \frac{1}{2}\frac{\vec{F}}{m}t^2. \tag{6.20}$$

Note that this equation is a vector equation, which will give rise to three equations for components.

$$x = x_0 + v_{0x}t + \frac{1}{2}\frac{F_x}{m}t^2.$$
$$y = y_0 + v_{0y}t + \frac{1}{2}\frac{F_y}{m}t^2.$$
$$z = z_0 + v_{0z}t + \frac{1}{2}\frac{F_z}{m}t^2.$$

Typically, if the force is constant, you would orient one of the Cartesian axes along the direction of the constant force vector. Suppose, you orient the x-axis in the direction of the force then y and z motions will be constant velocity the motions.

$$x = x_0 + v_{0x}t + \frac{1}{2}\frac{F_x}{m}t^2.$$
$$y = y_0 + v_{0y}t.$$
$$z = z_0 + v_{0z}t.$$

Example 6.4.6. Uniform circular motion. A stone of mass m is tied with a string and swung in a horizontal circle of radius R with a constant speed v. (a) What is the acceleration of the stone? (b) Which force is responsible for the acceleration? Find the magnitude and direction.

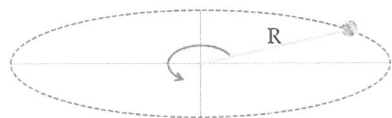

Figure 6.6: Example 6.4.6.

Solution. (a) A particle in a uniform circular motion, defined as a circular motion with constant speed, has the acceleration pointed towards the center of the circle with magnitude v^2/R.

(b) The force is the force of tension between the stone and the hand linked by the string. The tension force acting on the stone is directed towards the hand which is at the center of the circle. The magnitude of the tension force on the stone is $ma = mv^2/R$.

$$\vec{T} = -\left(\frac{mv^2}{R}\right)\hat{u}_r.$$

Note that this force is not a constant force! Why? Although the magnitude is constant, the direction is changing, since, in order to be always pointed towards the center of the circle, the direction in space would have to change as the position of the particle changes around the circle.

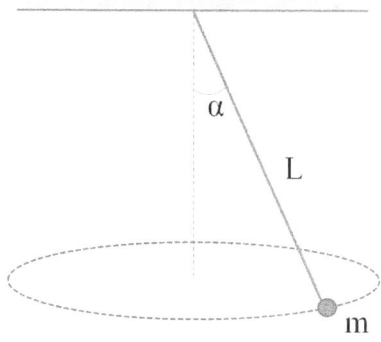

Figure 6.7: Example 6.4.7.

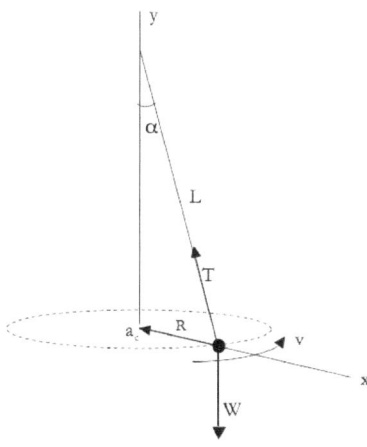

Figure 6.8: Example 6.4.7. The forces on a conical pendulum and choice of axes. The figure is drawn at an instant when the ball is crossing the x-axis so that the centripetal acceleration is pointed towards the negative x-axis.

Example 6.4.7. Conical pendulum. A ball of mass m is tied to a string of length L and suspended from the ceiling. The ball is then swung smoothly with a constant speed in a horizontal circle so that the point of suspension and the horizontal circle form a cone of apex angle α (Fig. 6.7). Find the tension in the string and the speed of the ball in terms of m, L, α, and g.

Solution. Let v be the constant speed of the ball. Since speed of the ball is constant, its acceleration is all centripetal. That is, the acceleration of the ball is pointed towards the center of the circle of motion and has the magnitude given by v^2/R, where $R(= L \sin\alpha)$ is the radius of the circle as seen from Fig. 6.8.

There are only two forces on the ball: the weight W of the ball and the tension T in the string, ignoring minor effects of air resistance. At any instant, the forces fall in one plane. We examine the equation of motion at an arbitrary instant when the ball is crossing x-axis such that the two forces have only x and y-components non-zero, and the acceleration has only x-component non-zero. Organizing the information about forces and their components in a table is quite helpful.

Force name	x-component	y-component
$\vec{W}$	0	$-mg$
$\vec{T}$	$-T \sin\alpha$	$T \cos\alpha$
$\vec{F}_{\text{net}}$	$-T \sin\alpha$	$T \cos\alpha - mg$

The corresponding components of the acceleration are:

$$a_x = -\frac{v^2}{R} = -\frac{v^2}{L \sin\alpha}, \quad a_y = 0.$$

Here a_x is negative since the acceleration at the instant is pointed towards the negative x-axis. Now equating the x-component of force to mass times x-component of acceleration, and similarly for the y-components, we find

$$-T \sin\alpha = m \left(-\frac{v^2}{L \sin\alpha}\right) \quad (6.21)$$

$$T \cos\alpha - mg = 0. \quad (6.22)$$

From Eq. 6.22 we find that the magnitude of tension in the string is

$$T = \frac{mg}{\cos\alpha}. \quad (6.23)$$

Now, putting T from Eq. 6.23 into Eq. 6.21 and solving for speed v we find

$$v = \sqrt{g L \sin\alpha \tan\alpha}.$$

It is interesting to note that the speed of the ball is independent of its mass. Therefore, all conical pendula of the same length L will have an

6.4. PROBLEMS CONCERNING MOTION OF PARTICLES

identical period ($2\pi R/v = 2\pi L \sin\alpha/v$) if swung with the same angle α regardless of their masses.

Example 6.4.8. Plane pendulum - An example of a motion in a vertical circle.

A ball of mass m tied to a string of length L is suspended from the ceiling. The ball is pulled to one side so that the string makes an angle θ_0 with the vertical and let go from rest. The ball swings in an arc of total angle $2\theta_0$ in a circle of radius L such that its speed is varying with time. Find relations between the tension in the string, the speed of the ball, and the rate at which the speed of the ball is changing with time at an arbitrary instant when the string makes an angle θ.

Solution. Let v be the speed of the ball when the string makes an angle θ with the vertical. The acceleration of the ball has both a radial component v^2/L pointed towards the suspension point, and a tangential component dv/dt pointed towards increasing speed tangential to the circle of motion. The two forces on the ball are its weight $\vec{W}$ and the tension $\vec{T}$ in the string which varies with time and depends on the instantaneous angle θ of suspension.

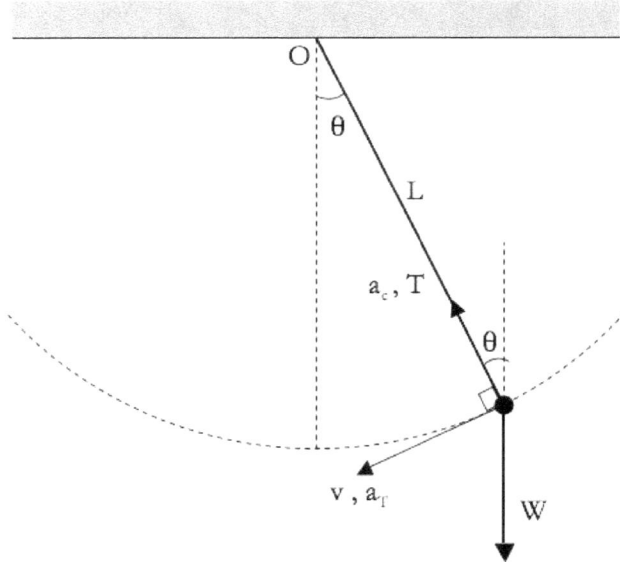

Figure 6.9: The free-body diagram of forces on a plane pendulum. The acceleration has both radial and tangential components. Weight vector $\vec{W}$ also has both radial and tangential components, while tension vector $\vec{T}$ has only a radial component.

Again, we use a table to organize the information about forces and their components. Here we use radial and tangential directions for components.

Force name	radial	tangential
$\vec{W}$	$-mg\ \cos\theta$	$mg\ \sin\theta$
$\vec{T}$	T	0
$\vec{F}_{net}$	$T - mg\ \cos\theta$	$mg\ \sin\theta$

The corresponding components of the acceleration are:

$$\text{Radial inward:}\quad a_c = \frac{v^2}{L}, \quad \text{Tangential:}\quad a_T = \frac{dv}{dt}.$$

Equating the radial and tangential components of the net force to mass times the corresponding components of acceleration we find the required relations.

$$T - mg\ \cos\theta = m\frac{v^2}{L},$$
$$mg\ \sin\theta = m\frac{dv}{dt}.$$

6.5 PROBLEMS CONCERNING MOTION OF EXTENDED BODIES

6.5.1 General Considerations

We have laid out the foundation for how to apply Newton's law for motion of individual particles. An extended body can be modeled as consisting of numerous particles, each obeying Newton's laws of motion.

Motion of a two particle system

To be specific, consider a simple system consisting of only two particles of masses m_1 and m_2 that interact with each other and with bodies external to the system. Let $\vec{F}_1$ and $\vec{F}_2$ be net forces on the two particles, and $\vec{p}_1$ and $\vec{p}_2$ be the corresponding momenta. Each particle may have more than one force on it, and the forces, $\vec{F}_1$ and $\vec{F}_2$ represent the vector sums of all forces on each particle. Then, assuming the masses to be constant, Newton's second law says that each particle will accelerate according to force on that particle.

$$\text{Particle 1:}\quad \vec{F}_1 = m_1 \vec{a}_1. \quad (6.24)$$
$$\text{Particle 2:}\quad \vec{F}_2 = m_2 \vec{a}_2. \quad (6.25)$$

Just as we had done in the last chapter, we separate the forces on each particle of the system into two categories: internal and external. For instance, the net force $\vec{F}_1$ on particle 1 is a vector sum of forces on particle

1 by particle 2, which we call internal force on particle 1 or $\vec{F}_1^{int}$, and forces on particle 1 from objects that are outside the system, which we call external force on particle 1 or $\vec{F}_1^{ext}$.

$$\vec{F}_1 = \vec{F}_1^{int} + \vec{F}_1^{ext}.$$

Then, equations of motion of the two particles become

$$\text{Particle 1:} \quad \vec{F}_1^{int} + \vec{F}_1^{ext} = m_1 \vec{a}_1. \tag{6.26}$$

$$\text{Particle 2:} \quad \vec{F}_2^{int} + \vec{F}_2^{ext} = m_2 \vec{a}_2. \tag{6.27}$$

Adding Eqs. 6.26 and 6.27 leads to the cancellation of internal forces because they are equal in magnitude and opposite in direction. We find that the net external force gives rise to mass weighted acceleration.

$$\vec{F}_1^{ext} + \vec{F}_2^{ext} = m_1 \vec{a}_1 + m_2 \vec{a}_2. \tag{6.28}$$

The left side of this equation is the net external force $\vec{F}_{net}^{ext}$ on all particles of the system. The net external force is a vector sum of all forces regardless of which particle a particular force acts. To add the forces we draw all the external forces coming out of one point rather than the point they may actually be acting. The figure collecting all forces coming out of one point is called a **free-body diagram** of the system.

$$\vec{F}_{net}^{ext} = \vec{F}_1^{ext} + \vec{F}_2^{ext}. \tag{6.29}$$

The right side of Eq. 6.28 is equal to sum of mass times accelerations of each particle. We can write the right side as total mass M $(m_1 + m_2)$ of the particles times acceleration of a space point, called center of mass of the system. We will have a more extended discussion of center of mass in the next chapter. We will be content that the right-side of equation, can be written as total mass times an acceleration simply on the dimensional ground.

$$M\vec{A} = m_1 \vec{a}_1 + m_2 \vec{a}_2. \tag{6.30}$$

The acceleration $\vec{A}$ is sum of the mass-weighted accelerations of the parts.

$$\vec{A} = \left(\frac{m_1}{M}\right) \vec{a}_1 + \left(\frac{m_2}{M}\right) \vec{a}_2. \tag{6.31}$$

The acceleration $\vec{A}$ is called the acceleration of the Center of Mass (CM) as we will see in the next chapter. Therefore, the net external force on a system of two particles is equal to the product of the total mass and the acceleration of the CM.

$$\vec{F}_{net}^{ext} = M\vec{A}. \tag{6.32}$$

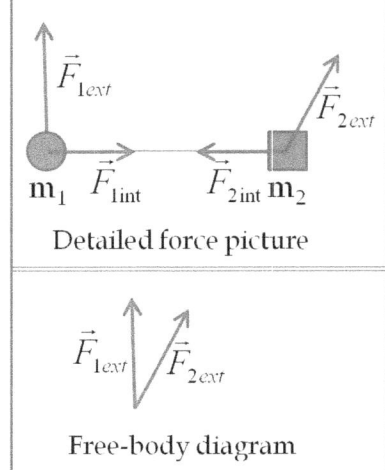

Figure 6.10: We collect all the external forces on various parts of the body in a separate diagram, called the free-body diagram, so that their vector sum can be worked out systematically.

Although we have deduced this equation from a consideration of a two-particle system, the result is valid for a system of any number of particles. Some special situations of interest are:

Case 1: All particles having the same acceleration

All particles of a rigid body that moves in a straight line will have the same acceleration, $\vec{a}$. In this case, the acceleration $\vec{A}$ of the center of mass will be same as $\vec{a}$ as we can prove easily by putting $\vec{a}_1 = \vec{a}_2 = \vec{a}$ in Eq. 6.30.

$$\vec{A} = \left(\frac{m_1}{M}\right)\vec{a}_1 + \left(\frac{m_2}{M}\right)\vec{a}_2$$
$$= \left(\frac{m_1}{M}\right)\vec{a} + \left(\frac{m_2}{M}\right)\vec{a}$$
$$= \left(\frac{m_1 + m_2}{M}\right)\vec{a} = \vec{a}.$$

Therefore, if all points of a body have the same acceleration, then $\vec{A}$ will be the same as the acceleration of any of the particles.

Case 2: Different particles with different accelerations

In this general situation, different parts of the body will move with different accelerations, and depending upon whether the body was rigid or not the behavior will be different. For a rigid body, different accelerations of different parts will give rise to the rotation of the body as well the translational motion of the body as a whole. We will study this motion in the chapter on rotation.

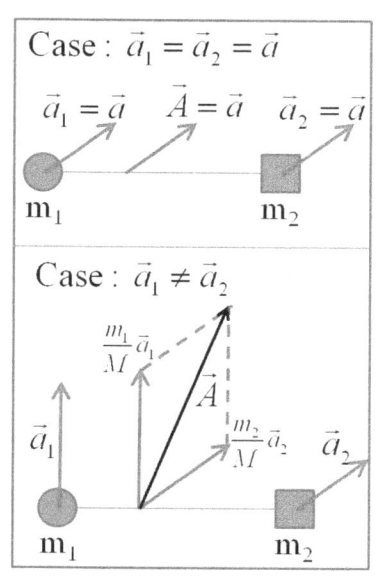

Figure 6.11: The upper figure shows that center of mass acceleration is same as the acceleration of the parts if all parts have the same acceleration. This would be the case of a rigid body executing only a translational motion. The lower figure illustrates the general case of unequal accelerations of the parts: the figure shows that the center of mass acceleration is mass-weighted acceleration of the parts.

6.5.2 Examples of Translational Motion of one Body

This section contains a variety of standard physics problems. Some books emphasize step-by-step approach to solving physics problems, implying that there is some standard method of solving problems, while there is no such thing. Every problem requires you to think through the problem and understand what is given and what is required. We also restrict our examples to the physical situations where the net torque about the center of mass is zero. With a balanced torque about the center of mass, the system under study will not start rotating about the center of mass or tumble over. We will discuss non-zero torque situations when we study rotation in a later chapter.

Example 6.5.1. A box pushed on a horizontal surface. A box of mass m is pushed with a constant horizontal force of magnitude F on a flat horizontal surface so that the box slides on a surface. If the coefficient of kinetic friction between the surface and the bottom of the box is μ_k, find the acceleration of the box in terms of m, g, F, and μ_k.

Solution. We draw a figure of the physical situation first, and identify all the external forces on the box. We find that the external forces on the box are its weight $\vec{W}$ with magnitude mg, the normal $\vec{N}$ from the table, whose magnitude is not known, the kinetic friction $\vec{F}_k$ from the table, whose magnitude will be written in terms of the magnitude of the normal force, $F_k = \mu_k N$, and the applied force $\vec{F}$ from an agent that has not been specified in the problem, as shown in Fig. 6.12. For brevity we will choose symbol N for the normal force rather than the symbol F_N.

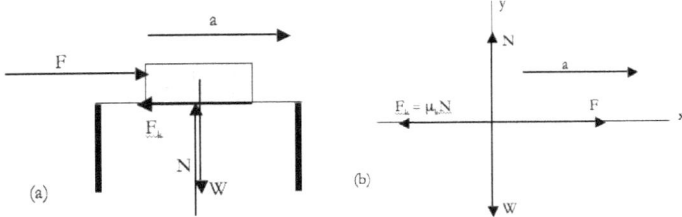

Figure 6.12: (a) Forces on a rectangular box pushed along on a horizontal surface and the resulting acceleration. (b) The free-body diagram, the direction of acceleration, and the Cartesian coordinates.

Next, we draw a free-body diagram, where we indicate the expected direction of the acceleration which helps us pick the direction of one of the Cartesian axes. Note that, effectively, all the forces are in one plane, which we will choose to be the xy-plane with x-axis pointed towards the expected acceleration.

With this choice of coordinates, the z-components of the force will be zero, which would mean that if z-component of the velocity is zero at any time, it will remain zero all the time. This choice of coordinate system will let us ignore z-axis.

Next, we organize the information about the forces and their components in a table so that it is convenient for finding the net force on the box.

Force name	x-component	y-component
$\vec{W}$	0	$-mg$
$\vec{N}$	0	N
$\vec{F_k}$	$-\mu_k N$	0
$\vec{F}$	F	0
$\vec{F}_{net}$	$F - \mu_k N$	$N - mg$

Let a be the magnitude of the acceleration, then the acceleration vector has the following components: $a_x = a$, $a_y = 0$ in the present case. Now, equating each component of the net force to mass times the corresponding acceleration we find

$$F - \mu_k N = m\, a$$
$$N - m\, g = 0.$$

We can easily solve these equations for the magnitude a of the acceleration.

$$a = \frac{F}{m} - \mu_k\, g.$$

The magnitude of the acceleration is given by this equation and the direction of the acceleration is towards the positive x-axis as shown in the figure.

Example 6.5.2. A box pushed on a frictionless surface.

A 5-kg box is pushed with a constant horizontal force of 10 N on a flat frictionless surface. Find the displacement of the box in 5 sec starting from rest.

Solution. Since forces are constant, the acceleration will be constant also. Hence, we can use the constant acceleration formulas given in the chapters on kinematics to determine the displacement in 5 sec if we know the acceleration of the box. Therefore, our first task is to find the acceleration of the box, and then use one-dimensional kinematics to find the distance.

We start by listing the forces acting on the box. We find the following external forces on the box: the weight $\vec{W}$ of the box of magnitude mg, the normal force $\vec{N}$ from the floor which has unknown magnitude N, and the push $\vec{F}$ of magnitude 10 N as shown in Fig. 6.13.

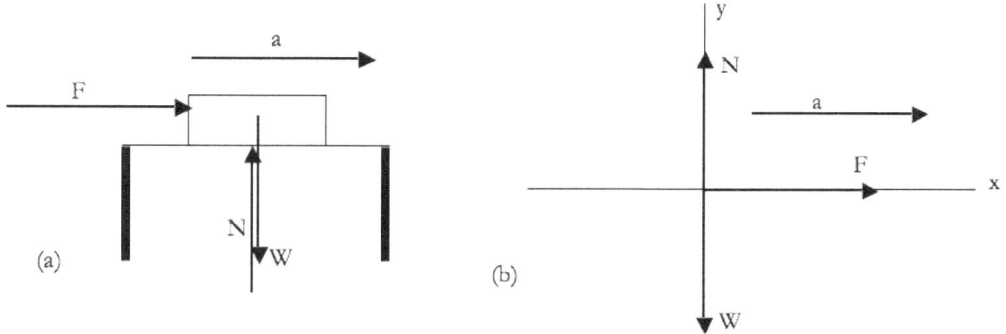

Figure 6.13: (a) Forces on a rectangular box pushed along on a frictionless horizontal surface and the resulting acceleration. (b) The free-body diagram, the direction of acceleration, and the Cartesian coordinates.

Since the acceleration is in the horizontal direction as indicated in the figure, the net force on the box must be in the horizontal direction also. The only force among the forces on the box that has any horizontal component is the horizontal push $\vec{F}$, therefore the acceleration of the box is from the horizontal component of this force alone.

$$x\text{-component:}\quad a_x = \frac{F_{\text{net},x}}{m} = \frac{F_x}{m} = \frac{10 \text{ N}}{5 \text{ kg}} = 2 \text{ m/s}^2.$$

Now, using the kinematics of 1-dimensional constant acceleration along x-axis we find the displacement of the box over the 5 *sec* interval.

$$\begin{aligned} x &= v_0\, t + \frac{1}{2} a_x\, t^2 \\ &= 0 + \frac{1}{2} \times (2 \text{ m/s}^2) \times (5 \text{ s})^2 \\ &= 25 \text{ m}. \end{aligned}$$

Since the box was at rest initially, the displacements along y and z-axes are zero. Therefore, the net displacement of the box is towards the positive x-axis with a magnitude 25 m.

Example 6.5.3. Sliding on an inclined plane.

A hockey puck of mass 160 grams is shot upwards on a slanted flat icy surface inclined at 10° from the horizontal direction. Immediately after leaving the stick, the puck has a speed of 20 m/s. If the coefficient of kinetic friction between the puck and the ice is 0.03, how far the hockey puck will go before coming to rest? Assume zero air resistance.

Solution. This example is similar to the last example except that we now have an incline and the frictional force to worry about. The trick here is to isolate the time interval of interest. The problem gives us the speed after the puck leaves the hockey stick, so we start the time from there. This means that the force of the hockey stick will not be on the puck during the time interval of our interest. The only external forces acting on the puck are the weight $\vec{W}$, the normal $\vec{N}$ from the incline, and the kinetic friction force $\vec{F}_k$, whose magnitude will be written in terms of the magnitude of the normal force as $F_k = \mu_k N$ as before, where μ_k is the coefficient of kinetic friction between the incline and the puck. In Fig. 6.14 we draw a free-body diagram of forces on the hockey puck after the puck has left the stick and before it has come to rest. In this diagram, we also show the direction of the velocity and acceleration of the puck at a representative instant on the motion of the puck up the incline.

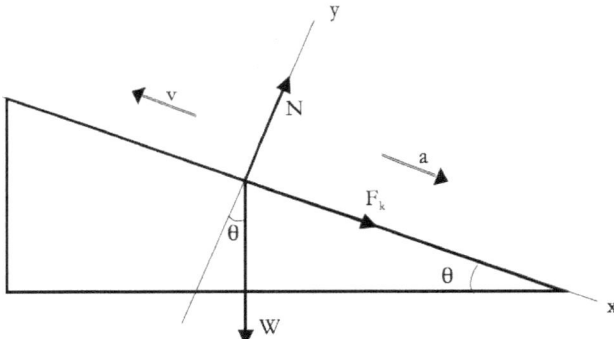

Figure 6.14: The free-body diagram of forces on a sliding box on an inclined plane. The kinetic friction force is pointed opposite to the velocity which is pointed up the incline since the puck is moving up. The expected direction of the acceleration is used to choose x-axis towards the direction of the acceleration.

We have also chosen Cartesian coordinates with the x-axis pointed in the direction of the acceleration. By choosing the x-axis in the direction of acceleration, we have made sure that y-component of acceleration is zero and the x and y-components of the second law for the hockey puck are

$$F_x^{net} = ma_x \tag{6.33}$$
$$F_y^{net} = 0 \tag{6.34}$$

Referring to Fig. 6.14, we can replace the left side of these equations with the components of the forces using their magnitudes and angles in the

xy-plane.

$$\mu_k N + W \sin\theta = ma_x, \quad (6.35)$$
$$N - W \cos\theta = 0, \quad (6.36)$$

where we have used $F_k = \mu_k N$. Solving for N in Eq. 6.36 and substituting it in Eq. 6.35 we find a_x to be:

$$a_x = (\mu_k \cos\theta + sin\theta)g = 2.0 \text{ m/s}^2.$$

Along the x-axis, the motion is that of constant acceleration in a straight line with zero final velocity. Therefore, we immediately obtain the distance traveled before stopping.

$$x = \frac{v_x^2 - v_{0x}^2}{2a_x} = \frac{0 - 20^2}{2 \times 2} = -100 \text{ m}.$$

The answer for the x-component of the displacement vector is negative since the puck has climbed up the incline in the negative x-axis direction. Since y and z-components are zero, the distance d traveled up the incline is

$$d = \sqrt{(\Delta x)^2 + (\Delta y)^2 + (\Delta z)^2} = 100 \text{ m}.$$

6.5.3 Coupled Systems

Often movements of two or more objects are linked in the sense that the coordinates of one are related to the coordinates of others. For instance, when you hang two masses on the two sides of a pulley connected by an inextensible cord, the distance traveled by one mass must equal the distance traveled by the other. Similarly, the motion of earth is coupled to the motion of sun because the gravitational force on the two bodies depends on the distance between the two.

These relations are called constraint equations, which usually relate displacements of objects in a coupled system. These constraints create a relation among the accelerations, which must be taken into account when we solve equations resulting from an application of Newton's second law to the individual objects in coupled systems. In this section, we will examine simple coupled systems, where we will see that the constraints and equations of motion of the individual masses can be solved simultaneously to gain information about the dynamics of a coupled system.

Example 6.5.4. Moving objects together. Perhaps the simplest coupled system consists of two masses that move together in the same direction. Consider two blocks of masses m_1 and m_2 that are next to each other on a table, which, we will assume to be frictionless for the sake of simplicity. What happens when a force $\vec{F}$ pushes on mass m_1?

Note that, in the physical situation shown in Fig. 6.15, mass m_1 cannot move without also moving m_2. Therefore, the accelerations of the

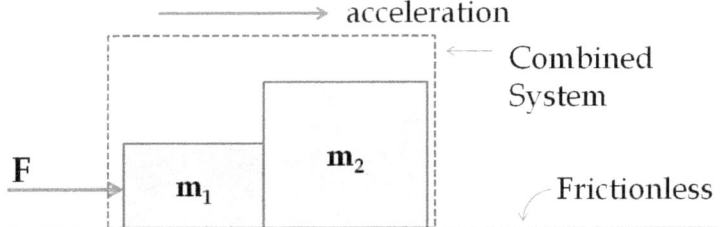

Figure 6.15: Coupled system of two blocks. Blocks m_1 and m_2 move together. The combined system moves with the same acceleration.

two masses for a right-moving horizontal motion must be equal. In this sense, we can treat both m_1 and m_2 as one system of total mass $m_1 + m_2$. On the combined system, there is only one horizontal force, $\vec{F}$. Therefore, the acceleration of both masses is simply given as

$$\vec{a} = \frac{\vec{F}}{m_1 + m_2}.$$

We can obtain the same result from examining one mass at a time. By treating one mass at a time, we obtain additional information than just an overall acceleration as we will see below. We will write equations of motion for the two masses separately and then, we will see what we can do with them to obtain the accelerations of the two.

Let us do the problem in a particular coordinate system. We choose x-axis to point in the direction of the external force $\vec{F}$. Let x_1 and x_2 be the x-coordinates of the center of masses of the two bodies so that the separation of the CM's of the two bodies is a constant D.

$$x_2 - x_1 = D. \tag{6.37}$$

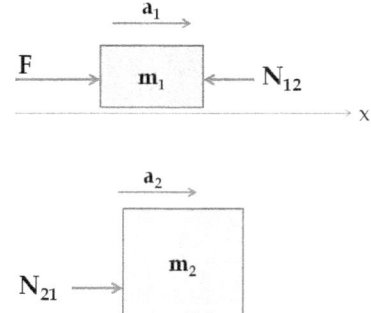

This equation expresses the constraint in the motion of the two bodies as given by the coupling of their x-coordinates. Now, looking at the horizontal forces on m_1 alone, we find that there are two forces on m_1, the applied force $\vec{F}$ and a force on m_1 by m_2 as shown in Fig. 6.16. Similarly, there is one force on m_2, which is the force from m_1 on m_2.

Note that the force $\vec{F}$ does not act on m_2- this force acts on m_1, not on m_2. When we look at the combined system, that is one containing both m_1 and m_2, then an external force acting on either m_1 or m_2 would also act on the combined system also. But since now, we are looking at masses separately, the force $\vec{F}$ acts only on m_1 and not on m_2.

Figure 6.16: Coupled system of two blocks - set up for separate equations of motion.

Let $\vec{a}_1$ and $\vec{a}_2$ be the accelerations of the two masses. Writing out the x-components of equations of motion of the two masses we obtain the following equations.

$$F - N_{12} = m_1 a_{1x} \tag{6.38}$$
$$N_{21} = m_2 a_{2x} \tag{6.39}$$

where the normal forces from 2 on 1 has magnitude N_{12} and the normal forces from 1 on 2 has magnitude N_{12}. Note that this normal force is different from the normal force between the blocks and the table - the normal forces in these equations are the normal forces between the two blocks. The two normal force magnitudes are equal. Therefore, we can use one symbol N_h for both of them. We use subscript h to distinguish this normal force from the normal force between the blocks and the table.

$$F - N_h = m_1 a_{1x} \tag{6.40}$$
$$N_h = m_2 a_{2x} \tag{6.41}$$

Taking two time derivatives of both sides in Eq. 6.37 we immediately see that x-components of the accelerations of the two blocks are equal, which we can replace with a simpler symbol, a_x.

$$a_{1x} = \frac{d^2 x_1}{dt^2} = \frac{d^2 x_2}{dt^2} = a_{2x} \equiv a_x. \tag{6.42}$$

Replacing the acceleration components in Eqs. 6.40 and 6.41 and adding the two equations we obtain.

$$a_x = \frac{F}{m_1 + m_2}$$

which gives the same acceleration as before since the y and z-components are zero.

$$\vec{a} = \frac{\vec{F}}{m_1 + m_2}.$$

However, we can get more information about the system from a treatment of the two masses separately - we can also obtain the force of one block on the other by calculating the normal forces in Eqs. 6.40 and 6.41. Thus, by using a_x we just found into any one of these equations, we find

$$N_h = \left(\frac{m_2}{m_1 + m_2}\right) F.$$

This shows that if $m_2 \gg m_1$, then the normal force N_h will be equal to the external force applied, which would balance the force $\vec{F}$ on m_1 resulting in zero acceleration of block m_1.

Example 6.5.5. Atwood machine.

As our first example of coupled systems consider two masses m_1 and m_2 tied with an unstretchable string that goes over a massless and frictionless pulley. We will find the acceleration of the masses and tension in the string.

Solution. Since all the forces are in the vertical direction, we can point one of the Cartesian axes vertically. It is customary to take y-axis pointed vertically upwards (Fig. 6.17).

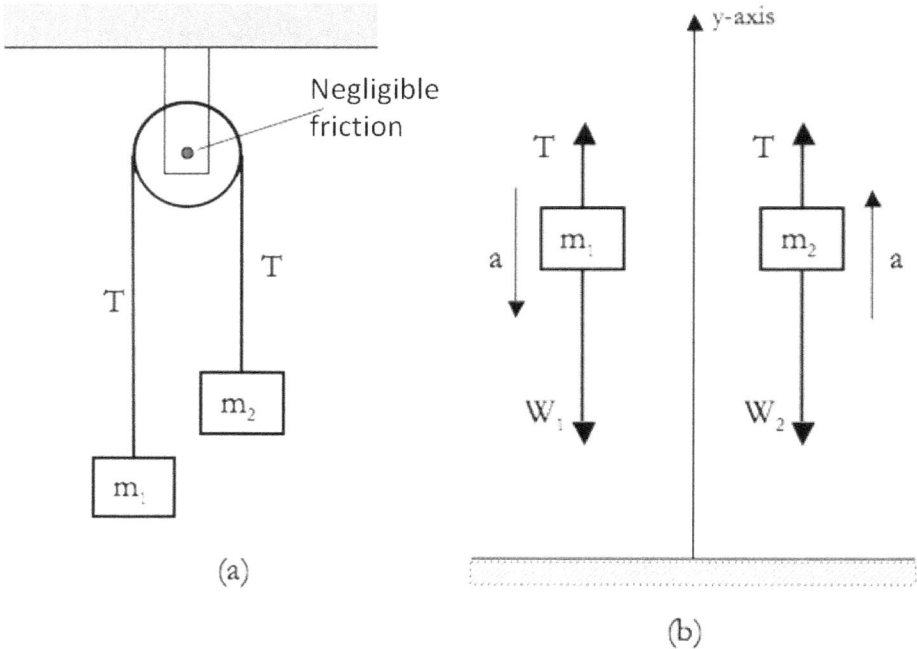

Figure 6.17: (a) An Atwood machine has two masses hanging on two sides of a pulley. Here we assume that the mass of the pulley negligible compared to the two masses, and the axle of the pulley is frictionless. (b) Separate free-body diagrams of the two masses, the directions of their accelerations, and the direction of the y-axis.

Let us first discuss the constraint in the motions of the two masses. The motions of m_1 and m_2 are constrained so that the change Δy_1 in y-coordinate of m_1 is negative of the change Δy_2 in y-coordinate of m_2.

$$\Delta y_2 = -\Delta y_1.$$

This means that y-components of their velocities are equal in magnitude but opposite in sign.

$$v_{2y} = -v_{1y}.$$

By taking time derivatives of both sides, we find that the y-component of their accelerations are equal in magnitude and opposite in sign.

$$a_{2y} = -a_{1y}. \tag{6.43}$$

Let a denote the common magnitude of the accelerations of the two blocks. Let $m_1 > m_2$, then we expect a_{2y} to be positive and a_{1y} negative. Equation 6.43 can be replaced by:

$$a_{1y} = -a \qquad (6.44)$$
$$a_{2y} = a \qquad (6.45)$$

Note: The tension in the string on the two sides of a massless and frictionless pulley have equal magnitude. If either of the two requirements, i.e. massless or frictionless, is not satisfied the tensions on the two sides may be different!

Before we write the separate equations of motion of the two masses it is worth noting another simplifying feature of the problem. When the mass of the pulley is negligible compared to m_1 and m_2, and the friction at the axle of the pulley has negligible torque compared to the torques from the tensions in the strings, then we can show that the tensions in the two parts of the string - tension $\vec{T}_1$ between the pulley and m_1 and tension $\vec{T}_2$ between pulley and m_2 have the same magnitude, and therefore, we can denote them by the same symbol T.

$$T_1 = T_2 = T \quad \text{(magnitude only)}. \qquad (6.46)$$

Now, we draw free-body diagram of the two masses and write down Newton's second law of motion for each (Fig. 6.17(b)) where we have already incorporated requirements of the constraint given in Eqs. 6.44 and 6.45.

$$\text{y-equation for } m_1: \ F_{1y}^{net} = m_1 a_{1y} \implies T - m_1 g = -m_1 a. \quad (6.47)$$
$$\text{y-equation for } m_2: \ F_{2y}^{net} = m_2 a_{2y} \implies T - m_2 g = m_2 a. \quad (6.48)$$

Solving Eqs. 6.47 and 6.48 for a and T we find

$$a = \left(\frac{m_1 - m_2}{m_1 + m_2}\right) g. \qquad (6.49)$$

$$T = \left(\frac{2 m_1 m_2}{m_1 + m_2}\right) g. \qquad (6.50)$$

The magnitude of the acceleration $a > 0$ if $m_1 > m_2$ as assumed. If $m_2 > m_1$, then we will find that $a < 0$, which would reverse the direction of the acceleration assumed in the calculation above.

Let us look at what these results say about a system where you hang two equal masses, $m_1 = m_2 = m$. As expected, Eq. 6.49 says that $a = 0$, and Eq. 6.50 says that the magnitude of the tension force in the string is $T = mg$, not $2mg$. Thus, even when you hang two masses on two sides of a pulley, the magnitude of the tension is equal to the weight of only one of the masses hanging from the ceiling. This can be seen from the second law on any one of the mass - since the tension force in a balanced situation is required to balance only the weight of one mass. Suppose you give an initial velocity, say velocity v_0 up to m_1 and v_0 down to m_2, the two masses will continue to move at constant velocity since the net force on each mass will be zero.

Example 6.5.6. Two masses on different inclines linked together. Two blocks of masses m_1 and m_2 are tied to a cord that goes over a

6.5. PROBLEMS CONCERNING MOTION OF EXTENDED BODIES

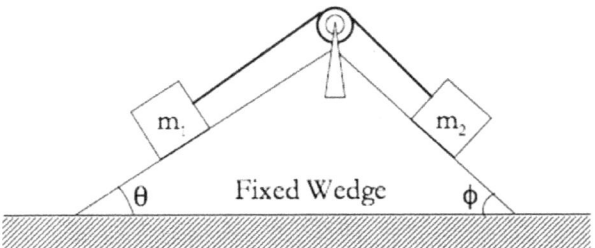

Figure 6.18: Two blocks hanging on two sides of a pulley on a fixed wedge. Here we assume that the mass of the pulley is negligible compared to the two masses, and the axle of the pulley is frictionless.

"massless and frictionless" pulley. The blocks move on two inclines of a fixed wedge as shown in the Fig. 6.18. Let the coefficient of kinetic friction for m_1 and m_2 be μ_1 and μ_2 respectively. Find the accelerations of the two blocks and the tension in the cord.

Solution. The motions of the two blocks are linked here such that the two blocks cover the same distance in the same time. Therefore, the magnitude of their velocities and accelerations are equal. Let us denote the magnitude of their accelerations by common symbol a. Note that their accelerations are in different directions. We also have another element that simplifies the situation: the tension in the entire cord on both sides of the pulley has the same magnitude since the pulley is assumed to be "massless and frictionless," although the direction of the tension force changes, being different on the two sides of the pulley. We draw the free-body diagrams for the two blocks separately in Fig. 6.19, and then write out the equations of motion for each of the masses separately since we have already determined the quantities for the two that are the same.

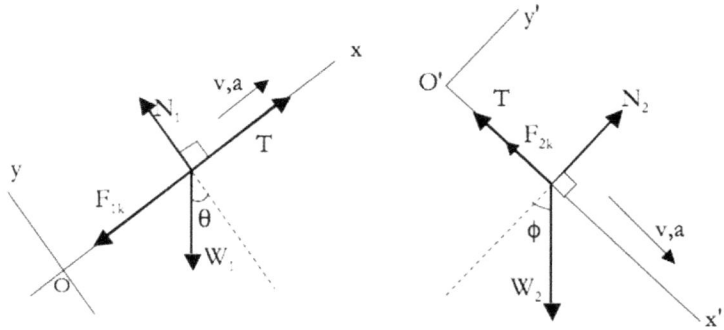

Figure 6.19: The free-body diagrams of the two masses, the assumed directions of their velocities and accelerations, and their separate Cartesian coordinates. Note that we do not need to use the same Cartesian axes to resolve the vector equations into components. The relations obtained are relations among magnitudes and angles which would be same in every coordinate system and therefore, the relations obtained from different coordinate systems can be solved together.

mass m_1:

$$T - \mu_1 N_1 - W_1 \sin\theta = m_1 a \quad (\text{using } F_{1k} = \mu_1 N_1) \quad (6.51)$$
$$N_1 - W_1 \cos\theta = 0. \quad (6.52)$$

mass m_2:

$$-T - \mu_2 N_2 + W_2 \sin\phi = m_2 a \quad (\text{using } F_{2k} = \mu_2 N_2) \quad (6.53)$$
$$N_2 - W_2 \cos\phi = 0. \quad (6.54)$$

We have four equations, Eq. 6.51 to Eq. 6.54, and four unknown a, T, N_1, and N_2. Now, we solve Eq. 6.52 for N_1, and Eq. 6.54 for N_2, and substitute them in Eq. 6.51 and Eq. 6.53, which can be solved for a and T.

$$a = \frac{W_2(\sin\phi - \mu_2 \cos\phi) - W_1(\sin\theta + \mu_1 \cos\theta)}{m_1 + m_2}. \quad (6.55)$$

$$T = m_1 a + W_1(\sin\theta + \mu_1 \cos\theta)$$
$$= \frac{m_1 W_2(\sin\phi - \mu_2 \cos\phi) + m_2 W_1(sin\theta + \mu_1 \cos\theta)}{m_1 + m_2}. \quad (6.56)$$

These results give the magnitudes of the acceleration of the blocks and the tension in the string joining them. The directions of the accelerations of the two masses and the tension force on the masses are as indicated in Fig. 6.19. Numerical values of W_1, W_2, ϕ and θ will give $a > 0$ or $a < 0$. If $a < 0$ then we will reverse the direction for the acceleration given in the figure.

Equations 6.55 and 6.56 are quite complicated, and it is instructive to check some limits. When both angles θ and ϕ are 90^0, then the system should become the same as the Atwood machine discussed above.

$\theta = 0, \phi = 0$:

$$a = \left(\frac{m_2 - m_1}{m_1 + m_2}\right) g.$$
$$T = \left(\frac{2m_1 m_2}{m_1 + m_2}\right) g.$$

Example 6.5.7. Block on a moving wedge. A block of mass m that slides on the frictionless incline of a wedge of mass M and angle of inclination α, which itself slides on a frictionless horizontal table. Find the accelerations of the block and wedge at an arbitrary time before the block reaches the bottom of the incline.

Solution. We first draw a figure to understand the physical situation better as shown in Fig. 6.20. Note that the wedge also accelerates in this problem. Therefore, we do not use a coordinate system attached to the wedge since we need an inertial frame for Newton's equation to be written as $\vec{F} = m\vec{a}$. A coordinate system fixed to the table as shown in Fig. 6.20 is convenient for this problem. We will generate equations of motion for

6.5. PROBLEMS CONCERNING MOTION OF EXTENDED BODIES

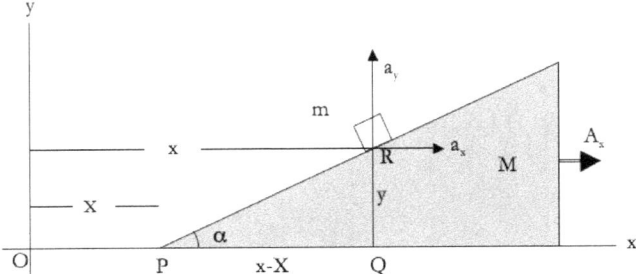

Figure 6.20: A block sliding on a wedge placed on a frictionless table. The block as well as the wedge will accelerate. A coordinate system attached to the wedge will not be an inertial frame. We will use a coordinate system attached to the fixed table as shown.

the block and wedge and their components with respect to this coordinate system.

The position of the wedge requires assignment of only one point on the wedge since every point of the wedge has the same movement on the table. We will choose the left corner of the wedge whose x-coordinate will be indicated by X.

The position of the block on the wedge requires both x and y-coordinate. By our choice, the z-coordinate is zero all the time. Let (x, y) be the x and y-coordinates of the block at a representative instant in its motion down the incline.

We note that X of the wedge and (x, y) of the block are not all independent variables. We will now work out a relation between them based on the geometry here. This is the constraint equation in this problem. From the triangle $\triangle PQR$ we have

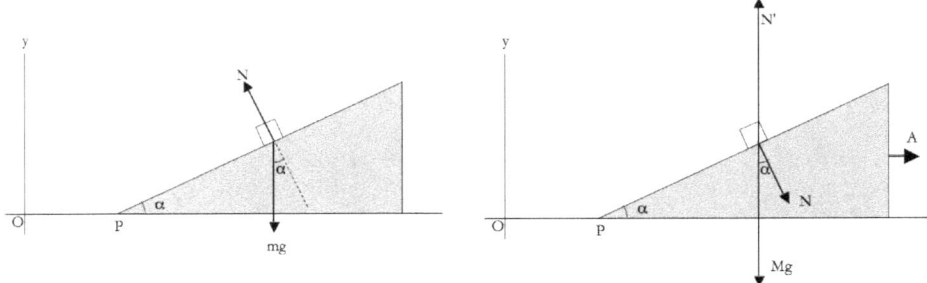

Figure 6.21: The free-body diagram of the block on frictionless incline (left) and the free-body diagram of the wedge (right). Here $\vec{N}$ is the normal force at the surface between the block and the wedge and $\vec{N}\,'$ is the normal at the surface between the wedge and the table.

$$\text{Constraint:} \quad \tan \alpha = \frac{y}{x - X} \implies y = (x - X)\tan \alpha. \qquad (6.57)$$

Taking successive derivatives with respect to time t, we conclude that the components of velocities and accelerations of the block are related to the corresponding quantities for the wedge. We continue to use small letters for the quantities of the block and capital letters for the wedge.

$$v_y = (v_x - V_x) \tan \alpha. \tag{6.58}$$

$$a_y = (a_x - A_x) \tan \alpha. \tag{6.59}$$

A point of confusion in this problem is that there are two surfaces of contact: the surface between the block and the wedge and the surface between the wedge and the table. Therefore, we will have two different normal forces and two different frictional forces in this problem. Frictional forces have been assumed to be negligible for both surfaces. Let N and N' be the magnitudes of the normal force at the surface between the wedge and the block and the wedge and the table respectively. As shown in the free-body diagrams, the block has one normal force but the wedge has two normal forces. From the free-body diagrams for the block and the wedge we find the following equations of motion.

Block:

$$F_x^{net} = m\, a_x \implies -N \sin \alpha = m\, a_x \tag{6.60}$$

$$F_y^{net} = m\, a_y \implies N \cos \alpha - m\, g = m\, a_y \tag{6.61}$$

Wedge:

$$F_x^{net} = M\, A_x \implies N \sin \alpha = M\, A_x \tag{6.62}$$

$$F_y^{net} = M\, A_y \implies N' - M\, g - N \cos \alpha = 0 \ (\text{since } A_y = 0.) \tag{6.63}$$

From Eqs. 6.60 and 6.62 we have

$$A_x = -\left(\frac{m}{M}\right) a_x. \tag{6.64}$$

Using N from Eq. 6.60 into Eq. 6.61 we have

$$a_y = -a_x \cot \alpha - g. \tag{6.65}$$

Now, substituting A_x and a_y from Eqs. 6.64 and 6.65 into the constraint equation, Eq. 6.59 we find

$$a_x = \frac{-g \cot \alpha}{\frac{m}{M} + \text{cosec}^2 \alpha}. \tag{6.66}$$

Then, by substituting a_x into Eqs. 6.64 and 6.65 we find

$$a_y = -\left(\frac{1 + \frac{m}{M}}{\frac{m}{M} + \text{cosec}^2 \alpha}\right) g \tag{6.67}$$

$$A_x = \left(\frac{\frac{m}{M} \cot \alpha}{\frac{m}{M} + \text{cosec}^2 \alpha}\right) g. \tag{6.68}$$

From the components of the acceleration of the block we determine the magnitude and direction of the acceleration of the block in the usual way.

$$a = \sqrt{a_x^2 + a_y^2} \tag{6.69}$$

$$\tan \theta = \frac{a_y}{a_x}, \tag{6.70}$$

where angle θ is measured counter-clockwise from the positive x-axis Once again, it is helpful to check some useful limits on our equations. One such limit is $M \to \infty$. In this limit, the wedge becomes fixed, and then the problem becomes that of a block sliding on a frictionless incline of angle α. In this limit, we find

$$a \xrightarrow{M \to \infty} g \sin \alpha, \text{ and} \qquad (6.71)$$
$$\theta \xrightarrow{M \to \infty} \pi + \alpha, \qquad (6.72)$$

which says that the acceleration is down the incline with magnitude $g \sin \alpha$ as expected.

6.6 EXERCISES

Calculating Momentum

Ex 6.6.1. Provide both the magnitude and direction of the momentum in each question. (a) Find the momentum of a 2 kg ball flying towards the East at a speed of 15 m/s. (b) Find the combined momentum of a 50 kg rider and a 10 kg bicycle moving to the North at a speed of 10 m/s. (c) Find the total momentum of two cars, each of mass 2500 kg, one moving with speed 28 m/s towards the East and the other with speed 29 m/s towards the West. (d) Find the net momentum of a car of mass 3000 kg moving towards the North at a speed of 25 m/s and a truck of mass 10,000 kg moving towards the East at a speed of 8 m/s. (e) Find the net momentum of three rockets, each of mass 200 kg, one moving horizontally towards the East at a speed of 100 m/s, the second moving towards the North at a speed of 150 m/s and the third is moving straight up at a speed of 75 m/s.

Ex 6.6.2. (a) The Earth (mass $\approx 6.0 \times 10^{24}$ kg) makes a full revolution around Sun in almost a circular path in approximately 365 days. Assuming circular orbit of radius 1.5×10^{11} m, find the magnitude of momentum of the Earth with respect to the Sun. (b) Draw momentum vectors at two points on the "circular" orbit that is separated by six months in time. (c) Is the momentum of Earth constant? Why or why not? (d) What is the momentum of Earth with respect to Earth itself?

Ex 6.6.3. A box has a momentum 500 kg.m/s pointed towards the East. The box crashes into a wall and after some time comes to rest. What is the change in momentum of the box? Provide both magnitude and direction.

Ex 6.6.4. A baseball of mass 0.145 kg moving at a speed of 100 miles per hour (mph) is struck by a bat. After the hit the ball reverses direction and moves in the opposite direction with a speed of 120 mph. (a) Use a Cartesian coordinate and write the components of momenta before the hit and after the hit in the chosen coordinates. (b) What is the magnitude and direction of the change in momentum of the ball?

Ex 6.6.5. A proton ($m = 1.67 \times 10^{-27}$ kg) is moving towards the East with the speed 5×10^6 m/s. The proton runs into a thin gold foil and emerges on the other side at $30°$ to the original direction but having the same speed. Determine the change in momentum of the proton. Provide both the magnitude and the direction. Note: You may find it useful to choose a coordinate system with original direction as the x-axis and the proton emerging in the xy-plane. Use this coordinate system to compute the components of the momenta to find their difference.

Free-Body Diagrams, Axes, Components of $\vec{F} = m\vec{a}$

Ex 6.6.6. Fig. 6.22 shows four different situations of forces on a particle in (a)-(d). Determine the magnitude and direction of the net force in each case. If the mass of the particle is 0.1 kg, what are the accelerations of the particle in the four cases.

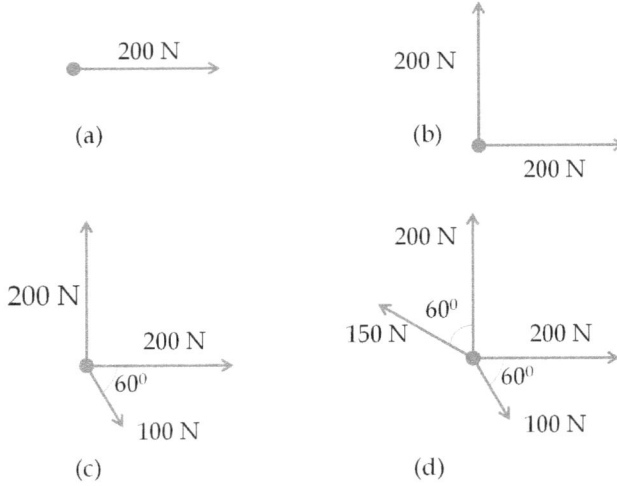

Figure 6.22: Exercise 6.6.6.

Ex 6.6.7. A projectile of mass m is flying freely with only the force of gravity for the Earth acting on the projectile. (a) Draw a free-body diagram for the projectile. (b) Choose a Cartesian coordinate system and set up the Newton's second law expression in the component form. (c) Find the magnitude and direction of acceleration of the projectile.

Ex 6.6.8. A child of mass m in a sleigh sleds down a snowy hill at a constant incline of angle θ that has virtually no friction. Ignore the air resistance also. (a) Draw a free-body diagram for the forces on the child. Use symbols to denote forces. (b) Choose a Cartesian coordinate system and set up Newton's second law in the component form for the constant mass case. (c) Find the magnitude and the direction of acceleration of the projectile. You would get the magnitude of acceleration in terms of g and θ. (d) Find the normal force from the incline on the child in terms of m, g and θ. Do not use a known formula for the magnitude of the normal force, but deduce the magnitude of the normal force from the equations of motion you have written down.

Ex 6.6.9. A truck slides down an icy hill, which slopes at a constant angle θ (Fig. 6.23). (a) Ignoring the friction from the road and the air resistance, determine the instantaneous acceleration of the truck at an arbitrary time during the slide in terms of g and θ. (b) Assuming the forces on the truck remain constant throughout the slide on the incline, predict the speed with which the truck will move after sliding for a distance D on the hill if it was moving down the hill with a speed v_0 at rest at the starting point.

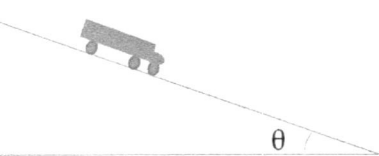

Figure 6.23: A truck sliding down an incline frictionlessly. Exercise 6.6.9

Ex 6.6.10. In a physics experiment designed for studying kinetic friction coefficient a student releases a 500-gram block from rest on an incline. He finds that it takes 1 second to cover a distance of 1 meter on the incline sloped at 15°. Ignoring the effect of air resistance but not that of kinetic friction of the incline, find the coefficient of kinetic friction as follows: first find the magnitude and direction of the acceleration from the given kinematic data assuming the acceleration down the incline to be constant, and then, set up Newton's second law to determine the normal and kinetic friction forces by using the known acceleration. Use the techniques of free-body diagram and choice of axes to work with components.

Ex 6.6.11. A hockey puck of mass 160 grams is shot with a hockey stick on a flat concrete surface. The coefficient of kinetic friction between the bottom of the puck and the concrete surface is 1.5. After the puck is no longer in contact with the hockey stick, the puck moves on the flat surface in a straight line with decreasing speed and comes to rest after traveling a distance of 50 meters. From the given data we wish to find the speed with which the puck leaves the hockey stick and it instantaneous acceleration before coming to rest by utilizing the following steps. (a) Draw a free-body diagram for the forces on the puck at an arbitrary time after the puck has left the hockey stick and before it has come to rest. Use symbols to denote forces. (c) Choose a Cartesian coordinate system and set up Newton's second law in the component form. (d) Find the magnitude and direction of acceleration of the puck. (d) Assuming the acceleration of the puck to be constant, determine the speed of the puck immediately after it leaves the hockey stick.

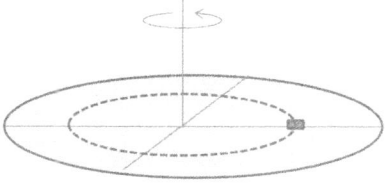

Figure 6.24: Exercise 6.6.12.

Ex 6.6.12. A block of mass M is resting on a rotating platform so that it rotates in a circle of radius R. The block starts to slide on the platform if the platform rotates faster than N turns per second. (a) Find the magnitude of the centripetal acceleration of the block when the platform rotates at a rate of one turn per second. (b) Suppose you use a coordinate system whose origin is at the center of the platform and the platform is in the xy-plane. The platform rotates but the axes remain fixed in space. Now, pick an instant when the block is crossing the x-axis. Set up equations of motion in the component form at this instant using symbols for various forces and the expression for the acceleration you found in part (a). (c) Solve the equations of motion to find the coefficient of static friction between the block and the platform.

Ex 6.6.13. A flat curve on a road has a radius of curvature of 200 m. If the coefficient of static friction between the tires of a car and the road is as low as 0.5, what speed limit in km/h should be posted there so that cars do not slide on the curve. Hint: the static friction force has a maximum value.

Ex 6.6.14. A toy car of mass 250 grams is shot on a horizontal track which then loops in a circle of radius 20 cm. Note that magnitude of centripetal acceleration is still given by the formula v^2/R since the motion is circular, even though v now varies in time, but since the speed of the car in the vertical circle will not be constant, there will be a tangential acceleration in addition to the centripetal acceleration. Suppose that the toy is moving at a speed of 2.4 m/s when it is rounding the track at the top. Find the force with which the car presses the track when it is at the top of the circular path. Hint: Set up $\vec{F} = m\vec{a}$ for the instant when the car is at the top of the circular path.

Ex 6.6.15. In the previous exercise, find the minimum speed with which the car must travel at the top so that it rounds the circle. Hint: The non-zero normal from the track adds to the net centripetal force.

Ex 6.6.16. At an air show a stunt pilot of mass 60 kg flies an air plane in a vertical loop of radius 100 m such that his head always faces the center of the circle. (a) At the top of the loop the pilot feels that he is no longer putting any weight on his seat, i.e. he is weightless. Find his speed when he is at the top of the vertical circle. (b) When he is at the bottom of the circle he is flying at a speed of 400 km/h with what force he must be pushing his seat when he is at the bottom of the circle? Hint: this exercise is similar to the toy car in the vertical circle.

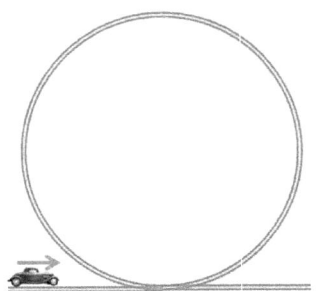

Figure 6.25: Exercise 6.6.14.

Integrating $\vec{F} = m\vec{a}$

Ex 6.6.17. A particle of mass m is subject to a constant force of magnitude $\vec{F}_0$ and direction towards the West. Choose a Cartesian coordinate system and write the components of $\vec{F} = m\vec{a}$ or $\vec{F} = m d\vec{v}/dt$ to answer the following questions. (a) what is the change in velocity of the particle between $t = 0$ and $t = t_1$? (b) What is the displacement of the particle between $t = 0$ and $t = t_1$?

Ex 6.6.18. A particle of mass m is subject to a force that is always pointed towards the East but whose magnitude changes linearly with time. The magnitude of the force is given as $F = bt$. Let x-axis point towards the East. (a) Find the change in the x-component of velocity between $t = 0$ and $t = t_1$. (b) Find the change in x-coordinate of the particle between $t = 0$ and $t = t_1$.

Ex 6.6.19. A particle of mass m is subject to a force that is always pointed towards the North but whose magnitude changes quadratically with time. The magnitude of the force is given as $F = bt^2$. (a) Find the change in the velocity between $t = 0$ and $t = t_1$. (b) Find the displacement of the particle between $t = 0$ and $t = t_1$. Hint: Take y-axis to point towards the North, and work with the y-components.

Ex 6.6.20. A particle of mass m is subject to a force that is always pointed towards East or West but whose magnitude changes sinusoidally with time. With the positive x-axis pointed towards the East, the x-component of the force is given as $F_x = F_0 \cos(\omega t)$, where F_0 and ω are constant. At $t = 0$ the particle is at $x = 0$ and has the x-component of the velocity, $v_x = 0$. (a) Find the x-component of velocity at time t. (b) Find the x-coordinate of the particle at time t.

Ex 6.6.21. A particle of mass m is subject to a force that is always pointed down but whose magnitude changes with time as given by the function, $F = mg\left[1 - \exp\left(-ct\right)\right]$, where c is a constant which takes only positive values. (a) Find the change in velocity between $t = 0$ and $t = t_1$. (b) Find the displacement of the particle between $t = 0$ and $t = t_1$. Hint: Take y-axis to point down, and work with the y-components.

Ex 6.6.22. A particle of mass m is subject to a force that is proportional to the speed and pointed opposite to the direction of velocity: $\vec{F} = -b\vec{v}$, where b is a constant which takes only positive values. Let x-axis be the direction of velocity. (a) Find the change in the velocity with time if the initial velocity has the magnitude u and points towards the East direction. (b) Find the change in the position with time.

Motion of Coupled Systems

Ex 6.6.23. Two large crates of masses m_1 and m_2 are sitting side by side on the floor. (a) You push horizontally on one crate with a force of magnitude F but the crates do not move. Find the net frictional force on the two crates while you are pushing on the first crate crate. (b) You increase your force to F' and find that crates start to slide while they did not slide when the force was less than F'. What is the value of average coefficient of static friction between the surface of crates and the floor? (c) What is the acceleration of the masses when your force is F'' if the coefficient of kinetic friction between m_1 and m_2 and the table is μ_k (d) With what force does the second crate push back on the first crate?

If you like numbers try these: $m_1 = 1000$ kg, $m_2 = 1200$ kg, $F = 2000$ N, $F' = 3300$ N, $F'' = 3400$ N, and $\mu_k = 0.1$.

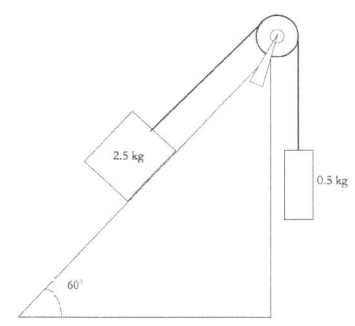

Figure 6.26: Exercise 6.6.25.

Ex 6.6.24. Two masses m_1 and m_2 are connected using a "massless and frictionless" pulleys as shown in Fig. 6.27. When the masses are let go from rest, the mass m_1 moves on a frictionless surface and the mass m_2 moves down towards the floor. Find the time it would take for the mass m_2 to drop a distance h before the mass hits the floor in terms of m_1, m_2, g, and h.

Ex 6.6.25. Two blocks of masses 0.5 kg and 2.5 kg are tied with a string which goes over a massless and frictionless pulley as shown in Fig. 6.26. The 2.5-kg block moves over an incline of angle of inclination 60°. The coefficient of kinetic friction between the 2.5-kg block and the surface of the incline is 0.08. Find the accelerations of the two blocks.

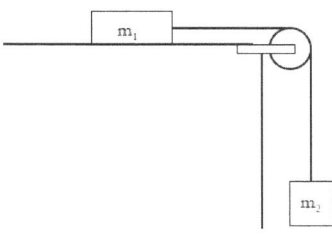

Figure 6.27: Exercise 6.6.24.

Ex 6.6.26. A pulley has masses m_1 and m_2 around it. The support of the pulley is pulled up at a constant velocity V (Fig. 6.28). (a) Find the constraint equation relating Δy_1, Δy_2, and Δy_p. (b) Find the tension in the string and the acceleration of the masses. Assume the pulley to be massless and frictionless.

Ex 6.6.27. In a three-pulley system shown Fig. 6.29, three masses and one of the pulleys are allowed to move. (a) Find the constraint equation(s)

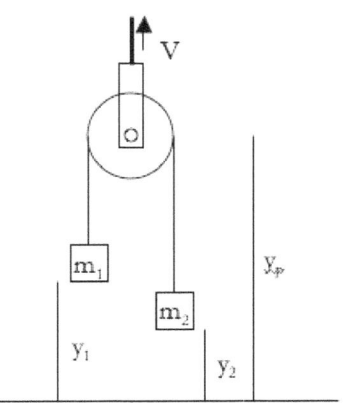

Figure 6.28: Exercise 6.6.26.

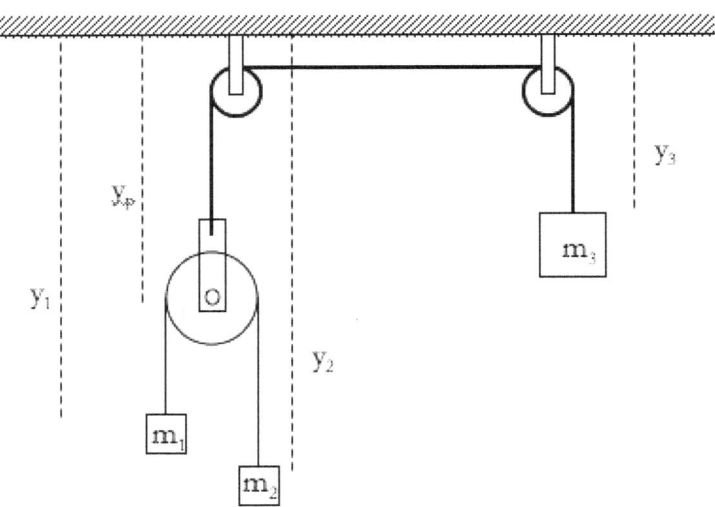

Figure 6.29: Exercise 6.6.27.

among a_1, a_2, and a_3 by first figuring out a relation among y_1, y_2, y_3, and y_p. (b) Find the tension in the strings and the acceleration of each mass. Assume pulley to be "ideal".

Ex 6.6.28. Two masses m_1 and m_2 are connected using two massless and frictionless pulleys as shown in Fig. 6.30. The mass m_1 moves on a frictionless surface and the mass m_2 moves vertically. The masses are let go from rest. (a) Find the values of the magnitudes of accelerations of the

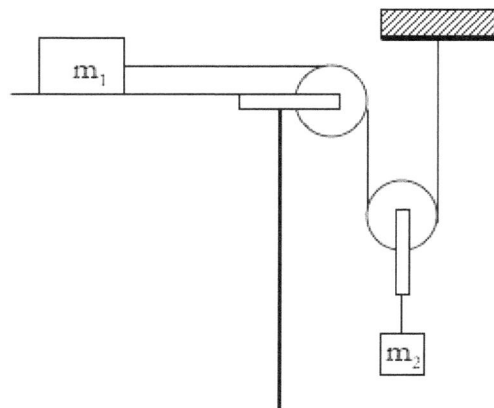

Figure 6.30: Exercise 6.6.28

masses m_1 and m_2 after a time $t = T$ has elapsed. (b) Find the speeds v_1 and v_2 of the masses m_1 and m_2 when m_2 has dropped a distance h.

Assume when $t = T$, the mass m_1 has not hit the pulley at the end of the table of the mass m_2 has not hit the floor. If you like numbers, try these $m_1 = 5$ kg, $m_2 = 0.4$ kg, and $h = 5$ cm.

6.7 PROBLEMS

Problem 6.7.1. Many amusement parks have a ride called antigravity where riders are lined up against a wall of a large cylindrical drum, which is spun to some final angular speed when the bottom of the floor is removed. The riders however remain stuck to the wall without falling in the pit. If the final angular speed is ω and the radius of the drum R, what must be the minimum static frictional coefficient μ_s so that a rider does not slide vertically?

Figure 6.31: Problem 6.7.1.

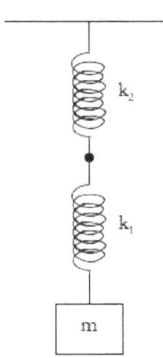

Figure 6.32: Problem 6.7.2.

Problem 6.7.2. A block of mass m is hung from two springs of spring constant k_1 and k_2 attached as shown in the figure. Each spring stretches to different amounts with the total stretch equal to L. Find the stretch of each spring in terms of m, k_1, k_2, L and g. Assume the mass to be at rest when hanging.

Problem 6.7.3. Roads are usually banked at the curves to prevent car from sliding out. If the coefficient of static friction between the road surface and the tires is 1.5, what must be the minimum angle of banking for a bend in a road of radius of curvature of 75 m so that a car traveling at 65 km/h can safely make the turn if maximum static friction acts on the tires.

Problem 6.7.4. A ball of mass m is tied to a vertical revolving bar by two strings of equal length L so that it revolves with a uniform speed of v in a horizontal circle as shown Fig. 6.33. Find tensions in the two strings.

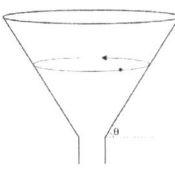

Figure 6.34: Problem 6.7.5.

Problem 6.7.5. A particle of mass 50 grams slides in a circle without friction inside a funnel as shown in Fig. 6.34 at a constant speed of 10 m/s. Find the radius of the circle.

6.7. PROBLEMS

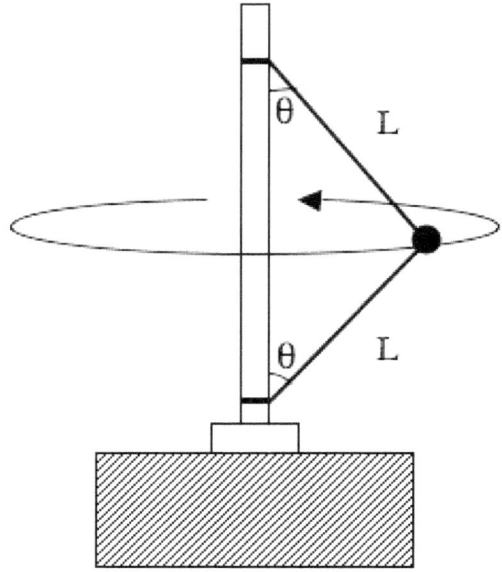

Figure 6.33: Problem 6.7.4.

Problem 6.7.6. A bead of mass m slides on a ring of radius R meter without any friction. When the ring rotates about a vertical axis at a rate of ω radians per sec, the bead is found to be stable in a circular motion of non-zero radius. Find the radius of the circle for the motion of the bead when it is not sliding in the ring in terms of the given quantities and g.

Problem 6.7.7. A block of mass M is tied to a massless string which is passed through a small hole in a smooth table as shown in Fig. 6.36. A pebble of mass m is attached to the other end of the string. When block moves in a circle of radius R on the surface of the table at a particular speed v the pebble does not move. Assuming the table to be frictionless find the speed v of the block.

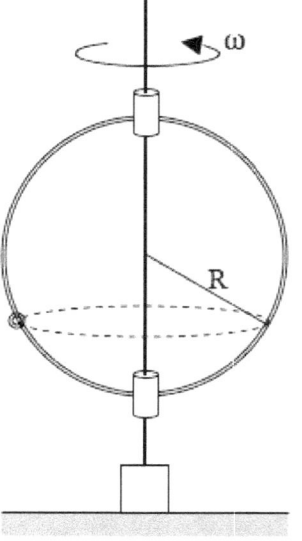

Figure 6.35: Exercise 6.7.6.

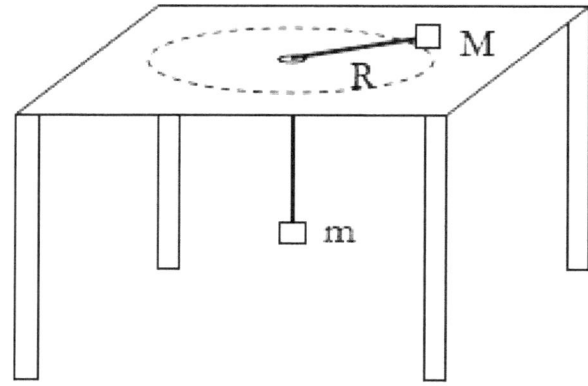

Figure 6.36: Pr. 6.7.7.

Problem 6.7.8. Three masses are connected to two fixed pulleys and a moving pulley as shown in Fig. 6.37. Assume all pulleys massless and frictionless and all strings massless. Find the accelerations of the three masses.

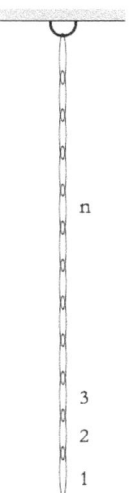

Figure 6.38: Problem 6.7.9.

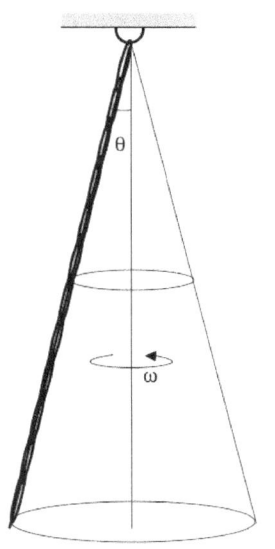

Figure 6.39: Problem 6.7.10.

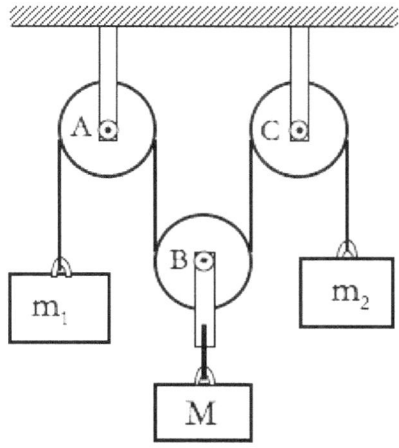

Figure 6.37: Pr. 6.7.8.

Problem 6.7.9. A long chain consists of links of mass m each. The chain is hung from a tower. What is the tension in the chain at the nth link from the bottom?

Problem 6.7.10. A rope of mass M and length L is swung uniformly with the angular speed ω in a circle that makes an angle θ with the vertical. What is the tension in the rope at a distance b from the top? Note: each element of the rope moves in a horizontal circle with uniform circular motion.

Chapter 7

IMPULSE AND COLLISION

Contents

- **7.1 IMPULSE** 264
  - 7.1.1 Impulse of a Constant Force 264
  - 7.1.2 Impulse of a Time Dependent Force 265
  - 7.1.3 Area Under the Curve Method 268
- **7.2 IMPULSE AND CHANGE OF MOMENTUM** 270
- **7.3 TRANSLATIONAL MOTION OF MULTIPAR-
TICLE SYSTEMS** 271
  - 7.3.1 Two-Particle System 271
  - 7.3.2 Generalization 273
  - 7.3.3 Motion of Center of Mass 273
- **7.4 CALCULATIONS OF CENTER OF MASS** . 275
  - 7.4.1 Examples of Discrete Masses 275
  - 7.4.2 Examples of Continuous System 278
- **7.5 ISOLATED SYSTEMS AND CONSERVATION
OF MOMENTUM** 282
- **7.6 COLLISIONS** 286
- **7.7 COLLISIONS IN THE CM FRAME** 291
- **7.8 SYSTEMS WITH TIME VARYING MASS** . 294
  - 7.8.1 Rocket Motion With no External Force ... 295
  - 7.8.2 Rocket Motion With Constant External Force . 299
- **7.9 EXERCISES** 300
- **7.10 PROBLEMS** 306

7.1 IMPULSE

7.1.1 Impulse of a Constant Force

When a constant force $\vec{F}$ acts for a duration Δt, it exerts an **impulse** $\vec{J}$ equal to $\vec{F}\Delta t$.

$$\boxed{Impulse,\ \vec{J} = \vec{F}\Delta t.} \tag{7.1}$$

Impulse is a vector and has the same direction as the force itself. Therefore, if you pull a cart with a constant force of magnitude 15 N for 2 sec, then you would impart an impulse of 30 N.s to the cart in the direction of the force. Same impulse of 30 N.s can be imparted by a smaller force, say 1 N in the same direction, that acts for 30 sec instead of 15 N for 2 sec. Note that the unit of impulse in the SI system of units is the same as the unit of momentum:

$$\text{Unit of impulse} = \text{N.s} = \frac{\text{kg.m}}{\text{s}^2}\text{s} = \frac{\text{kg.m}}{\text{s}} = \text{Unit of momentum}.$$

Impulses from different forces add vectorially. Suppose force $\vec{F}_1$ acts for a time Δt_1 and $\vec{F}_2$ acts for a time Δt_2, then the net impulse $\vec{J}_{net}$ will be the vector sum of the impulses $\vec{J}_1$ and $\vec{J}_2$ of the two forces.

$$\vec{J}_{net} = \vec{J}_1 + \vec{J}_2, \tag{7.2}$$

where $\vec{J}_1 = \vec{F}_1 \Delta t_1$ and $\vec{J}_2 = \vec{F}_2 \Delta t_2$.

Example 7.1.1. Impulse as a vector. A box is on a level surface. A force of magnitude 3 N acts on the box in the direction towards the East for 10 sec and another force of magnitude 2 N acts on the box in the direction towards the North for 5 sec. What is the net impulse?

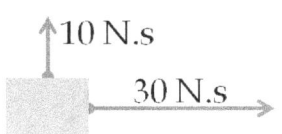
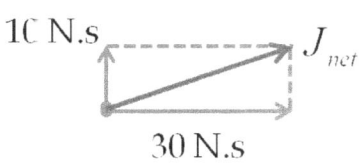

Figure 7.1: Adding impulse vectors, Example 7.1.1.

Solution. We will work this problem in the analytic picture for vectors. For that purpose, we will choose a coordinate system whose x-axis is pointed towards the East and the y-axis is pointed towards the North. We do not need the z-coordinate for this problem since the all the forces are in one plane, which is the xy-plane. The x and y-components of the two impulses and their sum are collected in the following table.

Impulse name	x-component (N.s)	y-component (N.s)
$\vec{J}_1$	30	0
$\vec{J}_2$	0	10
$\vec{J}_{net}$	30	10

The net impulse is now constructed from its components. We obtain the following magnitude and direction of the net impulse. Let us write J_x and J_y for the x and y-components of $\vec{J}_{net}$.

Magnitude:

$$J_{net} = \sqrt{J_x^2 + J_y^2} = 32 \text{ N.s},$$

Direction:

Since the vector is in the xy-plane, the angle of the vector with the positive x-axis can be used to indicate the direction.

$$\theta = \arctan\left(\frac{10}{30}\right) = 18°,$$

Therefore, the direction of $\vec{J}_{net}$ is $18°$ North of East.

7.1.2 Impulse of a Time Dependent Force

Often, impressed forces on a body change with time. For instance, when you hit a base ball, the force between the bat and the ball varies with the time - having a zero magnitude before coming into contact, rising to a maximum magnitude after coming to contact, and then decreasing to zero at which time the two are no longer in contact. A representation of the magnitude of the force between the bat and the ball is shown in Fig. 7.2.

How can we calculate the impulse of the force of the bat on the ball? This is similar to how we extended the formulas for the constant acceleration for the case of varying acceleration. The general strategy is to think in terms of small steps of time over which a varying property can be replaced by an average value and then combine the effects from each subinterval.

Thus, we start by dividing the the total time, say from $t = 0$ to $t = T$, into smaller segments of time. We will then replace the force in each interval by the average force during that interval. This would give us a collection of constant forces acting on the body in each subinterval from which we can calculate the impulses in each subinterval. Finally, we can add the impulses in each subinterval to obtain the net impulse for the total time.

Let $\vec{F}_1, \vec{F}_2, \cdots, \vec{F}_N$ be the average forces in the N subintervals, $\Delta t_1, \Delta t_2, \cdots, \Delta t_N$, respectively. Each subinterval of time will generate an impulse vector. Let $\vec{J}_1, \vec{J}_2, \cdots, \vec{J}_N$ be the corresponding impulse vectors.

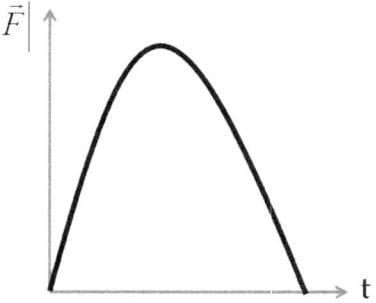

Figure 7.2: Magnitude of the force on a base ball by the bat changes with time.

$$\vec{J}_1 = \vec{F}_1 \Delta t_1$$
$$\vec{J}_2 = \vec{F}_2 \Delta t_2$$
$$\vdots$$
$$\vec{J}_N = \vec{F}_N \Delta t_N, \tag{7.3}$$

with the total time T:

$$T = \Delta t_1 + \Delta t_2 + \cdots + \Delta t_N.$$

The impulse in different subintervals must now be added to obtain the net impulse.

$$\vec{J}_{net} = \vec{J}_1 + \vec{J}_2 + \cdots + \vec{J}_N. \tag{7.4}$$

The sum of vectors in Eq. 7.4 can be done either geometrically or analytically. Typically, an addition of vectors is more easily done in the analytic picture for vectors. For this purpose, we choose a coordinate system with respect to which we can determine the components of the forces, obtain the components of each impulse, and then add the components separately to obtain the components of the net impulse. This exercise results in the following set of equations.

$$J_x^{net} = F_{1x}\Delta t_1 + F_{2x}\Delta t_2 + \cdots + F_{Nx}\Delta t_N$$
$$J_y^{net} = F_{1y}\Delta t_1 + F_{2y}\Delta t_2 + \cdots + F_{Ny}\Delta t_N$$
$$J_z^{net} = F_{1z}\Delta t_1 + F_{2z}\Delta t_2 + \cdots + F_{Nz}\Delta t_N.$$

As long as the subintervals $\{\Delta t_i, i = 1, 2, \cdots, N\}$ are finite intervals, no matter how small, we will only get an approximation of the exact value of the impulse of a time-dependent force. However, the difference between the exact answer and the approximate answer can be minimized to an arbitrary precision by making the sizes of the subintervals arbitrarily small, and consequently, making the number of subintervals infinitely large. The result is then written using the integral sign.

$$J_x^{net} = \int_0^T F_x(t)dt \qquad (7.5)$$

$$J_y^{net} = \int_0^T F_y(t)dt \qquad (7.6)$$

$$J_z^{net} = \int_0^T F_z(t)dt. \qquad (7.7)$$

A definite integral such as in Eqs. 7.5 - 7.7 have also the meaning of the "area under the curve" with area being positive if function is positive and negative if function is negative as we have encountered before. Therefore, we can use area under the curve for each of the components of forces if the forces are given as plots rather than as analytic functions. Formally, we can write the three equations for the components in vector notation.

$$\boxed{\vec{J}_{net} = \int_0^T \vec{F}(t)dt.} \qquad (7.8)$$

Example 7.1.2. Impulse of a sinusoidally varying force. The time dependence of a sinusoidally varying force can be written as a sine or cosine of time. For instance, when an electron is illuminated by a monochromatic light, the force on the electron can be represented by a cosine or sine function. Suppose the force on an electron at some time t has the following form with respect to a coordinate system, $\vec{F} = A\cos(t)\hat{u}_x + B\sin(t)\hat{u}_y$, where $\hat{u}_x$ and $\hat{u}_y$ are the unit base vectors along the x and y-axes respectively, and A and B are some constants. What will be the impulse on the electron during an interval $t = 0$ to $t = T$?

Solution. Since the force is already given in an analytic form, we will express the impulse in the same coordinate system. Using Eq. 7.8 and

7.1. IMPULSE

writing out the integrals for each component, we find

$$\vec{J}_{net} = A\left[\int_0^T \cos(t)dt\right]\hat{u}_x + B\left[\int_0^T \sin(t)dt\right]\hat{u}_y$$
$$= A\sin(T)\hat{u}_x + B\left[1 - \cos(T)\right]\hat{u}_y.$$

Further Remarks:

Is the magnitude of impulse equal to the magnitude of a force times the duration? The answer depends on whether or not the force is changing with time.

1. When a single constant force acts on the system, then the impulse will have the same direction as the direction of the force, and the magnitude of the impulse will be simply the product of the amplitude of the force and the duration over which the force has acted on the system.

2. When multiple constant forces act on the system, then the net force will also be constant. This situation is same as the one force case with one force now being the net force. Therefore, the impulse will have the same direction as the direction of the net force, and the magnitude of the impulse will be the product of the amplitude of the net force and the duration over which the force has acted on the system.

3. When a time varying force acts on the system, the situation is more complicated. As a concrete example of this situation, let us examine in detail the case of a force of constant magnitude but time-dependent direction. Let $F_x(t)$ and $F_y(t)$ be the non-zero components of a force whose magnitude is constant in time.

$$F = \sqrt{F_x(t)^2 + F_y(t)^2} = \text{constant, even if the components are not.}$$

For instance, if $F_x(t) = F_0\cos(bt)$ and $F_y(t) = F_0\sin(bt)$, then the magnitude will be constant, even when the components are time-dependent. The components of impulse will be

$$J_x^{net} = \int_0^T F_x(t)dt.$$

$$J_y^{net} = \int_0^T F_y(t)dt.$$

This gives the magnitude of the impulse

$$J_{net} = \sqrt{\left(J_x^{net}\right)^2 + \left(J_y^{net}\right)^2}.$$

The question now becomes: Can this J be written as F times T? The answer is no, since F_x and F_y are time-dependent.

$$J_{net} \neq FT.$$

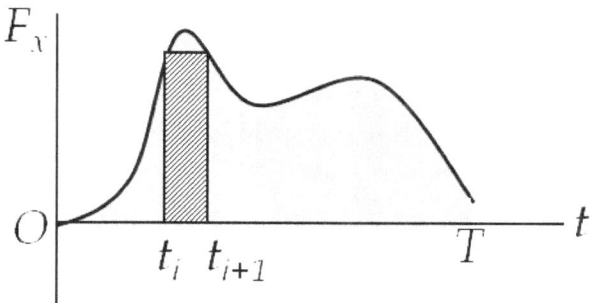

Figure 7.3: Area under curve shown shaded is evaluated from the sum of the areas of rectangles of the area. In the limit of infinitesimal t subintervals the area obtained from summing the areas of the rectangles approaches the exact area under the curve.

Therefore, when the force varies with time, you cannot simply take the magnitude of the force and multiply it by the duration to obtain the impulse. You will have to calculate the magnitude and direction of the impulse vector from its components, which would be separately obtained from the time-varying force components.

7.1.3 Area Under the Curve Method

Often the data for force with respect to time is in terms of a plot of force versus time as shown in Fig. 7.2. If you can write the data as an analytic function of time, then you could do the integrals of Eqs. 7.5 - 7.7 to obtain the components of the impulse vector.

Alternately, we can utilize another interpretation of the definite integral of one variable: a definite integral of a function $f(t)$ of one independent variable t is equal to the area under the curve of $f(t)$ versus t plot. The area for a curve above the abscissa gives a positive area and the area for a curve below the abscissa has a negative area. The integral value for integration from $t = t_1$ to $t = t_2$ is the sum of all areas under the curve between the limits, $t_1 \leq t \leq t_2$.

Let us look at one of the Cartesian components, say the x-component. A plot of x-component of a force is shown in Fig. 7.3.

The shaded part has an "area under the curve" of $F_x(t_i)\Delta t$, which is the contribution to the x-component of the impulse for the time segment t_i to t_{i+1} of duration Δt. You can easily see that the area $F_x(t_i)\Delta t$ is not really equal to the "true" area under the curve - the rectangle misses the curve by some amount depending upon the size of Δt and how curved the curve is at that point in the plot. However, as you make Δt progressively smaller, the difference between the exact area and the approximate area evaluated by rectangles becomes progressively smaller, and the two become more and more equal to each other. By choosing small enough Δt we can

7.1. IMPULSE

make the error less than any desired value.

To obtain an exact value of the integral of F_x from $t = 0$ to $t = T$ from a plot of F_x versus t, we divide the total interval, $0 \leq t \leq T$, into subintervals of size Δt, and sum over the areas of the rectangles. If Δt is small enough, then the value obtained by summing the areas gives us the exact value of the integral.

$$\int_0^T F_x dt = \lim_{\Delta t \to 0} [F_x(0)\Delta t + F_x(\Delta t)\Delta t + F_x(2\Delta t)\Delta t + \cdots] \quad (7.9)$$
$$= \text{"Area under } F_x \text{ vs } t \text{ from } t = 0 \text{ to } t = T\text{"}.$$

The same procedure will work for the y and z-components also. The "area under the curve" method provides an alternate way of evaluating components of impulse vector given in Eqs. 7.5 - 7.7.

$$x\text{-component:} \quad J_x^{net} = \int_0^T F_x(t)dt = \text{"Area under } F_x \text{ vs } t\text{"} \quad (7.10)$$

$$y\text{-component:} \quad J_y^{net} = \int_0^T F_y(t)dt = \text{"Area under } F_y \text{ vs } t\text{"} \quad (7.11)$$

$$z\text{-component:} \quad J_z^{net} = \int_0^T F_z(t)dt = \text{"Area under } F_z \text{ vs } t\text{"} \quad (7.12)$$

Example 7.1.3. Impulse of a time-varying force The force between a baseball and a bat is an example of a time-varying force. The magnitude of the force varies with time in a complicated way as shown in Fig. 7.2. A simpler picture of the magnitude of the force is shown in Fig. 7.4. What is the impulse of the force given in Fig. 7.4?

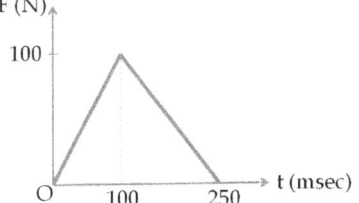

Figure 7.4: Magnitude of the force on a base ball by the bat changes with time.

Solution. Figure 7.4 gives us the magnitude of the force on the ball. Although, the variation of the magnitude of the force is an important information for impulse, but we also need information about the direction of the force to compute the impulse vector. Here, the force on the ball has the same direction during the entire time the bat is in contact with the ball. Therefore, if we choose the x-axis to point in the direction from the bat to the ball, then Fig. 7.4 is same as the x-component of the force as shown in Fig. 7.5.

The area under the curve in Fig. 7.5 is the area of the triangle under the curve. The formula for the area of a triangle is base times height divided by 2. We convert the unit of time from millisecond to sec before putting the numbers in the area formula. This gives the x-component of impulse to be

$$J_x = \frac{1}{2} \times 100 \text{ N} \times 0.25 \text{ s} = 12.5 \text{ N.s},$$

which is equal to the magnitude of the impulse since the y and z-components are zero, and the direction of the impulse is from the bat towards the ball.

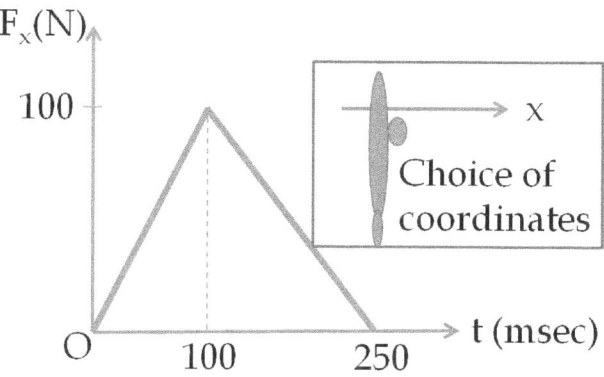

Figure 7.5: x-component of force on the baseball.

7.2 IMPULSE AND CHANGE OF MOMENTUM

Newton's second law says that at each instant the net force on a body is equal to the rate of change of momentum of the body.

$$\vec{F} = \frac{d\vec{p}}{dt}. \tag{7.13}$$

Suppose you push a box on a floor with a net average force $\vec{F}_{ave}$ for a time Δt. Then, the change in momentum will be

$$\Delta \vec{p} = \vec{F}_{ave} \Delta t, \tag{7.14}$$

which means that the change in momentum is actually equal to the average impulse imparted to the body. For a time-varying force, the impulse will be an integral as shown above. If a time varying force acts from time t_1 to t_2, then the approximate equation, Eq. 7.14 must be replaced by the following more complete statement of the change in momentum over the interval.

$$\boxed{\vec{p}(t_2) - \vec{p}(t_1) = \int_{t_1}^{t_2} \vec{F}(t)dt.} \tag{7.15}$$

Example 7.2.1. Change of momentum of a golf ball. A player hits a 46-gram golf ball with an average force of 4,600 N while the golf club is in contact with the ball for 0.5 msec. Find the speed of the ball immediately after leaving the club if the golf ball was at rest before being it was struck.

Solution. We can find the speed of the ball by first calculating the change in momentum. Now, a change in momentum can be obtained if we know the momentum at the end and at the beginning. But, here, we only know the momentum at the beginning, which is zero. Our discussion in this section says that, according to Newton's second law, you can use impulse to deduce the resulting change in momentum also. For a calculation of the impulse, we need the force and the duration for which the force has acted, both of which are given in the problem. Therefore, the impulse on the ball is given by

$$\text{Magnitude: } J = 4600 \text{ N} \times 0.5 \times 10^{-3} \text{ s} = 2.3 \text{ N.s.}$$
$$\text{Direction: in the direction of the force.} \quad (7.16)$$

This is equal to the change in momentum of the ball. Dividing by mass we find the velocity at the end of the interval since velocity is zero at the beginning of the interval.

$$\vec{v} = \{50 \ m/s, \text{from club towards ball}\}.$$

Therefore, the speed of the ball after the hit will be 50 m/s.

7.3 TRANSLATIONAL MOTION OF MULTI-PARTICLE SYSTEMS

We call a group of particles a multiparticle system when we are interested in describing the motion of the group as a whole. A baton, for instance, with two masses connected by a light rod, as a whole, moves as a two-particle system. When you throw a baton in the air, the two masses tumble about each other and move under the influence of gravity of the Earth. Although the motion of each mass is quite complicated, we are often interested in an overall motion of the baton.

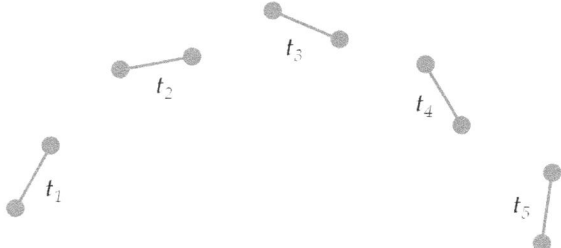

Figure 7.6: A baton with two masses moves as a two-particle system. In this sectio we will find that the total momentum of the two masses is affected only by the external force on the two masses.

The physics of multipoarticle systems are based on the physics of individual particles of the system. When we apply Newton's laws of motion to each particle we obtain the rate of change of momentum of each particle. These rules for individual particles can be combined to deduce the overall motions, such as the overall translational and the rotational motion of the multiparticle system. In this section, we will see how the total momentum of a multiparticle system can be studied. The rotational motion of multiparticle systems will be studied in a later chapter.

7.3.1 Two-Particle System

As a an example of a multiparticle system, let us examine a two-particle system first. We take a familiar example from Astronomy and study the motion of the Earth and the Moon together as one multiparticle system. To keep the physical situation simple we will ignore other planets and consider only the Sun as the sole external object with which the Earth and the Moon can interact as shown in Fig. 7.7 in addition to their interaction with each other.

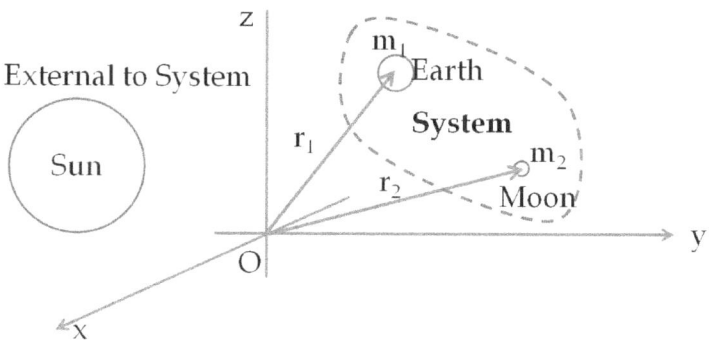

Figure 7.7: In the three-body world of the Earth, the Moon and the Sun, when we wish to study the motions of the Earth and the Moon, then they form a two-particle system and the Sun becomes an external body to the system. Had we decided to study the motion of the Earth only, then the Earth would be a one-body system and the Moon and the Sun would have been the external bodies to the system. Had we decided to study the motions of all three bodies shown, then the system would be a three-body system, and since there is nothign external to this system, the system would have been an isolated system

To simplify the notation we will use subscript 1 for the Earth, 2 for the Moon, and ext for the Sun in our calculations below. Let us start with the separate equations of motion of the two objects in the system as given by Newton's second law of motion.

$$\text{Earth:} \quad \frac{d\vec{p}_1}{dt} = \vec{F}_1^{\text{ext}} + \vec{F}_{12} \tag{7.17}$$

$$\text{Moon:} \quad \frac{d\vec{p}_2}{dt} = \vec{F}_2^{\text{ext}} + \vec{F}_{21} \tag{7.18}$$

where $\vec{F}_{12}$ is the force on the Earth by the Moon, $\vec{F}_1^{\text{ext}}$ is the force of the Sun on the Earth, $\vec{F}_{21}$ is the force on the Moon by the Earth, and $\vec{F}_2^{\text{ext}}$ is the force of the Sun on the Moon. Note that forces of the Moon on the Earth and that of the Earth on the Moon are internal forces for the system since the Earth and Moon belong to the multiparticle system. Only the forces by Sun on the Earth and Moon are external to the Earth-Moon system.

According to Newton's third law, the internal forces $\vec{F}_{12}$ and $\vec{F}_{21}$ are equal in magnitude but act in opposite directions. Hence their vector sum must be zero.

$$\vec{F}_{12} + \vec{F}_{21} = 0. \tag{7.19}$$

This suggests that we should add equations 7.17 and 7.18, which will result

7.3. TRANSLATIONAL MOTION OF MULTIPARTICLE SYSTEMS

in an equation without any reference to the internal forces.

$$\frac{d(\vec{p}_1 + \vec{p}_2)}{dt} = \vec{F}_1^{ext} + \vec{F}_2^{ext}, \qquad (7.20)$$

which can be re-written more compactly as

$$\frac{d\vec{P}_{system}}{dt} = \vec{F}_{net}^{ext}. \qquad (7.21)$$

where I have written $\vec{P}_{system}$ for the total momentum $\vec{p}_1 + \vec{p}_2$ and $\vec{F}_{net}^{ext}$ for the sum of all the external forces on the Earth-Moon system. Thus, we find that the rate of change of total momentum of a composite system depends only on the external forces, and completely independent of the internal forces. Therefore, even if you do not know anything about the internal forces of a composite system, you can still predict the translational motion of the system as a whole by looking at the total momentum vector.

7.3.2 Generalization

From the procedure outlined above for a two-particle system it is clear what to expect for the total momentum of an arbitrary number N of particles instead of just two. The following is a summary of results one can easily obtain.

Let $\vec{P}_{system}$ be the total momentum of the system of N particles, i.e.,

$$\vec{P}_{system} = m_1 \vec{v}_1 + m_2 \vec{v}_2 + \cdots + m_N \vec{v}_N.$$

The rate of change of the total momentum will be equal to the net external force on the system.

$$\boxed{\frac{d\vec{P}_{system}}{dt} = \vec{F}_{net}^{ext},} \qquad (7.22)$$

where $\vec{F}_{net}^{ext}$ is the vector sum of external forces on all particles of the multiparticle system.

$$\boxed{\vec{F}_{net}^{ext} = \vec{F}_1^{ext} + \vec{F}_2^{ext} + \cdots + \vec{F}_N^{ext}.}$$

7.3.3 Motion of Center of Mass

Another useful interpretation of the rate of change of total momentum equation, Eq. 7.22, is that it represents the motion of a mathematical particle of mass equal to the total mass of the system located at a special point in space called the **center of mass** or **CM** of the system as we will see below. The center of mass point does not have to be anywhere in the body or bodies making up the multiplarticle system. For instance, the center of mass of a ring, whose mass is spread out uniformly around the ring, is at the center of the ring where no particle of the ring is located.

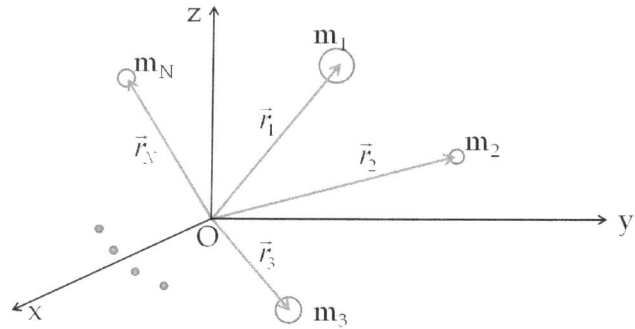

Figure 7.8: The location of the center of mass (CM) of an N-particle system is obtained mass-weigting the positions of each vector as shown in Eq. 7.24.

Consider a system consisting of N particles of masses m_1, m_2, $\cdots$, m_N, whose position vectors with respect to some origin are given by $\vec{r}_1$, $\vec{r}_2$, $\cdots$, $\vec{r}_N$ respectively as shown in Fig. 7.8. Let M be the total mass of the system,

$$M = m_1 + m_2 + \cdots + m_N.$$

The position vector of center of mass (CM) of the system, to be denoted by $\vec{R}_{\text{cm}}$, is defined by mass-weighting the position vectors of all the particles.

$$M\vec{R}_{\text{cm}} = m_1\vec{r}_1 + m_2\vec{r}_2 + \cdots + m_N\vec{r}_N, \tag{7.23}$$

or,

$$\boxed{\vec{R}_{\text{cm}} = \frac{m_1\vec{r}_1 + m_2\vec{r}_2 + \cdots + m_N\vec{r}_N}{M}.} \tag{7.24}$$

Taking derivative with respect to time on both sides of Eq. 7.23 we see that the momentum of a particle of mass M moving with R_{cm} is equal to the total momentum of the system.

$$M\frac{d\vec{R}_{\text{cm}}}{dt} = m_1\frac{d\vec{r}_1}{dt} + m_2\frac{d\vec{r}_2}{dt} + \cdots + m_N\frac{d\vec{r}_N}{dt}, \tag{7.25}$$

or,

$$M\vec{V}_{\text{cm}} = m_1\vec{v}_1 + m_2\vec{v}_2 + \cdots + m_N\vec{v}_N, \tag{7.26}$$

where we have defined the center of mass velocity by the rate of change of the center of mass position vector.

$$\vec{V}_{\text{cm}} = \frac{d\vec{R}_{\text{cm}}}{dt}. \tag{7.27}$$

Define the center of mass momentum $\vec{P}_{cm}$ as the momentum of a fictitious particle of mass equal to the total mass of the system and moving with center of mass velocity.

$$\vec{P}_{\text{cm}} = M\vec{V}_{\text{cm}}. \tag{7.28}$$

Then, we have the result that the center of mass momentum is equal to the sum of the momenta of all particles in the system.

$$\vec{P}_{\text{cm}} = \vec{p}_1 + \vec{p}_2 + \cdots + \vec{p}_N = \vec{P}_{\text{system}} \tag{7.29}$$

The acceleration of the center of mass is defined by the derivative of the velocity of the CM.

$$\vec{A}_{\text{cm}} = \frac{d\vec{V}_{\text{cm}}}{dt}. \tag{7.30}$$

Using the result given in Eq. 7.22, we can now write the equation of motion of the center of mass as follows.

$$\boxed{\vec{F}_{\text{net}}^{\text{ext}} = \frac{d\vec{P}_{\text{cm}}}{dt},} \tag{7.31}$$

which for a constant mass system becomes

$$\boxed{\vec{F}_{\text{net}}^{\text{ext}} = M\vec{A}_{\text{cm}}.} \tag{7.32}$$

Equations 7.31 and 7.32 give the translational motion of the system as a whole since the system has been replaced by one point particle of total mass moving with the center of mass. The detailed information about the motions of individual particles has been lost since the motion of any particle depends on both the external forces and the internal forces on that particle.

For instance, if a system consists of two particles of equal mass moving in opposite directions, then we will find that center of mass is at rest. In this example, just because the center of mass is at rest does not imply no motion is taking place in the system. The detailed motion of each particle requires the study of the equations of motion of each particle separately.

Example 7.3.1. CM of an exploding shell

As an example consider the motion of pieces of an exploded shell in free-fall. Although the pieces fly away in various directions, the CM falls as if no explosion had taken place since during the explosion all forces are internal to the system (see Fig. 7.9).

The motion of CM is that of a single particle with total mass of the system. The fictitious particle at the CM falls freely with acceleration equal to g until one of the pieces hits the ground. At that instant there would be an additional external force on the CM due to the normal force on the fallen particle from the ground. During the time when the fallen part is coming to rest there is an upward force on the system. As a result, there would be additional force in the equation for the CM which would decelerate the CM.

7.4 CALCULATIONS OF CENTER OF MASS

7.4.1 Examples of Discrete Masses

The calculation of CM of discrete masses is based on applying the definition given in Eq. 7.24 to the given system as the following examples illustrate.

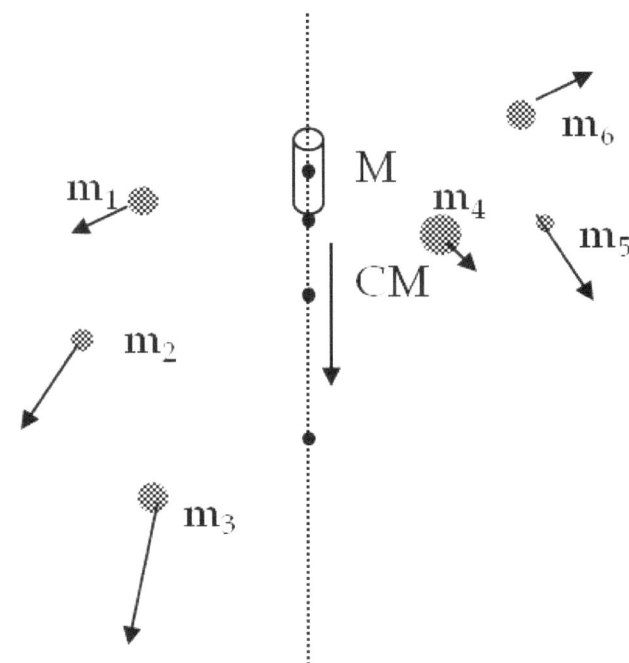

Figure 7.9: A falling explosive explodes but its CM falls as if nothing happened. This continues till the first piece hits the ground.

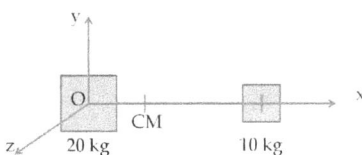

Figure 7.10: Example 7.4.1.

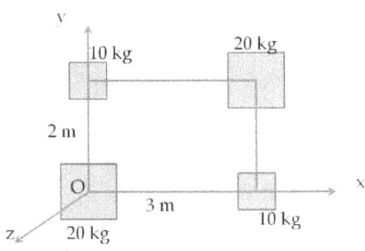

Figure 7.11: Example 7.4.2.

Example 7.4.1. CM of a two-mass system. Two blocks of masses 10 kg and 20 kg are attached at the ends of a 1-meter rod of negligible mass. Where is the CM located?

Solution. The calculation of the CM of a multiparticle system consisting of discrete masses is usually analytically by choosing a convenient set of Cartesian axes. In the present case, since we have only two masses, we place one of the masses at the origin and the other mass along the x-axis. Suppose we place the 20-kg block at the origin and the 10-kg block at $x = 1$ m.

Now, we use the x-component of Eq. 7.24 to find the x-coordinate of the CM as

$$X_{\text{cm}} = \frac{(15 \text{ kg})(1 \text{ m}) + (20 \text{ kg})(0)}{30 \text{ kg}} = \frac{1}{3} \text{ m}.$$

Therefore, the CM is located $\frac{1}{3}$ m from the 20-kg block towards the other block. Note that the units of mass appear both in the numerator and the denominator, therefore it is not necessary to change the unit of mass in these calculation as long as you keep the same unit for all the masses.

Example 7.4.2. Masses at the corners of a rectangle. Find the location of the CM of the four masses given in Fig. 7.11.

Solution. Again, we perform the calculations analytically. Let us choose Cartesian coordinates so that the centers of the four masses fall in the xy-plane with one of the masses at the origin as indicated in the figure. In this coordinate system, the coordinates of the masses are organized in

7.4. CALCULATIONS OF CENTER OF MASS

the following table.

Mass	x (m)	y (m)	z (m)
20 kg	0	0	0
10 kg	3	0	0
20 kg	3	2	0
10 kg	0	2	0

Here the z-coordinate is given for the sake of completeness since based on the choice of the coordinate system, it was already clear that the z-coordinates of all masses will be zero, and hence the z of CM will be zero also.

The x and y-coordinates of the CM are:

$$X_{cm} = \frac{10 \times 3 + 20 \times 3}{60} = 1.5 \text{ m}$$
$$Y_{cm} = \frac{10 \times 2 + 20 \times 2}{60} = 1.0 \text{ m}$$

This says that the CM of the system is actually at the center of the rectangle. This makes sense if you pair up the masses: the CM of the two 20-kg blocks will be at center and the CM of the two 10-kg blocks will also be at the center. The CM calculations can be done in steps: you can find CM of some parts of the total system and replace those parts by one point mass at the CM of those parts, successively simplifying the whole system.

In this example, we could have found the CM of the two 20-kg blocks and replace them by a 40-kg point mass at the center of the diagonal line joining them. Similarly, we could have replaced the two 10-kg blocks by one point mass of mass 20 kg at the center of the other diagonal. Since the centers of the two diagonals are at the same point we would obtain the center of mass at the center. This happened here because of the symmetry in the problem. Suppose the four masses were different, say 10 kg, 20 kg, 30 kg and 40 kg, would the CM be still at the center? The answer is no. Check it out.

Example 7.4.3. Masses at the corners of a triangle. Find the location of the CM of the three equal masses placed at the corner of a triangle.

Solution. Suppose we place masses m at the corners of a triangle as shown in Fig. 7.12. Now, we replace the two masses on the base by a $2m$ point particle at the center of that side of the triangle. This leaves a two mass system of masses $2m$ and m along the dashed line in the figure. The CM of $2m$ and m system is at a distance $\frac{1}{3}^{rd}$ of the distance between the two masses as measured from the $2m$. Therefore, the CM of the three-mass system is at the centroid point of the triangle as shown in the figure. Again, if the masses were different, then the CM will not be at

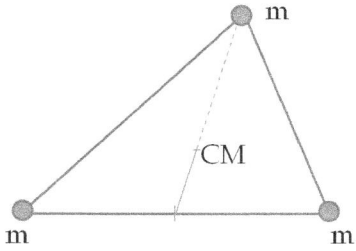

Figure 7.12: Example 7.4.3.

the **centroid**. For instance, if the mass at the top vertex was $2m$, then the CM will be half-way along the dashed line and not at the centroid.

7.4.2 Examples of Continuous System

Most of the time we will be concerned with regular solids of uniform density whose CM can be easily found by inspection and an appeal to symmetry. We will show by explicit calculations that these intuitive feelings are indeed correct.

The general strategy of finding the center of mass of a continuous body is to first discretize the body by conceptually "breaking" it up into small cells. The shapes of the cells are arbitrary and are chosen to exploit the symmetry or other simplifying features of the body. Each cell is then replaced by its mass Δm at the centers of the cells. This process converts the original continuous body problem into an equivalent system of point masses, where we use the procedure of CM for point masses to obtain the CM of the original body. With infinitesimal cells, the sum becomes integral over the body. In the following examples we will demonstrate the use of this general strategy.

Example 7.4.4. CM of a rod of uniform density. The simplest system of continuous mass is a uniform rod of length L and area of cross-section A. Where is the CM of the rod located?

Solution. Let M be the mass of the rod. The density of the rod ρ is then given by

$$\rho = \frac{M}{AL}$$

In calculations concerning a uniform rod, the area of cross-section usually plays a passive role and it is useful to define another density, called **linear density**, which is mass per unit length, denoted by μ.

$$\mu = \frac{M}{L},$$

which is related to the mass per unit volume by

$$\mu = \rho A.$$

To implement this strategy for a uniform rod, let us place the rod along x-axis with one end at the origin at the other end at $x = L$ as shown in Fig. 7.13.

If we divide the rod into N cells, each of length $\Delta x = L/N$, then the procedure gives us a discrete system of masses $m_1, m_2, \cdots, m_N$ at the

7.4. CALCULATIONS OF CENTER OF MASS

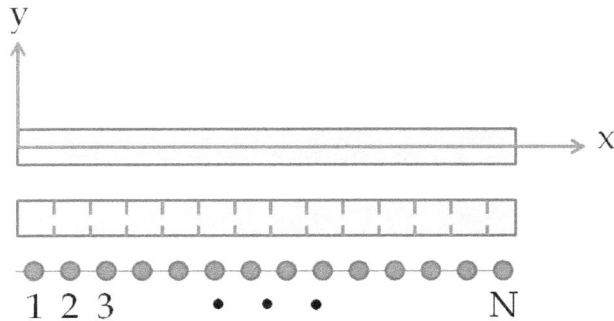

Figure 7.13: Calculating CM of a rod by partitioning the rod into smaller cells.

following places.

$$m_1 \text{ at } x_1 = \Delta x/2$$
$$m_2 \text{ at } x_2 = (3/2)\Delta x$$
$$\vdots$$
$$m_N \text{ at } x_N = (L - 1/2\Delta x)\Delta x$$

Therefore an approximate formula for X_{cm} will be given by applying definition of CM for discrete masses to $m_1, m_2, \cdots, m_N$.

$$X_{\text{cm}} = \frac{1}{M}\left[m_1 \frac{1}{2}\Delta x + m_2 \frac{3}{2}\Delta x + \cdots + m_N\left(L - \frac{1}{2}\Delta x\right)\right]. \qquad (7.33)$$

As the cell size is made progressively smaller, the limit of the sum gives the exact value of X_{cm}.

The path to infinitesimal cell sizes is better described by the following procedure, we will call **the method of infinitesimals**. Consider a representative cell between x and $x + \Delta x$ is shown in Fig. 7.14. The mass Δm in this cell can be written using the linear density μ as

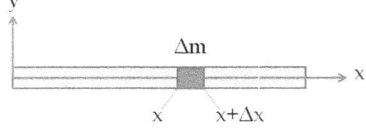

Figure 7.14: Calculating CM of a rod - infintesimals.

$$\Delta m = \mu \Delta x. \qquad (7.34)$$

Then, the sum over cells given in Eq. 7.33 can be formally written as

$$X_{\text{cm}} = \frac{1}{M} \sum_{\text{cells}} x \Delta m,$$

In the limit of infinitesimal cells, this sum becomes an integral.

$$\boxed{X_{\text{cm}} = \frac{1}{M} \int_{\text{rod}} x\, dm.} \qquad (7.35)$$

The integral in Eq. 7.35 is often called a "conceptual" integral since we have not actually specified the limits for the integration variable. Integration over rod is better thought of in terms of the length of the rod. If we can convert the conceptual integral over dm into an integration over the x-coordinate of the cells, then we can specify the limits of integration of the x variable more easily, as from $x = 0$ to $x = L$. We change from the "conceptual integration variable" m to the operational variable x by replacing dm using the infinitesimal form of Eq. 7.34.

$$dm = \mu\, dx, \qquad (7.36)$$

which gives the following integral to evaluate for the X_{cm}.

$$X_{\text{cm}} = \frac{1}{M} \int_0^L x\mu\, dx. \qquad (7.37)$$

If mass density is constant throughout, as is the case in this example, the density $\mu = M/L$ will come out of the integral and the integral would be easily evaluated and simplified to give the X_{cm}.

$$X_{\text{cm}} = \frac{L}{2} \quad \text{(Uniform rod)} \qquad (7.38)$$

Of course, $Y_{\text{cm}} = 0$ and $Z_{\text{cm}} = 0$ here if the x-axis goes right through the middle of the rod. Therefore, the CM of the uniform rod is at the mid-point in the rod as expected.

Example 7.4.5. CM of a triangle of uniform density

Consider a plate of uniform density and uniform thickness cut in the shape of right-angled triangle with base b and height h. Where is the center of mass?

Solution. Let thickness of the plate be t. Along the thickness of the plate, the CM will be in the plane half-way between the two faces of the plate. So, we only need to find the center of mass coordinates on the triangular shape surface.

Let us place the triangular face in the xy-plane as shown in Fig. 7.15. We will proceed with the method of infinitesimals and will try to find the

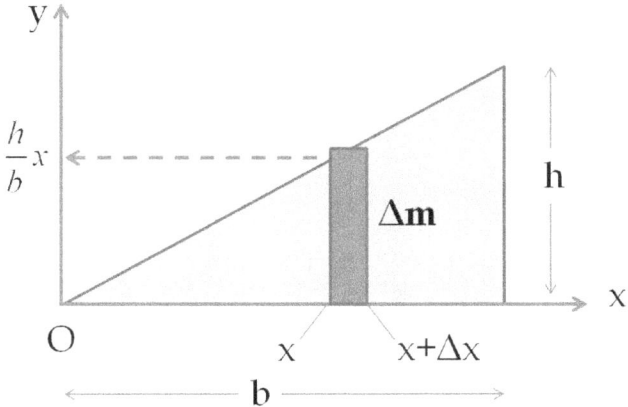

Figure 7.15: Calculating CM of right-angle triangle.

definite integral for the X_{cm} first. Once we complete our calculation for

7.4. CALCULATIONS OF CENTER OF MASS

X_{cm} we will guess the answer for Y_{cm}. For the calculation of X_{cm}, the "conceptual integral" is

$$X_{cm} = \frac{1}{M}\int_{plate} x\, dm. \tag{7.39}$$

Here dm is the mass of an element in a cube-shaped element of thickness t and area $dxdy$ located at the point (x,y) in the xy plane. The volume of the element is $tdxdy$. Let ρ be the density of the material, then we have

$$dm = \rho\, tdxdy \tag{7.40}$$

Therefore Eq. 7.39 can be transformed in operational form, where we need the limits of x and y variables that go over the plate.

$$X_{cm} = \frac{\rho\, t}{M}\int_{plate} x\, dxdy. \tag{7.41}$$

The integral in Eq. 7.41 is a double integral: there is an integration over x and another one over y, but the two integrations are linked since we must restrict the points to the points on the surface of the triangular plate. Suppose we do the integration over y first, keeping fixed at some value x, then we see from Fig. 7.15 that the limit will be from $y = 0$ to $y = y_{max}$, where y_{max} is the value of y at the hypotenuse. Writing the equation of the line for the hypotenuse we find that

$$y_{max} = \frac{h}{b}x.$$

After we have done the integration over y, the value of x goes over the entire base to include all the elements of the triangle. Therefore, Eq. 7.41 becomes

$$X_{cm} = \frac{\rho\, t}{M}\int_0^b dx\left[x\int_0^{y_{max}} dy\right]. \tag{7.42}$$

Doing the integration over y leaves x integral to do in the following

$$X_{cm} = \frac{\rho\, t\, h}{M\, b}\int_0^b x^2 dx. \tag{7.43}$$

Finally, we obtain the following answer for X_{cm} which can be simplified.

$$X_{cm} = \frac{\rho\, t\, h}{M\, b}\frac{b^3}{3}. \tag{7.44}$$

Note that the volume of the plate is $t \times$ area of the surface, or $tbh/2$, which gives $M = \rho \times$ Volume $= \rho \times tbh/2$. Therefore,

$$X_{cm} = \frac{2}{3}b. \tag{7.45}$$

This says that X_{cm} is two-third of the way along the base from the tip, or one-third of the distance from the edge opposite to the tip.

To obtain the Y_{cm} we do not need to do another calculation. We can make use of the result for the X_{cm} calculation, which says that that the

CM in this direction will be located $\frac{2}{3}^{rd}$ from the tip or $\frac{1}{3}^{rd}$ from the side. In the choice of coordinates in Fig. 7.15, the Y_{cm} is from the edge. Therefore,

$$Y_{cm} = \frac{1}{3}h. \tag{7.46}$$

Example 7.4.6. CM of a composite system. CM of composite systems are found by using the CM of individual parts. Where is the CM of a composite system of a rectangle of mass M_1 and a right angle triangle of mass M_2, both of uniform mass density of dimensions and separation shown in Fig. 7.16.

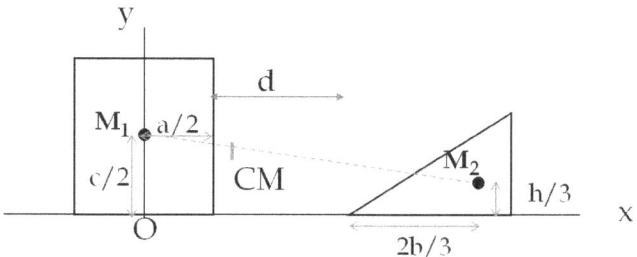

Figure 7.16: CM of a composite system.

Solution. To find the CM of a composite system the best strategy is to replace each object with a point masses at its center of mass. This process turns a complex problem into a problem of point mass. In the present case, the problem turns into a mass M_1 at $(0, c/2)$ and M_2 at $(a/2 + c + 2b/3, h/3)$. Therefore, the coordinates of the CM are

$$X_{cm} = \frac{M_2}{M_1 + M_2}\left(\frac{1}{2}a + d + \frac{2}{3}b\right)$$

$$Y_{cm} = \frac{1}{M_1 + M_2}\left(\frac{1}{2}M_1 c + \frac{1}{3}M_2 h\right)$$

7.5 ISOLATED SYSTEMS AND CONSERVATION OF MOMENTUM

We found above that the change in the total momentum of any system, whether it contains a single particle or several particles, is caused only by the external forces on the system, and the rate of that change at any instant is exactly equal to the net external force on all parts of the system.

$$\frac{d\vec{P}_{system}}{dt} = \vec{F}_{net}^{ext}. \tag{7.47}$$

Therefore, if the net external force on a system is zero, then the total momentum cannot change. That is, the magnitude and direction of the momentum of a system that does not interact with anything else will remain fixed in time, i.e. conserved. We will call such systems **isolated**

7.5. ISOLATED SYSTEMS AND CONSERVATION OF MOMENTUM

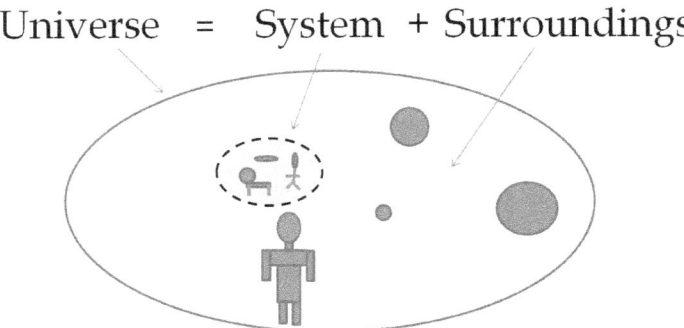

Figure 7.17: Division of the universe into system and surroundings takes place when we focus on a part of the universe for study. The objects of interest make up the system and everything else is the surroundings or external to the system. We usually draw a dashed boundary line to indicate the objects that are within the system from those which are external to the system.

systems. This result is also called **the principle of conservation of momentum**.

By a system we usually mean any physical body of interest. A choice of a physical body for study leads to an artificial division of the entire universe into a system and the rest, called the surroundings or the external world, as illustrated in Fig. 7.17. We study the overall motion of the system by examining the forces on the system by the external world. The surroundings is external to the system and influences the changes in momentum of the system. Similarly, the system is external to the surroundings and therefore, influences the changes in the momentum of the surroundings. The system and surroundings will obey their own equations of motion which can be written as:

$$\text{System:} \quad \frac{d\vec{P}_{\text{system}}}{dt} = \vec{F}_{\text{net}}^{\text{on system}} \tag{7.48}$$

$$\text{Surroundings:} \quad \frac{d\vec{P}_{\text{surroundings}}}{dt} = \vec{F}_{\text{net}}^{\text{on surroundings}} \tag{7.49}$$

Now, when we add the two equations we obtain the equations of motion of the universe. Since the forces from the surroundings on the system are equal in magnitude but opposite in directions to the forces from the system on the surroundings, the net external force on the universe is zero.

$$\boxed{\frac{d\vec{P}_{\text{Univese}}}{dt} = 0.} \tag{7.50}$$

This makes sense, since the larger system, which we have called the universe above, is an isolated system. Since there is nothing external to the larger system, the total momentum of the system and surroundings together would not change with time.

Principle of conservation of momentum

The total momentum of a system and the rest of the universe does not change in time. We will call the rest of the universe "the surroundings" or "the environment". The force on the system by the objects in the environment changes the momentum of the system and the force on the objects of the environement by the system changes the momentum of the environment. But, since the forces are equal in magnitude and opposite in directions, the impulse on the system imparted by the environment in any interval must be equal in magnitude and opposite in direction to the impulse imparted to the environment by the system. Consequently, any change in the momentum of the system is accompanied by a change in the momentum of the environment of an equal magnitude and of opposite direction.

$$\Delta \vec{P}_{\text{System}} = -\Delta \vec{P}_{\text{Environment}}. \tag{7.51}$$

Since, momentum is a vector quantity, the conservation of momentum applies independently for each direction in space. Decomposing these vectors in x, y and z-components, we find that changes in different components of the momenta of the system are accompanied by equal changes of opposite sign in the changes in corresponding components of the momenta of the surroundings.

$$\Delta P_x^{\text{System}} = -\Delta P_x^{\text{Environment}} \tag{7.52}$$

$$\Delta P_y^{\text{System}} = -\Delta P_y^{\text{Environment}} \tag{7.53}$$

$$\Delta P_z^{\text{System}} = -\Delta P_z^{\text{Environment}} \tag{7.54}$$

When you bring the changes in momenta of the system and the environment on one side of the equation, we find that the net change in any component of the total momentum of the system and the environment together is always zero, i.e the total momentum is conserved component-by-component. The decomposition in three Cartesian components makes it clear that the momentum component in any direction is independently conserved from the momentum component in any other direction that is perpendicular to it.

For instance, when a football is kicked at an angle to the ground, the momentum in the horizontal direction is conserved independently of whatever happens in the momentum component in the vertical direction. Actually, although the horizontal component of the momentum of a football in free flight does not change, the absolute value of the vertical component changes with time, decreasing in the upward part of the flight, becoming momentarily zero at the top of the flight, and increasing afterwards until the ball hits the ground. The change in the vertical component of the momentum happens because of the external force on the ball in that direction, which is the gravitational force of the Earth on the ball.

The principle of conservation of momentum provides a powerful tool for solving problems by giving us a conservation equation for the direction in which there are no external forces. It is easy to peel off that component of the motion that will have conserved momentum and depending upon the question to be investigated, that may be enough for the problem.

Example 7.5.1. Momentum of ice skaters. Khalil (mass 60 kg) and Mary (50 kg) are ice skating on a smooth surface. They start from rest by pushing on each other with an average force of magnitude 30 N and directed horizontally for 0.5 sec. Assume the skating surface to be horizontal and frictionless. Find (a) the total momentum of the two skaters at $t = 0.2$ sec, (b) the momentum of each skater at $t = 0.5$ sec and the total momentum, (c) the velocity of each skater at $t = 0.5$ sec, and (d) Mary's momentum at $t = 0.6$ sec.

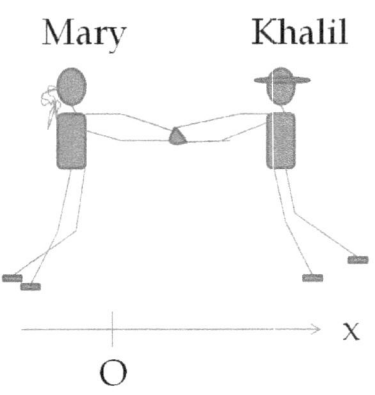

Figure 7.18: Example 7.5.1.

Solution. Since the value of momentum depends upon the reference frame, we will pick a reference frame before starting on the solution of the problem. Furthermore, since the principle of conservation of momentum requires an inertial reference frame, we will look for an inertial frame in the physical setting of the problem. In the present case, a frame attached to the ice surface would be convenient choice for an inertial frame.

(a) Now, if we consider the two skaters together as our multi-particle system. Then, their forces on each other will be an internal force of the system. The only other forces on the two bodies consisting of the system are the gravitational pull of the Earth and the normal force from Earth's surface. Both of these external forces are in the vertical direction. Hence, from the conservation of momentum, the horizontal component of the total momentum of the skaters will not change. As for the vertical component of the momentum, we note that since there is no acceleration in that direction, the vertical forces must be balanced. Therefore, the vertical component of the total momentum is also fixed in time. Since, the two skaters started out with zero momentum and their momentum cannot change by the argument presented, therefore their net momentum will remain zero.

(b) Although the total momentum of the two skaters together remains zero, the momentum of each skater can change since individually they are not isolated systems. **To find the change in the momentum of a particular skater we focus on that skater alone as a new system.** The vertical forces, i.e. weight of the skater and the normal force are still balanced, but the horizontal force from the other skater is not zero any more.

To be specific, let us calculate the change in momentum of Khalil over the time interval [0, 0.5 s] using the analytic approach for vectors. Let Mary and Khalil move on the x-axis so that the force from Mary is towards positive x-axis as shown in the Fig. 7.18. Let P_x be the x-component of Khalil's momentum at time $t = 0.5$ sec and P_{0x} be his momentum at $t = 0$. Then the x momentum equation for Khalil's momentum will be

$$\text{Khalil:}\quad P_x - P_{0x} = 30 \text{ N} \times 0.5 \text{ s} \implies P_x = 15 \text{ kg.m/s}.$$

The change in momentum of Mary must be equal in magnitude but opposite in direction.

$$\text{Mary:}\quad P_x = -15 \text{ kg.m/s}.$$

The total momentum of the two skaters together, of course, does not change since the net external force on the two skaters together as one system is zero.

(c) The velocity of each skater can be obtained by dividing the corresponding momentum by the appropriate mass. Just as the momentum, the only non-zero component of the velocity is the x-component here.

$$\text{Khalil:} \quad v_x(0.5\text{s}) = \frac{15 \text{ kg.m/s}}{60 \text{ kg}} = 0.25 \text{ m/s}.$$

$$\text{Mary:} \quad v_x(0.5\text{s}) = -\frac{15 \text{ kg.m/s}}{50 \text{ kg}} = -0.30 \text{ m/s}.$$

(d) Since there is no net force on Mary during [0.5 s, 0.6 s] interval, Mary herself is an isolated system during this interval with zero net external force. Therefore, Mary's momentum at $t = 0.6$ s will be equal to her momentum at $t = 0.5$ s, which is already given in part (b).

7.6 COLLISIONS

In a two-body collision, two objects collide with one another. In a collision, each body applies an impulse on the other body. Therefore, the momenta of the two bodies change in a collision.

However, since the collision event involves only internal forces between the two bodies, the total momentum of the colliding bodies after the collision will be equal to their total momentum before the collision, both in magnitude and direction.

$$\boxed{\left(\vec{P}_{\text{total}}\right)_{\text{after}} = \left(\vec{P}_{\text{total}}\right)_{\text{before}} \quad \text{(bodies isolated)}} \qquad (7.55)$$

If two colliding objects 1 and 2 have momenta $\vec{p}_1$ and $\vec{p}_2$ immediately before the collision, and $\vec{p}_1'$ and $\vec{p}_2'$ immediately after the collision, the momentum conservation equation across the collision process produces the following vector equation.

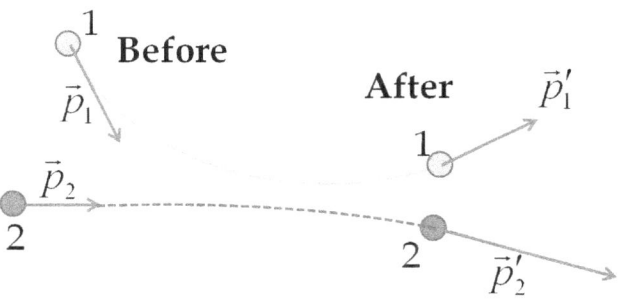

Figure 7.19: Collision process leads to transfer of momentum among bodies such that the total momentum before the collision is equal to the total momentum after the collision: $\vec{p}_1 + \vec{p}_2 = \vec{p}_1' + \vec{p}_2'$.

7.6. COLLISIONS

$$\boxed{\vec{p}_1 + \vec{p}_2 = \vec{p}_1\,' + \vec{p}_2\,'.} \qquad (7.56)$$

Therefore, during a collision, a transfer of momentum among the bodies takes place such that the total momentum of bodies before the collision is equal to the total momentum after the collision. Both the direction and the magnitude of the momenta of the two colliding objects may change, subject to the constraint that the total momentum remains fixed at the value immediately prior to the collision. Note that Eq. 7.56 is a vector equation. Often, it is helpful to work with the Cartesian components of this equation.

$$p_{1x} + p_{2x} = p'_{1x} + p'_{2x} \qquad (7.57)$$
$$p_{1y} + p_{2y} = p'_{1y} + p'_{2y} \qquad (7.58)$$
$$p_{1z} + p_{2z} = p'_{1z} + p'_{2z} \qquad (7.59)$$

If the incoming and outgoing momenta lie in one plane, then we usually take the plane to be the xy plane, and work with only the x and y-components.

The change in momentum of each object is due to the impulse by the other object. Suppose the collision lasts for a duration Δt when the two bodies are interacting and suppose that there is no interaction between the two bodies before or after the collision, then the change in momentum of each body during the collision gives us an idea of the average force between the bodies during the collision.

$$\boxed{\text{Change in momentum of 1: } \vec{p}_1\,' - \vec{p}_1 = \vec{F}_{\text{by 2 on 1}} \Delta t} \qquad (7.60)$$

$$\boxed{\text{Change in momentum of 2: } \vec{p}_2\,' - \vec{p}_2 = \vec{F}_{\text{by 1 on 2}} \Delta t} \qquad (7.61)$$

If we know the momentum change of one colliding body and have a good estimate of the duration Δt for which the two objects were in contact, these equations can be used to obtain an estimate of the average force $\vec{F}_{ave}$ between the two bodies.

Example 7.6.1. One dimensional collision. Two carts of masses 200 g and 300 g are moving towards each other on a straight track with speeds of 10 m/s and 15 m/s respectively. Immediately after the collision, the 200-g cart reverses direction and moves with a speed of 8 m/s. What is the velocity (i.e. speed and direction) of the 300-gram cart after collision?

Solution. To make the collision event clearer, we normally draw two figures, one for the situation **before the collision** and one for the situation **after the collision** as shown in Fig. 7.20. The direction of the 300-gram cart after the collision is not known yet, so we arbitrarily pick a direction. If we find that x-component of the momentum for 300-gram is positive then our chosen direction would be the correct one, otherwise the correct direction would be just the opposite of our choice. For component calculations, we do not need to know the position of the origin, we need only

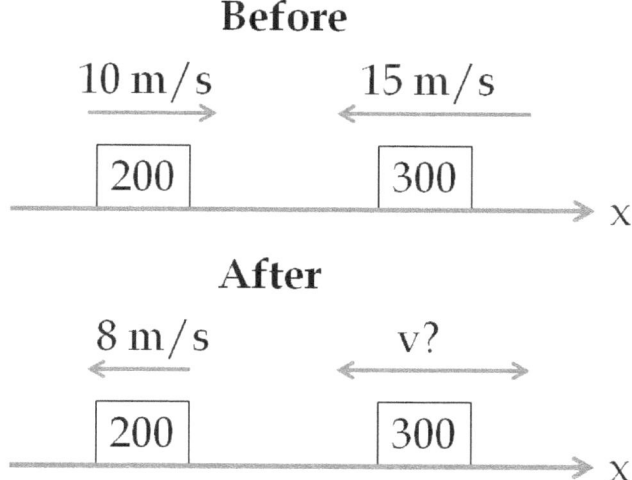

Figure 7.20: Example 7.6.1. Normally two figures are drawn to understand the collision process, one for the physical situation before the collision and the other after the collision.

the directions of the axes. As shown in Fig. 7.20, we pick the x-axis to be the axis along which the collision takes place in this problem. This choice will give zero y and z-components and we will not worry about then any further.

Let us label the 200-gram cart as "1" and the 300-gram cart as "2". Then we have the following quantities known from the statement of the problem when referred to the figure drawn. [To keep the expressions simple I will leave out the units from calculations, and put the units in at the end only. I will also drop the subscript x from the x-component of velocity in the notation.]

Before collision:

$$p_{1x} = +(0.2)(10) = +2;$$
$$p_{2x} = (0.3)(-15) = -4.5.$$

After collision:

$$p'_{1x} = (0.2)(-8) = -1.6;$$
$$p'_{2x} = +(0.3)(v) = 0.3v.$$

The conservation of the x-component of the total momentum across the collision yields an equation for the unknown v.

$$p_{1x} + p_{2x} = p'_{1x} + p'_{2x}$$
$$+2 - 4.5 = -1.6 + 0.3\ v, \implies v = -3\ m/s.$$

Therefore, we conclude that the 300-gram cart continues to move towards the negative x-axis with a speed of 3 m/s, the same direction as the direction before the collision.

7.6. COLLISIONS

Example 7.6.2. Two-dimensional collision. Two balls A and B of masses 0.4 kg and 0.45 kg are moving towards each other with speeds 5 m/s and 8 m/s respectively and collide at an angle of 60°, which is the angle between their velocities before the collision. After the collision, ball A continues in the direction that is 15° to its original direction at a speed of 10 m/s. What is the velocity of ball B after the collision?

Solution. We usually draw four figures when the collision takes place in two or more dimensions. These consist of a schematic diagram for the situation before, another for the situation after the collision, and two diagrams with coordinate choices to help with the components of vectors as shown in Fig. 7.21.

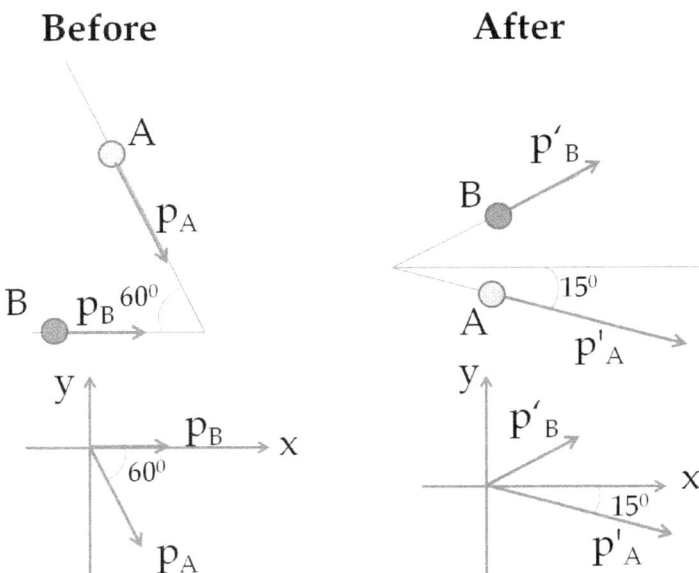

Figure 7.21: Example 7.6.2.

Just as before, when dealing with the addition or subtraction of vectors, it is helpful to organize the information about the components in a table format. Using the magnitudes from the given data and the direction with respect to the axes shown in the figure, we work out the components of various momenta. Since we do not know either the magnitude or the direction of the momentum of B, we will leave the momentum in terms of the x and y-components. Furthermore, to simplify the writing, we will denote the components of the unknown momentum by p_x and p_y in place of p'_{Bx} and p'_{By}, respectively, since the later notation is too cumbersome. We put the units in the column heads in the table.

Vector	x-component (kg.m/s)	y-component (kg.m/s)
$\vec{p}_A$	$0.4 \times 5\cos(60°) = 1$	$-0.4 \times 5\sin(60°) = -1.732$
$\vec{p}_B$	$0.45 \times 8 = 3.6$	0
$\vec{p}_A'$	$0.4 \times 10\cos(15°) = 3.86$	$-0.4 \times 10\sin(15°) = -1.04$
$\vec{p}_B'$	p_x	p_y

The conservation of the x and y-components of momenta give the following values for p_x and p_y.

$$p_x = 0.74 \text{ kg.m/s}$$
$$p_y = -0.70 \text{ kg.m/s}$$

The magnitude of momentum of ball B after collision is

$$p_B = \sqrt{p_x^2 + p_y^2} = 1.0 \text{ kg.m/s},$$

The direction of the momentum of ball B is in the fourth quadrant of the Cartesian coordinates in Fig. 7.21. The clockwise angle from the positive x-axis would suffice to determine the direction of the momentum vector. This angle is given by

$$\theta = \arctan\left(\frac{-0.70}{0.74}\right) = -43°,$$

where the negative value indicates that the direction of the vector is at an angle of 43° clockwise from the positive axis towards the negative y-axis. The velocity of ball B is in the same direction as the momentum and has the magnitude

$$\text{Magnitude of velocity, } v = \frac{1 \text{ kg.m/s}}{0.45 \text{ kg}} = 2.2 \text{ m/s}.$$

Example 7.6.3. Colliding bodies stuck together. Two balls A and B of mass 0.4 kg and 0.45 kg are moving towards each other with speeds 5 m/s and 8 m/s respectively, and then they collide at an angle of 60° as in the last example. Upon collision, they stick together, and then move as one object after the collision. Find the speed and direction in which the pair moves.

Solution. See Fig. 7.21 for the physical situation except that after the collision, now, the two bodies stick together and move in some direction to be determined. Let the components of the two bodies together after the collision be P_x and P_y. The values for components of momenta before and after the collision are

Vector	x-component (kg.m/s)	y-component (kg.m/s)
$\vec{p}_A$	1	-1.732
$\vec{p}_B$	3.6	0
$\vec{P}$	P_x	P_y

The conservation of momentum gives the values of P_x and P_y.

$$P_x = 4.6 \text{ kg.m/s}; \quad P_y = -1.732 \text{ kg.m/s}.$$

The magnitude of velocity is equal to the magnitude of momentum divided by the mass of the two bodies moving together.

$$\text{Magnitude of velocity: } v = \frac{\sqrt{4.6^2 + 1.732^2} \text{ kg.m/s}}{0.85 \text{ kg}} = 5.8 \text{ m/s}.$$

The direction of the velocity is in the fourth quadrant, and can again be given by the clockwise angle θ from the positive x-axis.

$$\theta = \arctan\left(\frac{-1.732}{4.6}\right) = -21°.$$

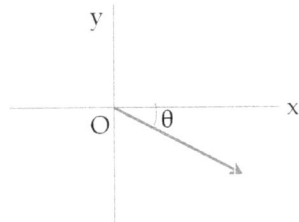

Figure 7.22: Example 7.6.3.

Example 7.6.4. A projectile breaking up in mid air. A projectile of mass 2 kg moving at 50 m/s breaks up in mid air into two pieces. The larger piece comes out at 60 m/s in the forward direction, and the smaller piece at 100 m/s in the backward direction. What are the masses of the two pieces?

Solution. Since an explosion is the opposite of a collision, only internal forces are involved in explosion. Therefore, the total momentum of the multiparticle system after explosion must equal the total momentum of the exploding body before the explosion.

The given situation has a one-dimensional scenario. Hence, we will use only one of the Cartesian axes for the analysis. Let the positive x-axis be pointed in the forward direction. Let m be the mass of the piece in the forward direction. Then, the x-component of momentum equation conservation for the problem yields the following condition.

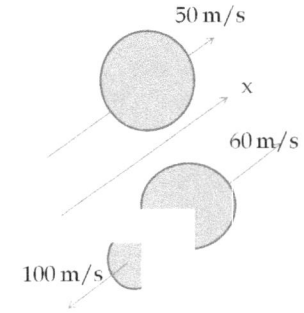

Figure 7.23: Example 7.6.4.

$$2 \text{ kg} \times 50 \text{ m/s} = m60 \text{ m/s} - (2-m) \times 100 \text{m/s}.$$

Solving this equation for m we find that $m = 1.875 \ kg$. Therefore, the larger piece has a mass of $1.875 \ kg$ and the smaller piece $0.125 \ kg$.

7.7 COLLISIONS IN THE CM FRAME

If the colliding bodies together form an isolated system, then, a coordinate system in which the CM of the colliding bodies is fixed at the origin is very helpful in the analysis of the collision. This frame of reference, calles the **CM frame**, is also an inertial frame since the CM would not have an acceleration if the system is an isolated system. In this section we examine the collision process from the CM frame and discuss the relation to the collision as observed in the laboratory frame.

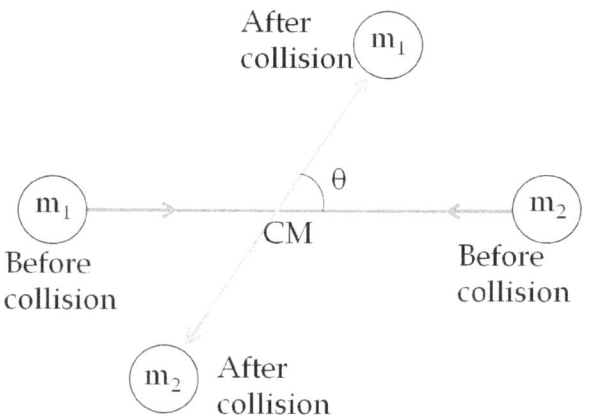

Figure 7.24: Collision in CM frame.

Recall that the total momentum of a system in any frame is equal to the total mass of the system times the velocity of CM in that frame. In the CM frame, the velocity of the CM is zero, by definition. Therefore, the total momentum of the system would be zero in the CM frame.

$$\vec{P}_{\text{System}} = 0 \quad \text{(CM frame)} \tag{7.62}$$

Let us choose a notation for the symbols for quantities with respect to the CM frame. We will attach a subscript or superscript "CM" to the symbol to indicate the quantity with respect to the CM frame. The standard frame in the laboratory is also called **LAB frame**, and sometimes we will attach a subscript or superscript "LAB" to the symbol to emphasize the reference frame.

Let $\vec{P}_1^{\text{CM}}$ and $\vec{P}_2^{\text{CM}}$ be the momenta of the two masses in the CM frame. Then, as stated above, their vector sum will be zero.

$$\boxed{\vec{P}_{\text{System}}^{\text{CM}} = \vec{P}_1^{\text{CM}} + \vec{P}_2^{\text{CM}} = 0.} \tag{7.63}$$

Therefore, in the CM frame, the magnitudes of the momenta of two colliding particles must be equal and directions must be opposite.

$$\vec{P}_1^{\text{CM}} = -\vec{P}_2^{\text{CM}}. \tag{7.64}$$

As shown in Fig. 7.24, before the collision, the colliding bodies approach each other toward the CM, where the origin of the CM frame is located.

Let $\vec{P}_1'^{\text{CM}}$ and $\vec{P}_2'^{\text{CM}}$ be the momenta of objects 1 and 2 after the collision. Since, the conservation of momentum applies to all inertial frames, and since the CM-frame of an isolated system is also an inertial frame, the sum of the momenta of the two objects would also zero after the collision since their sum was zero before the collision in this frame.

$$\vec{P}_1^{\text{CM}} + \vec{P}_2^{\text{CM}} = 0 = \vec{P}_1'^{\text{CM}} + \vec{P}_2'^{\text{CM}}. \tag{7.65}$$

Therefore, the momenta of the two objects after the collision are also collinear. The colliding objects after the collision move away from each

7.7. COLLISIONS IN THE CM FRAME

other with equal magnitude of momentum! In the CM frame, the line joining the colliding objects rotates by an angle θ, called the **scattering angle** (Fig. 7.24). The scattering angle can be obtained by taking a dot product of $\vec{P}_1^{CM}$ and $\vec{P}_1'^{CM}$.

$$\cos\theta = \frac{\vec{P}_1^{CM} \cdot \vec{P}_1'^{CM}}{\left|\vec{P}_1^{CM}\right|\left|\vec{P}_1'^{CM}\right|}. \tag{7.66}$$

Relation between the CM frame and the LAB frame

It is easy to relate the momentum an object in the CM-frame to the momentum of the same object the LAB-frame. Let m_1 and m_2 be the masses of the two bodies, and their velocities be $\vec{v}_1^{LAB}$ and $\vec{v}_2^{LAB}$ in the LAB-frame. From the definition of the CM we know the velocity of the CM of the two masses will be

$$\vec{V}_{CM}^{LAB} = \frac{m_1 \vec{v}_1^{LAB} + m_2 \vec{v}_2^{LAB}}{m_1 + m_2}$$

Writing this equation in terms of the momenta of the two bodies in the LAB frame we have

$$\vec{V}_{CM}^{LAB} = \frac{\vec{p}_1^{LAB} + \vec{p}_2^{LAB}}{m_1 + m_2}.$$

The momentum of body 1 in the CM frame will be obtained by multiplying the velocity of this body with respect to the CM and its mass.

$$\vec{P}_1^{CM} = m_1 \left(v_1^{LAB} - V_{CM}^{LAB}\right) = \vec{p}_1^{LAB} - m_1 \left(\frac{\vec{p}_1^{LAB} + \vec{p}_2^{LAB}}{m_1 + m_2}\right).$$

Simplifying, we obtain,

$$\vec{P}_1^{CM} = \frac{m_2 \vec{p}_1^{LAB} - m_1 \vec{p}_2^{LAB}}{m_1 + m_2}.$$

Similarly, you can show the following for the momentum of body 2 in the CM frame.

$$\vec{P}_2^{CM} = -\left(\frac{m_2 \vec{p}_1^{LAB} - m_1 \vec{p}_2^{LAB}}{m_1 + m_2}\right).$$

Example 7.7.1. Scattering of alpha particles from gold nucleus. Alpha particles of mass 4 amu are incident on gold nucleus of mass 197 amu [atomic mass unit]. Before the collision, an alpha particle is moving with a speed of 2×10^5 m/s in the LAB-frame. After the collision, the alpha particle comes out with a speed of 1.5×10^5 m/s at an angle of $10°$ from the original direction in the LAB frame. Find the angle between the incoming and outgoing directions in the CM-frame. (1 amu = 1.66053×10^{-27} kg)

Solution. Let us orient our axes of the LAB and CM frames so that their axes are parallel, and initially the motion is along the x-axis. The velocity of the origin of the CM-frame with respect to the LAB-frame is along the x-axis.

$$\vec{V}_{CM}^{LAB} = V_x^{LAB} \hat{u}_x = \frac{4 \times 2 \times 10^5 + 0}{4 + 197} \hat{u}_x = 4 \times 10^3 \hat{u}_x \text{ m/s}.$$

Using the velocity of CM, we can obtain the velocity of the colliding particles as seen from the CM-frame. Let $\vec{v}_1^{\text{CM}}$ and $\vec{v}_1'^{\text{CM}}$ be velocities of the alpha particle in the CM-frame before and after the collision, respectively. We use the same letters for the velocities in the LAB frame.

$$\begin{aligned} \vec{v}_1^{\text{CM}} &= \vec{v}_1^{\text{LAB}} - \vec{V}_{\text{CM}}^{\text{LAB}} = \left[2 \times 10^5 - 4 \times 10^3\right] \hat{u}_x \text{ m/s}. \\ \vec{v}_1'^{\text{CM}} &= \vec{v}_1'^{\text{LAB}} - \vec{V}_{\text{CM}}^{\text{LAB}} \\ &= \left[1.5 \times 10^5 \cos(10°)\hat{u}_x + 1.5 \times 10^5 \sin(10°)\hat{u}_y\right] \text{ m/s} \\ &\quad - 4 \times 10^3 \hat{u}_x \text{ m/s}. \end{aligned}$$

Therefore, the momenta of the alpha particle before and after collision are:

$$\begin{aligned} \vec{p}_1^{\text{CM}} &= m_1 \vec{v}_1^{\text{CM}}. \\ \vec{p}_1'^{\text{CM}} &= m_1 \vec{v}_1'^{\text{CM}}. \end{aligned}$$

Therefore the angle between the incoming direction and outgoing direction of alpha particles in the CM-frame.

$$\cos\theta = \frac{\vec{P}_1^{\text{CM}} \cdot \vec{P}_1'^{\text{CM}}}{\left|\vec{P}_1^{\text{CM}}\right|\left|\vec{P}_1'^{\text{CM}}\right|} = 0.986 \implies \theta = 9.6°.$$

Note that mass and common constant factors cancels out in the calculation of the angle.

7.8 SYSTEMS WITH TIME VARYING MASS

In many situations of practical interest we find that the mass of the body of interest changes with time. For instance, in the motion of a rocket, the mass of the rocket decreases as burnt fuel is ejected from the rocket. Although the total mass of the rocket and the ejected fuel is constant in this situation, we are normally interested in describing the motion of the rocket only. Similarly, when rain drops fall, they gather more water molecules thus increasing their mass with time. These systems are called syterms with **time-varying mass** or **open systems**.

These problems are best handled by using the complete form of Newton's second law in terms of the change of momentum rather the constant-mas formulation of Newton's second law. In this section, we will show how to adapt Newton's second law for systems with varying mass.

To see the special problem that arises in systems with time-varying mass substitute $\vec{p} = m\vec{v}$ in the full Newton's second law and carry out the derivative with respect to time to obtain the following.

$$\vec{F} = \left(\frac{dm}{dt}\right)\vec{v} + m\vec{a}. \tag{7.67}$$

7.8. SYSTEMS WITH TIME VARYING MASS

Let us rearrange the terms in this equation so that we keep $m\vec{a}$ on the one side of the equation and other terms on the other side

$$m\vec{a} = \vec{F} - \left(\frac{dm}{dt}\right)\vec{v}. \qquad (7.68)$$

This says that a system whose mass varies with time can have a non-zero acceleration even when there is no external force ($\vec{F} = 0$). The inertial term, $[-\left(\frac{dm}{dt}\right)\vec{v}]$, acts as a force on the system as far as acceleration is concerned. This force is called **thrust** when these equations are applied to the motion of rockets, where it is the force applied on the main body of the rocket by the exiting fuel.

To further elaborate the use of Newton's complete expression for the second law of motion to varying mass systems, we now apply these considerations to the motion of rockets. One complicating and sometimes confusing feature of the calculations for a rocket motion is that the velocity of the ejected fuel is commonly given with respect to the body of the rocket which is not an inertial frame. Since Newton's second law is usually written in an inertial frame, we will need to convert all given data in terms of an inertial frame.

The general strategy for an analysis of a time-varying system involves the following standard steps.

1. Pick an inertial coordinate system.

2. Write down the momenta of all masses at two nearby instants: an arbitrary time t and an infinitesimally instant later $t + dt$. This will be momenta of rocket of mass m at time t and the rocket of mass $m - dm$ and fuel of mass dm at time $t + dt$.

3. Equate rate change in momentum to average force.

7.8.1 Rocket Motion With no External Force

Consider a rocket in outer space with total mass M kg at $t = 0$. The rocket ejects burnt fuel at a constant rate of α kg/s from its rear at a constant speed u m/s with respect to the rocket. We wish to find the speed of the rocket at an arbitrary time t sec before the rocket runs out of fuel.

To deuce the equation of motion of the rocket that has no external force, we compare the momenta of the rocket and the burnt fuel at time t and $t + \Delta t$. To help us do this calculation, we make use of Figure 7.25.

Let m be the mass of the rocket plus remaining fuel at time t, and let v_y be the y-component of its velocity with respect to the inertial frame. The x and z-components of all quantities are zero. Let $|\Delta m|$ be the mass of the ejected burnt fuel between time t and $t + \Delta t$, so that at time

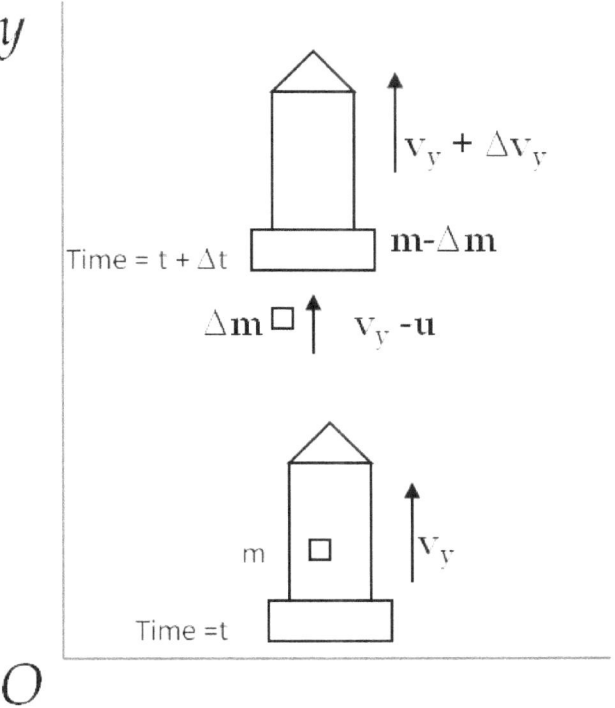

Figure 7.25: Rocket at two instants in time. The burnt fuel of mass Δm which is inside the rocket at time t is ejected from the back of the rocket at speed u with respect to the rocket. The positive y-axis of an inertial frame is pointed towards the forward direction of the rocket. In this frame, the y-component of the velocity of the ejected fuel is $v_y - u$ where v_y is the y-component of the velocity of the rocket at tine t.

$t + \Delta t$, the y-component of the velocity of the rocket is $v_y + \Delta v_y$ and the y-component of the velocity of the burnt fuel is $-u$ with respect to the rocket. Therefore, the y-component of the velocity of the burnt fuel with respect to the inertial observer will be $v_y - u$ by simple vector addition rules. Now, we summarize the y-components of momenta of rockets and fuels at the two nearby instants with respect to the inertial observer are:

At time t: $\quad p_y(t) = mv_y$ \hfill (7.69)

At time $t + \Delta t$:

$$p_y(t + \Delta t) = (m - |\Delta m|)(v_y + \Delta v_y) + |\Delta m|(v_y - u) \quad (7.70)$$

Here I have put absolute sign around the change in mass so that we do not need to worry about the negative sign at this point. Since there is no external force on the combined system of the rocket and the ejected fuel, the total momentum is conserved. Therefore, we equate the momentum of the system at the two instants.

$$mv_y = (m - |\Delta m|)(v_y + \Delta v_y) + |\Delta m|(v_y - u), \quad (7.71)$$

which can be simplified to

$$m\Delta v_y - |\Delta m|\Delta v_y - |\Delta m|\, u = 0. \quad (7.72)$$

7.8. SYSTEMS WITH TIME VARYING MASS

We need to take the limit of this equation as the duration Δt becomes small. In this limit, we keep only the dominant terms. To see the dominant term, it is best to compare dimensionless quantities. Therefore, let us rearrange the first two terms and combine them to get

$$m\Delta v_y \left(1 - \frac{|\Delta m|}{m}\right) - |\Delta m|\, u = 0. \tag{7.73}$$

For most of the operation of a rocket, the ratio $|\Delta m|/m \ll 1$, therefore, we can ignore the second term within the parenthesis. Now, taking the infinitesimal limit, the change in y-component of velocity Δv_y is replaced by dv_y and the change in mass $|\Delta m|$ is replaced by $-dm$ since dm is negative.

Using the infinitesimal notation we write this equation as follows.

$$\boxed{m\, dv_y + u\, dm = 0.} \tag{7.74}$$

This equation is called the **rocket equation**. Here, the mass m and the y-components of velocity v_y are functions of time.

To predict the velocity of the rocket at an arbitrary time we divide Eq. 7.74 by m, which separates the equation into two terms, one has only v_y and the other has the constant u and the variable m. This type of equation can be integrated. The v_y integration has limits in v_y and m integration has limits in m. Let M be the mass and v_{0y} be the y-component of the velocity at time $t = 0$, then we obtain

$$\int_{v_{0y}}^{v_y(t)} dv_y + u \int_{M}^{m(t)} \frac{1}{m} dm = 0, \tag{7.75}$$

which gives

$$v_y(t) - v_{0y} = u \ln\left(\frac{M}{m(t)}\right). \tag{7.76}$$

The ratio the mass at some time t and the initial mass of the rocket is usually written in terms of a quantity, called the **burn rate**, which is defined as

$$\boxed{\text{Burn rate: } \alpha = -\frac{dm}{dt}.} \tag{7.77}$$

There are two types of burn rates of particular practical and theoretical interest.

1. Constant burn rate

If the rate of fuel burn is constant, i.e. if α is constant, Eq. 7.77 can be integrated to yield the following result for the mass of the rocket at time t if it was M at time $t = 0$.

$$m(t) = M - \alpha t. \tag{7.78}$$

Note that $m(t)$ cannot be negative. If the body of the rocket has mass M_R, then the rocket will run out of fuel at time $t = t_f$ given by

$$M_R = M - \alpha t_f \implies t_f = \frac{M - M_R}{\alpha} \tag{7.79}$$

When we substitute Eq. 7.78 in Eq. 7.76 we find the y-component of velocity at an arbitrary time to be

$$v_y(t) - v_{y0} = u \ln\left(\frac{M}{M - \alpha t}\right), \quad (0 \leq t \leq t_f), \qquad (7.80)$$

which says that the motion of the rocket is not a constant acceleration motion since velocity does not change linearly in time. Figure 7.26 shows

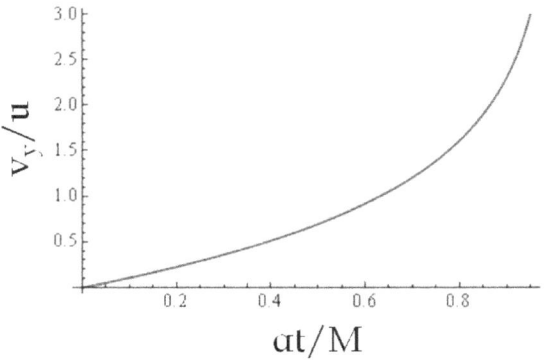

Figure 7.26: Velocity of the rocket as a function of time for constant burn rate case.

a plot of the rocket velocity as a function of time. Initially the velocity of the rocket increases quite linearly but when the rocket has less fuel on board the velocity of rocket rises sharply.

2. Variable burn rate

If the burn rate is proportional to the mass of the rocket, say $\alpha = \beta m$, then

$$\frac{dm}{dt} = -\beta m. \qquad (7.81)$$

This equation leads to

$$m(t) = M \exp(-\beta t). \qquad (7.82)$$

The final time when the rocket runs out of fuel will be $t = t'_f$. At that time, the mass of the rocket will be only the mass M_R of the rocket body.

$$M_R = M \exp(-\beta t'_f). \qquad (7.83)$$

For this type of burn rate, the mass of the rocket drops exponentially. By substituting Eq. 7.82 into Eq. 7.76 you can easily show that the rocket moves at a constant acceleration.

$$v_y(t) - v_{y0} = (u\beta)\, t, \quad 0 \leq \leq t'_f \qquad (7.84)$$

which shows the y-component of the acceleration is a constant, $a_y = u\beta$.

7.8.2 Rocket Motion With Constant External Force

Rocket motion near the surface of the Earth surface is subject to approximately constant gravitational force of the Earth. In this subsection we examine the vertical component of the motion of the rocket near the Earth, for instance, when a rocket is launched as shown in Fig. 7.27. Let the fuel be ejected at a constant speed u with respect to the rocket. We set up the problem in the same way as we did for the rocket in space, and add the force of gravity to the equation to obtain the following equation of motion of the y-coordinate of the rocket.

$$\boxed{m\, dv_y + u\, dm = -mg\, dt.} \qquad (7.85)$$

Dividing this equation by m and combining dv_y and dt terms we obtain a simpler form.

$$d\,(v_y + gt) + \frac{u}{m}\, dm = 0. \qquad (7.86)$$

Let us write

$$w \equiv v_y + gt, \qquad (7.87)$$

so that Eq. 7.86 looks simpler.

$$dw + \frac{u}{m}\, dm = 0. \qquad (7.88)$$

We integrate this equation from $t = 0$ to $t = t$ to obtain

$$w(t) - w(0) = u \ln\left(\frac{M}{m(t)}\right), \qquad (7.89)$$

where M is the mass of the rocket at $t = 0$. Putting the velocity back in we get

$$v_y(t) - v_{y0} = -gt + u \ln\left(\frac{M}{m(t)}\right). \qquad (7.90)$$

For a constant burn rate α, we replace $m(t)$ by $M - \alpha t$ to obtain the following for the y-component of the velocity.

$$v_y(t) - v_{y0} = -gt + u \ln\left(\frac{M}{M - \alpha t}\right),\ \ 0 \le t \le t_f, \qquad (7.91)$$

where $t_f = (M - M_R)/\alpha$ is the time when the rocket runs out of fuel and left with the rocket body only with mass M_R.

Figure 7.27: The launch of the Zvezda service module of the International Space Station on a Russian Proton-K rocket. Photcredit: NASA/Wikicommons.

7.9 EXERCISES

Impulse and Momentum

Ex 7.9.1. A ball of mass 0.5 kg is thrown in the air. What is the magnitude and direction of the impulse of the weight of the ball during the 5 sec interval the ball is in flight? Ans: 24.6 N.s pointed down.

Ex 7.9.2. A man pushes a shopping cart with a constant force of 5 N pointed towards North for 3 sec. What is the magnitude and direction of the impulse of the push? Ans: 15 N.s North.

Ex 7.9.3. A large crate is pushed by two men. One of them applies a constant force of 100 N due North and the other applies a constant force of magnitude 120 N and direction due East. What is the net impulse over a period of 3 sec? Ans: 469 N.s 39.8° North of East.

Ex 7.9.4. The x-component of a force on a 46-gram golf ball by a 7-iron versus time is plotted in Fig. 7.28. Find the x-component of the impulse during the intervals (i) [0, 50 msec], and (ii) [50 msec, 100 msec]. (b) Find the change in the x-component of the momentum during the intervals (i) [0, 50 msec], and (ii) [50 msec, 100 msec]. Ans: (a) (i) 0.75 N.s, (ii) 1.5 N.s. (b) (i) 0.75 kg.m/s, (ii) 1.5 kg.m/s.

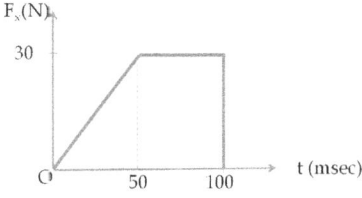

Figure 7.28: Exercise 7.9.4.

Ex 7.9.5. The y-component of a force on 2-kg box varies with time as shown in the Fig. 7.29. Find the y-component of the impulse during the intervals (i) [0, 2 sec], (ii) [2 sec, 4 sec], (iii) [4 sec, 8 sec], (iv) [8 sec, 10 sec], and (v) [0, 10 sec]. (b) Find the change in the y-component of the velocity during the intervals in part (a).

Ans: (a)(i)3 N.s, (ii)6 N.s, (iii) 2 N.s, (iv) -2 N.s, (v) 9 N.s.

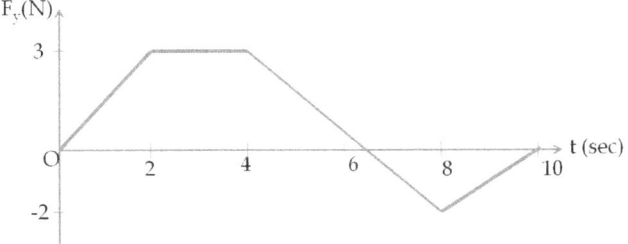

Figure 7.29: Exercise 7.9.5.

Ex 7.9.6. A batter strikes a ball of mass 300 grams with a bat of mass 3 kg. Before the hit the ball is moving horizontally towards the bat at a speed of 50 m/s. After the hit, the velocity of the ball changes direction by 180° and leaves the bat at 60 m/s speed. The impact of the bat on the ball lasts only 2 msec. (a) Find the magnitude and direction of the change in momentum of the ball. (b) What is the magnitude and direction of the impulse of the bat on the ball? (c) What is the magnitude and direction of the impulse of the ball on the bat? (d) What is the magnitude and direction of the average force of the bat on the ball? (e) What is the magnitude and direction of the average force of the ball on the bat?

Ans: (a) 33 kg.m/s pointed from the bat towards the ball, (d) 16,500 N, towards ball.

Ex 7.9.7. A batter strikes a ball of mass 300 grams with a bat of mass 3 kg. Before the hit the ball is moving horizontally towards the bat at a speed of 50 m/s. After the hit, the velocity of the ball changes direction by 120° but still pointed in the horizontal plane containing the velocity before the hit and velocity after the hit, and leaves the bat at 60 m/s speed (Fig. 7.30). The impact of the bat on the ball lasts only 2 msec. (a) Find the magnitude and direction of the change in momentum of the ball. (b) What is the magnitude and direction of the impulse of the bat on the ball? (c) What is the magnitude and direction of the impulse of the ball on the bat? (d) What is the magnitude and direction of the average force of the bat on the ball? (e) What is the magnitude and direction of the average force of the ball on the bat?

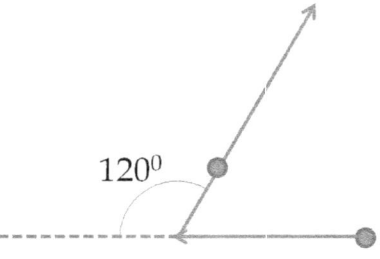

Figure 7.30: Exercise 7.9.7.

Ans: (a)28.6 kg.m/ at 33° with respect to the positive x-axis. (b) Same as (a). (c) Same as (b) but in the opposite direction. (d) 14,300 N in the same direction as (a). (e) Same as (d) but in opposite direction.

Ex 7.9.8. A car of mass 2,800 kg moving at 40 m/s strikes a stationary truck of mass 8,000 kg. After the collision, the car comes to rest within 2.5 seconds of the hit. (a) What is the magnitude and direction of the average force on the car? (b) Compare this force to the force on the truck by the car.

Ans: (a) 44,800 N in the opposite direction of the initial momentum of the car. (b) 44,800 N in the direction of the initial momentum of the car.

Ex 7.9.9. Two cars, of mass 3000 kg each, moving in the opposite directions on a straight road collide. Before the collision, each car had a speed of 65 miles per hour with respect to the ground, and after the collision, they move in the backward direction with a speed of 20 miles per hour. If the cars were in contact for 1.5 sec, what was the average force on each car? Hint: Find the change in momentum of one of the cars.

Ans: Magnitude of change in momentum 114,000 kg.m/s. Magnitude of average force 76,000 N.

Ex 7.9.10. A car of mass 2,000 kg and a truck of mass 10,000 kg start from rest at an intersection of two perpendicular roads. The car has a constant velocity of 10 m/s due East while the truck has a constant velocity of 25 m/s due North. Find the total momentum of the two vehicles at $t = 0$ and $t = 1$ sec.

Ans: Zero at $t = 0$ and magnitude 251,000 kg.m/s direction 85° N of E at $t = 10$ sec.

Center of Mass

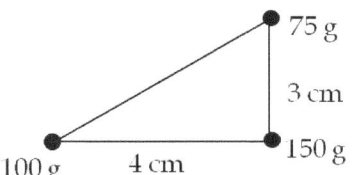

Figure 7.31: Exercise 7.9.11.

Ex 7.9.11. Three point masses are placed at the corners of a triangle as shown in Fig. 7.31. Find the center of mass of the three-mass system.

Ans: Xcm = -1.23 cm and Ycm = 0.69 cm with 150 gram mass at the origin.

Ex 7.9.12. Two particles of masses m_1 and m_2 separated by a horizontal distance D are let go from the same height h at the same time. Find the vertical position of the center of mass at a time before the two particles strike the ground. Assume no air resistance. If you like numbers try: m_1 = (0.2 kg), m_2 = (0.25 kg), D = (10 cm), and h = (30 m).

Ans: $X_{cm} = 5.6\ cm$; $Y_{cm} = 30m - 4.9(m/s^2)t^2$.

Ex 7.9.13. Two particles of masses m_1 and m_2 separated by a horizontal distance D are let go from the same height h at <u>different times</u>. The particle 1 starts at $t = 0$, and the particle 2 is let go at $t = T$. Find the vertical position of the center of mass at a time before the first particle strikes the ground. Assume no air resistance.

Ans: if m_1 = (0.2 kg), m_2 = (0.25 kg), D = (10 cm), and h = (30 m), $T = 0.2$ sec, then $Y_{cm}(t) - Y_{cm}(0.2sec) = (0.87)(t-0.2) - (9.8)(t-0.2)^2$.

Ex 7.9.14. Two particles of masses m_1 and m_2 move uniformly in different circles of radii R_1 and R_2 about origin in the xyplane. The x- and y-coordinates of the center of mass and that of particle #1 are given as follows where length is in meters and t in seconds.

$$X_{\text{CM}}(t) = 3\cos(2t); Y_{\text{CM}}(t) = 3\sin(2t)$$
$$x_1(t) = 4\cos(2t); y_1(t) = 4\sin(2t)$$

(a) Find the radius of the circle in which particle #1 moves. (b) Find the x and y-coordinates of particle #2 and the radius of the circle this particle moves.

Ans: (a) 4 m. (b) $R_2 = \dfrac{1}{m_2}\sqrt{9M^2 + 16m_1^2 + 24m_1 M}$ with $M = m_1 + m_2$.

Ex 7.9.15. Two particles of masses m_1 and m_2 move uniformly in different circles of radii R_1 and R_2 about the origin in xy plane. The coordinates of the two particles in meters are given as follows ($z = 0$ for both). Here t is in seconds.

$$x_1(t) = 4\cos(2t); y_1(t) = 4\sin(2t)$$
$$x_2(t) = 2\cos\left(3t - \frac{\pi}{2}\right); y_2(t) = 2\sin\left(3t - \frac{\pi}{2}\right)$$

(a) Find the radii of the circles of motion of particles #1 and #2. (b) Find the x and y-coordinates of the center of mass. (c) Decide if the center of mass moves in a circle by plotting the trajectory of CM.

Ans: (a) $R_1 = 4$ m and $R_2 = 2$ m. (b) $X_{cm} = \dfrac{m_1 x_1 + m_2 x_2}{m_1 + m_2}$, $Y_{cm} = \dfrac{m_1 y_1 + m_2 y_2}{m_1 + m_2}$. (c) Yes, moves in a circle of radius $R = \dfrac{1}{M}\sqrt{16m_1^2 + 4m_2^2}$.

7.9. EXERCISES

Ex 7.9.16. Find the center of mass of a one-meter long rod, made of 50-cm long iron (density 8 g/cc) and 50-cm long aluminum (density 2.7 g/cc) rods.

Ans: 12.6 cm from the center of the iron rod towards the aluminum rod.

Ex 7.9.17. Find the center of mass of a rod of length L whose mass density changes from one end to the other quadratically. That is, if the rod is laid out along the x-axis with one end at the origin and the other end at $x = L$, the density is given by $\rho(x) = \rho_0 + (\rho_1 - \rho_0)(x/L)^2$, where ρ_0 and ρ_1 are constant values.

Ans: $\frac{3}{4}L\left(\frac{\rho_1+\rho_0}{\rho_1+2\rho_0}\right)$.

Ex 7.9.18. Find the center of mass of a rectangular block of length a and width b that has a non-uniform density such that when the rectangle is placed in the xy-plane with one corner at the origin and the block placed in the first quadrant with the two edges along the x and y-axes, the density is given by $\rho(x,y) = \rho_0 x$, where ρ_0 is a constant.

Ans: $(2/3a, b/2)$.

Ex 7.9.19. Find the center of mass of a rectangular material of length a and width b made up of a material of non-uniform density. The density is such that when the rectangle is placed in the xy-plane as in Exercise 7.9.18, the density is given by $\rho(x,y) = \rho_0 xy$.

Ans: $(2a/3, 2b/3)$.

Ex 7.9.20. Find the center of mass of a sphere of mass M and radius R and a cylinder of mass m, radius r and height h arranged as shown in Fig 7.32(a) and (b). Express your answer in a coordinate system that has the origin at the center of the cylinder.

Ans: (a) $Y_{cm} = \frac{M(\frac{1}{2}h+R)}{m+M}$. (b) $Y_{cm} = \frac{M(r+R)}{m+M}$.

Ex 7.9.21. Find the center of mass of the structure given in Fig. 7.33. Assume uniform thickness of 20 cm, and density to uniform having magnitude 1 g/cm^3.

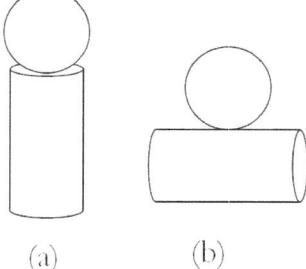

Figure 7.32: Exercise 7.9.20.

Conservation of Momentum - Single Body

Ex 7.9.22. A car of mass 2,000 kg is moving with a constant velocity of 10 m/s due East. What is the momentum of the car at $t = 0$ and $t = 1$ sec? Ans: 20,000 kg.m/s due East.

Ex 7.9.23. A skater of mass 40 kg is carrying a box of mass 5 kg. The skater has a speed of 5 m/s with respect to the floor and is gliding without any friction on a smooth surface. (a) Find the momentum of the box with respect to the floor. (b) Find the momentum of the box with respect to the floor after she puts the box down on the frictionless skating surface.

Ans: (a) Magnitude: 25 kg.m/s. (b) Same as (a).

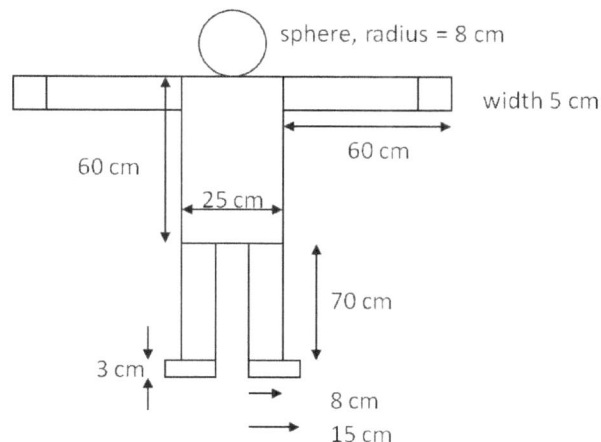

Figure 7.33: Exercise 7.9.21.

Ex 7.9.24. A hockey puck mass 150 gram is sliding horizontally on a frictionless table with a speed of 10 m/s. Suddenly a constant force of magnitude 5 N and direction due North is applied vertical to its initial direction, which was due East, for 1.5 seconds. (a) Which component(s) of momentum is(are) conserved during the 1.5 second interval and which is(are) not conserved? (b) Find the components of the momentum along its original direction and in the direction vertical to it at the end of 1.5 second duration.

Ans: (a) Only the component of momentum in the direction of North (or in the direction of South) is conserved. (b) Let positive x-axis be in the direction of the original momentum. Then $p_x = 1.5$ kg.m/s and p_y =7.5 kg.m/s.

Ex 7.9.25. A ball of mass 250 grams is thrown with an initial velocity of 25 m/s at an angle of 30° with the horizontal direction. Ignore air resistance. (a) After the ball leaves the hand of the thrower, which component(s) of the momentum will be conserved and which will not be conserved? (b) What is the momentum of the ball after 0.2 seconds? Do this problem by finding the components of the momentum first, and then constructing the magnitude and direction of the momentum vector from the components.

Ans: (a) horizontal component conserved. (b) 6 kg.m/s at 26°.

Conservation of Momentum - Collision

Ex 7.9.26. A bullet of mass 200 grams traveling horizontally towards East with speed 400 m/s strikes a block of mass 1.5 kg at rest on a frictionless table. After striking the block the bullet is imbedded in the block and the block and the bullet move together as one unit. (a) What is the magnitude and direction of the velocity of the block/bullet combination immediately after the impact? (b) What is the magnitude and direction of the impulse

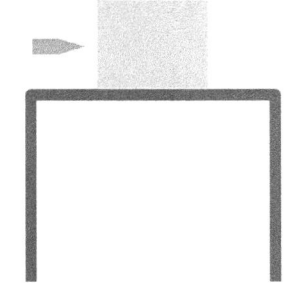

Figure 7.34: Exercise 7.9.26.

7.9. EXERCISES

by the block on the bullet. (c) What is the magnitude and direction of the impulse from the bullet on the block? (d) If it took 3 msec for the bullet to change the speed from 400 m/s to the final speed after impact, what is the average force between the block and the bullet during this time?

Ans: (a) 47 m/s in the bullet to block direction, (b) 70.6 $N.s$, towards bullet, (c) 70.6 $N.s$, towards block, (d) magnitude: 23,500 N.

Ex 7.9.27. A billiard ball, labeled 1, moving horizontally strikes another ball, labeled 2, at rest. Before the impact, ball 1 was moving at a speed of 3 m/s and after the impact it is moving at 1 m/s at 150° from the original direction. If the two balls have equal masses of 300 grams, what is the velocity of the ball 2 after the impact.

Ans: 3.9 m/s, 7° below the horizontal direction.

Ex 7.9.28. A projectile of mass 2 kg is fired in the air at an angle of 40° to the horizon at a speed of 50 m/s. At the highest point in its flight the projectile breaks into three parts of mass 1 kg, 0.7 kg and 0.3 kg. The 1-kg part falls straight down after breakup with an initial speed of 10 m/s, the 0.7-kg part moves in the original forward direction, and the 0.3-kg part goes straight up. (a) Find the speeds of 0.3-kg and 0.7-kg pieces immediately after the break-up. (b) How high from the break-up point does the 0.3-kg piece go before coming to rest? (c) Where does the 0.7-kg piece land relative to where it was fired from?

Ans: (a) 33.3 m/s and 109.4 m/s, (b) 56.5 m, (c) 482 m.

Ex 7.9.29. Two projectiles of mass m_1 and m_2 are fired in the opposite directions from two launch sites separated by a distance D. They both reach the same spot in their highest point and strike there. As a result of the impact they stick together and move as a single body afterwards. Find the place they will land.

Ans: $|m_1 - m_2|D/[2(m_1 + m_2)]$ from the mid point.

Time-Dependent Mass

Ex 7.9.30. Grains from a grain hopper falls at a rate of 10 kg/sec vertically onto a freight car that is moving horizontally at a constant speed 2 m/s on a straight track. What force is needed to keep the freight car moving at a constant velocity.

Ans: 20 N, horizontal.

Ex 7.9.31. A crate of mass 50 kg containing bananas is sliding in a straight line on a frictionless surface at a constant velocity of 10 m/s. A monkey of mass 45 kg hanging from the ceiling times it correctly and jumps vertically down landing in the crate. At the time monkey lands in the crate his velocity was 12 m/s vertically downward. (a) What is the velocity of the crate with monkey afterwards? (b) Compare the momenta of the care plus monkey afterwards to that of the crate plus the monkey separately.

Is the total momentum of the crate plus monkey system conserved? Why or why not?

Ans: 5.3 m/s horizontally in the original direction.

Ex 7.9.32. In the previous problem, the monkey starts to consume banana and throws away the peels each having mass of 200-grams at a constant rate of one banana per 40 seconds and a speed of 5 m/s with respect to the crate. The peels are being thrown in the forward direction of the motion of the cart. If there are 100 bananas, what will be the final velocity of the monkey and crate?

Ans: $v_N = v_0 + m_b u \sum_{j=1}^{N} \left[\frac{1}{M+(N-j+1)m_b} \right]$.

Ex 7.9.33. Find $v_x(t)$ from the following equation if $v_x(0) = v_{x0}$. Here, M, u, and α are constants.

$$\frac{dv_x}{dt} = \frac{u\alpha}{M - \alpha t}.$$

Ans: $u \ln\left(\frac{M}{M-\alpha t}\right)$.

7.10 PROBLEMS

Problem 7.10.1. A cube of side a is cut out of another cube of side b as shown in Fig. 7.35. Find the location of the CM of the structure. Hint: Think of the missing part as a negative mass overlapping a positive mass.

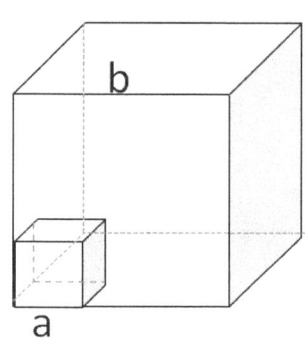

Figure 7.35: Problem 7.10.1.

Problem 7.10.2. Find the CM of cone of uniform density that has a radius R at the base, height h and mass M.

Ans: $Z_{cm} = \frac{1}{4}h$.

Problem 7.10.3. Find the CM of a thin wire of mass m and length L bent in a semicircular shape.

Ans: $Y_{cm} = \frac{2}{\pi^2}L$.

Problem 7.10.4. Find the CM of a uniform thin semicircular plate.

Ans: $Y_{cm} = \frac{4}{3\pi}R$.

Problem 7.10.5. A gun of mass 1.5 kg fires bullets of mass 100 grams each. The bullets come out at a speed of 400 m/s. If each firing lasts 200 msec, what is the average recoil force on the gun?

Ans: 200 N, towards gun. The **recoil force** is the force applied by the exiting bullet on the gun.

Problem 7.10.6. A tennis ball machine of mass M initially at rest on a frictionless surface shoots tennis balls of mass m. When a ball comes out of the muzzle the ball has a velocity of magnitude v_0 with respect to the ball machine and directed at an angle θ with respect to the horizontal direction as shown in Fig. 7.36. Find the velocity of the ball machine immediately after the ball leaves. (Be careful in this problem since the velocity of the ball is given in relation to an accelerating object.)

Ans: Speed $= \left(\frac{M}{M-m}\right) v_0 \cos\theta$.

Problem 7.10.7. There are N cannons on a railway car. The total mass of the car and the cannons is M. The cannons can fire cannon balls of mass m at a speed u with respect to the car regardless of the state of motion of the car. Initially the car is at rest. If cannons are fired horizontally in the direction of the back of the car, the car moves forward. (a) What is the speed of the car when all cannons are fired at the same time? (b) What is the final speed of the car if cannons are fired one after another?

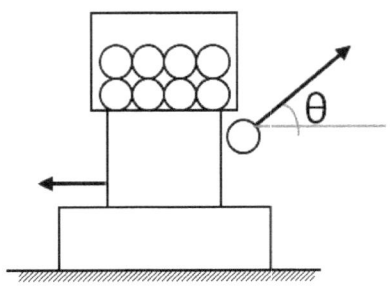

Figure 7.36: Problem 7.10.6.

Problem 7.10.8. A 65-kg person jumps from first floor window of a burning building and lands almost vertically on the ground with the horizontal component of the velocity of 3 m/s and the vertical component of velocity of -9 m/s with the axes pointed horizontally in the plane of the motion and vertically up. Upon impact with the ground he is brought to rest in a short time. The force experienced by his feet are different depending on whether he keeps his knees stiff or bends them. Find the forces on his feet in the two cases. (a) First find the impulse on the person from the impact on the ground. Calculate both its magnitude and direction. (b) Find the average force on the feet if the person keeps his leg stiff and straight and his CM drops by only 1 cm vertically and 1 cm horizontally during the impact. You will need to find the time the impact lasts by making reasonable assumptions about the deceleration. Although the force is not constant during the impact, working with constant average force for this problem is acceptable. (c) Find the average force on the feet if the person bends his legs throughout the impact so that his CM drops by 50 cm vertically and 5 cm horizontally during the impact. (d) Compare the results of part (b) and (c), and draw conclusions about which way is better.

Ans: (a) 703 N.s, 106°. (b) $|F_x| = 29{,}100$ N, $|F_y| = 307{,}000$ N. (c) $|F_x| = 5{,}850$ N, $|F_y| = 6{,}080$ N.

Problem 7.10.9. A spherical rain drop of radius R and mass M falls vertically through a cloud layer. The drop enters the cloud layer at a height H from the ground and exits the cloud at a height h. While inside the cloud the drop accumulates water molecules so that its mass and size grow with time. The mass of the rain drop at time t when inside the cloud is given as $m(t) = M + \alpha t$, with $t = 0$ at the time the drop enters the cloud. Find the speed of the rain drop as a function of time when it is inside the cloud if it has a speed of v_0 immediately before entering the cloud.

Problem 7.10.10. A chain of mass M and length L is suspended vertically with its bottom touching a pan of mass m on a scale. The chain is then let go from rest. As chain falls, the scale reads the force on the pan needed to balance the force of the chain that is on the pan and the chain link that is impacting the pan at that instant. Assume each chain link is infinitesimally small and comes to rest instantaneously upon impact so that at each instant the entire momentum of an infinitesimal part of the chain link falling on the pan is transferred to the pan. (a) Find an expression of the reading of the scale when a length y has fallen. (b) What is the reading on the scale at the instant the whole chain has fallen to the fan.

Ans: (a) $mg + 3\frac{xM}{L}g$, (b) $mg + 3Mg$.

Problem 7.10.11. Two blocks of mass m each are attached to the ends of a spring of spring constant k. They are placed on a frictionless surface and pushed against a wall with a horizontal force of magnitude F till the spring is compressed by a distance D. The masses are then released (Fig. 7.37). (a) What is the magnitude of the force F applied to hold the masses in place before they are released? (b) What is the amount of normal force from the wall immediately before the masses are released? (c) What are the forces acting on the masses immediately after the masses are released? (d) When does the normal force from the wall stop acting on the mass? (e) Find the position of the CM of the two blocks as a function of time taking the zero of time at the instant they are released.

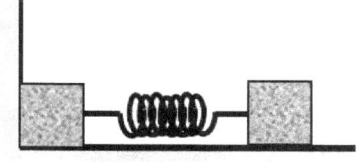

Figure 7.37: Problem 7.10.11.

Ans: (a) kD. (b) kD. (c) Spring, Normal from wall, Weight, Normal from floor. (d) At the instant the spring is not compressed or extended. (e) Hint: After the block leaves the wall the CM moves at a constant velocity.

Chapter 8

WORK, ENERGY AND POWER

Contents

8.1	WORK	310
	8.1.1 General Formula for Work	313
8.2	WORK-ENERGY THEOREM	319
	8.2.1 Integrating Equation of Motion of One Particle in One Dimension	319
	8.2.2 Integrating Equation of Motion of One Particle in Three Dimensions	321
	8.2.3 Work-Energy Theorem Applied to Multiparticle Systems	322
	8.2.4 Uses of Work-Energy Theorem	324
	8.2.5 Work-Energy Theorem For Conservative Forces	325
8.3	POTENTIAL ENERGY FUNCTIONS	327
	8.3.1 Potential Energy Associated with Gravity	328
	8.3.2 Potential Energy Associated with the Spring Force	331
	8.3.3 Potential Energy Associated With the Universal Gravitational Force	333
	8.3.4 Force and Potential Energy Function	336
8.4	CONSERVATION OF ENERGY	336
8.5	POWER	339
8.6	ELASTIC COLLISIONS	340
8.7	EXERCISES	343
8.8	PROBLEMS	348

CHAPTER 8. ENERGY

8.1 WORK

Work has a different meaning in physics than it has in everyday life. In physics, we say that a force has done work on an object over some interval if the object is displaced in that time. A weightlifter lifting a barbell above his head, a lift pulling passengers up to higher floors, a tugboat pulling a ship, a student lifting a backpack full of books upon his shoulders, etc, in each case, a work is done when a force is exerted on an object to cause its displacement.

As we have seen in previous chapters, the displacement of a particle is caused by the net force on the particle, the direction of the displacement may or may not in the same direction as one of the forces that make up the net force. Therefore, only the component of a force in the direction of the displacement or in the opposite direction to the displacement can be thought of as being involved in the displacement.

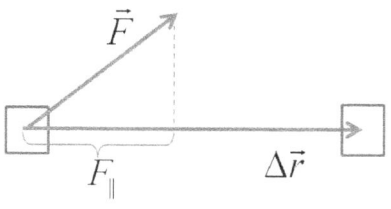

Figure 8.1: Work by a force is projection of force ($F_\parallel$) times the magnitude of displacement.

Mathematically, the projection of the force vector in the direction of displacement captures this physical reality. Therefore, quantitatively, we define the work of a force $\vec{F}$ during an interval Δt in which the particle has the displacement $\Delta \vec{r}$ by the product of projection of the force on the displacement and the magnitude of the displacement.

$$\text{Work} = (\text{Projection of } \vec{F} \text{ on } \Delta \vec{r}) \times (\text{Magnitude of } \Delta \vec{r})$$
$$= F_\parallel \, |\Delta \vec{r}| \tag{8.1}$$

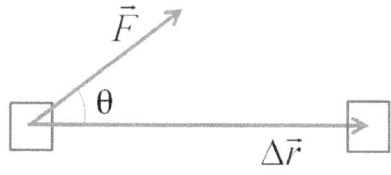

Figure 8.2: Work by a force is $W = \vec{F} \cdot \Delta \vec{r}$.

Clearly, no work is done when there is no displacement. For instance, in the picture shown in Fig. 8.4, when the model holds the barbell in any one position, she applies a force on the barbell to balance the weight of the barbell, but no work is done by this force. Only during the moving of the barbell does she any work on the barbell.

Recall that the scalar product between two vectors is also given by the product of the magnitude of one vector and the projection of the other vector. Therefore, the scalar product of the force vector and the displacement vector is also equal to the work by the force.

$$\boxed{W = \vec{F} \cdot \Delta \vec{r} = |\vec{F}||\Delta \vec{r}| \cos \theta,} \tag{8.2}$$

where θ is the angle between vectors $\vec{F}$ and $\Delta \vec{r}$.

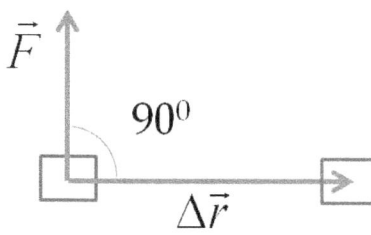

Figure 8.3: No work is done if $\vec{F} \perp \Delta \vec{r}$.

The scalar product shows clearly that no work would be done by a force that is directed perpendicularly to the displacement since $\theta = 90°$ in that case would give $\cos(90°) = 0$ which would make work equal to zero. Work is positive if force is in the same direction as the displacement, since $\theta = 0°$ and $\cos 0° = 1$, while work is negative if the force is oppositely directed to the displacement, since then $\theta = 180°$ and $\cos(180°) = -1$.

8.1. WORK

Figure 8.4: The girl in this picture does work on the barbell when she moves the weight from one position to another. When she is just holding the barbell in either position, she does not do work. Picture credits: Microsoft office clip-art

The analytic form of this definition with respect to a particular coordinate system can be written with the components of the force and the displacement.

$$W = F_x \Delta x + F_y \Delta y + F_z \Delta z. \tag{8.3}$$

From Eqs. 8.2 and 8.3, we note that the SI unit of work is N.m, which is also called Joule (J).

Example 8.1.1. Work done in pulling a cart. A cart of mass M is pulled a distance D on a flat horizontal surface by a constant force F that acts at an angle of θ with the horizontal direction. The other forces on the object during this time are gravity (F_w), Normal forces (F_{N1}) and (F_{N2}), and rolling frictions F_{r1} and F_{r2} as shown in Fig. 8.5. What is the work done by each force?

Solution. We can find the work done by each force by using either of the two ways of writing the dot product given in Eq. 8.2 and 8.3.

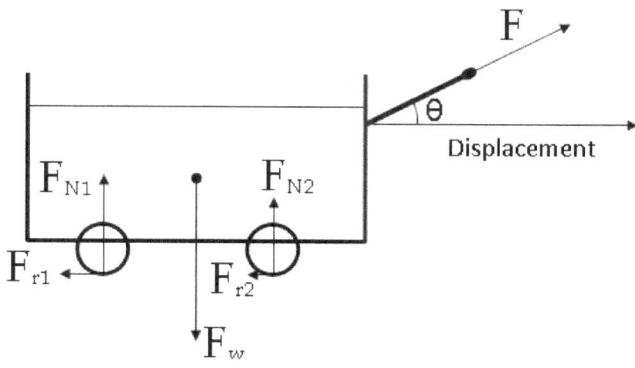

Figure 8.5: Example 8.1.1

Using Eq. 8.2 for the calculation of work

We can organize the information of magnitude and directions for each force with respect to the displacement vector in a table, and also present the answer for the work by the force.

Force	Magnitude	Angle with displacement	Work
$\vec{F}$	F	θ	$FD\cos\theta$
$\vec{F}_w$	Mg	$90°$	0
$\vec{F}_{N1}$	F_{N1}	$90°$	0
$\vec{F}_{N2}$	F_{N2}	$90°$	0
$\vec{F}_{r1}$	F_{r1}	$180°$	$-F_{r1}D$
$\vec{F}_{r2}$	F_{r2}	$180°$	$-F_{r2}D$

Using Eq. 8.3 for the calculation of work

We need directions of a Cartesian coordinate system to work out the components. We use the horizontal direction to the right in Fig. 8.5 to be positive x-axis and vertical direction up to be positive y-axis. The z-components of all forces and displacement will then be zero. Since, the displacement vector has only x-component non-zero, Eq. 8.3 simplifies to the following in this case.

$$W = F_x \Delta x = F_x D.$$

Therefore, if a force has a zero x-component, then the work will be zero. Once again, we can organize the information in a table.

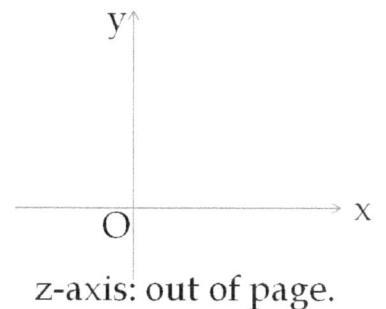

z-axis: out of page.

Figure 8.6: The coordinates for Example 8.1.1.

Force	x-component	Work
$\vec{F}$	$F\cos\theta$	$FD\cos\theta$
$\vec{F}_w$	0	0
$\vec{F}_{N1}$	0	0
$\vec{F}_{N2}$	0	0
$\vec{F}_{r1}$	$-F_{r1}$	$-F_{r1}D$
$\vec{F}_{r2}$	$-F_{r2}$	$-F_{r2}D$

8.1.1 General Formula for Work

Work is a scalar number. This means that work over various displacements of an object will add as simple numbers. Fig. 8.7 shows work by a force on an arbitrary path of motion of an object can be obtained by "breaking up" the path into tiny straight displacements $\Delta \vec{r}_n$ and evaluating the work on each segment. The work from all infinitesimal displacements can be added to obtain the work over the whole path. Suppose the path of the object is divided up into N small straight displacements of length, $\Delta \vec{r}_1$, $\Delta \vec{r}_2$, $\cdots$, $\Delta \vec{r}_N$, then the net work is given by the following ordinary sum.

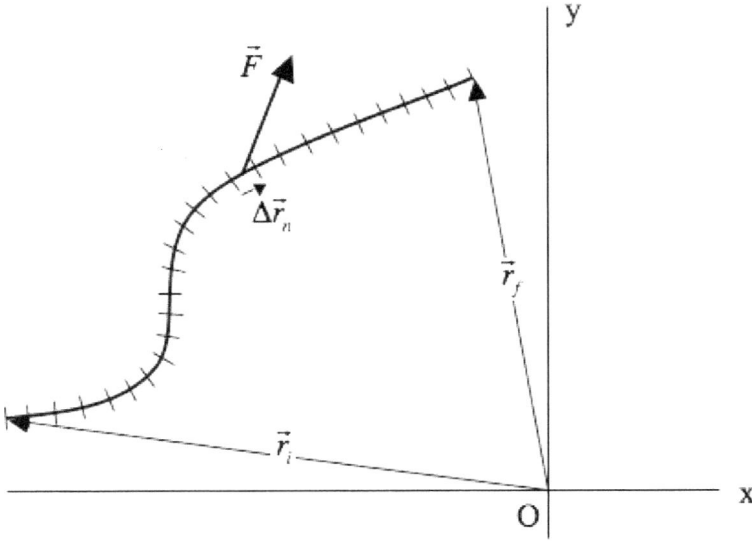

Figure 8.7: Work done by a force is the sum of scalar product of force with displacements on the segments of the path between initial and final positions.

$$W_{if} = \sum_{n=1}^{N} \vec{F} \cdot \Delta \vec{r}_n. \tag{8.4}$$

When the segments of the displacement are made smaller indefinitely, the result is written as an integral.

$$\boxed{W_{if} = \int_{i,\text{path}}^{f} \vec{F} \cdot d\vec{r},} \tag{8.5}$$

where subscript "path" has been added to the symbol of the integral to indicate the role of path in the definition, and i stands for the initial point of the path with coordinates (x_i, y_i, z_i) and f for the final point with coordinates (x_f, y_f, z_f). Often, the analytic form of Eq. 8.5 is useful for calculations. Using the analytic form of the infinitesimal displacement $d\vec{r}$ in Cartesian coordinates,

$$d\vec{r} = dx\,\hat{u}_x + dy\,\hat{u}_y + dz\,\hat{u}_z, \tag{8.6}$$

where $\hat{u}_x$, $\hat{u}_y$, and $\hat{u}_z$ are the unit base vectors towards the positive axes, and the components of the force vector we can rewrite Eq. 8.5 as:

$$W_{if} = \int_{i,path}^{f} (F_x dx + F_y dy + F_z dz), \tag{8.7}$$

which is a collection of three one-variable integrals but the integrations have to be done on the path of motion of the object, which constrains the independent variables x, y and z to the coordinates on the path only. To perform these integrals, we need to specify a path and the end points on the path in addition to the integrands, which are specified as functions of position, i.e, $F_x(x,y,z)$, $F_y(x,y,z)$, and $F_z(x,y,z)$. This type of integration is variously called a line or path integral.

We will now have some practice of doing these types of integrals below. Note that the subscripts x, y, and z refer to the components and the same symbols in the argument (x,y,z) refer to the coordinates of the points on the path over which we seek to compute the work integral.

Further Remarks

For situations where the force is constant, i.e. if the force does not depend on the position of the object, Eqs. 8.2 and 8.3 are more suitable for a calculation of work. However, if the force depends on the position, such as the spring force, then we must use Eq. 8.5 or 8.7 for a calculation of the work.

Example 8.1.2. Work done by a constant force - path along an axis. Consider a particle on which several forces act, one of which is known to be constant in time: $\vec{F}_1 = 3$ N $\hat{u}_x + 4$ N $\hat{u}_y$. As a result, the particle moves along the x-axis from $x = 0$ to $x = 5$ m position in some time interval. What is the work done by $\vec{F}_1$?

Solution. Note that, since force is constant, it is easier to use Eq. 8.2 or 8.3 to calculate the work. Here the displacement is along the x-axis, therefore, the work done will only use the x-component of the force. Therefore, the work will be equal to $W = 3$ N $\times$ 5 m $= 15$ N.m.

Example 8.1.3. Using path integral. Do the example above by Eq. 8.7.

Solution. The path here is along the x-axis as shown in Fig. 8.8.

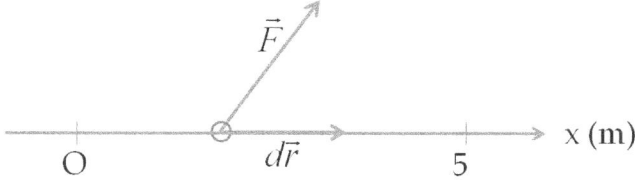

Figure 8.8: Example 8.1.3.

8.1. WORK

Therefore, $d\vec{r} = dx\hat{u}_x$, which means that the integral will be

$$W_{if} = \int_0^{5\text{ m}} (3\text{ N})dx = 15 \text{N.m}.$$

Example 8.1.4. Work done by a constant force - path in a plane.

Consider a particle on which several forces act, one of which is known to be constant in time: $\vec{F}_1 = 3\text{ N }\hat{u}_x + 4\text{ N }\hat{u}_y$. As a result, the particle moves in the xy-plane of a Cartesian coordinate system, first on the x-axis from $A(0,0)$ to $B(5\text{ m},0)$ and then parallel to the y-axis from $B(5\text{ m},0)$ to $C(5\text{ m},6\text{ m})$ as shown in Fig. 8.9. What is the work done by $\vec{F}_1$?

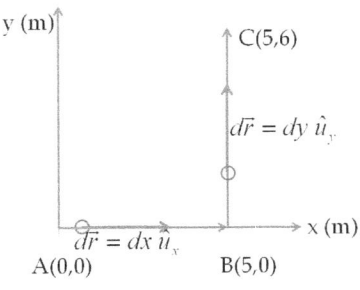

Figure 8.9: Example 8.1.4.

Solution. Since the force is constant, it is easier to work out the work done by the force using Eq. 8.2 or 8.3 on each path. As work is a scalar number, the total work on the full path is equal to the scalar sum of the parts. Let W_{AB}, W_{BC} and W_{ABC} denote work from A to B, B to C, and for the full path from A to B to C on the given path. Following the calculation presented in Example 8.1.2 it is readily seen that

$$W_{AB} = 15 \text{ N.m}$$
$$W_{BC} = 24 \text{ N.m}.$$

Therefore, the net work is

$$W_{ABC} = W_{AB} + W_{BC} = 39 \text{ N.m}.$$

This result can also be obtained by setting up the path integral over the full path with different expressions for the infinitesimal displacement $d\vec{r}$ on the two segments, $d\vec{r} = dx\ \hat{u}_x$ for the path parallel to the x-axis, and $d\vec{r} = dy\ \hat{u}_y$ for the path parallel to the y-axis. Performing the resulting integral on the x and y variables on each segment is simple also, and we leave them for the student to work out.

Example 8.1.5. Work done by a constant force - path in a plane.
The same force as the last example, but now the path is the straight path from $A(0,0)$ to $C(5\text{ m},6\text{ m})$. What is the work done by $\vec{F}_1$ now?

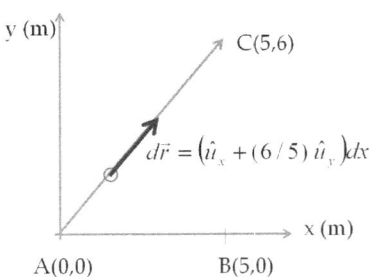

Figure 8.10: Example 8.1.5.

Solution. Again, since the force is constant, it is easier to work out the work done by the force using Eq. 8.2 or 8.3. We have components of the force and the displacements, viz., $\vec{F}_1 = (3\text{ N}, 4\text{ N})$ and $\Delta \vec{r} = (5\text{ m}, 6\text{ m})$. Therefore work done is

$$W_{AC\text{-direct}} = F_x\Delta x + F_y\Delta y = (3\text{ N}) \times (5\text{ m}) + (4\text{ N}) \times (6\text{ m}) = 39 \text{ N.m}.$$

How would we do the path integral (Eq. 8.7) for the direct path AC? Now, the path is not along the axes, but in the xy-plane. Furthermore, we have a relation between the x and y on the points on the path given by

$$y = \frac{6}{5}x.$$

Therefore, the infinitesimals dx and dy are related on the path by

$$dy = \frac{6}{5}dx.$$

We use this to simplify the infinitesimal displacement vector

$$d\vec{r} = \left(\hat{u}_x + \frac{6}{5}\hat{u}_y\right) dx.$$

Therefore the dot product $\vec{F} \cdot d\vec{r}$ becomes

$$\vec{F} \cdot d\vec{r} = \left((3 \text{ N}) \times 1 + (4 \text{ N}) \times \frac{6}{5}\right) dx,$$

which can be integrate from $x = 0$ to $x = 5$ m to give the result, $W = 39$ N.m.

Further Remarks

Work done by a constant force $\vec{F}_1$ in going from A to C directly is same as going by way of A-B-C worked out in Example 8.1.4. We got the same result because the force was constant. For every constant force, such as gravity near Earth, this will be the case.

Many forces, some even non-constant forces, have this feature that the work done by them is independent of path; when that is the case, we say that the force is a conservative force.

Example 8.1.6. Work done by a position-dependent force. Consider a particle on which several forces act, one of which is known to be a force whose magnitude and direction depends on the position of the particle. In a particular coordinate system, the force is given by $\vec{F}_2(x,y) = 2y\hat{u}_x + 3x\hat{u}_y$, where x and y are the coordinates of the position of the particle.

(a) Find the work done by this force when the particle moves from the origin to a point $A(5 \text{ m}, 0)$ on the x-axis. (b) Find the work done by this force, if the particle first moves on the y-axis from origin to a point $B(0, 3 \text{ m})$, turns 90°, moves parallel to the x-axis from point B to point $C(5 \text{ m}, 3 \text{ m})$, moves parallel to the y-axis from C to A, and finally ends up at A.

Solution. This example illustrates additional feature present in the evaluation of path integral given in Eqs. 8.5 and 8.7. So far, we have studied constant forces and shown that the infinitesimal displacement on paths assume different forms when get restricted to the path. A position-dependent force is usually given for an arbitrary position (x, y, z) which must be also restricted to the path. For instance, here $F_{2x}(x, y, z) = 2y$, which says that $F_{2x} = 0$ for a path on x-axis since on that path $y = 0$. Let us work out the work on each segment.

(a) On path OA the displacement of the particle is on the x-axis. Therefore, the displacement vector element will be $d\vec{r} = dx\,\hat{u}_x$, with $y = 0$.

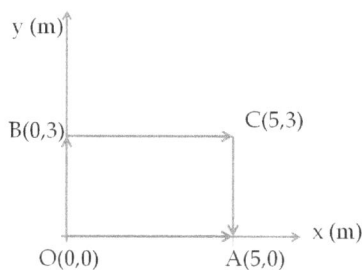

Figure 8.11: Example 8.1.6.

8.1. WORK

We now take the dot product of the force $\vec{F}_2$ with $d\vec{r}$, and set $y = 0$ in the result to restrict the integration to this path.

$$\vec{F}_2 \cdot d\vec{r} = 2ydx\Big|_{y=0} = 0.$$

The work done by $\vec{F}_2$ when the particle moves from origin to $x = 5$ m on the x-axis is zero.

(b) There are three path segments: OB, BC, CA.

SEGMENT OB:

Work done on OB, $W_{OB} = 0$ since $F_{2y}(x = 0) = 0$.

SEGMENT BC:

Displacement vector infinitesimal element, $d\vec{r} = \hat{u}_x \, dx$.

Other condition on the path: $y = 3$ m.

Dot product of the displacement vector element with force: $\vec{F}_2 \cdot d\vec{r} = 2ydx$.

Evaluating this for $y = 3$ m we obtain $\vec{F}_2 \cdot d\vec{r} = 6dx$.

Now, we integrate $\vec{F}_2 \cdot d\vec{r}$ for the segment. The integration variable is x, which changes from 0 to 5 m over the segment. Therefore, the work on BC segment is

$$W_{BC} = \int_0^{5 \text{ m}} 6dx = 30 \text{ N.m}.$$

··· continued

SEGMENT CA:

We repeat the process for this segment.

Displacement vector infinitesimal element, $d\vec{r} = dy \, \hat{u}_y$.

Other condition on the path: $x = 5$ m.

Dot product of displacement vector element with force: $\vec{F}_2 \cdot d\vec{r} = 3xdy$.

Evaluating this for $x = 5$ m we obtain $\vec{F}_2 \cdot d\vec{r} = 15 \, dy$.

Now, we integrate this for the segment. The integration variable is y, which changes from $3 \, m$ to 0 over the segment. Therefore, work on BC segment is

$$W_{CA} = \int_{3m}^{0} 15dy = -45 \text{ N.m}.$$

The net work done on the path O-B-C-A path is

$$W_{OBCA} = W_{OB} + W_{BC} + W_{CA} = -15 \text{ N.m}.$$

We find that work done on path OA is different from the work on path $OBCA$. If the work of a force depends on the path we say that the force is non-conservative. Therefore, the force $\vec{F}_2$ given here is an example of a non-conservative force.

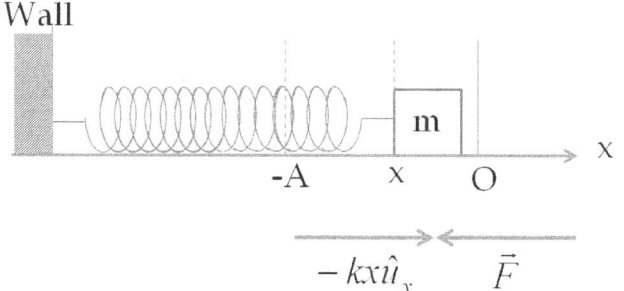

Figure 8.12: Example 8.1.7.

Example 8.1.7. Work against a spring force

A block is attached to a spring of spring constant k and placed on a horizontal table. The other end of the spring is attached to a support. A force $\vec{F}$ is applied to compress the spring by a distance A. What is the work done?

Solution. To compress the spring we need to apply a force that overcomes the spring force by an infinitesimal amount. We will assume that the applied force has magnitude equal to the spring force and has an opposite direction. Fig. 8.12 shows the forces on the block at an arbitrary instant during the compression when the spring is compressed by a distance $|x|$.

When dealing with spring force, you need to be careful with the direction of the vectors and signs of components. The spring force is always pointed towards the point when the spring is unstretched. Since the coordinate x of the point in Fig. 8.12 is negative, the spring force on the block will be equal to $-kx\hat{u}_x$. With the minus sign, the spring force would be pointed towards origin, which is towards the positive x-axis. The external force will be pointed such that the forces will add to give zero result for the x-components of the forces on the block.

$$\vec{F} - kx\ \hat{u}_x = 0,$$

and the infinitesimal displacement vector (setting $y = z = 0$ in the general formula for $d\vec{r}$)

$$d\vec{r} = dx\ \hat{u}_x.$$

Therefore, the work integral for x to go from $x = 0$ to $x = -A$ is given by

$$W = \int_i^f \vec{F} \cdot d\vec{r} = \int_0^{-A} kx\, dx = \frac{1}{2}kA^2.$$

Example 8.1.8. Work by a varying direction between displacement and force. In the motion of a pendulum the direction between the weight and the displacement of pendulum changes along the path of the pendulum. Consider a pendulum of mass m and length L. Find the work done by the force of gravity (i.e. weight) of the pendulum for a displacement of the pendulum from a position where the suspending thread makes an angle θ_0 with respect to the vertical axis to the point where the pendulum is hanging vertically.

Solution. Although weight mg of the pendulum is constant, it does different amount of work on different segments $d\vec{r}$ along the path of the pendulum since the angle between weight vector and $d\vec{r}$ changes along the path. Therefore, we must use the fundamental definition of work given in Eq. 8.5.

To set up the work integral, we look at an infinitesimal displacement at an arbitrary displacement in its motion, say between the suspension angles θ and $\theta + d\theta$, where $d\theta$ is negative for the motion under study. The displacement vector has the magnitude of the corresponding arc, $-Ld\theta$, which is positive since $d\theta < 0$. Since the direction of the displacement is tangent to the circle of motion, the angle between the weight and the displacement vector is equal to $(90° - \theta)$ as shown in Fig. 8.13.

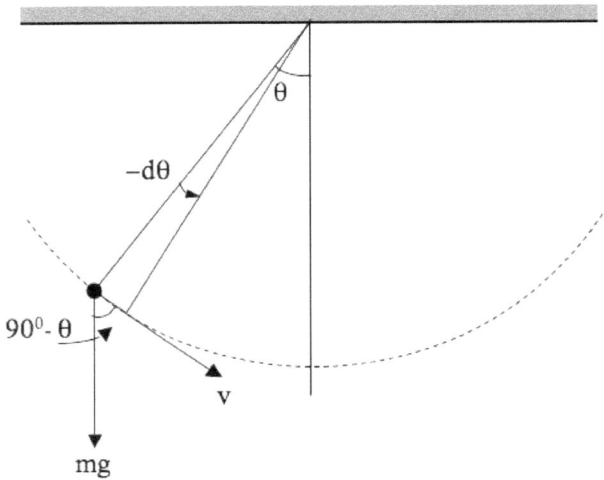

Figure 8.13: Example 8.1.8

Therefore, the work done by the weight during the infinitesimal displacement under consideration is

$$dW = (mg)(-Ld\theta)\cos(90° - \theta) = -mgL\sin\theta\, d\theta.$$

Now, we integrate from $\theta = \theta_0$ to $\theta = 0$ to obtain the required work.

$$W = \int_{\theta_0}^{0} [-mgL\sin\theta]\, d\theta = mgL(1 - \cos\theta_0).$$

8.2 WORK-ENERGY THEOREM

8.2.1 Integrating Equation of Motion of One Particle in One Dimension

Consider a one-dimension motion of a particle of mass m restricted to the x-axis. Newton's second law says that the x-coordinate of the mass will obey the following equation.

$$F_x^{net} = m\frac{dv_x}{dt}, \qquad (8.8)$$

where F_x^{net} is the x-component of the net force on the mass. Suppose we treat F_x^{net} as a function of the coordinate x of the particle and integrate with respect to x, corresponding to a particular path between x_i and x_f, so that the left side of this equation gives the work done by the net force or the net work done on the body.

$$\int_{x_i}^{x_f} F_x^{net} dx = m \int_{x_i}^{x_f} \frac{dv_x}{dt} dx. \qquad (8.9)$$

The right side is a complicated integral. It appears that, to perform this integral, we need x-component of acceleration as a function of x. However, a change of variable from x to t transforms the integral into an integral over v_x as we will see now. We note that

$$dx = \left(\frac{dx}{dt}\right) dt = v_x dt. \qquad (8.10)$$

Replacing dx on the right side of Eq. 8.9 by $v_x dt$ we find

$$\text{Right side of Eq.8.9} = m \int_{t_i}^{t_f} \frac{dv_x}{dt} v_x dt, \qquad (8.11)$$

where the limits on the integration have been changed to the corresponding time values. In Eq. 8.11, we can cancel out dt to obtain an integration over v_x instead of over dt, and the change in the limits accordingly.

$$\text{Right side of Eq.8.9} = m \int_{v_{ix}}^{v_{fx}} v_x dv_x, \qquad (8.12)$$

which can be performed to yield.

$$\text{Right side of Eq.8.9} = \frac{1}{2}mv_{fx}^2 - \frac{1}{2}mv_{ix}^2. \qquad (8.13)$$

Therefore, an integration of the equation of motion Eq. 8.8 for a one-dimensional motion gives the following equality.

$$\int_{x_i}^{x_f} F_x^{net} dx = \frac{1}{2}mv_{fx}^2 - \frac{1}{2}mv_{ix}^2. \qquad (8.14)$$

Since the motion is restricted to one dimension, the y and z-components are zero, and we can replace the square of the x-components of velocity by square of speed at those instants. The quantity $\frac{1}{2}mv^2$ is called **kinetic energy**, which will be denoted by K. In three-dimensions, kinetic energy can be written in terms of all three components of velocity, (v_x, v_y, v_z).

$$\boxed{K = \frac{1}{2}mv^2 = \frac{1}{2}m\left(v_x^2 + v_y^2 + v_z^2\right).} \qquad (8.15)$$

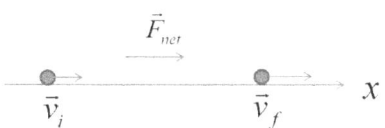

Figure 8.14: Work-energy theorem for a particle of mass says that the net work done by all forces equals the change in the kinetic energy of the particle, $W_{if}^{net} = \frac{1}{2}mv_f^2 - \frac{1}{2}mv_i^2$.

8.2. WORK-ENERGY THEOREM

The left side of Eq. 8.14 gives the net work done on the particle by all forces and the right side is the difference in kinetic energy, $K_f - K_i$.

$$\boxed{W_{if}^{\text{net}} = K_f - K_i, \quad \text{for a particle.}} \quad (8.16)$$

This result is called the **Work-Energy Theorem**. This equation also shows that the units of work and energy are same. In the SI system of units the unit of work is $N.m$ and the unit of energy is $kg.m^2/s^2$ as seen from the definition of kinetic energy. The two units are equal and are also known by another name, Joule (J).

$$1 \text{ N.m} = 1 \text{ kg.m}^2/\text{s}^2 = 1 \text{ J}.$$

Therefore, we can say that a 1-kg object moving at speed 1 m/s has 1 J of kinetic energy. Also, if 1 N force acts in the direction of the displacement of 1 m, then the work done by that force will be 1 J.

Although, we have derived the work-energy theorem using a motion in one dimension, it is true more generally. We will next illustrate this assertion by explicitly working out the calculation in three dimensions.

8.2.2 Integrating Equation of Motion of One Particle in Three Dimensions

The general equation of motion of a particle of mass m subject a net force $\vec{F}$ is given as

$$\vec{F}_{\text{net}} = m \frac{d\vec{v}}{dt}. \quad (8.17)$$

Let us think of $\vec{F}_{\text{net}}$ as a function of position $\vec{r}$ rather than a function of time t as we have done in previous chapters. Taking the dot product of both sides with $d\vec{r}$ and integrating from the initial position to the final position we notice that the left side corresponds to the net work done on the particle and the right side can be transformed into the change in kinetic energy as we have seen in the simpler situation of one-dimensional motion.

$$\int_{\vec{r}_i,\text{path}}^{\vec{r}_f} \vec{F}_{\text{net}} \cdot d\vec{r} = m \int_{\vec{r}_i,\text{path}}^{\vec{r}_f} \frac{d\vec{v}}{dt} \cdot d\vec{r}. \quad (8.18)$$

The left side of this equation is W_{if}^{net}. Now, we perform a series of manipulations, similar to the one done for the one-dimensional motion, to show that the right side of Eq. 8.18 is actually equal to the change in kinetic energy.

$$\begin{aligned}
\text{Right side of Eq. 8.18} &= m\int_{\vec{r}_i,\text{path}}^{\vec{r}_f} \frac{d\vec{v}}{dt}\cdot d\vec{r} = m\int_{t_i}^{t_f}\frac{d\vec{v}}{dt}\cdot\frac{d\vec{r}}{dt}dt \\
&= m\int_{t_i}^{t_f}\frac{d\vec{v}}{dt}\cdot\vec{v}dt = m\int_i^f \vec{v}\cdot d\vec{v} \\
&= m\left(\int_{v_{ix}}^{v_{fx}} v_x dv_x + \int_{v_{iy}}^{v_{fy}} v_y dv_y + \int_{v_{iz}}^{v_{fz}} v_z dv_z\right) \\
&= \frac{1}{2}m\left(v_{fx}^2+v_{fy}^2+v_{fz}^2\right)-\frac{1}{2}m\left(v_{ix}^2+v_{iy}^2+v_{iz}^2\right) \\
&= \frac{1}{2}mv_f^2 - \frac{1}{2}mv_i^2 = K_f - K_i
\end{aligned}$$

This shows that work-energy theorem given in Eq. 8.16 is applicable for a general motion of a particle in three dimensions.

8.2.3 Work-Energy Theorem Applied to Multiparticle Systems

A multiparticle system consists of two or more particles. Every macroscopic body, whether the body is a person, a car, a train, the entire planet or a star, is a multiparticle system. The work-energy theorem obtained in Eq. 8.16 apply to each particle of the multiparticle system. The work in the work-energy system consists of the work by every force on the particle, whether the force is by other particles of the same system or by some other external bodies.

In the case of a system consisting of two or more particles we have found that a separation of forces on the particles as internal forces and external forces is very helpful as far as the translation motion of the entire system is concerned. For instance, the acceleration of the center of mass depends on the external forces only. In the next chapter, we will see that the rotational motion of rigid bodies also depends on only the external forces.

Does a similar simplification occur when you combine the the work done on all the particles of a multiparticle system? We will answer this question here by actually writing out the work done on each particle and combining them.

Consider a system containing N particles of masses $m_1, m_2, \cdots, N$ at positions $\vec{r}_1, \vec{r}_2, \cdots, \vec{r}_N$, respectively, with respect to a reference point O. Let $\vec{F}_1^{\text{net}}, \vec{F}_2^{\text{net}}, \cdots, \vec{F}_N^{\text{net}}$ be the net forces on the particles $1, 2, \cdots, N$ respectively. The net force on each particle is a sum of external forces and internal forces. For instance, the net force on particle #1 is sum of the external force $\vec{F}_1^{\text{ext}}$ and the internal forces $\vec{F}_{12}, \vec{F}_{13}, \cdots, \vec{F}_{1N}$, where $\vec{F}_{1k}$ is the force on particle #1 by particle #k.

$$\vec{F}_1^{\text{net}} = \vec{F}_1^{\text{ext}} + \left(\vec{F}_{12} + \vec{F}_{13} + \cdots + \vec{F}_{1N}\right) \tag{8.19}$$

8.2. WORK-ENERGY THEOREM

The work done on each particle will separate into work by external forces and internal forces. The work done by all forces on particle #1 will be

$$W_1 = W_1^{\text{ext}} + W_1^{\text{int}}. \tag{8.20}$$

The work-energy theorem applied to particle #1 gives the net work on particle #1 equal to change in kinetic energy of particle #1. Let v_{1i} and v_{1f} be speed of particle #1 at the beginning and the end of the interval of interest. Then, we have

$$W_1^{\text{ext}} + W_1^{\text{int}} = \frac{1}{2}m_1 v_{1f}^2 - \frac{1}{2}m_1 v_{1i}^2. \tag{8.21}$$

The sum of work on by all forces, external as well internal give the net change in kinetic energy.

$$\sum_{k=1}^{N} W_k^{\text{ext}} + \sum_{k=1}^{N} W_k^{\text{int}} = \sum_{k=1}^{N} \frac{1}{2}m_k v_{kf}^2 - \sum_{k=1}^{N} \frac{1}{2}m v_{ki}^2. \tag{8.22}$$

The right side of this equation is the change in kinetic energy of all the particles and the left is the sum of **work done by the external and internal forces**. We can write this equation in simpler notation.

$$\boxed{W_{\text{net}}^{\text{ext}} + W_{\text{net}}^{\text{int}} = K_f - K_i.} \tag{8.23}$$

We find that, unlike the combining of forces on all particles, the separation of forces in the external and internal forces does not remove the necessity of knowing the internal forces. Therefore, we can conclude the it is much harder to use the work-energy theorem for calculations if internal forces can also do work on the system. There are two special cases in which internal work may vanish and then we will need to bother with external work only.

$$\boxed{W_{\text{net}}^{\text{ext}} = K_f - K_i, \quad \text{if } W_{\text{net}}^{\text{int}} = 0} \tag{8.24}$$

Case 1: $\vec{F}_{ij} \perp \Delta \vec{r}_i$

If internal forces are perpendicular to the displacements of the particles, then, the work done by the internal forces will be zero.

$$W_{\text{net}}^{\text{int}} = 0, \text{ if } \vec{F}_{ij} \perp \Delta \vec{r}_i, \text{ for all}\{ij\}$$

Case 2: $\vec{r}_i$ independent of $\{i\}$

If the displacement of every particle is same, then, the the work done by the internal forces will also be zero as we will show with a two-particle system.

$$W_{\text{net}}^{\text{int}} = 0, \text{ if } \Delta \vec{r}_1 = \Delta \vec{r}_2 = \cdots.$$

Two-particle System

Let us apply these equations for an N-particle system to a two-particle system to gain a better understanding of their content. The work by external forces and change in kinetic energy do not require more discussion since they are similar to the terms for a one-particle system. The new term is the work by internal forces. In the case of a two-particle system, the internal forces are a force on particle #1 by particle #2, denoted by $\vec{F}_{12}$ and a force on particle #2 by particle #1, denoted by $\vec{F}_{21}$. The force $\vec{F}_{12}$ will do work on particle #1 and the force $\vec{F}_{21}$ will do work on particle #2. Their works in a time interval when the displacements of the two particles are $\Delta\vec{r}_1$ and $\Delta\vec{r}_2$ respectively are given by

$$W_1^{\text{int}} = \vec{F}_{12} \cdot \Delta\vec{r}_1 \tag{8.25}$$

$$W_2^{\text{int}} = \vec{F}_{21} \cdot \Delta\vec{r}_2 \tag{8.26}$$

Therefore, the net work by internal forces is

$$W_{\text{net}}^{\text{int}} = \vec{F}_{12} \cdot \Delta\vec{r}_1 + \vec{F}_{21} \cdot \Delta\vec{r}_2 \tag{8.27}$$

From the third law of motion, the two forces have equal magnitude but are in opposite direction. That is,

$$\vec{F}_{12} = -\vec{F}_{21}. \tag{8.28}$$

Using this in Eq. 8.27 we obtain the following simpler result for the net work done by the internal forces.

$$W_{\text{net}}^{\text{int}} = \vec{F}_{12} \cdot (\Delta\vec{r}_1 - \Delta\vec{r}_2) \tag{8.29}$$

We find that, although the net internal force, $\vec{F}_{\text{net}}^{\text{int}} = \vec{F}_{12} + \vec{F}_{21} = 0$, the net internal work is not zero unless the displacements of the two particles are equal.

$$W_{\text{net}}^{\text{int}} = 0, \quad \text{if } \Delta\vec{r}_1 = \Delta\vec{r}_2. \tag{8.30}$$

We can generalize this result systems having any number of particles. We note that if all particles of a system move with the same displacement, then, we need work by only the external forces.

8.2.4 Uses of Work-Energy Theorem

The work-energy theorem equates the change in kinetic energy of a system to the net work done by all forces on all particles of the system. One practical use of the work-energy theorem is to find change in speed from given forces and displacements.

But, we have a major problem here: in the section on work we found that work integral may sometimes depend on the path of the particle. This means that, in order to find the change in kinetic energy, we need to know the actual path of the particle and the force at each instant on the path. This kind of detailed information required for using the work-energy theorem makes the theorem useless for general applications.

8.2. WORK-ENERGY THEOREM

However, there are many important forces, such as gravity, spring, etc, whose work integral does not depend on path but only on the end points of the motion. These forces are called **conservative forces**. For conservative forces, the work-energy theorem becomes very useful for relating the positions and speeds at different points in time without having to know anything about the full motion as we will see below.

$$\boxed{W_{if}^{\text{Conservative force}} : \text{Independent of path.}} \tag{8.31}$$

There is another situation where work-energy theorem becomes useful. Suppose the object is constrained to move in a pre-determined trajectory. For instance, when a ball is forced to move in a track, we know the CM of the ball must follow the path of the track. To constrain an object to a particular trajectory one requires a force that acts on the body to change the direction of the velocity by acting perpendicular to the velocity. Since constraining forces are directed perpendicular to the velocity, they are also perpendicularly to the infinitesimal displacement. This makes the work integral for constraining forces identically zero.

$$\boxed{W_{if}^{\text{Constraining force}} = 0.} \tag{8.32}$$

The forces which are neither conservative nor constraining are called non-conservative forces. The friction force is an example of a **nonconservative force**, so is the force of pull and push. In setting up problems that involve non-conservative forces, we need either a complete information about the path or some simplifying assumptions, such as constancy of force, if we wish to apply work-energy theorem to that problem.

8.2.5 Work-Energy Theorem For Conservative Forces

The work done by many commonly encountered forces such as push, pull, friction, etc, tend to depend on the actual path the system takes between the initial and final positions. However, work by many other forces, such as gravity, spring, and all fundamental forces, are path independent - the work by these forces depend only on the initial and final positions of the path. These forces are called **conservative forces**. Work-energy theorem is a very useful tool for an analysis of the motion when only conservative forces act on the body as we will see in this section.

Since the work done by a conservative force does not depend on the path but depends only on the position of the end points of the path, the result of the work integral can also be written as a difference of a function evaluated at the end points of the path. Denoting the conservative force by $\vec{F}_c$, we write the results of the work integral for this force as

$$\int_{\vec{r}_i}^{\vec{r}_f} \vec{F}_c \cdot d\vec{r} = \text{function of } (\vec{r}_f) - \text{function of } (\vec{r}_i), \tag{8.33}$$

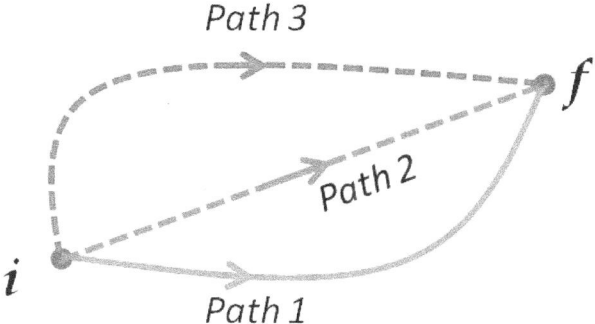

Figure 8.15: The work done by conservative force for displacement of the system between two points in space is independent of the path. The work integral $\int_i^f \vec{F}_c \cdot d\vec{r}$ for a conservative force $\vec{F}_c$ is same for all paths between points marked i and f.

Note that we have omitted mentioning the path for the work integral here since, for a conservative force, the integral does not depend on the path. The function on the right side of Eq. 8.33 is usually designated by letter U. We also include a negative sign for future convenience.

$$\int_{\vec{r}_i}^{\vec{r}_f} \vec{F}_c \cdot d\vec{r} = -\left[U(\vec{r}_f) - U(\vec{r}_i)\right]. \tag{8.34}$$

For the sake of brevity we will write this equation as

$$\boxed{\int_{\vec{r}_i}^{\vec{r}_f} \vec{F}_c \cdot d\vec{r} = -U_f + U_i.} \tag{8.35}$$

The individual quantities U_i and U_f are called the **initial and final potential energies** of the body that is subject to the conservative force $\vec{F}_c$. At each point in space the body will have potential energy contributions from every conservative force acting on the body. For instance, if two conservative forces, say gravity and spring force act on a body, then the body will have two contributions to the potential energy, one due to the gravity force and another due to the spring force.

$$\boxed{U^{\text{net}} = U^{\text{gravity}} + U^{\text{spring}} + \cdots}$$

Supposing only conservative and constraint forces act on the body, the net work will equal the work by conservative forces only because the work by constraint forces are zero as explained above. Therefore, we can write the net work on the body as a change in the net potential energy from all conservative forces. We denote the net potential energy by the same symbol, U_i and U_f.

$$W_{if}^{\text{net}} = U_i - U_f \quad \text{(Conservative and constraint forces only.)} \tag{8.36}$$

Now, when we replace W_{if}^{net} in the work energy theorem by $U_i - U_f$, we obtain the following important relation.

$$U_i - U_f = K_f - K_i \quad \text{(Conservative and constraint forces only.)} \tag{8.37}$$

This equation can be rearranged so that all quantities for the initial state are one side of the equation and those of the final state on the other side.

$$\boxed{K_i + U_i = K_f + U_f} \quad \text{(Conservative and constraint forces only.)} \tag{8.38}$$

The combination $(K + U)$ is called mechanical energy, or simply energy E of the system.

$$\boxed{E = K + U.} \tag{8.39}$$

Eq. 8.38 says that mechanical energy is conserved if only conservative and constraint forces act on the system. This statement is referred to as the principle of conservation of energy.

8.3 POTENTIAL ENERGY FUNCTIONS

We saw in the last section that the work done by a conservative force on a particle is accounted by a change in the potential energy of the particle. The contribution of the work done by the conservative forces can be included either as work by the force or by the potential energy change.

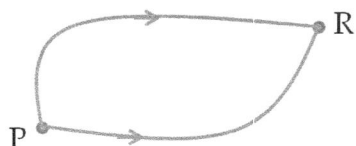

Figure 8.16: The value of the potential energy function at a space point P is obtained by work integral from P to a reference point R by any path.

To facilitate the use of conservation of energy given in Eq. 8.38 and other formulas containing potential energy terms, we now work out the formulas for the potential energies of point particles when they are subject to conservative forces. Since the work integral gives the potential energy difference, we usually write the potential energy as a function of coordinates of space by evaluating the potential difference between the space point of interest and a carefully chosen reference point in each case. These functions are called **potential energy functions** and usually denoted by letter U.

Thus, the potential energy function for a point particle placed at the space point $P(x, y, z)$ due to a conservative force is be obtained by performing the work integral W_{PR} for moving the particle from point P to an arbitrarily chosen reference point R, and equating the work done to the change in the potential energy $U_P - U_R$.

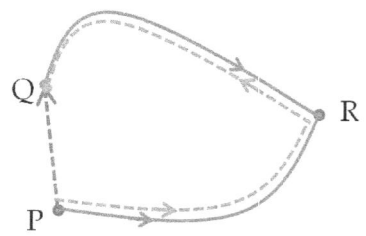

Figure 8.17: The difference in potential energy between points P and Q is independent of the choice of reference point R. Work from P to Q directly is equal to work from P to R then from R to Q.

$$U_P = U_R + W_{PR}, \tag{8.40}$$

The work integral W_{PR} can be done on any convenient path between points P and R as shown in Fig. 8.16 since the work integral for conservative forces are independent of the path. We choose the value of the arbitrary constant U_R and the location of the reference site R so that the potential energy function is well-defined in the region of interest and the formula is simplest as we will illustrate in the examples below. We will often write U_P as $U(x, y, z)$ to indicate the coordinates of point P in the formula for the potential energy function.

The arbitrariness of the reference energy U_R and the reference point R does not affect the difference in potential energy between any two points which is where the physical information resides. For instance, the difference $U(x,y,z) - U(x',y',z')$ of potential energies between points $P(x,y,z)$ and $Q(x',y',z')$ is independent of U_R and R.

$$\begin{aligned} U(x,y,z) - U(x',y',z') &= [U_R + W_{PR}] - [U_R + W_{QR}] \\ &= W_{PR} - W_{QR} \\ &= \int_P^R \vec{F}_c \cdot d\vec{r} - \int_Q^R \vec{F}_c \cdot d\vec{r} \\ &= \int_P^Q \vec{F}_c \cdot d\vec{r} \quad \text{(independent of } R \text{ and } U_R\text{).} \end{aligned}$$

Each conservative force has its own formula for the potential energy function obtained in this manner as we will see below. In the following, you will study examples illustrating a general procedure for finding the formula for the potential energy function corresponding to a given force. The common steps in the derivation of a potential energy function can be summarized as follows.

1. Pick an inertial coordinate system.

2. Write the force law for the conservative force $\vec{F}_c$ in that coordinate system.

3. Pick two points in space, one for the reference (R) and the other an arbitrary point (P). Use symbolic coordinates for R and P, with coordinates (x_R, y_R, z_R) for R and coordinates (x, y, z) for P. Use the analytic form of $\vec{F}_c$ to decide if you need a one-dimensional, a two-dimensional, or a three-dimensional consideration.

4. Perform the integral to find the work done for a path from P to R. While doing the integral, pick the path that simplifies the integral by using segments where the path and the force directions are aligned either parallel, anti-parallel or perpendicular to each other. Call the result symbolically as $U_P - U_R$.

5. The potential energy U_P is then obtained by carefully studying the equation produced, and making a choice for the coordinates of R and the value of U_R so that you get the simplest or the desired expression for U_P, which is now written as $U(x, y, z)$. The usual choice for U_R is zero, but any finite value for U_R will do just as well. The reference point R is usually placed at origin or infinity. However, sometimes, neither of these standard choices are allowed, in that case we pick an arbitrary point at a finite distance from the origin as the reference point.

8.3.1 Potential Energy Associated with Gravity

Consider a particle of mass m near the surface of earth. Since the weight of magnitude mg acts vertically at the particle, force of gravity will do work only for a vertical displacement. That is, the work by the force $\{mg,\text{pointed down}\}$ from any point on $y = a$ plane to any point in $y = b$ plane will be the same (Fig. 8.18). From the work done by the force of

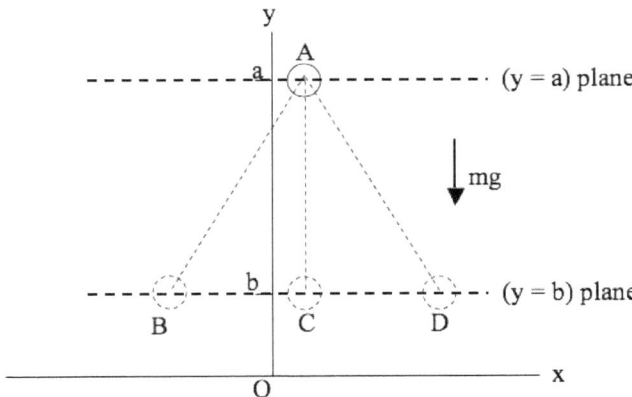

Figure 8.18: The work done by mg is same between points A and B, C or D since it depends only on the vertical displacement.

gravity we will deduce a formula for the potential energy as a function of height $y = h$ with respect to a reference at $y = y_R$.

Since weight is a constant force, we can find work easily by the dot product of the displacement vector with the constant force. With the y-coordinates pointed up we have the following vectors for the displacement from P to R and the weight vector.

$$\text{Displacement:}\quad \Delta\vec{r} = (y_R - h)\hat{u}_y.$$

$$\text{Force:}\quad \vec{F} = -mg\hat{u}_y.$$

Therefore, work done on the object when it is displaced from P to R is

$$W = \int_h^{y_R} F_y dy = -mg\left(y_R - h\right). \tag{8.41}$$

We now equate the work done to $U(h) - U(y_R)$ and obtain

$$U(h) = U(y_R) - mgy_R + mgh. \tag{8.42}$$

We can choose the reference point to be at the origin so that $y_R = 0$, and choose the reference value of potential energy at the origin to be zero, i.e. set $U(y_R) = 0$ to simplify the formula for the potential energy function corresponding to the force of gravity.

$$\boxed{U(h) = mgh,\quad \text{Reference at } h = 0.} \tag{8.43}$$

This result is often called the **gravitational potential energy** of an object of mass m at coordinate (x, h, z) with a reference point at $y = 0$. Note that $h < 0$ for points below the $y = 0$ plane so that gravitational potential energy is negative for $h < 0$ and positive for $h > 0$.

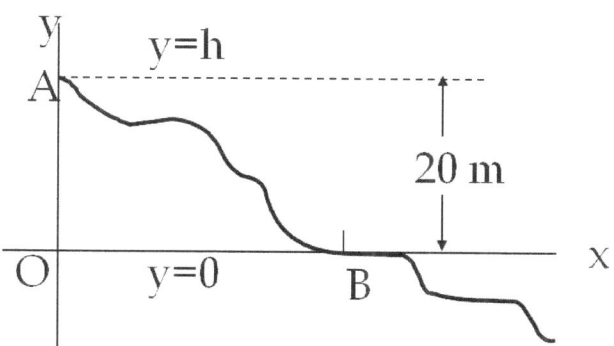

Figure 8.19: Example 8.3.1.

Example 8.3.1. Gravitational potential energy of a skier A skier starts from rest and slides downhill as shown in Fig. 8.19. What will be the speed of the skier if he drops by 20 meters in going from A to B? Ignore any air resistance (which will, in reality, be quite a lot), and any friction between the skis and the snow.

Solution. The motion of the skier occurs on a complicated surface. To apply $\vec{F} = m\vec{a}$ we would need the force on the skier at each instant on the path, which is impossible to obtain in this case. However, since the force of gravity is a conservative force, we can use the principle of the conservation of energy, which can relate quantities such as the position and speed at one instant to the position and speed at another instant.

When applying the principle of the conservation of energy to problems, the algebra is simpler if we choose the initial or the final point to be the reference point. In the case of gravity, it is customary to choose the lowest point in the motion as the reference point. This choice makes the gravitational potential energies at all other points of interest come out positive.

Therefore, we will choose point B as the reference point here, rather than the ground. Then, the height h above B in the gravitational potential energy formula will be positive for points above B and negative for points below B. The conservation of energy between points A and B takes the following form here.

$$K_A + U_A = K_B + U_B,$$

which gives the following equation

$$0 + mgh = (1/2)mv_B^2 + 0$$

solving for v_B we obtain

$$v_B = \sqrt{2gh},$$

where we keep only the positive root since the speed is a positive number. Now, we can plug in the values given in the problem to obtain the value of v_B:

$$v_B = \sqrt{2 \times (9.8 \ m/s^2) \times (20 \ m)} = 20 \ m/s.$$

8.3. POTENTIAL ENERGY FUNCTIONS

Note that, the mass of the skier canceled out in this problem. This happened because the gravitational potential energy U was proportional to mass. Therefore, all skiers, regardless of their masses, will have the same motion if we ignore the air resistance and ground friction on them.

8.3.2 Potential Energy Associated with the Spring Force

To find the formula for the potential energy of a particle subject a spring force we will carry out the same steps of calculations as we did for the particle subject to gravity. Consider a block of mass m attached to a spring of spring constant k. Without any loss of generality we can choose the origin of the coordinate system at the point in Fig. 8.20 when the spring is neither stretched nor compressed from its original length. We will also choose the x-axis to be the line in which the mass moves as shown in the figure. Let $x(t)$ be the x-coordinate of the block at an arbitrary

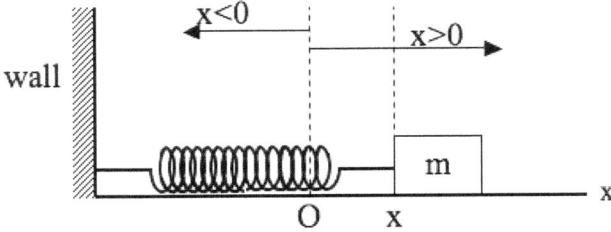

Figure 8.20: The set-up for calculation of the potential energy stored in the spring.

time t. The spring force on the block is given by the Hooke's law whose mathematical expression written using the coordinate system given in Fig. 8.20 is as follows.

$$\vec{F} = -k\, x\, \hat{u}_x. \tag{8.44}$$

The infinitesimal displacement vector along the x-axis is

$$d\vec{r} = dx\, \hat{u}_x, \tag{8.45}$$

which is obtained by setting $dy = dz = 0$ in $d\vec{r} = dx\, \hat{u}_x + dy\, \hat{u}_y + dz\, \hat{u}_z$. The work integral from point $P(x,0,0)$ to a reference point $R(x_R,0,0)$ gives

$$\begin{aligned} W &= \int_P^R \vec{F}\cdot d\vec{r} = -\int_x^{x_R} kx\, dx \\ &= \frac{1}{2}kx^2 - \frac{1}{2}kx_R^2 \end{aligned} \tag{8.46}$$

This is the change in the potential energy, $U(x) - U(x_R)$, of the block due to the spring force.

$$U(x) - U_R = \frac{1}{2}k\, x^2 - \frac{1}{2}k\, x_R^2. \tag{8.47}$$

Suppose, we choose the reference of the potential energy to be zero when $x = 0$, i.e. when the spring is neither stretched nor compressed.

Choice of reference: $x_R = 0$ and $U_R = 0$.

Then, the potential energy of the block due to the spring force when the spring is stretched or compressed by an amount $|x|$ is given by a simple formula.

$$\boxed{U(x) = \frac{1}{2}k\,x^2, \quad \text{Reference at } x = 0.} \tag{8.48}$$

The reference is the state when the spring is neither stretched nor compressed. Sometimes, this potential energy is called the **spring energy** since this energy is contained in the spring since the spring is deformed compared to the original length. When the spring comes back to the original configuration, the energy in the spring goes towards speeding up the objects connected at the two ends. In the case of one end connected to an immovable support and the other end to a movable block, the energy in the spring foes towards changing the kinetic energy of the block.

Example 8.3.2. Mass attached to a spring on a frictionless table. Consider a block of mass 0.2 kg attached to a spring of spring constant 100 N/m. The block is placed on a frictionless table, and the other end of the spring is attached to the wall so that the spring is level with the table. The block is then pushed in so that the spring is compressed by 10 cm. Find the speed of the block as it crosses (a) the point when the spring is not stretched, (b) 5 cm to the left of point in (a), and (c) 5 cm to the right of point in (a).

Solution. Since there is no friction, only the spring force does work on the mass and since spring force is a conservative force, the mechanical energy is conserved. Here the potential energy is due to the spring force only. Therefore,

$$(K + U)_i = (K + U)_f.$$

$$\implies \left(\frac{1}{2}mv^2 + \frac{1}{2}kx^2\right)_i = \left(\frac{1}{2}mv^2 + \frac{1}{2}kx^2\right)_f.$$

(a) In this part, the initial speed $v_i = 0$ and the initial position $x_i = -10\ cm = -0.1\ m$, and the final speed v_f is unknown and the final position $x_f = 0$. Therefore,

$$0 + \frac{1}{2}kx_i^2 = \frac{1}{2}mv_f^2 + 0$$

Ignoring the negative root, since v is speed and it is positive, we obtain

$$v_f = \sqrt{\frac{k}{m}}|x_i| = \sqrt{\frac{100\ N/m}{0.2\ kg}} \times 0.1\ m = \sqrt{5}\ m/s.$$

(b) In this part, we can choose i and f as follows. $x_i = 0$ (at origin); $x_f = -5$ cm to the left of origin. We know the following: $v_i = \sqrt{5}$ m/s, $x_i = 0$, $v_f =$ unknown, and $x_f = -5$ cm. Therefore, writing the conservation equation in symbols before we plug in numbers.

$$\frac{1}{2}mv_i^2 + 0 = \frac{1}{2}mv_f^2 + \frac{1}{2}kx_f^2$$

We solve this equation for v_B.

$$v_B = \sqrt{v_A^2 - \frac{k}{m}x_B^2} = \sqrt{3.75} \text{ m/s}.$$

(c) This part will give the unknown $v = \sqrt{3.75}$ m/s. The detail is left as an exercise for the student.

8.3.3 Potential Energy Associated With the Universal Gravitational Force

The universal gravitational force between any two objects is given by the Newton's law of gravitation. The law states that the gravitational force between two objects is proportional to the product of their masses and inversely proportional to the square of the distance between them. The direction of the force on the two objects are determined by the attractive nature of the force. Thus, if we place mass M at the origin, then force on mass m at a distance r is

$$\vec{F}_{on\ m} = G_N \frac{mM}{r^2}(-\hat{u}_r), \tag{8.49}$$

where $\hat{u}_r$ is a unit vector radially outward from the origin, and the negative sign reflects the fact that the force on m is towards M. Newton's gravitational constant G_N has value 6.67×10^{-11} $N.m^2/kg^2$.

How much work does this force do when the distance between the masses change from r_0 to r_R? To find this work, we will imagine a process in which mass M is fixed at the origin and mass m is moved "virtually" by exerting an external force $\vec{F}_{appl}$ that balances the gravitational force at each point.

$$\vec{F}_{appl} = -\left[G_N \frac{mM}{r^2}(-\hat{u}_r)\right], \tag{8.50}$$

The work done by $\vec{F}_{appl}$ will be equal to the negative of the work done by the gravitational force on m. By the radial nature of the force, it is clear that work done on m between any points of a sphere of radius r_0 to any point on a sphere of radius r_R will be same. To be concrete, let $r_R > r$ and let us choose point P on the spherical surface of radius r_0 and point R on the spherical surface of radius r_R that are on the same radial line from the origin (Fig. 8.21). Then an infinitesimal displacement between $\vec{r}$ and $\vec{r} + d\vec{r}$ on this path is

$$d\vec{r} = dr\ \hat{u}_r, \tag{8.51}$$

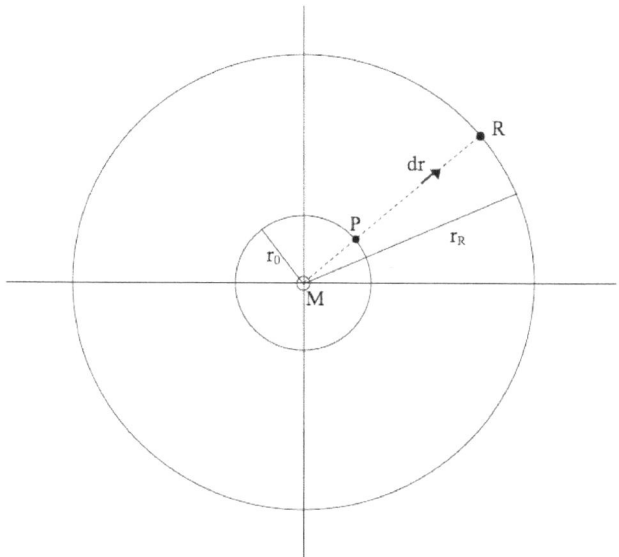

Figure 8.21: The set-up for calculation of the gravitational potential energy.

From Eqs. 8.50 and 8.51 we find the work done by the applied force for the infinitesimal displacement to be

$$dW = \vec{F}_{appl} \cdot d\vec{r} = G_N \frac{mM}{r^2} dr, \tag{8.52}$$

where we have used the identity $\hat{u}_r \cdot \hat{u}_r = 1$, since $\hat{u}_r$ is a unit vector. Performing the integral we obtain the work done by the applied force to be

$$W = G_N \frac{mM}{r_0} - G_N \frac{mM}{r_R}. \tag{8.53}$$

This is equal the negative of the work done by the gravitational force on m and can be equated to the potential energy difference $U(r_0) - U_R$ for the gravitational force, as we have done for other forces.

$$U(r_0) - U_R = G_N \frac{mM}{r_R} - G_N \frac{mM}{r_0}. \tag{8.54}$$

Now, we choose the values of r_R and U_R so that the potential energy $U(r_0)$ is simplest and has a well-defined value. Clearly $r_R = 0$ will not be an acceptable reference point since the first term on the right side of Eq. 8.54 is not defined for $r_R = 0$. A good reference here is $r_R = \infty$ and the choice of $U_R = 0$ further simplifies the formula. With these choices the gravitational potential energy of m when the separation between the masses is r_0 is given by

$$U(r_0) = -G_N \frac{mM}{r_0} \quad [\text{Reference: } r_R = \infty \text{ and } U_R = 0]. \tag{8.55}$$

Since there is nothing special about the distance r_0, we use Eq. 8.55 to write the gravitational potential energy function $U(r)$ for m when it is at a distance r from M.

$$\boxed{U(r) = -G_N \frac{mM}{r} \quad [\text{Reference: } r_R = \infty \text{ and } U_R = 0].} \tag{8.56}$$

8.3. POTENTIAL ENERGY FUNCTIONS

Note that gravitational potential energy of m is a negative number for all finite separation of the two masses. That is, the gravitational potential energy of m increases as the distance between the masses increases. This is consistent with our intuition that to pull m away from M an agent will have to do work on m thereby increasing the energy of the later. Note also that the gravitational potential energy is not defined when the masses are on top of each other.

Example 8.3.3. Escape speed If an object is fired with large enough speed it will escape the gravitational pull of the Earth. The minimum speed needed to escape the pull of the Earth is called the escape velocity or more appropriately **escape speed**. Find the escape speed from the Earth. Assume the Earth to be a sphere of mass M_E and radius R_E.

Solution. Let v_e denote the escape speed of an object of mass m when the object is at the surface of the Earth. When the object has escaped far away, the speed there will be taken to be zero since we are looking for the minimum speed of escape. If the air resistance can be ignored, the energy of the fired object will be conserved since it has only the gravitational force of the Earth acting on it, which is a conservative force. The conservation of energy of the object between when it is at the surface of the Earth and when it at a far away point, which will be taken to be $r = \infty$, yields the following equation.

$$E_{\text{on earth}} = \frac{1}{2}mv_e^2 + \left(-G_N \frac{mM_E}{R_E}\right)$$

$$E_{\text{far away}} = 0$$

$$\text{Therefore, } \frac{1}{2}mv_e^2 + \left(-G_N \frac{mM_E}{R_E}\right) = 0$$

$$\Rightarrow \quad v_e = \sqrt{\frac{2G_N M_E}{R_E}} \quad \text{(positive root since its a speed)}.$$

Since the mass of the fired object cancels out from the equation, the escape speed would be same for all objects regardless of the mass. Putting the values for G_N, M_E and R_E, the value of the escape speed from the Earth turns out to be $v_e \approx 112$ km/s, which is approximately $400{,}000$ km/h or $250{,}000$ mph.

Note that the calculation given above is with respect to a frame fixed to the center of Earth. A frame on Earth will be moving towards East with a speed equal to product of the angular rotation speed of the Earth and the distance from the axis of rotation of the Earth. Therefore, firing the projectile horizontally towards East will add the velocity of Earth-surface based frame to the launch velocity. This will result in a speed of launch needed to reach the escape speed to be considerably less than 112 km/s if you launch the projectile towards East. Also, firing from a place near the equator will give the largest distance from the axis of rotation, and hence the Earth-surface based frame with the largest speed with respect to the Earth-center based frame. That is why the space rocket launches in the

United States are done from the state of Florida and not from the state of Alaska.

8.3.4 Force and Potential Energy Function

We have found above that the change in potential energy of an object is given by the negative of the work done by the conservative force $\vec{F}_c$. That is, the difference of potential energy between when the object is at P and when it is at Q is

$$U(\text{at Q}) - U(\text{at P}) = -\int_P^Q \vec{F}_c \cdot d\vec{r}. \tag{8.57}$$

It is possible to invert this equation and obtain the conservative force if we know the potential energy function U. Consider a one-dimensional situation, and let us apply Eq. 8.57 to an infinitesimal displacement between x and $x + \Delta x$.

$$U(x + \Delta x) - U(x) = -\int_x^{x+\Delta x} F_x dx \approx -F_x \Delta x, \tag{8.58}$$

where F_x is the x-component of the conservative force $\vec{F}_c$. Now divide both sides of Eq. 8.58 by Δx and take $\Delta x \to 0$ limit to obtain

$$F_x = -\lim_{\Delta x \to 0} \frac{U(x + \Delta x) - U(x)}{\Delta x} = -\frac{dU}{dx} \tag{8.59}$$

Thus, for a one-dimensional displacement along the $x$axis, the x-component of the force is equal to the negative of the x-derivative of the potential energy function $U(x)$.

$$F_x = -\frac{dU}{dx} \tag{8.60}$$

We now state without proof that if you have a three-dimensional situation, then the x, y, and z-components of the force will be obtained from the partial derivatives of the potential energy function.

$$\boxed{F_x = -\frac{\partial U(x,y,z)}{\partial x}; \quad F_y = -\frac{\partial U(x,y,z)}{\partial y}; \quad F_z = -\frac{\partial U(x,y,z)}{\partial z}} \tag{8.61}$$

In these equations we have used partial derivatives. Partial derivatives of a function of more than one independent variable refers to the derivative with respect to one variable while keeping the other variables fixed. That is, while performing the partial derivative with respect to x, we will keep y and z fixed here. Similarly, for y and z derivatives.

8.4 CONSERVATION OF ENERGY

The principle of conservation of energy has a broader scope than stated above in Eq. 8.38 as we explain in this section. The forces in the work-energy theorem can also be divided into internal forces and external forces

8.4. CONSERVATION OF ENERGY

as opposed to conservative and non-conservative forces. For instance, suppose you are pulling a cart, then if we can consider you and the cart as one system, then the force of the Earth on you and the cart will be external forces while the force on the cart by you and its pair force, i.e. the force of cart on you, will be the internal forces. The work energy theorem then becomes.

$$W_{\text{int}} + W_{\text{ext}} = K_f - K_i. \tag{8.62}$$

A force can be either conservative or non-conservative. As we have seen above that we can write the work done by a conservative force as causing a change in the potential energy of the object upon which the force acts. Therefore, another way of writing Eq. 8.62 is in terms of conservative and non-conservative forces.

$$W_{\text{int}}^{\text{cons}} + W_{\text{int}}^{\text{non-cons}} + W_{\text{ext}}^{\text{cons}} + W_{\text{ext}}^{\text{noncons}} = K_f - K_i. \tag{8.63}$$

We now combine all the work by non-conservative forces and write the result as W_{NC} and combine all the work by conservative e forces and write the result as change in potential energy $U_i - U_f$ to obtain

$$W_{\text{NC}} + U_i - U_f = K_f - K_i. \tag{8.64}$$

This equation can be rearranged to obtain the change in $(K + U)$ as

$$[K_f + U_f] - [K_i + U_i] = W_{\text{NC}}, \tag{8.65}$$

which is written more simply as

$$\boxed{E_f - E_i = W_{\text{NC}},} \tag{8.66}$$

Therefore, the energy of a system does not change in the absence of non-conservative forces.

Example 8.4.1. Two bodies interacting by a conservative force Show that the mechanical energy of an isolated system consisting of two bodies interacting with a conservative force is conserved.

For discussion, let us label two objects labelled 1 and 2. Since the system is isolated, there is no external force on the system and we have

$$W_{\text{int}} = (K_f - K_i)_1 + (K_f - K_i)_2. \tag{8.67}$$

The work W_{int} is the work by force by 1 on 2 and force by 2 on 1. Since these forces are are assumed to be conservative, their works can be written in terms of changes in the potential energies of objects 1 and 2 respectively.

$$W_{\text{on } 1} = (U_i - U_f)_1 \tag{8.68}$$
$$W_{\text{on } 2} = (U_i - U_f)_2. \tag{8.69}$$

Using Eqs. 8.67, 8.68 and 8.69 we find

$$[E_f - E_i]_1 = -[E_f - E_i]_2, \Leftrightarrow \boxed{\Delta E_1 = -\Delta E_2,} \tag{8.70}$$

where $E = K+U$ for each particle, the subscripts i and f refer to the initial and final states respectively, and ΔE is the change in energy. Therefore, if there are no external forces on a system so that $W_{\text{ext}} = 0$, i.e. when a system is isolated, and all the internal forces are conservative, then the total energy of the system cannot change.

It is interesting to note that for two bodies in this example, the work on 1 by force from 2 and the work on 2 by force from 1 are conduits for the transfer of energy from one to the other. The energy of the parts of a system, such as the energy of object 1 or the energy of object 2 may change with time due to the interaction between them, but the increase in energy of one will be accompanied by an equal decrease in the energy of the other.

Further Remarks

Since all the fundamental forces in nature are conservative forces, at the most basic level all systems are conservative and energy of an isolated system is always conserved.

Principle of conservation of energy: The energy of an isolated system cannot change with time.

The principle of conservation of energy is one of the most fundamental laws of nature that holds true even when Newton's laws of motion fail although we have arrived at it through the use of Newton's laws.

Example 8.4.2. Skier down hill with air resistance. Use the same problem as in Example 8.3.1 above, but this time, suppose that the work done by the air-resistance cannot be ignored. Let the work done by the air resistance when the skier goes from A to B along the given hilly path be -2000 Joules. The work done by the sir resistance is negative since the air resistance acts in the opposite direction to the displacement. Suppose the mass of the skier is 50-kg, what is the speed of the skier at point B?

Solution. Note that mechanical energy is not conserved in this situation since we have non-conservative force on the system. Therefore, we must use the full work-energy theorem given in Eq. 8.66.

$$W_{\text{NC}} = (K_f + U_f) - (K_i + U_i).$$

$$W_{\text{NC}} = \left(\frac{1}{2}mv_B^2 + 0\right) - (0 + mgh)$$

Solve for v_B and then plugging in the numbers to find v_B.

$$v_B = \sqrt{\frac{2(W_{\text{NC}} + mgh)}{m}} = 18 \text{ m/s}.$$

The speed v_B is less when the work by air resistance is taken into account. The mechanical energy of the skier is lost to the air particles and the internal energy of the molecules of the skier. We say that the energy of the system is converted to the thermal energy.

8.5 POWER

The rate at which a force does work is called the **power**. Thus, if a force $\vec{F}$ does work ΔW is done in time Δt then average power P_{ave} over that period of time is given as

$$\boxed{P_{\text{ave}} = \frac{\Delta W}{\Delta t}.} \tag{8.71}$$

If the force varies during the interval, then we will use the average force F_{ave}. Writing ΔW in terms of the average force and the displacement $\Delta \vec{r}$, we find that the average power is given by

$$P_{\text{ave}} = \frac{\vec{F}_{\text{ave}} \cdot \Delta \vec{r}}{\Delta t} = \vec{F}_{\text{ave}} \cdot \frac{\Delta \vec{r}}{\Delta t}. \tag{8.72}$$

Writing the average velocity v_{ave} for $\frac{\Delta \vec{r}}{\Delta t}$.

$$\boxed{P_{\text{ave}} = \vec{F}_{\text{ave}} \cdot \vec{v}_{\text{ave}}.} \tag{8.73}$$

Larger force and rapid displacement would correspond to more power. A person with stronger muscles normally will be able to lift weights more quickly than the same weight lifted by someone with weaker muscles. Consequently, you might say that building muscles gives you more power!

The dimensions of power can be found from the dimensions of work and time in Eq. 8.71.

$$[P] = \frac{[M][L]^2/[T]^2}{[T]} = \frac{[M][L]^2}{[T]^3}. \tag{8.74}$$

Therefore, the SI unit of power would be kg.m^2/s^3. This unit is given the name Watt (W) after James Watts who is given the credit for the invention of the steam engine. One Watt is also 1 Joule/second or J/s.

$$1 \text{ W} = 1 \frac{\text{kg.m}^2}{\text{s}^3} = 1 \frac{\text{J}}{\text{s}}. \tag{8.75}$$

A common unit of power in the United States is **Horsepower** (hp). One horsepower equals 746 W. Thus, an engine rated at 300 hp will deliver work at a rate of (300 hp $\times$ 746 W/hp) = 223,800 W, or, 223,800 J/s.

Often, the power is used to denote the rate at which a device uses up energy. For instance, a 60-W electric bulb uses 60 Joules of energy per second, and a microwave oven rated at 1100-W uses 1100 Joules of energy per second.

If we take the limit of infinitesimal time interval in Eq. 8.72, we obtain instantaneous power $P(t)$ at instant t.

$$P(t) = \lim_{\Delta t \to 0} \left[\vec{F}_{\text{ave}} \cdot \frac{\Delta \vec{r}}{\Delta t} \right] = \vec{F} \cdot \vec{v}. \tag{8.76}$$

This gives instantaneous power as the dot product of the force $\vec{F}$ and the velocity $\vec{v}$ at instant t.

$$P(t) = \vec{F} \cdot \vec{v}. \tag{8.77}$$

The work done over a finite interval, then, can be obtained by integrating the instantaneous power over the time interval. Therefore, the work W_{12} over an interval from $t = t_1$ to $t = t_2$ would be

$$W_{12} = \int_{t_1}^{t_2} P(t)dt. \tag{8.78}$$

Example 8.5.1. Power calculation An agent applies a constant force of 5 N horizontally on a 2 kg box sliding on a smooth frictionless horizontal surface. (a) What is the instantaneous power of the agent when the velocity of the box is 2 m/s in the same direction as the force? (b) What is the work done by the agent in 15 sec?

Solution. (a) Using the formula for the power of a force given as $P = \vec{F} \cdot \vec{v}$ we find that
$$P = (5 \text{ N})(2 \text{ m/s}) \cos(0°) = 10 \text{ W}.$$

(b) Since the power is constant, the work done will be $W = P\Delta t$. Putting the numerical values we find that the work done is equal to 150 N.m, or 150 J.

8.6 ELASTIC COLLISIONS

Collisions are typically categorized as elastic or inelastic depending upon whether or not kinetic energy is conserved in the collision. If total kinetic energy is conserved, then the collision is said to be elastic. This can happen if the colliding bodies do not stick together, change shape, or break apart, or lose energy in heat and vibration. For instance, the collision of billiard balls and metal balls are elastic to a large extent. The motion of planets around the sun can be examined from this perspective also.

In an elastic collision, you have two conservation conditions, the conservation of momentum and the conservation of kinetic energy, which must hold independently of each other.

Elastic Collisions:
$$(\vec{p}_{\text{total}})_{\text{before}} = (\vec{p}_{\text{total}})_{\text{after}}. \tag{8.79}$$
$$(K_{\text{total}})_{\text{before}} = (K_{\text{total}})_{\text{after}}. \tag{8.80}$$

Note that, while the conservation of momentum is a vector equation, the conservation of kinetic energy is a scalar equation.

The elasticity of collision between two objects in a one-dimensional collision is also expressed by a quantity called the restitution. In a one-dimensional collision, the restitution is defined as the ratio of relative

8.6. ELASTIC COLLISIONS

speed of separation after collision to the relative speed of approach before collision.

$$\epsilon = \frac{(v_{\text{rel}})_{\text{after}}}{(v_{\text{rel}})_{\text{before}}} \quad (8.81)$$

For a perfectly inelastic collision the two particles will stick to each other upon collision, then the restitution will equal zero. For a perfectly elastic collision the relative speed of separation after the collision is equal to the relative speed of approach before the collision, and therefore restitution is equal to 1.

Perfectly inelastic, $\epsilon = 0$; Perfectly elastic, $\epsilon = 1$.

Example 8.6.1. One-dimensional elastic collision. Consider a ball of mass m_1 traveling at speed v_1 towards another ball of mass m_2 at rest. As a result of the collision ball 1 is found to bounce back with speed v_1' and ball 2 moves with speed v_2' in ball 1's original direction. If the collision is elastic, find v_1' and v_2' in terms of m_1, m_2 and v_1.

Solution. We have two conditions here: (1) Conservation of momentum and (2) conservation of kinetic energy due to elasticity of the collision. To proceed analytically, we choose axis system so that the original direction of ball 1's motion be along the positive x-axis. Then, x-component of momentum will be conserved.

$$m_1 v_1 + 0 = -m_1 v_1' + m_2 v_2'$$
$$\frac{1}{2} m_1 v_1^2 + 0 = \frac{1}{2} m_1 v_1'^2 + \frac{1}{2} m_2 v_2'^2$$

Solving these two equations for v_1' and v_2' we find the following.

$$v_1' = \left(\frac{m_2 - m_1}{m_1 + m_2}\right) v_1$$
$$v_2' = \left(\frac{2 m_1}{m_1 + m_2}\right) v_1$$

Note that when $m_1 = m_2$, then after the collision ball 1 comes to rest and ball 2 moves with the speed of ball 1. On the other hand, if m_2 is very large, as would be the case when you bounce a ball off the floor, then m_1 bounces off m_2 with the same speed as before, unless it is deformed during the collision and the collision is not perfectly elastic.

Example 8.6.2. Two dimensional elastic collision. A particle A of mass m_1 is subject to a repulsive force from another particle B of mass m_2. Initially particle A is traveling towards particle B which is at rest. After interacting with particle B, particle A is found to move with speed v_1' at an angle θ with its original direction, and B with speed v_2' at an angle ϕ from the original direction of A. Angle θ is called the angle of scattering of particle A while angle ϕ is called the angle of recoil of the target particle. Assuming the motion to be in the xy-plane and the collision elastic, find speeds v_1', v_2' and the scattering angle θ in terms of m_1, m_2, v_1, and ϕ.

Solution. We draw a figure to make the situation clearer and choose the Cartesian axes for calculations. As shown in Fig. 8.22, particle 1 is approaching particle 2 on the x-axis. After collision, the particle 1 is scattered in the direction specified by counterclockwise angle θ from the positive x-axis and the particle 2 recoils in the direction given by the clockwise angle ϕ from the positive x-axis.

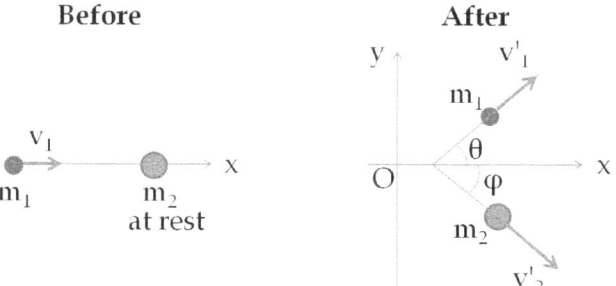

Figure 8.22: Particle 1 in motion collides with particle 2 at rest. After the collision particle 1 moves in the direction at angle θ while particle 2 recoils in the direction ϕ.

Since, the motion of the particles is confined to a plane, we will get two conditions, one for the x-components and the other for the y-components, from the conservation of momentum across the collision process. In addition, since the collision is assumed to be elastic, we will get one more condition from the conservation of kinetic energy.

x-components of momentum: $m_1 v_1 + 0 = m_1 v_1' \cos\theta + m_2 v_2' \cos\phi$

y-components of momentum: $0 + 0 = m_1 v_1' \sin\theta - m_2 v_2' \cos\phi$

Elastic collision: $\frac{1}{2} m_1 v_1^2 = \frac{1}{2} m_1 v_1'^2 + \frac{1}{2} m_2 v_2'^2$

One can solve these three equations for at most three unknowns. Suppose m_1, m_2, v_1 and one of the after-collision variables, say the recoil angle ϕ are given, then one can solve for v_1', v_2', and ϕ in terms of the given quantities. Let us write c for m_1/m_2 to simplify the answer.

$$v_1' = \left(\frac{\sqrt{1 - 2c \cos(2\phi) + c^2}}{1 + c} \right) v_1$$

$$v_2' = \left(\frac{2 \cos\phi}{1 + c} \right) v_1$$

$$\theta = \arctan\left(\frac{\sin(2\phi)}{c - \cos(2\phi)} \right)$$

The calculation is tedious and not very illuminating. To facilitate the calculation, it helps to introduce the ratios of speeds and masses, namely $a = v_1'/v_1$, $b = v_2'/v_1$ and $c = m1/m2$, and make use of $\sin^2\theta + \cos^2\theta = 1$ to eliminate θ and solve for a and b. A student is encouraged to carry out the necessary calculations as an exercise.

8.7 EXERCISES

Work

Ex 8.7.1. A boy pulls a 5 kg cart with a 20 N force at an angle of 30° with the horizon for some time. Over this time the cart moves a distance of 12 m on a horizontal floor. (a) Find the work done on the cart by the boy. (b) What will be the work done by the boy if he pulled with the same force horizontally instead of at an angle 30° with the horizon over the same distance? Ans: (a) 208 N.m; (b) 240 N.m.

Ex 8.7.2. A crate of mass 200 kg is to be brought from a site on the ground floor to a third floor apartment. The workers know that they can either use the elevator first, then slide it on the third floor to the apartment or first slide the crate to another location marked C in the Fig. 8.23, and then take the elevator to the third floor and then slide it on the third floor a shorter distance.

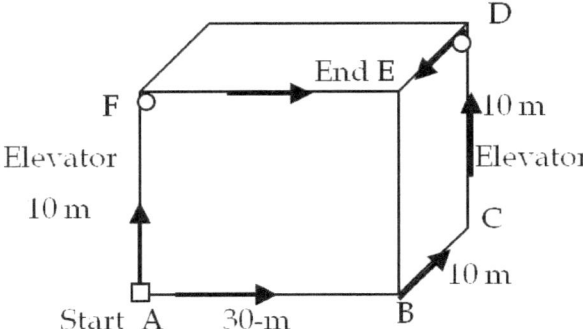

Figure 8.23: Exercise 8.7.2.

The trouble is the third floor is very rough compared to the ground floor. Given that the coefficient of kinetic friction between the crate and the ground floor is 0.1 and between the crate and the third floor surface is 0.3, find the work needed by the workers for each path. Assume that the force workers need to do is just enough to slide the crate at constant velocity at zero acceleration. Note: The work by the elevator against the force of gravity is not done by the workers. Ans: $W_{ABCDE} = 13,720\ J$, $W_{AFE} = 17,640\ J$.

Ex 8.7.3. A hockey puck is shot across a rough floor with the roughness different at different place, which can be described by a position-dependent coefficient of kinetic friction. For a puck moving along the x-axis, the coefficient of kinetic friction is the following function of x when x is in m, $\mu(x) = 0.1 + 0.05x$. Find work done by the kinetic frictional force on the hockey puck when it has moved (a) from $x = 0$ to $x = 2$ m, and (b) from $x = 2$ m to $x = 4$ m. Ans: (a) -0.9 N.m; (b) -1.5 N.m.

Conservation of Energy Involving Weight

Ex 8.7.4. A mysterious constant force 10 N acts horizontally on everything. The direction of the force is found to be always pointed towards a wall in a big hall. Find the potential energy of a particle due to this force when it is at a distance x from the wall assuming the potential energy at the wall to be zero. Ans: $10x$ with x-axis pointed away from the wall and origin at the wall.

Ex 8.7.5. In an amusement park a car rolls in a track as shown in Fig. 8.24. Find the speed of the car at A, B and C. Note that the work done by the rolling friction is zero since the displacement of the point at which the rolling friction acts on the tires is momentarily at rest and therefore has a zero displacement. Ans: $v_A = 24$ m/s; $v_B = 14$ m/s; $v_C = 31$ m/s.

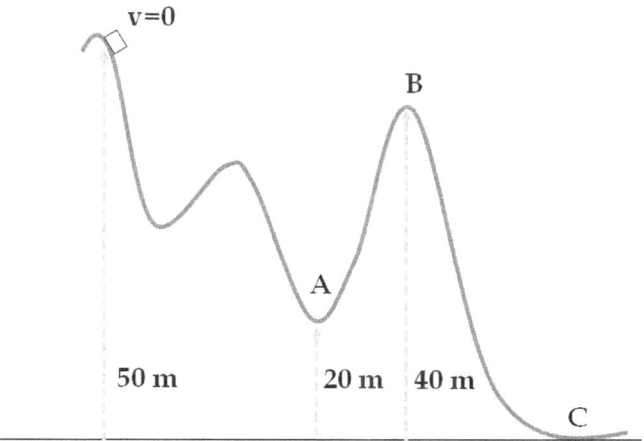

Figure 8.24: Exercise 8.7.5.

Ex 8.7.6. A 200-g steel ball is tied to a 2-m "massless" string and hung from the ceiling to make a pendulum, and then, the ball is brought to a position making a 30° angle with the vertical direction and released from rest. Ignoring the effects of the air resistance, find the speed of the ball when the string (a) is vertically down, (b) makes an angle of 20° with the vertical and (c) makes an angle of 10° with the vertical. Ans: (a) 2.29 m/s. (b) 1.70 m/s. (c) 2.16 m/s.

Ex 8.7.7. How would your answer(s) in the Ex. 8.7.6 change if the string had a mass of 100 g distributed uniformly along the string? (Hint: No new calculations necessary.)

Conservation of Energy Involving Spring Force

Ex 8.7.8. A block of mass 300 g is attached to a spring of spring constant 100 N/m. The other end of the spring is attached to a support while the block rests on a smooth horizontal table and can slide freely without any friction. The block is pushed horizontally till the spring compresses by

12 cm, and then the block is released from rest. (a) How much potential energy was stored in the block-spring-support system when the block was just released? (b) Determine the speed of the block when it crosses the point when the spring is neither compressed nor stretched. (c) Determine the speed of the block when it has traveled a distance of 20 cm from where it was released. Ans: (a) 0.72 J. (b) 2.2 m/s. (c) 1.6 m/s.

Ex 8.7.9. A block of mass 500 g is attached to a spring of spring constant 80 N/m. The other end of the spring is attached to a support while the mass rests on a smooth surface inclined at angle of 30° (Fig. 8.25). The block is pushed along the surface till the spring compresses by 10 cm and then released from rest. (a) How much potential energy was stored in the block-spring-support system when the block was just released? (b) Determine the speed of the block when it crosses the point when the spring is neither compressed nor stretched. (c) Determine the position of the block where it just comes to rest on its way up the incline. Ans: (a) 0.4 J. (b) 0.79 m/s. (c) 6.8 cm.

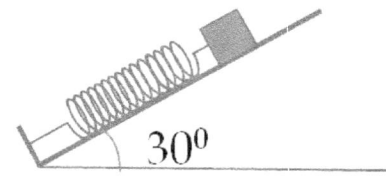

Figure 8.25: Exercise ??.

Ex 8.7.10. A block of mass 200 g is attached at the end of a massless spring of spring constant 100 N/cm. The other end of the spring is attached to the ceiling and the mass is brought to rest. Let us mark this point as O. Suppose, this point is taken to be the zero of the potential energy of the block, both from the weight and the spring force. The mass hangs freely and the spring is in a stretched state. The block is then pulled downward by another 5 cm and released from rest. (a) What is the net potential energy of the block at the instant the block is at the lowest point? (b) What is the net potential energy of the block at the instant the block returns to the point marked O? (c) What is the speed of the block as it crosses the point marked O? (d) How high above the point marked O does the block rise before coming to rest again? Ans: (a) 12.4 J, (b) U = 0, (d) 11.1 m/s, (d) 4.96 cm.

Conservation of Energy Involving Universal Gravitational Force

Ex 8.7.11. Determine the escape speed of a projectile from the Moon. Ans: 2,380 m/s.

Ex 8.7.12. Determine the escape speed of a projectile from the planet Jupiter. Ans: 60,000 m/s.

Ex 8.7.13. A satellite of mass m is to be placed in a circular orbit of radius $2R_E$, where R_E is the radius of the Earth. With what speed with respect to the fixed center of the Earth should the satellite be launched from a point on the equator of the Earth? Ans: 9200 m/s.

Ex 8.7.14. The Earth revolves around the Sun in an elliptical orbit with the Sun at one of the foci. In its orbit, the Earth makes a closest approach to the Sun, called the perihelion, and a farthest distance from the Sun, called the aphelion. The closest approach in the year 2000 took place on January 3, 2000 when the Earth-Sun distance was 1.46×10^{11} m and the farthest distance took place on July 4, 2000 with the Earth-Sun distance of 1.51×10^{11} m. Find the difference in the speed of Earth at the two locations. Ans: $v_2 = \sqrt{\dfrac{2\left[U(r_2) - U(r_1)\right] r_1^2}{m(r_2^2 - r_1^2)}}$, $(r_1 \neq r_2)$, $v_1 r_1 = v_2 r_2$.

Ex 8.7.15. Suppose your friend claims to have discovered a mysterious force in nature that acts on all particles in the three-dimensional space. He tells you that the force is always pointed towards a definite point in space, which we can call the **force center**. The magnitude of the force turns out to be proportional to $1/r^3$, where r is the distance from the force center to any other point. Let b be the proportionality constant, i.e. the magnitude of the force on a particle can be written as b/r^3, when the particle is at a distance r from the force center. (a) Find the potential energy of a particle when it is at a distance D from the force center, assuming the potential energy to be zero when the particle is at infinity. (b) Explain why the reference of the potential energy cannot be chosen at the force center. (c) Formally, write the potential energy if the magnitude of the force is an arbitrary function $f(r)$? Ans: (a) $-b/r^2$, (b) $U(0)$ not defined, (c) $U(r) = \int_{\text{ref}}^{r} f(r) dr$.

Nonconservation of Energy

Ex 8.7.16. In Exercise 8.7.5, the track was frictionless. Now, suppose 10% of the change in gravitational energy due to weight is lost due to friction. What will be the speeds at A, B, and C? Ans: $v_A = 23$ m/s.

Ex 8.7.17. A sky diver exits an air plane at almost zero speed. He falls a distance of 50 m. His speed at that point is 10 m/s. Show calculations to decide if his energy is conserved or not. Ans: Energy not conserved.

Ex 8.7.18. A hockey puck is shot across a iced pond. Before the hockey puck was hit the puck was at rest. After the hit, the puck has a speed of 40 m/s. The puck comes to rest after going a distance of 30 m. (a) Describe how the energy of the puck changes over time, giving the numerical values of any work or energy involved. (b) Find the magnitude of the net friction force. Ans: (a) Loss of energy = 240 N.m, (b) $F = 8$ N.

Ex 8.7.19. A bottle rocket is shot straight up in the air with a speed 30 m/s. If the air resistance is ignored, the bottle would go up to a height of approximately 46 m. But, the rocket goes up to only 35 m before returning to the ground. What happened? Explain, giving numerical values of any work or energy involved.

Power

Ex 8.7.20. A crate is being pushed across a rough floor surface. If no force is applied on the crate, the crate will slow down and come to a stop.

8.7. EXERCISES

If the crate of mass 50 kg moving at speed 8 m/s comes to rest in 10 seconds, what is the rate at which the frictional force on the crate takes energy away from the crate? Ans: 160 J/s.

Ex 8.7.21. In the exercise above, suppose that a horizontal force of 20 N in the direction of the velocity is required to maintain the speed of the crate constant at 8 m/s. (a) What is the power of this force? (a) Note that the acceleration of the crate is zero despite the fact that 20 N force acts on the crate horizontally. What happens to the energy given to the crate as a result of the work done by this 20-N force? Or, does this force do any work? Decide and give reason for your answer. Ans: (a) 160 W.

Ex 8.7.22. Grains from a hopper falls at a rate of 10 kg/sec vertically onto a conveyor belt that is moving horizontally at a constant speed 2 m/s. (a) What force is needed to keep the conveyor belt moving at the constant velocity? (b) What is the minimum power of the motor driving the conveyor belt? Ans: (a) 20 N; (b) 40 W.

Ex 8.7.23. A cyclist in a race must climb a 5° hill at a speed of 8 m/s. If the mass of the bike and the biker together is 80 kg, what must be the power output of the biker to achieve the goal? Ans: 547 W.

Elastic Collision

Ex 8.7.24. An alpha particle (which is the nucleus of a Helium atom) of mass 4 u (1 u = 1.66×10^{-27} kg) at a speed of 1×10^4 m/s strikes a gold nucleus (m = 197 u) at rest head on and is repelled <u>elastically</u> backward. Find the final velocities of the alpha particle and the gold nucleus. Ans: Speeds: 9,600 m/s and 392 m/s.

Ex 8.7.25. A billiard ball of mass m moving with speed 2 m/s strikes another billiard ball of the same mass at rest. After the collision, the incoming ball is deviated in the direction 15° from the original direction and moves with speed u while the struck ball is sent with speed w in the direction of 10° direction from the original direction of the first ball. (a) What are the speeds of the two balls after the collision? (b) Is the collision elastic? Why or why not. Ans: (a) $0.82\ m/s$ and $1.22\ m/s$; (b) not elastic.

Ex 8.7.26. A neutron (mass = 1 u, with u given in Exercise 8.7.24) moving at a speed of 3×10^5 m/s strikes a stationary hydrogen atom (mass = 1 u) head on. After the collision the stationary atom is found to move with a speed of 2×10^5 m/s in the direction of the neutron's original direction. (a) Find the velocity of the neutron after the collision. (b) Find the restitution and determine if the collision is elastic. Ans: (a) $1 \times 10^5\ m/s$, forward; (b) restitution = 1/3.

Ex 8.7.27. A particle of mass m moving at the speed v strikes head on another particle of mass $2m$ at rest in an elastic collision. Find the speeds of the two particles after the collision. Ans: $1/3\ v$ and $2/3\ v$.

Ex 8.7.28. A particle of mass m ($= 1$ gram) moving with a speed v strikes a stationary particle of mass $2m$ head on. After the collision, the heavier particle breaks up into two parts of mass $0.5\,m$ and $1.5\,m$. The initial particle comes to rest as a result of the collision, but $0.5\,m$ particle moves in the direction of $30°$ from the original direction of m with a speed of $0.2\,v$. (a) What is the direction and speed of $1.5\,m$ particle after the collision. (b) Is the collision elastic? Why or why not. Ans: (a) $0.61\,v$, $0.31°$, (b) not elastic.

8.8 PROBLEMS

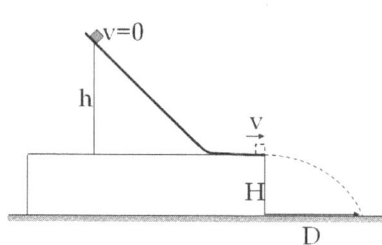

Figure 8.26: Problem 8.8.1.

Problem 8.8.1. A block leaves a frictionless inclined surface horizontally after dropping off by a height h (Fig. 8.26). Find the horizontal distance D where it will land on the floor in terms of h, H and g. Ans: $D = 2\sqrt{hH}$

Problem 8.8.2. A block of mass m after sliding down a frictionless incline strikes another block of mass M that is attached to a spring of spring constant k (Fig. 8.27). The blocks stick together upon impact and travel together.

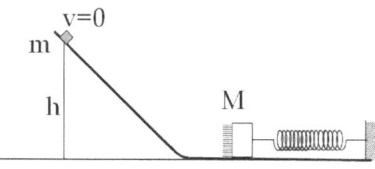

Figure 8.27: Problem 8.8.2.

(a) Find the compression of the spring in terms of m, M, h, g and k when the combination comes to rest. (b) The loss of kinetic energy as a result of the bonding of the two masses upon impact is stored in the so-called binding energy of the two masses. Calculate the binding energy. Ans: (a) $\sqrt{\frac{2m^2 gh}{k(m+M)}}$; (b) $\frac{mMgh}{m+M}$.

Problem 8.8.3. A block of mass m slides in a frictionless track that bends in a circular arc and then straightens out (Fig. 8.28). If the block is released from rest from an appropriate height h, the block will make it through the circular track without falling off; any less height and the block would not complete the circular path. Find the force that the block applies at positions marked A, B and C in terms of m, g, h and R. Ans: (a) $F_A = 6mg$, $F_B = 3mg$, $F_C = 0$, (b) $h = 2.5R$.

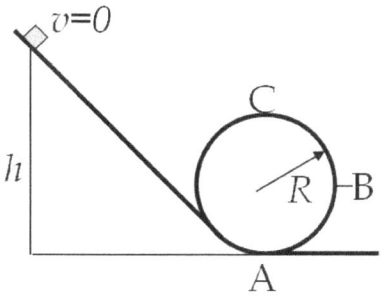

Figure 8.28: Problem 8.8.3.

Problem 8.8.4. A small block of mass m is sitting at the top of a sphere of radius R (Fig. 8.29). When the block is moved infinitesimally to one side, it slides on the spherical surface up to a point when it leaves the spherical surface. Assuming no friction between the block and the surface of the sphere find the point at which the block will fall off the surface. You can specify the location of the place on the sphere by an angle a line from the center of the sphere to that point will make with the vertical line. Hint: The instant after the block is no longer in contact, there is no normal force on the block, but the speed of the block is same as it was when it left the surface. Ans: $\cos^{-1}(2/3)$.

Problem 8.8.5. Ballistic pendulum. A bullet of mass m is fired horizontally with speed v in a block of M that is hanging by a massless string of length L from the ceiling (Fig. 8.30). After the impact the bullet is embedded in the block at its center and the two move together.

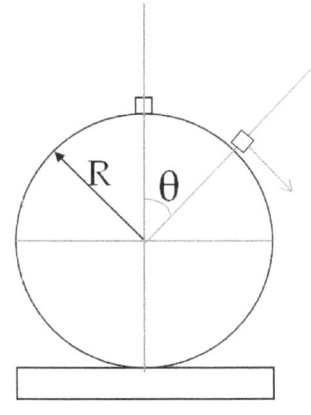

Figure 8.29: Problem 8.8.4.

Find the angle θ the string will make with the vertical when the bullet/block system comes to rest. Give your answer in terms of quantities given and g. Ans: $\cos\theta = \left(1 - \frac{m^2v^2}{2gL(m+M)^2}\right)$.

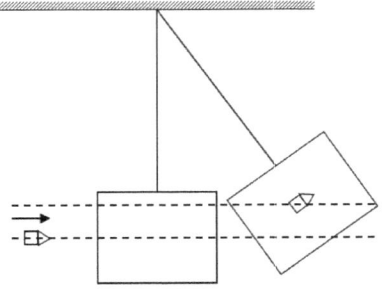

Figure 8.30: Problem 8.8.5.

Problem 8.8.6. A particle of mass m and speed v_0 collides head on with another particle of mass M and an unknown speed v. After the collision, particle m travels with speed $\frac{3}{4}v_0$ in the vertical direction to its original path and M moves at angle ϕ with respect to the original path of m but with an unknown speed v'. Find v and v' in terms of m, M, ϕ and v_0. Ans: $v = \left(\frac{mv}{M}\right)\left(1 - \frac{3}{4}\cot\phi\right)$; $v' = \frac{3mv}{4M}\cosec\phi$. Answer check: For $m = M$ and $\phi = 45°$, $v = v_0/4$ and $v' = (3\sqrt{2}/4)v_0$.

Problem 8.8.7. A ball bounces off the floor inelastically so that each bounce takes the ball to a lower height than the previous bounce. Eventually the ball comes to rest. (a) If the coefficient of restitution is 0.95, what is the speed of the ball immediately after the first bounce if it was dropped at rest from a height of 3 meters? (b) How long does it take for the ball to come to rest? Ans: (a) 7.3 m/s; (b) 30.6 sec.

Problem 8.8.8. Two boys, Jack and Majid, want to compare powers of their muscles by pulling up a bucket of water of mass M hanging from a frictionless pulley. Each pulls on the cord with a constant force and measures the time for the bucket to rise by a height h meter. (a) Find an expression of the average power delivered in terms of force $\vec{F}$ applied, mass M of the bucket, height h, and the acceleration due to gravity g. (b) With what force the cord must be pulled so that an average 1 horse power can be supplied in raising the bucket by 2 meters? Ans: (a) $F\sqrt{\frac{h}{2}\left(\frac{F}{M} - g\right)}$, (b) 217 N.

Problem 8.8.9. A pendulum bob of mass 0.250 kg is hanging from a ceiling with a 150 cm long massless cord (Fig. 8.31). When the pendulum is let go from rest at 60° from the vertical, it strikes a block of mass 0.500 kg when the pendulum is vertical. After the collision the block moves on a horizontal surface with decreasing speed due to the kinetic friction with the coefficient of kinetic friction equal to 0.2. If the collision between the pendulum bob and the block is elastic, find (a) the place the block comes to rest, and (b) the angle of maximum swing of the pendulum afterwards.

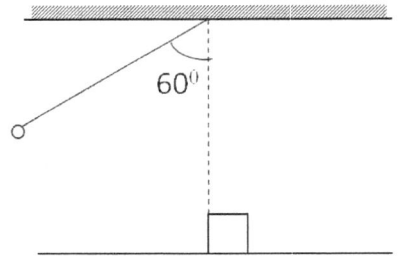

Figure 8.31: Problem 8.8.9.

Problem 8.8.10. A ring of mass M and radius R has two metal beads, each of mass m and negligible radius (Fig. 8.32). The beads can slide frictionlessly on the ring. The ring with the beads is hung from the ceiling with a string and the beads are brought at the top and released from rest. The beads then slide off on the two sides in a symmetric way, and when the beads reach a particular points on the ring, the ring tends to go up. Find this unique place and any conditions on the masses m and M for this phenomenon to occur.

Problem 8.8.11. A mysterious force acts on all particles along a particular line and always points towards a particular point (P) on the line.

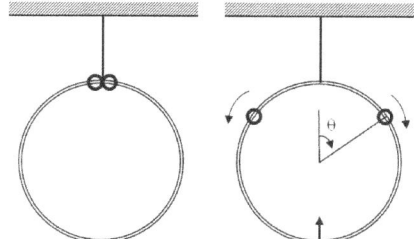

Figure 8.32: Problem 8.8.10.

The magnitude of the force on a particle increases as the cube of the distance from that point, that is, $F \propto r^3$, if the distance from the P to the position of the particle is r. Let b be the proportionality constant, and write the magnitude of the force as $F = br^3$. Find the potential energy of a particle subjected to this force when the particle is at a distance D from P assuming the potential energy to be zero when the particle is at P. Ans: $bD^4/4$.

Chapter 9

FIXED-AXIS ROTATION

Contents

9.1	KINEMATICS OF PURE ROTATION	352
9.2	ANGULAR MOMENTUM	363
	9.2.1 Angular Momentum of a Point Particle	363
	9.2.2 Angular Momentum of an Extended Body and the Moment of Inertia	366
	9.2.3 Calculations of Moments of Inertia	369
9.3	DYNAMICS OF FIXED-AXIS ROTATION	379
	9.3.1 Rotational Dynamics of a Single Particle	380
	9.3.2 Rotational Dynamics of Extended Bodies	381
	9.3.3 The Law for Fixed Axis Rotation	382
	9.3.4 Practice With Torque Calculations	382
	9.3.5 Example Problems - Single Rigid Bodies	383
	9.3.6 Example Problems - Coupled Systems	386
9.4	CONSERVATION OF ANGULAR MOMENTUM	391
	9.4.1 For One Particle	391
	9.4.2 Conservation of Angular Momentum For Extended Bodies	393
9.5	ROTATIONAL WORK AND KINETIC ENERGY	396
9.6	ROLLING MOTION IN A STRAIGHT LINE	397
9.7	EXERCISES	403
9.8	PROBLEMS	411

Rotation is an ubiquitous phenomena. In this chapter we will study fixed-axis rotation. A fixed-axis rotation is a rotation in which the direction of the axis of rotation is fixed in space. This covers a large variety of situations. For instance, if a wheel is rotating about a fixed axle, the rotation axis is fixed in space. The tires of a car also have a fixed axis of rotation as long as the car moves in a straight line - in this case, although the tire moves and the axis moves in space, the direction of the axis does not change as long as the tire is moving in a straight line. If the car takes a turn, then the direction of the axis changes, and we no longer have a fixed-axis rotation.

9.1 KINEMATICS OF PURE ROTATION

Angle and Axis of Rotation

The most basic aspects of a rotating body are the axis of rotation and the amount of rotation about that axis. Look at a rotating body and you will find that the particles of the body go around in circles about a line in space. The line about which the body rotates is called the axis of rotation. A rotating body changes orientation with respect to some fixed direction in space and therefore, we use an angle to measure the amount of rotation. The angle between a line from the axis to a point in the body and a fixed direction in space changes as the body rotates as shown in Fig. 9.1. Therefore, we use this angle as a quantitative measure of the

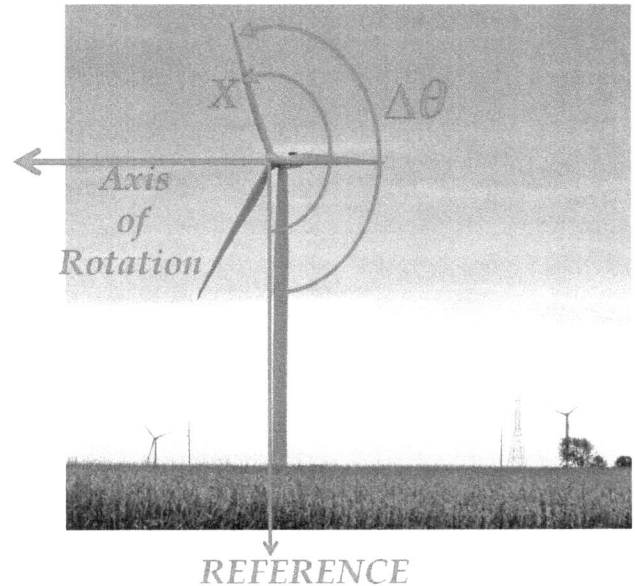

Figure 9.1: The axis and the angle of rotation. The angle of rotation is defined by the angle between a line from the axis of rotation to a point in the body and a fixed reference direction, which is shown as a vertical line in this figure. Since, every particle of a rigid body rotates at the same rate, you need only one angle to describe the rotation of the whole body.

9.1. KINEMATICS OF PURE ROTATION

orientation of the entire body in space and describe the rotation of the body in terms of this angle.

Often we assign a sense of rotation as being clockwise or counterclockwise. When you observe the rotating body from one side of the axis, the body appears to rotate in a clockwise or a counterclockwise sense. Of course, if you look at the same body from the other side the sense of rotation will be opposite. A right-hand rule given in Fig. 9.2 illustrates the direction from which we look when we state the sense of rotation. This is the same rule as the rule for a right-handed Cartesian coordinate system. Thus, if the rotation happens about the z-axis of a coordinate system in such a way that particles of the body go from positive x-axis towards positive y-axis, then the sense of rotation will be counterclockwise as observed from the side of positive z-axis.

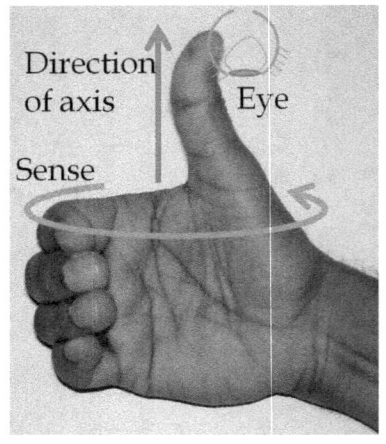

Figure 9.2: Right-hand rule. Sense of rotation and direction of the axis of rotation.

The rotation angle can be expressed in various units as shown in Fig. 9.3. The most common ones being, the number of revolutions, degrees, arc-minutes, arc-seconds, and radians. We will be mostly using radians to express the rotation angle. For instance, if a body goes around five times then we say that it has rotated by 5 revolutions or 10π radians or $1800°$.

Angular Displacement

The angular displacement $\Delta\vec{\theta}$ in a time interval Δt is defined as a vector whose magnitude is the angle of rotation and whose direction is the direction of the axis of rotation as given by the right-hand rule given in Fig. 9.2. Rather than denote the direction as along the axis or opposite to the axis, we also use the language of clockwise and counterclockwise senses of rotation using the right-hand rule.

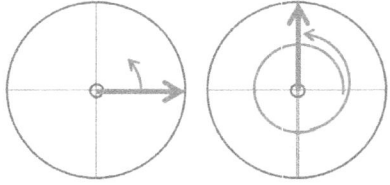

Figure 9.3: Angle of rotation = 1.25 rev, or $450°$, or $\frac{9}{4}\pi$ rad.

$$\Delta\vec{\theta} = \begin{cases} \text{magnitude} = \Delta\theta, \\ \text{direction} = \text{towards axis using right-hand rule.} \end{cases} \quad (9.1)$$

When you use the direction of the axis of rotation as the direction of the angular displacement vector, then you can make use of an analytic description of rotation just as for other vectors we have studied in this book. Suppose the axis of rotation is pointed in an arbitrary direction with respect to a fixed Cartesian axes (see Fig. 9.4). Using the Cartesian axes, the angular displacement vector $\Delta\vec{\theta}$ pointed along the axis of rotation can be decomposed into the x, y and z-components similar to other vectors in the three-dimensional space.

$$\Delta\vec{\theta} = \Delta\theta_x \hat{u}_x + \Delta\theta_y \hat{u}_y + \Delta\theta_z \hat{u}_z. \quad (9.2)$$

Another way to write the displacement vector is to use a unit vector $\hat{u}$ in the direction of the axis of rotation. The product of the magnitude of the angular displacement $\Delta\theta$ and the unit vector denotes the angular

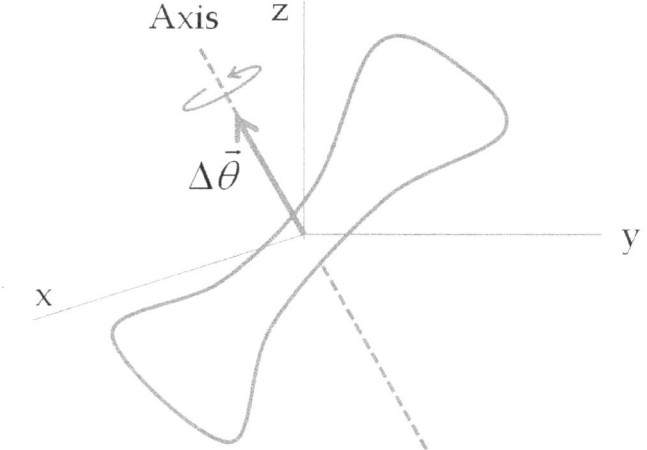

Figure 9.4: Angular displacement vector and Cartesian axes.

displacement vector.

$$\Delta\vec{\theta} = \Delta\theta\hat{u}, \qquad (9.3)$$

The two analytic forms given in Eqs. 9.2 and 9.3 are, of course, equivalent, although, the later gives the magnitude and direction of the vector directly.

The magnitude and direction of the angular displacement vector can be determined from its components in Eq. 9.2 as we have learned to do for other vectors. Thus, the magnitude of the angular displacement vector, which is equal to the angle of rotation about the axis, is related to the components as follows.

$$\text{Magnitude:} \quad |\Delta\theta| = \sqrt{\Delta\theta_x^2 + \Delta\theta_y^2 + \Delta\theta_z^2}. \qquad (9.4)$$

The direction of $\Delta\vec{\theta}$ is obtained from the angle the vector makes with respect to the Cartesian axes. If only one component is non-zero, we say that the situation is one-dimensional, and the sign of the component tells us the direction with respect to the axis for which the component is non-zero. If only two components are non-zero, then the situation is called two-dimensional, and the angle with respect to any one of the two axes can be worked out to determine the direction of the vector. If all three components are non-zero, then we need two angles in space to tell the direction - the two angles of a spherical coordinate system is often used for this purpose, although other choices are also possible.

For fixed-axis rotation, we can orient the fixed Cartesian axes so that one of the axes, usually the z-axis points in the direction of the axis of rotation. With the axis of rotation pointed towards the positive z-axis, the displacement vector has only the z-component non-zero. This simplifies the mathematical treatment of rotation considerably.

$$\text{Fixed-axis along } z\text{-axis:} \quad \Delta\vec{\theta} = \Delta\theta_z\hat{u}_z. \qquad (9.5)$$

Thus, the absolute value of $\Delta\theta_z$ is the angle rotated and the sign of $\Delta\theta_z$ together with the unit vector $\hat{u}_z$ gives us the direction of the rotation.

9.1. KINEMATICS OF PURE ROTATION

If the rotation is counterclockwise as seen from the positive z-axis, then $\Delta\theta_z > 0$ and $\Delta\vec{\theta}$ will point towards the positive z-axis. Similarly, the clockwise rotation corresponds to the negative z-component. Since, a fixed axis rotation is only one-component problem, we often draw analogy to the one-component problems of translational motion. Also, many authors drop the vector character of the angular displacement altogether and treat the problems of fixed-axis rotation in a non-vector language. We will not use that language and continue to use the vector language since angular displacement, whether one-component type or three-components type are at the core vector quantities.

Angular Speed

In a rotation of a rigid body about a fixed axis, every particle covers the same amount of angle in the same amount of time. Therefore, the rate of angle covered by each particle in their own circle of motion is a characteristic of the entire rigid body. This rate is called the angular speed of rotation.

As explained above, we measure the angle of rotation in a plane perpendicular to the axis of rotation. Let $\Delta\theta$ be the angle rotated in a time duration Δt, then the average angular speed ω_{ave} is given by

$$\boxed{\text{Average angular speed:} \quad \omega_{\text{ave}} = \frac{\Delta\theta}{\Delta t}.} \qquad (9.6)$$

The commonly used units for angular speed are revolutions per minute (rpm), degrees per second (deg/sec), and radian per second (rad/sec). We will mostly used rpm and rad/sec.

In the limit of infinitesimal durations, we obtain the instantaneous speed ω.

$$\boxed{\text{Instantaneous angular speed:} \quad \omega = \frac{d\theta}{dt}.} \qquad (9.7)$$

Example 9.1.1. Steady rotation of a computer hard disk.

A computer disk is rotating at a constant rate of 7200 rpm, where rpm stands for revolutions per minute. What is the angle in radians rotated in 2 msec?

Solution. Since the rate of rotation is constant, the angle rotated will simply be equal to the product of the speed and time. We do not need to place the axis of rotation along any Cartesian axis. We can work with the magnitudes alone in this problem. In this problem we need to convert units of the speed of rotation from revolutions per minute to radians per second.

$$\omega = 7200 \frac{\text{rev}}{\text{min}} \times \frac{2\pi \text{ rad}}{1 \text{ rev}} \times \frac{1 \text{ min}}{60 \text{ sec}} = 754 \text{ rad/sec}.$$

Now, we multiply the angular speed by the time (converted in sec) to obtain the angle rotated.

$$\Delta\theta = \omega t = 754 \frac{\text{rad}}{\text{sec}} \times 2 \times 10^{-3} \text{ sec} = 1.51 \text{ rad}.$$

Angular Velocity

The rate of change of angular displacement is called angular velocity. Thus, average angular velocity $\vec{\omega}_{ave}$ during an interval Δt in which angular displacement is $\Delta\vec{\theta}$ is given by the ratio of angular displacement to the interval.

$$\boxed{\vec{\omega}_{ave} = \frac{\Delta\vec{\theta}}{\Delta t}.} \tag{9.8}$$

The direction of the angular velocity is the same as the direction of the angular displacement. Thus, if the displacement vector is 200 rad in 20 sec in the counterclockwise sense, then the average velocity during this interval is 10 rad/sec in the counterclockwise sense.

In the limit of infinitesimal time interval, we obtain the instantaneous angular velocity $\vec{\omega}$ as the time derivative of the angular position vector, which is the angular displacement with respect to a fixed direction in a plane perpendicular to the axis of rotation.

$$\boxed{\vec{\omega} = \frac{d\vec{\theta}}{dt}.} \tag{9.9}$$

The magnitude of instantaneous angular velocity is equal to the angular speed of rotation and the direction is the direction of the axis of rotation as determined from using the right-had rule stated above.

Instantaneous angular velocity $\vec{\omega}$:

$$\boxed{\begin{array}{l}\text{Magnitude} = \text{Angular speed.} \\ \text{Direction} = \text{Towards the axis using the right-hand rule.}\end{array}} \tag{9.10}$$

Analytically, the angular velocity vector is treated in the same way as we have described for the angular displacement vector above. Suppose a unit vector $\hat{u}$ points towards the axis of rotation and the angular speed is ω, then analytically we write the angular velocity vector as

$$\vec{\omega} = \omega\hat{u}, \tag{9.11}$$

We can also express the angular velocity vector which is pointed in the direction of the axis of rotation in terms of its components along Cartesian axes.

$$\vec{\omega} = \omega_x \hat{u}_x + \omega_y \hat{u}_y + \omega_z \hat{u}_z. \tag{9.12}$$

The angular speed, ω, is as usual the magnitude of angular velocity vector,

$$\omega = \sqrt{\omega_x^2 + \omega_y^2 + \omega_z^2}. \tag{9.13}$$

9.1. KINEMATICS OF PURE ROTATION

Figure 9.5: Merry-go-round. Credits: Jebulon, commons.wikimedia.org

For fixed-axis rotation, if we choose to orient the z-axis along the axis of rotation, then only the z-component will be non-zero.

$$\text{Fixed axis along } z\text{-axis:} \quad \vec{\omega} = \frac{d\theta_z}{dt} \hat{u}_z \equiv \omega_z \hat{u}_z, \quad (9.14)$$

where the z-component of the angular velocity vector is

$$\boxed{\omega_z = \frac{d\theta_z}{dt}.} \quad (9.15)$$

Equation 9.15 can be inverted giving the z-component of the angular displacement in terms of integration of the z-component of the angular velocity.

$$\theta_z(t) - \theta_{0z} = \int_0^t \omega_z(t) dt, \quad (9.16)$$

where $\theta_z(t) - \theta_{0z}$ is the z-component of the angular displacement.

Example 9.1.2. A merry-go-round.

A merry-go-round is rotating about an axis that is pointed vertically up. The line from the center of the merry-go-round to a rider makes an angle ϕ with respect to a line pointing East from the center of the merry-go-round. The angle ϕ changes with time according to $\phi(t) = bt^2$, where b is a positive constant. What is the angular velocity at $t = T$?

Solution. Using Cartesian coordinates whose z-axis is pointed up we see that the angle $\phi(t)$ gives the z-component of angular displacement of the rotation of the Merry-go-round. Therefore, the z-component of angular velocity is obtained by taking the derivative of $\phi(t)$ with respect to time. For the fixed-axis rotation described in the problem statement, the other components of the angular velocity are zero.

$$\omega_z = \frac{d\theta_z}{dt} = \frac{d\phi}{dt} = 2bt,$$

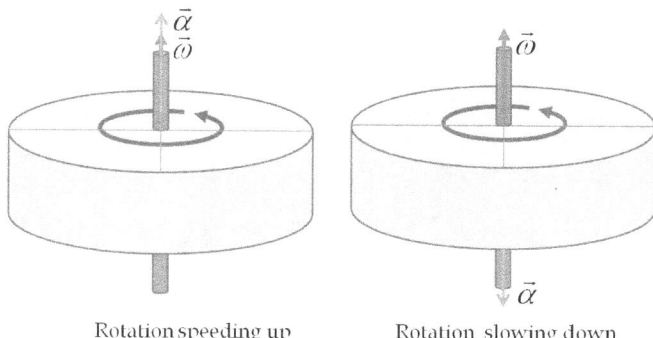

Figure 9.6: The angular acceleration and angular velocity may be in different directions. When the angular acceleration is in the same direction as the angular velocity, the rotational speed increases. When the angular acceleration is in the opposite direction to that of the angular velocity, the rotational speed decreases. In the later case, body will pick the rotational speed after the body has stopped rotating, but the rotation velocity thereafter will be in the same direction as the angular acceleration.

which is evaluated for $t = T$ to give $\omega_z = 2bT$. Therefore, the angular velocity has the magnitude $2bT$ and the direction is pointed up towards the z-axis as shown in the figure.

Angular Acceleration

The rate of change of angular velocity vector defines the angular acceleration of the body. Thus, if the angular velocity changes by $\Delta\vec{\omega}$ in a time interval Δt, the average angular acceleration $\vec{\alpha}_{ave}$ will be

$$\boxed{\vec{\alpha}_{ave} = \frac{\Delta\vec{\omega}}{\Delta t}.} \qquad (9.17)$$

When time duration is infinitesimal we obtain the instantaneous angular acceleration $\vec{\alpha}$.

$$\boxed{\text{Angular acceleration vector,} \quad \vec{\alpha} = \frac{d\vec{\omega}}{dt}.} \qquad (9.18)$$

Angular acceleration $\vec{\alpha}$ may or may not be in the same direction as the angular velocity $\vec{\omega}$. Suppose the axis of rotation is fixed and the body is rotating in a counterclockwise sense with increasing speed, then the angular acceleration would also be in the counterclockwise sense as shown in Fig. 9.6. However, if the body is rotating counterclockwise with decreasing speed, then the angular acceleration will be in the clockwise sense.

Analytically, we represent angular acceleration with the components along Cartesian axes. In the case of fixed axis rotation about the z-axis, only the z-component of the angular acceleration will be non-zero.

$$\vec{\alpha} = \alpha_z \hat{u}_z, \qquad (9.19)$$

9.1. KINEMATICS OF PURE ROTATION

where z-component of angular acceleration is the rate of change of the z-component of the angular velocity.

$$\alpha_z = \frac{d\omega_z}{dt}. \qquad (9.20)$$

The angular acceleration can also be written in terms of the angular displacement by replacing the z-component of the angular velocity by the time derivative of the z-component of the angular position vector, which is the angle rotated from the positive x-axis.

$$\vec{\alpha} = \frac{d^2\theta_z}{dt^2}\hat{u}_z. \qquad (9.21)$$

Equation 9.20 can be inverted giving the change in the z-component of the angular velocity in terms of integration of the z-component of the angular acceleration.

$$\boxed{\omega_z(t) - \omega_{0z} = \int_0^t \alpha_z(t)dt,} \qquad (9.22)$$

where $\omega_z(t)$ and ω_{0z} are z-component of angular velocity at time $t = t$ and $t = 0$ respectively.

Constant Angular Acceleration

Just as constant acceleration was an important case of study for the translational motion, constant angular acceleration case also plays an important role in the study of the rotational motion. Once again, let z-axis to be the axis of rotation, and to emphasize constant angular acceleration, we place a bar over the symbol for the z-component of the angular acceleration.

$$\boxed{\vec{\alpha}(t) = \bar{\alpha}_z \hat{u}_z} \qquad (9.23)$$

For constant angular acceleration, we expect the angular velocity to change linearly in time, which is validated by replacing $\alpha_z(t)$ by $\bar{\alpha}_z$ in Eq. 9.22 and carrying out the integration.

$$\boxed{\omega_z(t) = \omega_{0z} + \bar{\alpha}_z t} \qquad (9.24)$$

Now, using this expression for ω_z in Eq. 9.16, and carrying out the integration gives the following angular displacement.

$$\boxed{\theta_z(t) = \theta_{0z} + \omega_{0z}t + \frac{1}{2}\bar{\alpha}_z t^2} \qquad (9.25)$$

Equations 9.24 and 9.25 are the two main equations for the analysis of the constant angular acceleration motion. Just as in one-dimensional constant acceleration translational motion, we can eliminate time t from these two equations to obtain the following useful relation.

$$\boxed{\omega_z(t)^2 = \omega_{0z}^2 + 2\bar{\alpha}_z\left[\theta_z(t) - \theta_{0z}\right]} \qquad (9.26)$$

This equation does not have any information that is not already contained in Eqs. 9.24 and 9.25, but sometimes provides a short-cut way of solving some problems.

Table 9.1: Constant angular acceleration, $\alpha_z = \bar{\alpha}_z$

$\omega_z = \omega_{0z} + \bar{\alpha}_z t$
$\theta_z = \theta_{0z} + \omega_{0z}t + \frac{1}{2}\bar{\alpha}_z t^2$
$\omega_z^2 = \omega_{0z}^2 + 2\bar{\alpha}_z(\theta_z - \theta_{0z})$

Example 9.1.3. Constant acceleration rotation A wheel is mounted on an axle and placed in a support so that the wheel can rotate about an axis through the axle. The wheel is then rotated from rest with a uniform angular acceleration of 3 rad/sec^2 pointed so that the wheel rotates counterclockwise. (a) Find the angular velocity at t = 10 sec. (b) Find the total rotation in 10 sec.

Solution. Since the angular acceleration is constant, we can use the constant angular acceleration formulas. From the right-hand rule, we know that, if z-axis is the axis of rotation, then the counterclockwise corresponds to the direction of the vector towards positive z-axis. We will work with the components of the vectors in this problem.

(a) The z-component of the constant angular acceleration is $\alpha_z = 3$ rad/sec^2. The z-component of the angular velocity changes linearly for a constant angular acceleration. Therefore, we find the z-component of the angular velocity to be

$$\omega_z = \omega_{0z} + \bar{\alpha}_z t = 0 + 3 \frac{\text{rad}}{\text{sec}^2} \times 10 \text{ sec} = 30 \frac{\text{rad}}{\text{sec}}.$$

Since, other components of the angular velocity are zero, the magnitude of the angular velocity would be 30 rad/sec and the direction would be towards positive zR axis, or counterclockwise sense.

(b) The z-component of the net rotation angle in the given time is also readily obtained from the given formulas for the constant angular acceleration case.

$$\Delta \theta_z = \omega_{0z} t + \frac{1}{2} \bar{\alpha}_z t^2 = 0 + \frac{1}{2} 3 \frac{\text{rad}}{\text{sec}^2} \times (10 \text{ sec})^2 = 150 \text{ rad}.$$

This gives the total rotation angle to be 150 rad, which is approximately 24 revolutions.

Example 9.1.4. Constant acceleration rotation. A bicycle wheel rotates by an angle 2000 radians in 5 sec before coming to rest with a fixed axis of rotation. Assuming constant angular acceleration, what is the angular acceleration?

Solution. Just as the last example, we again work with the components of vectors along the axis of rotation. By using the right-hand rule on the initial motion, let us define the positive direction of the z-axis to be along the direction of the initial angular velocity. From the given information, we can set up the following two equations since the final angular velocity is zero.

$$0 = \omega_{0z} + \bar{\alpha}_z t$$
$$\Delta \theta_z = \omega_{0z} t + \frac{1}{2} \bar{\alpha}_z t^2$$

In these equations, we have only two unknowns, ω_{0z} and $\bar{\alpha}_z$. We can solve for ω_{0z} in the first equation and then substitute in the second equation.

9.1. KINEMATICS OF PURE ROTATION

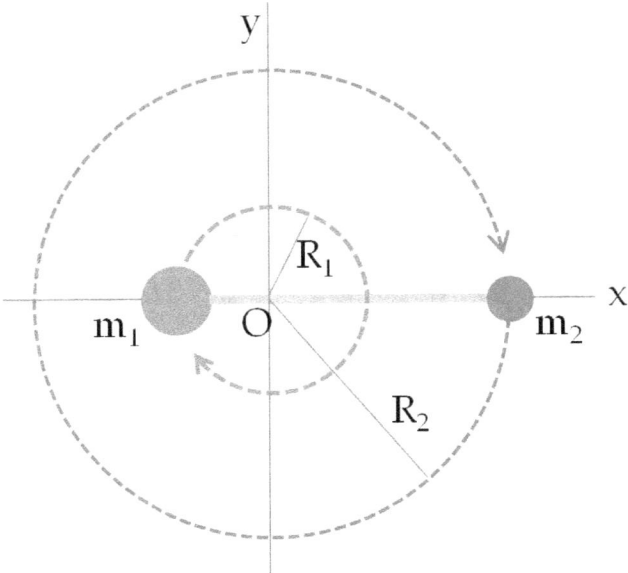

Figure 9.7: Two masses of a dumb-bell move in different circles when the dumb-bell rotates about an off-center point. The spatial speeds of masses are $v_1 = R_1\omega$ and $v_2 = R_2\omega$ if angular speed ω in rad/sec.

The resulting equation can be solved for $\bar{\alpha}_z$. Once we have solved for $\bar{\alpha}_z$ we can put the numerical values to find the value of the z-component of the angular acceleration. At that point we will state the magnitude and direction of this vector. From the first equation, $\omega_{0z} = -\bar{\alpha}_z t$, therefore,

$$\Delta\theta_z = -\bar{\alpha}_z t^2 + \frac{1}{2}\bar{\alpha}_z t^2 = -\frac{1}{2}\bar{\alpha}_z t^2.$$

This gives

$$\bar{\alpha}_z = -2\frac{\Delta\theta_z}{t^2} = -2 \times \frac{2000 \text{ rad}}{(5 \text{ sec})^2} = -160 \text{ rad/sec}^2.$$

Since the other components of the angular acceleration are zero, the negative sign of the z-component of the angular acceleration means that the angular acceleration is pointed towards the negative z-axis. Recall that the initial velocity was used to define the direction of the positive z-axis. Therefore, angular acceleration has the magnitude 160 rad/sec^2 and has the direction opposite to the direction of the initial angular velocity.

Spatial Velocity Versus Angular Velocity

You know that particles of a rotating rigid body move in circles about the axis of rotation. This is shown schematically in Fig. 9.7 for two masses of a dumb-bell that rotate about an axis through some point O in the rod that joins the two masses.

The mass m_1 moves in a circle of radius R_1 and the mass m_2 in a circle of radius R_2. In one revolution, the two masses move different distances,

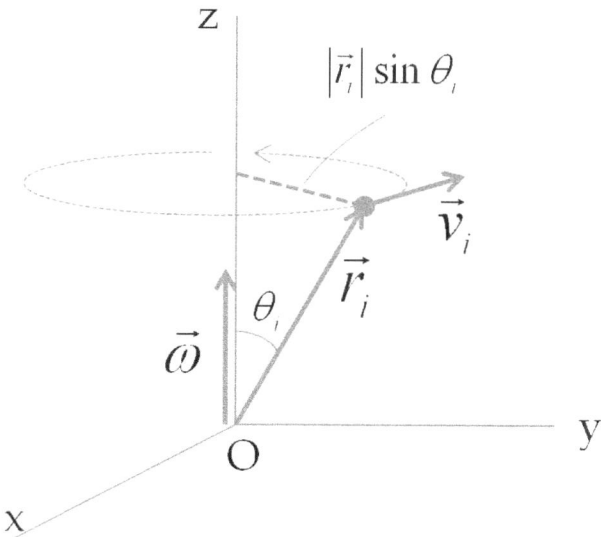

Figure 9.8: The spatial velocity $\vec{v}_i$ of a particle located at position $\vec{r}_i$ is equal to the vector product of the angular velocity $\vec{\omega}$ and the position vector, $\vec{v}_i = \vec{\omega} \times \vec{r}_i$.

one moves a distance equal to $2\pi R_1$ and the other $2\pi R_2$. Therefore, although, the two masses possess the same angular speed, they have different "real" or "spatial" speeds.

Let the common angular speed of the two masses in Fig. 9.7 be ω in radians/sec. Then, the time for one revolution T in sec will be

$$T = \frac{2\pi}{\omega}.$$

Now, we can deduce the spatial speeds v_1 and v_2 of the masses in terms of the radii R_1 and R_2 and the common angular speed ω.

$$v_1 = \frac{2\pi R_1}{T} = R_1 \omega$$
$$v_2 = \frac{2\pi R_2}{T} = R_2 \omega$$

The relation between the position and spatial velocity of a point of a rotating body are more generally related by a cross product as we show now. Consider a point particle at position $\vec{r}_i$ which is located at an angle θ_i from the axis of rotation, which will be taken to be the z-axis as shown in Fig. 9.8.

The point particle will move in a circle of radius $|\vec{r}_i| \sin \theta_i$ as shown. The physical distance covered by the particle will be equal to the product of the angle of rotation, expressed in radians, and the radius of this circle, as given from the arc-length and angle formula for a circle.

$$\boxed{\text{Arc length} = \text{Radius} \times \text{Angle in radians.}} \qquad (9.27)$$

Therefore, the magnitude of the spatial velocity will be equal to the product of the magnitude of the angular velocity and the radius of the circle of

9.2. ANGULAR MOMENTUM

rotation, and the direction of the spatial velocity will be in the direction of the tangent to the circle at that point in time.

Instantaneous spatial velocity $\vec{v}_i$:

$$\boxed{\begin{array}{l}\text{Magnitude} = \omega|\vec{r}_i|\sin\theta_i \\ \text{Direction} = \text{Tangent to the circle.}\end{array}} \qquad (9.28)$$

We can write the definition of the spatial velocity of the particle at $\vec{r}_i$ more compactly using the vector product notation.

$$\boxed{\vec{v}_i = \vec{\omega} \times \vec{r}_i.} \qquad (9.29)$$

9.2 ANGULAR MOMENTUM

9.2.1 Angular Momentum of a Point Particle

We start with the simplest case of a motion of a point particle of mass m. Let $\vec{r}$ be the position of the particle with respect to an origin of a given coordinate system, and $\vec{p}$ the momentum. The angular momentum $\vec{L}$ with respect to the origin is defined by the cross product of $\vec{r}$ and $\vec{p}$.

$$\boxed{\vec{L} = \vec{r} \times \vec{p}.} \qquad (9.30)$$

From this definition we see that the unit of angular momentum in meter-kg-sec system will be $kg.m^2/s$.

Example 9.2.1. Angular momentum of a particle moving in a circle. To get a feel for the angular momentum, we evaluate the angular momentum of particle of mass m that moves with a speed v in a circle of radius R shown in Fig. 9.9. Let the origin be at the center of the

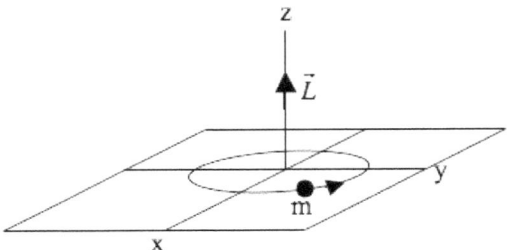

Figure 9.9: Example 9.2.1. The angular momentum of a particle moving in a circle is pointed perpendicular to the plane of the circle and has the magnitude $L = mvR$.

circle and the circle be in the xy-plane of a Cartesian coordinate system as shown in Fig. 9.9. Since the particle moves in a circle, the position vector $\vec{r}$ is perpendicular to the momentum vector $\vec{p}$, hence the magnitude of the angular momentum is

$$\text{Magnitude, } L = (R)(mv)(\sin 90°) = mvR$$

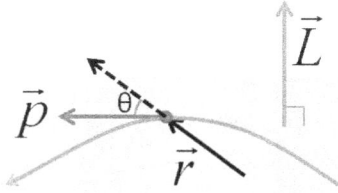

Figure 9.10: The direction of $\vec{L}$ is perpendicular to both $\vec{r}$ and $\vec{p}$ as given by the right-hand rule. If $\vec{r}$ and $\vec{p}$ are collinear, then the angular momentum is zero. When $\vec{r}$ and $\vec{p}$ are non-collinear, then they define a plane, and angular momentum is perpendicular to this plane.

The direction is obtained by using the right-hand rule for the cross-product. This gives the direction of angular momentum along the positive z-axis in the figure.

The rules for the cross-product tell us about the magnitude and direction of the angular momentum. The direction is given by applying the right-hand rule of cross-product on the vectors $\vec{r}$ and $\vec{p}$ as shown in Fig. 9.10.

Using the geometric definition of the cross product we can write angular momentum in terms of the magnitudes of the position and the momentum vectors and the angle θ between them.

$$\vec{L} = \vec{r} \times \vec{p} = \begin{cases} \text{Magnitude} = rp\sin\theta \\ \text{Direction: Use Right Hand rule} \end{cases} \quad (9.31)$$

The analytic method for evaluating a cross-product uses the decomposition of the two vectors into their Cartesian components and then computing the determinant.

$$\begin{aligned} \vec{L} = \vec{r} \times \vec{p} &= \begin{vmatrix} \hat{u}_x & \hat{u}_y & \hat{u}_z \\ x & y & z \\ p_x & p_y & p_z \end{vmatrix} \\ &= \hat{u}_x (yp_z - zp_y) + \hat{u}_y (zp_x - xp_z) + \hat{u}_z (xp_y - yp_x). \end{aligned} \quad (9.32)$$

How would you obtain the angular momentum of a multiparticle system? Well, you just add up the angular momenta of all particles vectorially to obtain the angular momentum of the whole. For example, the angular momentum of a system consisting of N particles will be

$$\boxed{\vec{L} \equiv \vec{L}_{\text{net}} = \sum_{i=1}^{N} \vec{L}_i = \sum_{i=1}^{N} (\vec{r}_i \times \vec{p}_i).} \quad (9.33)$$

Note the role of the reference point and reference frame in the definition of the angular momentum. Since the position vector $\vec{r}$ is measured from a reference point O, the position vector will change if you pick another

9.2. ANGULAR MOMENTUM

reference point, thereby changing $\vec{L}$. Similarly, since the momentum depends on the choice of the reference frame, the angular momentum will also depend on the choice of a reference frame.

Example 9.2.2. Angular momentum of a conical pendulum. A pendulum bob of mass m and length b can be rotated about the pivot so that the bob moves in a circular path with an angle of suspension θ as shown in Fig. 9.11. This arrangement is called a **conical pendulum**. Find the angular momentum of the pendulum (a) about the center P of the circle, and (b) about the pivot point O.

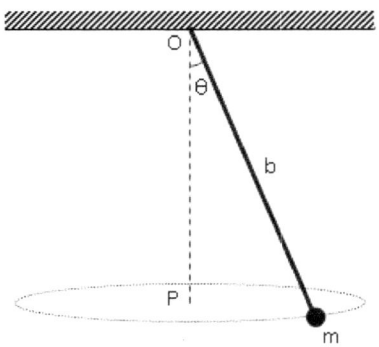

Figure 9.11: Example 9.2.2.

Solution. Since the angular momentum depends on the reference point we will organize the answer in the two parts as follows.

(a) $\vec{L}$ about P

To find the angular momentum about point P we need the position and momentum vectors with respect to P. The position vector of m from point P has the magnitude of the radius of the circle and a direction radially outward in the plane of the circle. The momentum of m has the magnitude mv and the direction tangent to the circle. Therefore, the position vector is perpendicular to the momentum vector. That means that the magnitude of the angular momentum will be equal to product of the radius R of the circle and the momentum mv. Since the radius of the circle about P is $R = b\sin\theta$, we have

The magnitude of $\vec{L}$ about P $= mvR = mvb\sin\theta$.

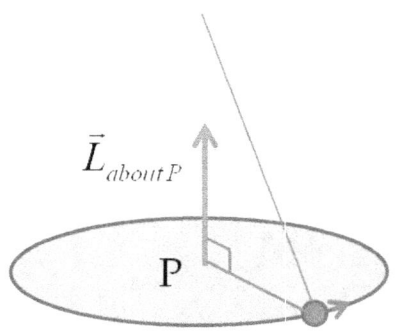

Figure 9.12: Example 9.2.2. The direction of the angular momentum about P.

The direction of the angular momentum is obtained by using the right-hand rule on vectors $\vec{r}$ and $\vec{p}$. This gives the direction of $\vec{L}$ to be perpendicular to the plane of the circle, pointed up as shown in Fig. 9.12.

(b) $\vec{L}$ about O

The position vector now has a magnitude b and pointed in the direction from O to m. Since the direction of the momentum is perpendicular to the plane containing points O, P, and m in the figure, the angle between the position vector and the momentum vector is 90°. The magnitude of angular momentum will therefore be simply the product of the magnitudes of position and momentum vectors. The position of the particle with respect to the reference point O has magnitude b now. Therefore, the angular momentum about O has the following magnitude.

The magnitude of $\vec{L}$ about O $= mvb$.

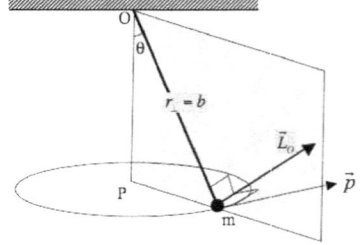

Figure 9.13: Example 9.2.2.

The direction of the angular momentum vector is again obtained by applying the right-hand rule to the position and the momentum vectors. The resulting angular momentum vector would be in the plane containing the points O, P, and m, in the direction shown in Fig. 9.13. Make sure that the angular momentum vector is perpendicular to both the position vector and the momentum vector.

9.2.2 Angular Momentum of an Extended Body and the Moment of Inertia

The angular momentum of an extended body can be constructed from the sum of angular momenta of particles that make up the body. Suppose an extended body is made up of N masses, $m_1, m_2, \cdots, m_N$, whose positions with respect to a reference point O be $\vec{r}_1, \vec{r}_2, \cdots, \vec{r}_N$, respectively, and the spatial velocity be $\vec{v}_1, \vec{v}_2, \cdots, \vec{v}_N$, respectively. The angular momentum

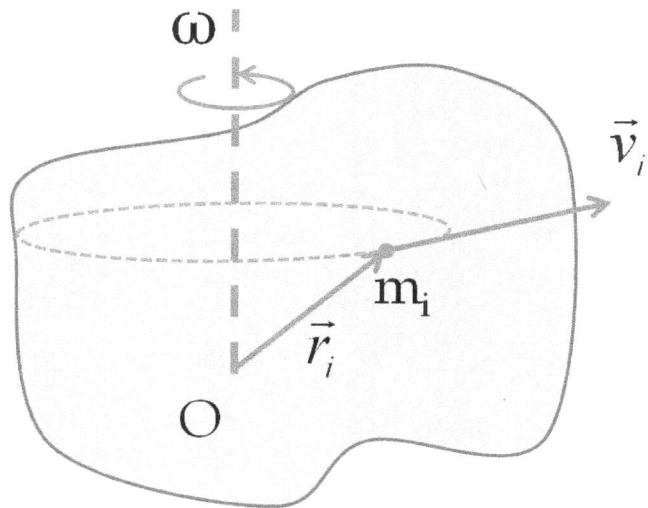

Figure 9.14: Angular momentum of an extended body is obtained by a sum of angular momenta of its parts. For mass m_i, the angular momentum is $\vec{L}_i = \vec{r}_i \times m\vec{v}_i$.

of ith mass $\vec{L}_i$ about point O can be written using the point mass formula given above.

$$\vec{L}_i = \vec{r}_i \times \vec{p}_i = \vec{r}_i \times m\vec{v}_i = m\vec{r}_i \times \vec{v}_i \quad \text{[About O]} \tag{9.34}$$

Now, we sum up vectors $\vec{L}_i$ for each particle to get the angular momentum $\vec{L}$ of the body as a whole about point O.

$$\vec{L} = \sum_{i=1}^{N} \vec{L}_i = \sum_{i=1}^{N} (m_i \vec{r}_i \times \vec{v}_i) \quad \text{[About O]} \tag{9.35}$$

Each particle's spatial velocity can also be expressed in terms of the angular velocity of the body using Eq. 9.29.

$$\boxed{\vec{L} = \sum_{i=1}^{N} [m_i \vec{r}_i \times (\vec{\omega} \times \vec{r}_i)] \quad \text{[About O]}} \tag{9.36}$$

This formula for the angular momentum of a body is very complicated for an arbitrarily rotating body [no kidding!]. If the axis of rotation is towards the z-axis, the formula simplifies considerably. In this case, we can

9.2. ANGULAR MOMENTUM

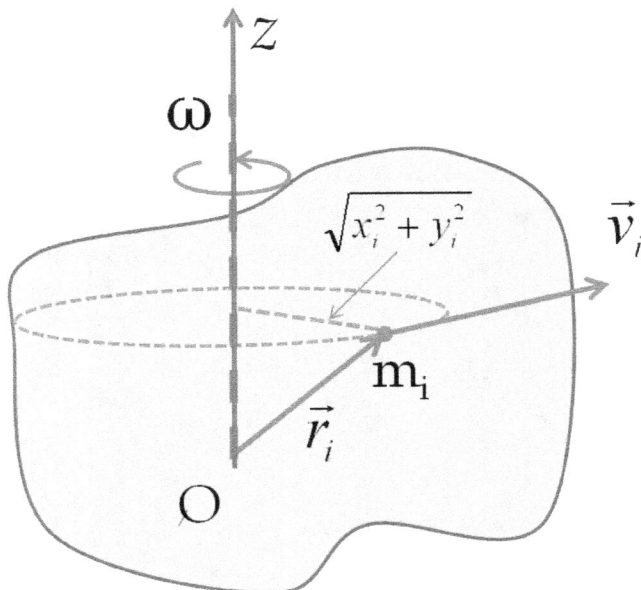

Figure 9.15: Choosing a Cartesian axis along the axis of rotation simplifies calculations for fixed axis rotations.

perform the cross products rather easily with the result that the angular momentum is also pointed along the z-axis. In the component form the result is:

$$\left\{\begin{array}{l} L_x = 0 \\ L_y = 0 \\ L_z = \left[\sum_{i=1}^{N} m_i \left(x_i^2 + y_i^2\right)\right] \omega_z \end{array}\right\} \quad \text{[Rotating about } z\text{-axis.]} \quad (9.37)$$

Proof: Let us write the vectors in component form so that we can perform the cross products in Eq. 9.36.

$$\vec{r}_i = x_i \hat{u}_x + y_i \hat{u}_y + z_i \hat{u}_z$$
$$\vec{\omega} = \omega_z \hat{u}_z \quad \text{[since rotating about the } z\text{-axis]}$$

The spatial velocity vector $\vec{\omega} \times \vec{r}_i$ is then obtained by a direct calculation.

$$\vec{\omega} \times \vec{r}_i = \omega_z x_i \hat{u}_y - \omega_z y_i \hat{u}_x.$$

Now we can take the cross product of $\vec{r}_i$ with this expression giving the following result.

$$\vec{r}_i \times (\vec{\omega} \times \vec{r}_i) = \omega_z \left(x_i^2 + y_i^2\right) \hat{u}_z.$$

Since ω_z and $\hat{u}_z$ are same for every particle, they come outside the sum giving the relation in Eq. 9.37.

The quantity in brackets in Eq. 9.37 is the coefficient by which we must multiply the z-component of angular velocity to obtain the z-component of the angular momentum. It is the zz component of a quantity called the moment of inertia, which is usually denoted by the letter I. We will denote the zz component of moment of inertial by I_{zz}.

$$I_{zz} = \sum_{i=1}^{N} m_i \left(x_i^2 + y_i^2 \right), \tag{9.38}$$

which can be used to write the z-component of the angular momentum more compactly as

$$L_z = I_{zz}\omega_z. \tag{9.39}$$

This equation shows that, for the fixed-axis rotation, the formula for the z-component of angular momentum is analogous to the definition of the z-component of the linear momentum $p_z = mv_z$. The moment of inertia reflects the geometrical distribution of masses in the body, and plays the role of inertia in rotation analogous to the role played by mass in the translational motion. Therefore, the fixed-axis rotation is analogous to translational motion in a straight line.

For a continuous body, the sum in Eq. 9.38 will turn into an integral. Suppose dm is a mass element located in a small volume at the coordinates (x, y, z), then the zz component of the moment of inertia cam be calculated by performing the following "conceptual" integral.

$$I_{zz} = \int_{\text{body}} \left(x^2 + y^2 \right) dm. \tag{9.40}$$

This integral is usually performed by first dividing the body into elements and then writing the mass of a representative element as a product of the density ρ and infinitesimal volume $dxdydz$ at the point (x, y, z).

$$dm = \rho \, dx \, dy \, dz. \tag{9.41}$$

The choice of the shape of the volume element depends on the symmetry of the situation as you will see in the examples below. For many cases of interest, the integral becomes an integration over only one variable if appropriately shaped elements are chosen for the calculation.

Further Remarks

If the axis of rotation is not along the z-axis, then, the angular velocity may have non-zero x, y and z-components and can be written as:

$$\text{Rotation about arbitrary axis: } \vec{\omega} = \omega_x \hat{u}_x + \omega_y \hat{u}_y + \omega_z \hat{u}_z. \tag{9.42}$$

Using Eq. 9.42 in Eq. 9.36 gives rise to the following components of angular momentum in terms of the components of the angular velocity that have nine coefficients, denoted by $\{I_{ij}, (i = x, y, z), (j = (x, y, z)\}$.

9.2. ANGULAR MOMENTUM

The values of the components I_{ij} tell us the way masses of the system are distributed with regard to the axis of rotation as captured by the Cartesian coordinates of the masses. For instance, if the body rotates about the z-axis, you need only I_{zz} which tells us about the distribution of masses of the system with regard to the z-axis.

$$L_x = I_{xx}\omega_x + I_{xy}\omega_y + I_{xz}\omega_z$$
$$L_y = I_{yx}\omega_x + I_{yy}\omega_y + I_{yz}\omega_z$$
$$L_z = I_{zx}\omega_x + I_{zy}\omega_y + I_{zz}\omega_z$$

These relations show that the x-component of the angular momentum depends not only on the x-component of the angular velocity but also on the y and z-components of the angular velocity. Therefore, in general, the angular momentum and angular velocity may be in different directions. This is very different than the relation between the momentum and velocity for translational motion, where the two are in the same direction.

In the case of a fixed-axis rotation, angular momentum and angular velocity are in the same direction since only one component of angular velocity is non-zero which can always be taken to be the z-axis. In this case $L_z = I_{zz}\omega_z$.

9.2.3 Calculations of Moments of Inertia

Moment of inertia of systems with point masses

The general formula for the moment of inertia component I_{zz} for discrete masses is given in Eq. 9.38.

$$I_{zz} = \sum_{i=1}^{N} m_i \left(x_i^2 + y_i^2\right)$$

where x_i and y_i are x and y-coordinates of the mass m_i. The quantity $x^2 + y^2$ can be replaced by the square of the distance r from z-axis.

$$\boxed{I_{zz} = \sum_{i=1}^{N} m_i r_i^2, \quad r_i = \sqrt{x_i^2 + y_i^2}.}$$

Now, we work out a few simple examples.

One particle of mass m

Consider a particle of mass m located at point P with Cartesian coordinates (x, y, z). This gives the three principal moments as follows. Note: we don't need any information about the rotation axis to calculate moments of inertia components; we only need the mass of the particle and its Cartesian coordinates.

$$I_{xx} = m\left(y^2 + z^2\right)$$
$$I_{yy} = m\left(z^2 + x^2\right)$$
$$I_{zz} = m\left(x^2 + y^2\right)$$

Two particles of masses m_1 and m_2

Let two point masses located at $P_1(x_1, y_1, z_1)$ and $P_2(x_2, y_2, z_2)$. We can immediately write down the three principal moments.

$$I_{xx} = m_1\left(y_1^2 + z_1^2\right) + m_2\left(y_2^2 + z_2^2\right)$$
$$I_{yy} = m_1\left(z_1^2 + x_1^2\right) + m_2\left(z_2^2 + x_2^2\right)$$
$$I_{zz} = m_1\left(x_1^2 + y_1^2\right) + m_2\left(x_2^2 + y_2^2\right)$$

Example 9.2.3. Moment of inertia of the methane molecule. To illustrate an application to molecular rotations consider determining the moment of inertia of the methane molecule for a coordinate system with the origin at the carbon atom and the z-axis through one of the C-H bonds.

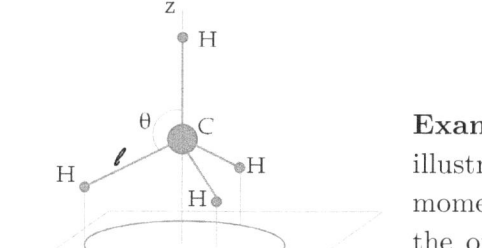

Figure 9.16: Methane molecule.

The methane molecule has four hydrogen atoms at the corners of a tetrahedron and one carbon atom at the center. Let l be the bond length between C and H. By symmetry all four bonds are of equal length. Let the angle between any two C-H bonds be θ.

We will calculate I_{zz} in the given coordinate system. For I_{zz} we need only the x and y-coordinates of the three H atoms that are not on the z-axis. By symmetry the three H atoms will fall on a circle of radius $l\sin\theta$ around z-axis separated by $120°$. Therefore,

$$I_{zz} = 3ml^2\sin^2\theta.$$

A uniform thin rod with the axis through center and perpendicular to the rod

Consider a uniform thin rod of mass M and length L. We assume that the area of cross-section of the rod is small and the rod can be thought of as a string of masses in one straight line. Suppose the axis of rotation is

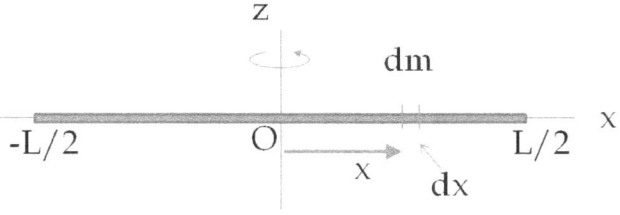

Figure 9.17: Calculation of I_{zz} of a thin rod about an axis through the center.

9.2. ANGULAR MOMENTUM

perpendicular to the rod and passes through the mid-point. We wish to find the moment of inertia about this axis. As before, we will orient the axes so that the z-axis is the axis of rotation. Since, we are dealing with a rod, and the integral is to go over the volume of the rod, it will be helpful to orient the x or y-axis in the direction of the rod. For definiteness, let the x-axis pass through the length of the rod as shown in fig. 9.17.

Let dm be a small element of mass located between $(x, 0, 0)$ and $(x + dx, 0, 0)$. The element has mass equal to ρdx, where ρ is mass per unit length, $\rho = M/L$.

$$dm = \frac{M}{L} dx, \tag{9.43}$$

which says that, to take into account the contribution from the entire rod, we need to integrate over the x variable. Now, using the coordinates $(x = x, y = 0)$ in Eq. 9.40 for the element and replacing dm by the expression for the rod given in Eq. 9.43, we obtain

$$I_{zz} = \int_{-L/2}^{L/2} \left(x^2 + 0\right) \frac{M}{L} dx.$$

This integral can be readily performed to yield the zz component of the moment of inertia of a thin rod about an axis through the center and perpendicular to the length of the rod.

$$\boxed{I_{zz} = \frac{1}{12} ML^2 \quad \text{[for rod about an axis through center]}}$$

A uniform thin rod with axis at the end and perpendicular to the length

The origin of the coordinate system will be at one end as shown in Fig 9.18. This changes only the range of integration. Now, the integration

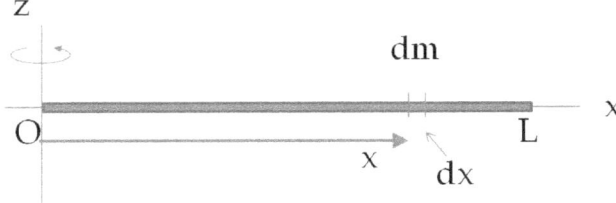

Figure 9.18: Calculation of I_{zz} of a thin rod about an axis at the edge.

will be from $x = 0$ to $x = L$, giving a different formula for I_{zz} of a rod.

$$I_{zz} = \int_0^L \left(x^2 + 0\right) \frac{M}{L} dx = \frac{1}{3} ML^2.$$

Performing the integration we find

$$\boxed{I_{zz} = \frac{1}{3} ML^2 \quad \text{[for rod about an axis through on end]}}$$

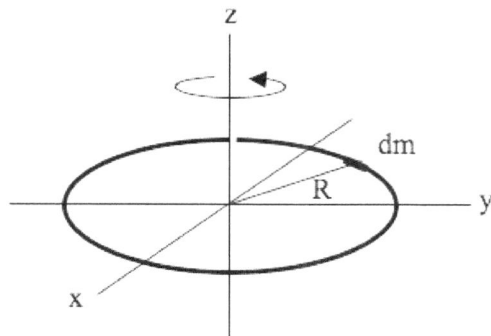

Figure 9.19: Calculation of I_{zz} of a ring.

This shows that the moment of inertia of a system depends on the axis of rotation. We have found that the I_{zz} for the same rod about an axis placed at the end of the rod is four times the moment of inertia when the axis passes through the center.

Why are the two I_{zz} for the same rod different? When the axis passes through the center, all mass elements of the rod rotate within a distance of $L/2$ from the axis, but when the axis passes through a point at the end, half of the masses of the rod rotate in larger circles, having distances between $L/2$ and L from the axis. Clearly, having masses more distant from the axis makes a big difference in rotation.

A uniform thin ring about an axis through the center and perpendicular to the ring

Consider a thin ring of mass M and radius R. We assume that the ring is so thin that we can place all masses on the ring at the same distance R from the center. To find the component I_{zz} of the moment of inertia for the rotation about the z-axis perpendicular to the ring and passing through the center, we place the ring in the xy-plane with the origin at the center as shown in Fig. 9.19.

Let dm be an element of the ring located at point $(x, y, 0)$ on the ring. Since $(x, y, 0)$ is located at the ring and the origin is at the center of the ring, the coordinates x and y are related by the radius R of the ring.

$$x^2 + y^2 = R^2.$$

Therefore, the formula for the conceptual integral for I_{zz} given in Eq. 9.40 simplifies and the integral can be done without any effort.

$$I_{zz} = \int_{body} \left(x^2 + y^2\right) dm = \int_{body} R^2 dm = R^2 \int_{body} dm.$$

The integration over dm is simply a sum of mass of all elements of the ring, which will give the total mass of the ring for the integral. Therefore, the result is

$$I_{zz} = MR^2 \quad \text{[Ring; Axis through the center and perpendicular to the ring]}$$

A uniform thin ring about an axis through the center and in the plane of the ring

Consider again a thin ring of mass M and radius R. Let us place the ring in the xy-plane so that z-axis passes through its center and is perpendicular to the ring (Fig. 9.20).

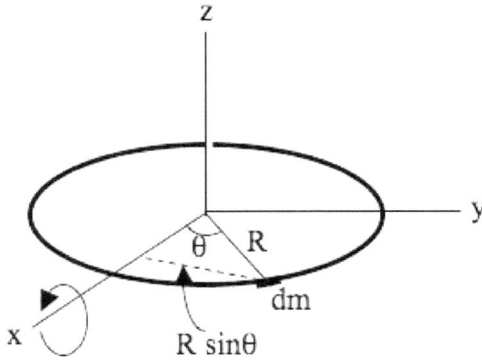

Figure 9.20: Calculation of I_{xx} of a ring.

From the symmetry, it is clear that rotation about x-axis is equivalent to a rotation about the y-axis. Therefore, the components I_{xx} and I_{yy} of the moment of inertia will be equal.

$$I_{xx} = I_{yy}.$$

The component I_{xx} has a similar expression as I_{zz} given above, specifically,

$$I_{xx} = \int_{body} \left(y^2 + z^2\right) dm.$$

When you rotate the ring about the x-axis, the mass elements move in circles in planes parallel to the yz plane. To calculate the moment of inertia component I_{xx} you can take the configuration at any instant. Here, we choose to work at the instant the ring is in $z = 0$ plane. With the ring in the $z = 0$ plane, I_{xx} simplifies to

$$I_{xx} = \int_{body} y^2 dm. \qquad (9.44)$$

The shape of the ring suggests that this integral will be easier to do if we formulate the problem in the polar coordinate. In the polar coordinate the radial variable is fixed to the radius of the ring.

$$r = R.$$

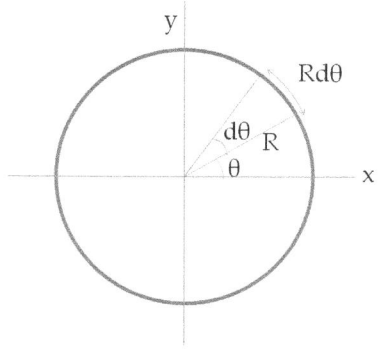

Figure 9.21: Calculation of dm.

The mass element has a shape of arc length $Rd\theta$, where $d\theta$ is the angle the arc element subtends at the center (Fig. 9.21). The mass dm in an arc element is

$$dm = \text{Mass per unit length} \times \text{Arc length} = \frac{M}{2\pi R} \times Rd\theta.$$

The y-coordinate written in terms of polar coordinate has the following form.

$$y = R\sin\theta.$$

Therefore, the integral in Eq. 9.44 transforms into an integral in one variable, the polar angle from 0 to 2π.

$$I_{xx} = \frac{MR^2}{2\pi} \int_0^{2\pi} \sin^2\theta d\theta = \frac{1}{2}MR^2.$$

Performing this integral we find

$$\boxed{I_{xx} = \frac{1}{2}MR^2 \quad \text{[Ring; Axis through the center and in plane of the ring]}}$$

A uniform thin disk about an axis through the center and perpendicular to the disk

Consider a thin disk of mass M and radius R. Since the disk is thin, its thickness can be ignored. For calculations, we place the disk in the xy-plane so that the z-axis is perpendicular to the disk and passes through its center. Now, we divide the disk into thin rings of various radii r and infinitesimal thickness dr.

One ring whose radius is less than R is shown in Fig. 9.22. The disk can be considered to be made up of rings like these from $r=0$ to $r=R$. The thin ring shown in Fig. 9.22 contains an infinitesimal mass given by

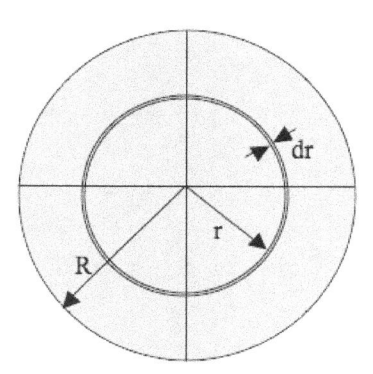

Figure 9.22: Calculation of I_{zz} of a disk.

$$dm = \frac{M}{\pi R^2} \times 2\pi r dr.$$

All masses in the infinitesimally thin ring are on circles of radius between r and $r+dr$. since dr is infinitesimal, it has a value that can be as small as you like, except zero. Therefore, we can say that all the masses in the infinitesimal ring between r and $r+dr$ are on a circle of radius r.

$$x^2 + y^2 = r^2. \quad \text{(Note: it is not } R^2\text{.)}$$

Using this expression in the definition of I_{zz} given in Eq. 9.44, we find that the conceptual integral for a disk becomes an integral over the radial coordinate. The range $r=0$ to $r=R$ includes all the infinitesimal rings that make up the disk.

$$I_{zz} = \int_0^R (r^2)\left(\frac{M}{\pi R^2}2\pi r dr\right) = \frac{2M}{R^2}\int_0^R r^3 dr = \frac{1}{2}MR^2.$$

After performing the integration, we find

$$\boxed{I_{zz} = \frac{1}{2}MR^2 \quad \text{[Disk; Axis through the center and perpendicular to the disk]}}$$

9.2. ANGULAR MOMENTUM

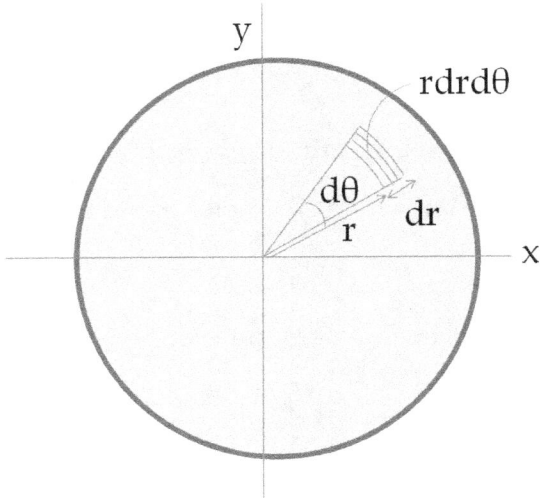

Figure 9.23: Calculation of I_{xx} of a disk.

A uniform thin disk about an axis through the center and in the plane of the disk

Once again consider a thin disk of mass M and radius R. From symmetry we note that rotation about x and y-axes are equivalent. Therefore, components I_{xx} and I_{yy} of the moment of inertia for rotations about the x and y-axes respectively, must be equal. We will perform the calculation for I_{xx} as we did for the ring above.

To make use of the cylindrical symmetry in the disk, we will do calculation in the polar coordinates. We focus on writing dm, and the x and y-coordinates of an element of the disk so that we can write the conceptual integral in Eq. 9.44 into a definite integral.

Consider a small element on the disk between r and $r + dr$ and θ and $\theta + d\theta$. The element has a length equal to the arc length $r d\theta$ and width dr. Therefore, the area of this element is $r dr d\theta$. The total mass M of the disk is spread over the total area πR^2 of the disk. Therefore, the mass in the mass element is

$$dm = \frac{M}{\pi R^2} r dr d\theta.$$

This says that we have to perform a two-dimensional integral, one for r and the other for θ. For I_{xx} calculation, the conceptual integral has the distance of the element from x-axis, which is $y^2 + z^2$. Here, the disk is placed in $z = 0$ plane, therefore, we have only y^2. Thus, we need y-coordinate of the element to go in the integral. Since the element is at the polar coordinates (r, θ), its y-coordinate is

$$y = r \sin \theta.$$

Therefore we find I_{xx} to be,

$$\begin{aligned} I_{xx} &= \int y^2 dm = \int\int (r\sin\theta)^2 \left(\frac{M}{\pi R^2} r dr d\theta\right) \\ &= \frac{M}{\pi R^2} \int_0^R r^3 dr \int_0^{2\pi} \sin^2\theta d\theta \end{aligned}$$

This gives the following After performing the integration, we find

$$\boxed{I_{zz} = \frac{1}{4}MR^2. \text{ [Disk; Axis through the center and in plane of the disk]}}$$

A uniform sphere about an axis through the center

Consider a sphere of mass M and radius R. Since all axes through the center of a sphere are equivalent, it suffices to work out only the component I_{zz} of moment of inertia for rotation about z-axis. It turns out that we

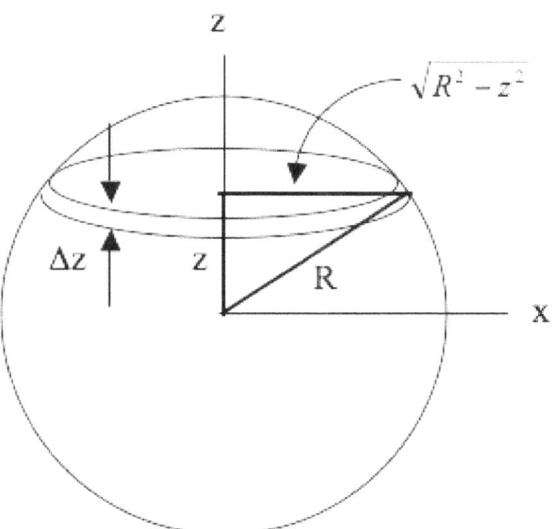

Figure 9.24: Calculation of I_{zz} of a sphere.

can make use of the results of the disk if we think of the sphere as thin disks of thickness dz stacked on top each other. The disk between z and z+ z has a radius of $\sqrt{R^2 - z^2}$. The moment of inertia of the thin disk between z and $z + dz$ is found as follows.

$$\begin{aligned} \text{Radius of the disk} &= \sqrt{R^2 - z^2} \\ \text{Volume of the disk} &= \pi\left(R^2 - z^2\right) dz \\ \text{Mass of the disk} = \text{density} \times \text{volume} &= \rho\pi\left(R^2 - z^2\right) dz \\ I_{zz} \text{ of the disk} &= \frac{1}{2}\left[\rho\pi\left(R^2 - z^2\right) dz\right]\left(R^2 - z^2\right). \end{aligned}$$

The moment of inertia of the sphere will be obtained by summing up moments of inertia of all the disks, which is obtained by integrating over

9.2. ANGULAR MOMENTUM

z from $-R$ to R.

$$I_{zz} = \int_{-R}^{R} \frac{1}{2}\left[\rho\pi\left(R^2 - z^2\right)dz\right]\left(R^2 - z^2\right) = \frac{8}{15}\pi\rho R^5.$$

This formula can be written by eliminating the density ρ by mass over volume.

$$\rho = \frac{M}{\frac{4}{3}\pi R^3}.$$

Hence, I_{zz} component of the moment of inertia of the sphere written in terms of its mass and radius is

$$\boxed{I_{zz} = \frac{2}{5}MR^2 \quad \text{[Sphere; Axis through the center}}$$

THE PARALLEL AXIS THEOREM

In calculating the standard formulas for the components of the moments of inertia for regular shapes, we usually place origin at the center of mass (CM) of the body. Often, the axis of rotation of interest in a particular problem does not go through the CM. The parallel axis theorem relates the moment of inertia corresponding to two parallel axes, one of which passes through the center of mass and the other through an arbitrary point in space.

To illustrate the relation we seek, we calculate I_{zz} about the two z-axes, one for a coordinate system with origin at the CM and the other with origin at some point O' on the x-axis as shown in Fig. 9.25. Let the the two z-axis be separated by a distance D. Let us denote the two moments of inertial by I_{zz} and I_{ZZ} respectively. Then we will prove that,

$$I_{ZZ} = I_{zz} + MD^2. \tag{9.45}$$

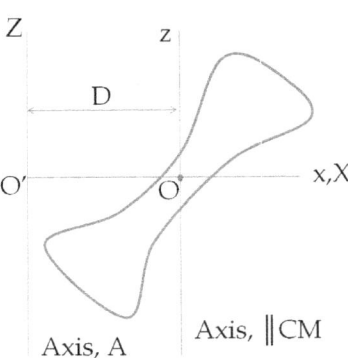

Figure 9.25: Parallel axis theorem. The origin O is at the CM.

Proof

We will prove the result by writing out I_{ZZ} from the definition of this component of moment of inertia and then use the following relation between the coordinates of any point in the two systems.

$$X = x + D \tag{9.46}$$
$$Y = y \tag{9.47}$$
$$Z = z \tag{9.48}$$

Hence,

$$\begin{aligned} I_{ZZ} &= \int \left(X^2 + Y^2\right) dm \quad \text{(Definition)} \\ &= \int \left[(x+D)^2 + y^2\right] dm \quad \text{(Eqs. 9.46 - 9.48)} \\ &= \int \left(x^2 + y^2\right) dm + \int D^2 dm + \int 2Dx\, dm \end{aligned}$$

The first term is I_{zz}, the second term evaluates to MD^2, and the last integral is zero since the origin of $Oxyz$ is the CM of the body.

$$\int x\,dm = 0 \quad \text{(using definition of } X_{cm}\text{)}$$

Therefore,

$$I_{ZZ} = I_{zz} + MD^2.$$

Often, we state this result in more "practical" language. The quantity I_{ZZ} is called the moment of inertia about axis A that does not pass through the CM. The quantity I_{zz} is called the moment of inertia about an axis parallel to axis A that passes through the CM. Using this nomenclature, we can write our result as

$$\boxed{I_{\text{Axis}} = I_{\|\text{CM}} + MD^2}$$

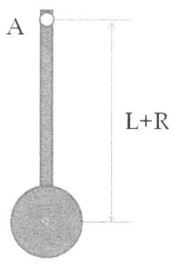

Figure 9.26: Application of parallel axis theorem.

Example 9.2.4. Moment of inertia about an axis through the end of a rod.

As an application of the parallel axis theorem let us work out the I_{zz} of a rod about an axis through its end. We have already calculated this I_{zz} above. Here, we show that from the answer for I_{zz} about the axis through the center, which was equal to $\frac{1}{12}ML^2$, we can find I_{zz} about the axis at the end.

To find I_{zz} about the axis at the end, we add MD^2 to the I_{zz} about the axis through the center, where D is the distance between the two parallel axes. In the present case, the distance between the axes is $\frac{1}{2}L$. Therefore from the parallel axis theorem, I_{zz} about the axis at the end of the rod is

$$I_{\text{Axis}} = I_{\|\text{CM}} + MD^2 = \frac{1}{12}ML^2 + M\left(\frac{L}{2}\right)^2 = \frac{1}{3}ML^2,$$

which is the answer we had found by a direct calculation.

Example 9.2.5. Moment of inertia of a spherical ball. We have found the moment of inertia of a spherical ball about an axis through the center of the ball. The answer was $I_{zz} = \frac{2}{5}MR^2$ for a ball of mass M and radius R. Suppose the ball is hung by attaching it to a light rod of length L and suspending the rod from the other end. Find the moment of inertia about an axis perpendicular to the plane shown in Fig. 9.26 with origin at the suspension point A.

Solution. Using the parallel axis theorem, the moment of inertia about an axis through A can be written in terms of the moment of inertia about another axis through the center of mass (CM) of the ball as long as the two axes are parallel to each other. Here, the distance between the two axes are $D = L + R$, therefore

$$I_{\text{Axis}} = I_{\|\text{CM}} + MD^2 = \frac{2}{5}MR^2 + M(L+R)^2.$$

RADIUS OF GYRATION

The moment of inertia has dimensions of mass times distance squared. If you have a single mass M at a distance b from the axis of rotation, then its moment of inertia about that axis would be simply Mb^2. Note: By the distance of a point from a line we mean the length of a line that is perpendicular from the point to the line (see Fig. 9.27).

For more complicated objects, different points of the object will be at different distances from the axis, and as a result the formula for I_{zz} or any other component of the moment of inertia that will contain a numerical pre-factor multiplying the mass and a length squared.

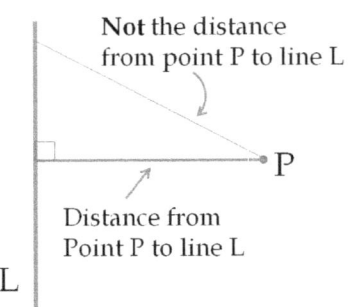

Figure 9.27: Meaning of distance between a point and a line.

The length in the moment of inertia formulas contains the information about the location of the axis relative to the body and the dimensions of the body. For instance, I_{zz} of a sphere with the origin at the center of the sphere is equal to $\frac{2}{5}MR^2$, or that of a sphere with the origin at the edge of the sphere is $\frac{7}{5}MR^2$.

We now ask: suppose you replace the extended object by a point mass, what distance will you have to place this point mass so that the moment of inertia of the point mass will be equal to the moment of inertia of the extended body? This distance is called the radius of gyration R_G of the body.

For instance, the radius of gyration of a sphere of radius R for the rotation about an axis through the center of the sphere can be obtained by equating I_{zz} of the point mass at a distance R_G from the axis to the I_{zz} of the sphere with respect to the same axis.

$$MR_G^2 = \frac{2}{5}MR^2,$$

which gives the following for the radius of gyration of sphere

$$R_G(\text{sphere}) = \sqrt{\frac{2}{5}}\,R.$$

The same body may have different radii of gyration depending on the axis. For instance, $R_G = \frac{1}{\sqrt{2}}R$ for a disk of radius R corresponding to I_{zz} and $R_G = \frac{1}{2}R$ corresponding to I_{xx}.

9.3 DYNAMICS OF FIXED-AXIS ROTATION

The laws of the dynamics of rotation are deduced from the fundamental laws of mechanics as given by the Newton's laws of motion. In this section we will first derive the relation governing the change of angular momentum of a single particle and then apply the result to multiparticle systems. We will find that the angular momentum of a body is essential for understanding the rotational dynamics. We will also discover that in a fixed-axis rotation we need study only one component of the angular momentum and the problem of rotation of a rigid body becomes equivalent to a one-dimensional problem.

9.3.1 Rotational Dynamics of a Single Particle

The dynamics of a single particle is completely given by Newton's laws of motion. By rotational dynamics of a single particle, we mean the study of the angular momentum of the particle. The second law of motion gives the rate at which momentum of a particle changes. Here, we ask: how can the angular momentum of a particle change? Recall that the angular momentum of a single particle about a reference at the origin is given by

$$\vec{L} = \vec{r} \times \vec{p},$$

where $\vec{r}$ is the position vector of the particle and $\vec{p}$ the momentum of the particle. We start by taking the derivative of both sides of this equation with respect to time t.

$$\frac{d\vec{L}}{dt} = \frac{d}{dt}(\vec{r} \times \vec{p}) = \frac{d\vec{r}}{dt} \times \vec{p} + \vec{r} \times \frac{d\vec{p}}{dt} \qquad (9.49)$$

The first term on the right side is $\vec{v} \times \vec{p}$. Since the velocity and momentum vectors point in the same direction, their cross product is equal to zero.

$$\text{First term in Eq.9.49:} \quad \frac{d\vec{r}}{dt} \times \vec{p} = \vec{v} \times m\vec{v} = m(\vec{v} \times \vec{v}) = 0. \qquad (9.50)$$

Using Newton's second law, we can replace $d\vec{p}/dt$ in the second term on the right side of Eq. 9.49 by the net force $\vec{F}_{net}$. This gives us the law for the rate of change of the angular momentum of a particle.

$$\boxed{\frac{d\vec{L}}{dt} = \vec{r} \times \vec{F}_{net}.} \qquad (9.51)$$

The right-hand side of this equation is the net torque $\vec{\tau}_{net}$ on the particle about the origin.

$$\vec{\tau}_{net} = \vec{r} \times \vec{F}_{net}. \qquad (9.52)$$

Note that both the torque and the angular momentum depend upon $\vec{r}$, which depends upon the choice of the reference point. The unit of torque in the SI system is kg. m^2/s^2, or N.m, which is the same as the unit of energy, Joule (J). The usual practice for the unit of torque is to use N.m rather than Joule.

Various techniques for the evaluation of torques have been discussed in Chapter 5. A student should review them before proceeding further in this chapter.

9.3.2 Rotational Dynamics of Extended Bodies

By now you must already know that we examine the dynamics of extended bodies by representing them as collections of particles. This makes sense since all physical bodies are eventually made up of discrete particles. Suppose, there are N particles that make up an extended body. A separate consideration of the dynamics of each particle will generate N vector equations of motion, one per particle. Let us label the particles as $1, 2, \cdots, N$. Let $m_1, m_2, \cdots, m_N$ be the masses of the particles and $\vec{r}_1, \vec{r}_2, \cdots, \vec{r}_N$ their position vectors with respect to the origin. Each particle of the body will have external forces from outside the system and internal forces from other particles. Let us label the external forces on individual particles by $\vec{F}_1^{\text{ext}}, \vec{F}_2^{\text{ext}}$, etc, and the internal forces between particles by $\vec{F}_{12}$ for force of 2 on 1, $\vec{F}_{13}$ for force of 3 on 1, $\cdots$, $\vec{F}_{ij}$ for force of j on i, etc. The rate of change of the angular momenta about the origin for each particle are given by by the following N vector equations.

$$\frac{d\vec{L}_1}{dt} = \vec{r}_1 \times \vec{F}_1^{\text{ext}} + \vec{r}_1 \times \vec{F}_{12} + \vec{r}_1 \times \vec{F}_{13} + \cdots + \vec{r}_1 \times \vec{F}_{1N}$$

$$\frac{d\vec{L}_2}{dt} = \vec{r}_2 \times \vec{F}_2^{\text{ext}} + \vec{r}_2 \times \vec{F}_{21} + \vec{r}_1 \times \vec{F}_{23} + \cdots + \vec{r}_1 \times \vec{F}_{2N}$$

$$\vdots$$

$$\frac{d\vec{L}_N}{dt} = \vec{r}_N \times \vec{F}_N^{\text{ext}} + \vec{r}_N \times \vec{F}_{N1} + \vec{r}_1 \times \vec{F}_{N2} + \cdots + \vec{r}_1 \times \vec{F}_{N-1,N}$$

Summing these equations gives an equation for the rate of change of the total angular momentum of the whole body, which can be separated into the torques from the external forces and the internal forces.

$$\frac{d\vec{L}}{dt} = \vec{\tau}_{\text{net}}^{\text{ext}} + \vec{\tau}_{\text{net}}^{\text{int}} \tag{9.53}$$

where

$$\vec{\tau}_{\text{net}}^{\text{ext}} = \vec{r}_1 \times \vec{F}_1^{\text{ext}} + \vec{r}_2 \times \vec{F}_2^{\text{ext}} + \cdots + \vec{r}_N \times \vec{F}_N^{\text{ext}},$$

and $\vec{\tau}_{\text{net}}^{\text{int}}$ is the sum of the torques from the internal forces. These internal torques have the following form for a force between arbitrary particle i and j.

$$\text{One term in } \vec{\tau}_{ij}^{\text{int}}: \quad (\vec{r}_i - \vec{r}_j) \times \vec{F}_{ij}. \tag{9.54}$$

If the internal forces between the particles of the system act only along the line joining the two particles, then these terms will all be zero since then $\vec{F}_{ij}$ will be either parallel to $(\vec{r}_i - \vec{r}_j)$ vector or antiparallel to it. This will leave only the torques from the external forces to generate the change in the net angular momentum of the body. Therefore, we find that the rate of change of the net angular momentum of an extended body is equal to the net external torque on all particles of the body.

$$\boxed{\frac{d\vec{L}}{dt} = \vec{\tau}_{\text{net}}^{\text{ext}}.} \tag{9.55}$$

This derivation is completely general and applies to the motion of any system as long as the forces between the particles act along the line between the particles. Care must be taken into computing the torque from external forces since each external force acts on individual particles, and the torque must be evaluated with respect to the same reference point as the angular momentum of the body.

9.3.3 The Law for Fixed Axis Rotation

In a fixed-axis rotation, the direction of the axis of rotation does not change with time. This simplifies the equations derived above since we do not need to be as general as Eq. 9.55. Suppose the fixed-axis of rotation is along the z-axis of a Cartesian coordinate system. Then, the angular momentum has only the z-component non-zero. Therefore, we need only the z-component of Eq. 9.55:

Simpler Notation:
$\frac{d}{dt}(I\omega) = \tau$
(Components along fixed axis understood)

$$\boxed{\text{Fixed Axis:} \quad \frac{dL_z}{dt} = \tau_{\text{net},z}^{\text{ext}}} \tag{9.56}$$

which can also be written in terms of the zz component of moment of inertia and the z-component of the angular velocity.

$$\boxed{\text{Fixed Axis:} \quad \frac{d}{dt}(I_{zz}\omega_z) = \tau_{\text{net},z}^{\text{ext}}.} \tag{9.57}$$

If the body does not change shape, then its moment of inertia remains unchanged, so that this equation can be written using z-component of the angular acceleration.

Simpler Notation:
$I\alpha = \tau$
(Components along fixed axis understood)

$$\boxed{\text{Fixed Axis/Rigid Body:} \quad I_{zz}\alpha_z = \tau_{\text{net},z}^{\text{ext}}.} \tag{9.58}$$

Equations 9.57 and 9.58 are analogous to the following two equations of the center of mass of an extended body.

$$\frac{d}{dt}\left(MV_z^{\text{CM}}\right) = F_{\text{net},z}^{\text{ext}} \tag{9.59}$$

$$MA_z^{\text{CM}} = F_{\text{net},z}^{\text{ext}} \tag{9.60}$$

where V_z^{CM} and A_z^{CM} are the z-components velocity and acceleration of the center of mass, respectively. Beware of the pitfalls of pushing the analogy between a translational motion in one dimension and a rotational motion about a fixed-axis too far since this superficial analogy does not go over to a more general motion of the bodies.

9.3.4 Practice With Torque Calculations

See section 5.5.

9.3.5 Example Problems - Single Rigid Bodies

Example 9.3.1. Torque on a rotating a wheel.

A wheel of mass 2 kg and radius 30 cm is rotating counterclockwise when viewed from one end of the axle. It is seen that the rotation speed is increasing uniformly by 1.5 rad/sec for every second. Find the net torque on the wheel. For the purposes of the moment of inertia calculations, assume the wheel to be a uniform thin disk.

Solution. The data given in the problem shows that the angular acceleration has magnitude 1.5 rad/sec^2 and direction same as the direction of the angular velocity since the rotation speed is increasing. The direction is given as counterclockwise sense of rotation when looked from a particular side of the axle. Suppose that side of the axle is pointed towards positive z-axis. Then we have the z-component of the angular acceleration as

$$\alpha_z = 1.5 \text{ rad/sec}^2$$

Now, we need the zz component of the moment of inertia. Using the disk formula for a disk for an axis through the center and perpendicular to the disk we find that,

$$I_{zz} = \frac{1}{2}MR^2 = \frac{1}{2}(2 \text{ kg})(0.30 \text{ m})^2 = 0.09 \text{ kg.m}^2.$$

Using the equation of motion for the rotation we find the z-component of the torque as

$$\tau_z = I_{zz}\alpha_z = 0.135 \text{ kg.m}^2/\text{s}^2.$$

Other components of the torque are zero here. Therefore, the torque has the magnitude 0.135 kg.m^2/s^2 or 0.135 N.m, pointed towards the positive z-axis, which is the direction of the particular side of the axle described in the question. Note: when the "unit" radian is multiplied by a dimensionful number such as a meter, the result is not radian-meter, but just meter since radian is a ratio of two lengths, and therefore, does not have a dimension.

Example 9.3.2. A simple plane pendulum.

In a simple pendulum a mass m is attached to the end of a "massless" string of length L which is then tied to a post so that the mass can swing in a vertical arc. In this example, we will work out the equation of motion of the pendulum. We will also solve the equation of motion by making an approximation for small angular displacement from the vertical.

From Fig. 9.28 of the pendulum, we note that the axis of the rotation of the pendulum passes through the suspension point O. The axis is fixed in time and perpendicular to the plane of the drawing. We have learned above that the signs of various quantities are easier to take into account if we work with with components of vectors in a particular coordinate system. Therefore, we start with a choice of coordinate directions. Let

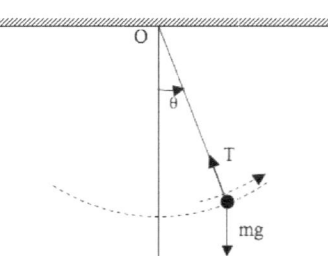

Figure 9.28: A plane pendulum.

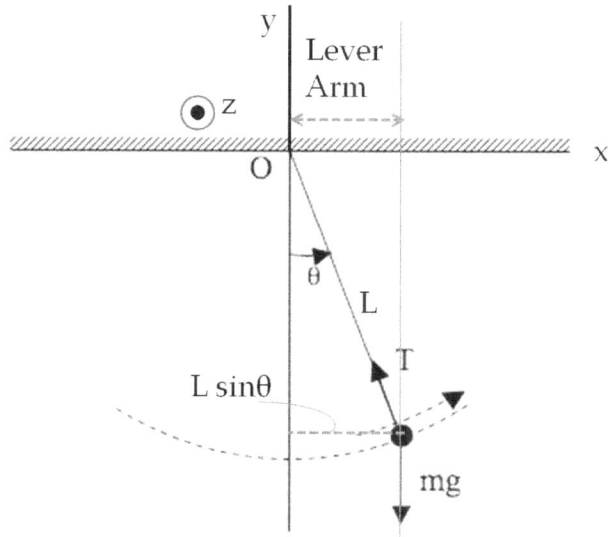

Figure 9.29: Calculation of torque about suspension point O. Note: z-axis is coming out of page.

the out of page be positive z direction and vertically up be positive y direction. The choices of coordinate system are shown in Fig. 9.29.

As shown in the figures, there are two forces on the mass: the tension from the string and the force of gravity from the Earth. The torque of the tension force about any axis through O is zero since the line of force goes through this point and the lever arm for this force will be zero. But the torque from the weight is not zero. Therefore, the net torque on the mass comes from only one force - its weight. At the instant of the motion shown in the drawing, the z-component of the torque has a clockwise sense, which means that the torque is pointed into the page or towards the negative z-axis. Therefore the z-component of torque will be negative

$$\text{Torque, } \tau_z = -mgL\sin\theta,$$

where $L\sin\theta$ is the lever arm of the weight about O. The zz component of moment of inertia of the mass m about O is mL^2. Now, the equation of motion for the rotation can be written to obtain the z-component of the angular acceleration from

$$mL^2\alpha_z = -mgL\sin\theta.$$

Writing the z-component of the angular acceleration in terms of the angle of rotation, this equation becomes

$$\frac{d^2\theta}{dt^2} = -\frac{g}{L}\sin\theta.$$

This relation describes the dynamics of a plane pendulum completely. However, this equation is difficult to solve in this form, but becomes easily solvable when we look at the motion for small angles. For small angle oscillations, we can make a linear approximation of $\sin\theta$ as $\sin\theta \approx \theta$ when the angle is expressed in radians. Therefore, the equation of motion takes the following simpler form.

$$\frac{d^2\theta}{dt^2} = -\frac{g}{L}\theta. \tag{9.61}$$

The solution of this equation is oscillatory in time and is given by a combination of sine and cosine functions of time with a well-defined period T.

$$\theta(t) = A\cos\left(\frac{2\pi t}{T}\right) + B\sin\left(\frac{2\pi t}{T}\right), \tag{9.62}$$

where A and B are constants of motion to be determined from the initial values of the angle θ and the z-component of the angular velocity ω_z. The time period T of the pendulum in this solution is given by

$$T = 2\pi\sqrt{\frac{L}{g}}.$$

This formula for the time period can be verified by inserting the solution given in Eq. 9.62 into the equation of motion, Eq. eq:simp-pendulum-1. This step is left an exercise for the student.

Example 9.3.3. Physical pendulum

A rigid body hung from a post swings just like a pendulum. Such oscillating bodies are called **physical pendulums**. Almost anything can be a physical pendulum. Find an expression for the time period of the oscillation of a physical pendulum.

Solution. An illustration of a physical pendulum and the forces on the body are shown in Fig. 9.30. Here the axis of rotation passes through the suspension point. Let M be the mass and D the distance between the axis of rotation and the center of mass (CM) of the body.

Let the z-axis be pointed out of the page. Since the force of gravity is proportional to the mass of the particles of the extended body, it can be shown that the torque on an extended body will equal to the entire weight acting on the CM only. The net torque on the physical pendulum is from the weight at the CM only. Let I_O is I_{zz} about the axis through the suspension point, i.e. the moment of inertia about an axis through O and perpendicular to the plane of the drawing. Therefore, the z-component of the equation of motion of angular momentum of the body gives the following equation for the angle θ made by the line between the suspension point and the CM with the vertical direction.

$$I_O \frac{d^2\theta}{dt^2} = -MgD\sin\theta.$$

To deduce the formula for the time period we need only the oscillations for small oscillations. For small oscillations, we use $\sin\theta \approx \theta$ to yield the

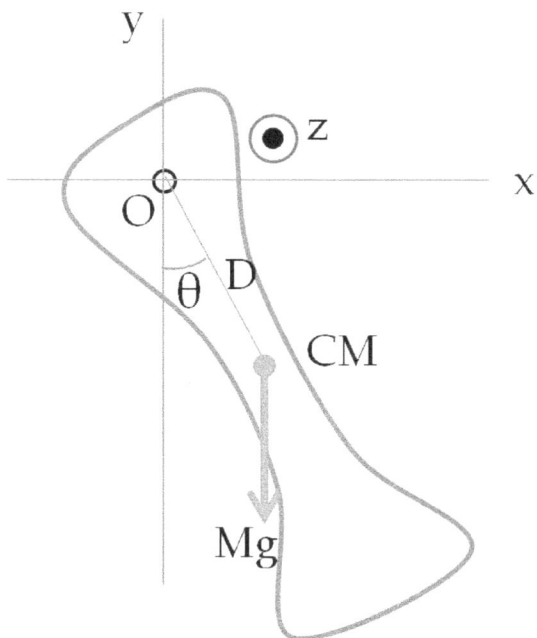

Figure 9.30: A physical pendulum can be any rigid body that can swing about an axis. Note z-axis is pointed out of page.

following.
$$\frac{d^2\theta}{dt^2} = -\frac{MgD}{I_O}\theta$$

This equation is similar to the equation for a plane pendulum, and can be solved by analogy. Therefore, we find that the period of oscillation T of a physical pendulum is given by:

$$T = 2\pi\sqrt{\frac{I_O}{MgD}}.$$

Suppose the physical pendulum is a rod of length L and mass M. Then, we will have
$$I_O = \frac{1}{3}ML^2$$
and
$$D = \frac{L}{2}$$

Therefore, the period of oscillation of the rod of length L suspended from one end would be

$$T = 2\pi\sqrt{\frac{I_O}{MgD}} = 2\pi\sqrt{\frac{2L}{3g}}.$$

9.3.6 Example Problems - Coupled Systems

Example 9.3.4. Unwinding a tape on an anchored wheel. Consider a wheel of mass M and radius R that is free to rotate about an axle through its center. Several turns of a thin light non-sticky tape, whose mass can

9.3. DYNAMICS OF FIXED-AXIS ROTATION

be neglected, are wound at the edge and a block of mass m is attached to the free end. When the block is released, the tape unwinds smoothly and the wheel rotates. Since the tape is thin, assume that the distance of the tape from the center of the wheel is approximately equal to the radius R of the wheel. Assuming the friction at the axle to be negligible, find the angular acceleration of the wheel and the tension in the tape.

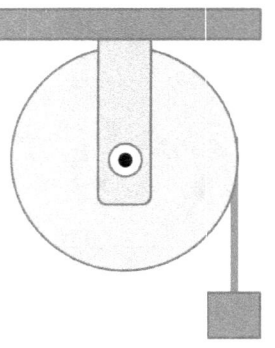

Figure 9.31: Example 9.3.4.

Solution. Here the translational motion of the block and the rotational motion of the wheel are coupled. For every radian of rotation, the tape unwinds by a distance R as given by the arc-radius-angle formula of a circle. Therefore, every radian of rotation is accompanied by a vertical displacement of the mass m by R. More generally, for an angular displacement of θ radians of the wheel, the vertical displacement of the block is $R\theta$.

Let the z-axis be point into the page and the y-axis pointed down as shown in Figure 9.33. Then, torque and angular acceleration of the wheel will be along the z-axis while the forces and acceleration of the block will be along the y-axis.

The vertical motion of the block

We start by drawing the free-body diagram for the translational motion of the block as shown in Fig. 9.32. From the free-body diagram, we get the following equation for the y-component of Newton's second law of motion for the block.

$$mg - T = ma_y \tag{9.63}$$

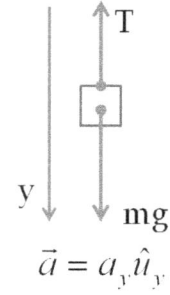

Figure 9.32: Free-body diagram of block.

The rotational motion of the wheel

For the rotational motion of the wheel, we examine torques on the wheel. Fig. 9.33 shows various forces that act on the wheel and we need to determine their torques about the center of the wheel.

The force F_A is a net force from the axle on the wheel, which acts all along the contact surface between the wheel and axle. We assume that the torque from this force is negligible as given in the problem statement.

The forces shown as F_{tp} are the forces from the tape pressing on the wheel. These forces act normally to the surface and can be assumed to have direction such that they go through the axis of rotation. This will give zero torque about the axis of rotation.

The force of gravity acts on all the particles of the wheel individually, but the torque from force of gravity on various particles is equal to the torque of the entire weight of the wheel placed at the CM of the wheel, which is a point on the axis of rotation here. Therefore, the weight also has a zero torque about the center of the wheel.

Thus, the tension in the tape is the only force that has a non-zero torque on the wheel about the z-axis through the center of the wheel. Since, the moment of inertia of the wheel about the axis is $I_{zz} = \frac{1}{2}MR^2$,

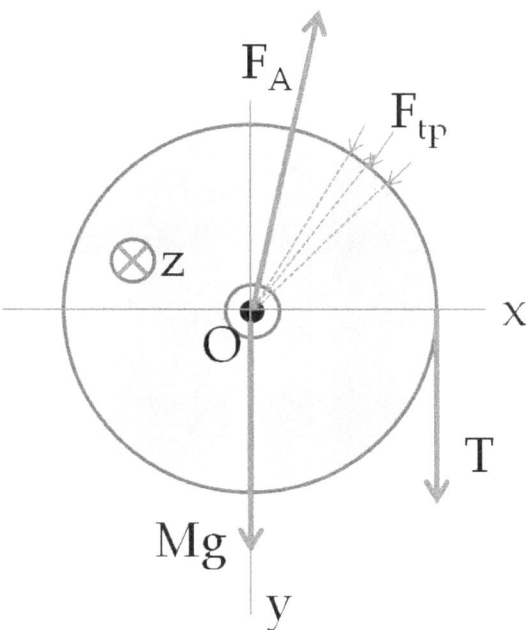

Figure 9.33: Forces on the wheel for torque calculation about O.

the z-component of the rotational equation of motion of the wheel is as follows.

$$TR = \frac{1}{2}MR^2\alpha_z, \qquad (9.64)$$

where α_z stands for the z-component of the angular acceleration of the wheel.

The constraint on the system

Finally, the coupling of the rotational motion of the wheel and the translational motion of the block gives rise to a relation between distance travelled by the block and angle of rotation of the wheel. In time dt the vertical displacement dy of the block and the angular displacement $d\theta_z$ of the wheel are related since the tape unwinds at the circumference of the wheel:

$$dy = Rd\theta_z$$

Dividing this relation by dt we find that the velocity of the block and the angular velocity of the wheel are related.

$$v_y = R\omega_z.$$

Taking a time derivative of this equation shows that the acceleration of the block and the angular acceleration of the wheel are related.

$$a_y = R\alpha_z. \qquad (9.65)$$

Now, we can solve Eqs. 9.63, 9.64 and 9.65 together to obtain expressions for T and α_z in terms of the masses and the acceleration due to gravity.

$$T = \left(\frac{M}{M+2m}\right)mg.$$

9.3. DYNAMICS OF FIXED-AXIS ROTATION

$$\alpha_z = \left(\frac{2m}{M+2m}\right)\frac{g}{R}.$$

The acceleration of the block can be obtained by multiplying the angular acceleration by R.

$$a_y = \left(\frac{2m}{M+2m}\right)g$$

Example 9.3.5. Atwood machine. In an Atwood machine two blocks connected by a light string hang from the two sides of a pulley. Suppose the pulley is frictionless but the mass of the pulley is not negligible compared to the masses of the blocks. Let M and R be the mass and radius of the pulley, which, for purposes of moment of inertia, can be assumed to be a uniform disk. Let m_1 and m_2 be the masses of the two blocks. What would be tensions in the string on the two sides and the acceleration of the masses and the angular acceleration of the pulley?

Solution. We have solved a similar problem before by assuming that the pulley was massless and frictionless. These assumptions about the pulley let us assert that the tensions on the two sides had the same magnitude. Here, we cannot make this assumption. We will see below that the equation of rotation of the pulley will become inconsistent if we assume that the tensions on the two sides have the same magnitude.

Let T_1 and T_2 be magnitudes of tension force on the two sides of the pulley.

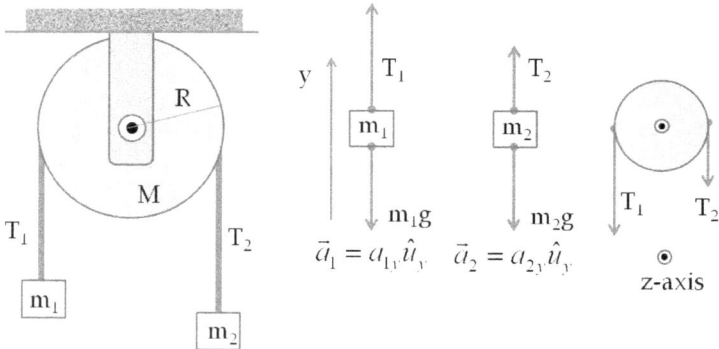

Figure 9.34: Atwood's machine. Free-body diagrams for the two masses and axes for computing the components. The two forces on the pulley that have non-zero torques about the center. Other forces on the pulley are not shown in this figure.

To solve this problem for accelerations and tensions we need to write out the equations of motion of the blocks and the pulley. We will not go into as much detail as was presented in the last example. A student who has skipped Example 9.3.4 given above is advised to go back and study that example first. Suffice to say that we need to identify the forces and relevant torques.

In Fig. 9.34 you will find the free-body diagrams for the forces on the two masses and the torque diagram for the pulley that helps us obtain the following components of the equations of motion in the coordinate system shown in the figure.

$$\text{Translation of } m_1 : T_1 - m_1 g = m_1 a_{1y} \qquad (9.66)$$

$$\text{Translation of } m_2 : T_2 - m_2 g = m_2 a_{2y} \qquad (9.67)$$

$$\text{Rotation of pulley} : T_1 R - T_2 R = \frac{1}{2} M R^2 \alpha_z \qquad (9.68)$$

The coupling of motion of the three moving bodies leads to the change in coodinates in time dt as follows.

$$\text{Translation of } m_1 \text{ and } m_2 : dy_1 = -dy_2 \qquad (9.69)$$

$$\text{Translation of } m_1 \text{ and rotation of pulley} : dy_1 = -R d\theta_z \qquad (9.70)$$

The relative signs are very important in these relations. You can check the signs as follows: let m_1 moving up, then $dy_1 > 0$. This will make m_2 go down, which makes $dy_2 < 0$ whose magnitude is equal to that of dy_1. When $dy_1 > 0$, the wheel rotates clockwise, which makes the angular displacement pointed towards the negative z-axis, and hence $d\theta_z < 0$. The string is at the edge of the pulley which is at the circumference of a circle of radius R, therefore, the absolute value of dy_1 must be R times absolute value of $d\theta_z$. These relations give rise to the following relations among the accelerations.

$$\text{Translation of } m_1 \text{ and } m_2 : a_{1y} = -a_{2y} \qquad (9.71)$$

$$\text{Translation of } m_1 \text{ and rotation of pulley} : a_{1y} = -R \alpha_z \qquad (9.72)$$

Therefore, we can write Eqs. 9.66-9.68 replacing all accelerations in terms of α_z.

$$T_1 - m_1 g = -m_1 R \alpha_z \qquad (9.73)$$

$$T_2 - m_2 g = m_2 R \alpha_z \qquad (9.74)$$

$$T_1 - T_2 = \frac{1}{2} M R \alpha_z \qquad (9.75)$$

It is elementary to solve these equations for α_z, T_1 and T_2. The results are:

$$R \alpha_z = - \left(\frac{m_1 + m_2}{m_2 - m_1 + M/2} \right) g \qquad (9.76)$$

$$T_1 = m_1 g - m_1 R \alpha_z \qquad (9.77)$$

$$T_2 = m_2 g + m_2 R \alpha_z \qquad (9.78)$$

The accelerations of the two blocks will be magnitude equal to R times the magnitude of α_z.

9.4 CONSERVATION OF ANGULAR MOMENTUM

9.4.1 For One Particle

It was shown above that the rate of change of the angular momentum of a particle is equal to the net torque on the particle.

$$\frac{d\vec{L}}{dt} = \vec{\tau}_{\text{net}}.$$

Therefore, if the net torque on a particle is zero, then the angular momentum cannot change. That is, both the magnitude and the direction of the angular momentum of the particle will remain fixed in time if the net torque on the particle is zero. We say that the angular momentum of the particle is conserved if there is no torque on the particle.

$$\boxed{\text{If } \vec{\tau}_{\text{net}} = 0, \text{ then } \frac{d\vec{L}}{dt} = 0 \implies \Delta \vec{L} = 0.}$$

A consequence of this result is that a particle with zero net torque about some point O will remain confined to a plane. This can be seen by examining the basic definition of angular momentum. The angular momentum of a particle located at the position $\vec{r}$ and having a momentum $\vec{p}$ is given by $\vec{L} = \vec{r} \times \vec{p}$.

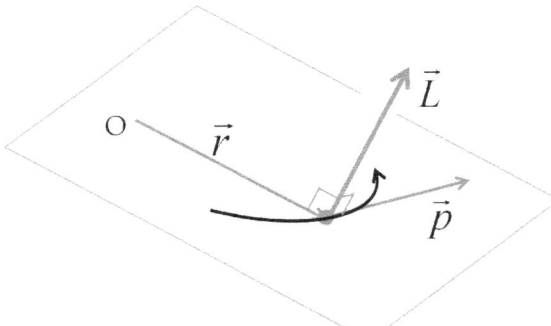

Figure 9.35: Angular momentum $\vec{L}$ is perpendicular to the plane containing $\vec{r}$ and $\vec{p}$.

When angular momentum is conserved, $\vec{r}$ and $\vec{p}$ are confined to a fixed plane in space since the direction of angular momentum, which is perpendicular to this plane, is constant. Therefore, a particle with a conserved angular momentum will remain confined to a planar motion.

Example 9.4.1. Equal area in equal time

The conservation of angular momentum also provides a fundamental reason for Kepler's second law of planetary motion, which states that the

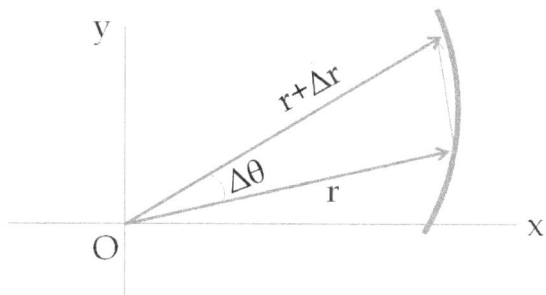

Figure 9.36: Set-up for equal area in equal time calculation.

line from the sun to the planets sweep out equal area in equal time. Since the gravitational force on a planet is pointed towards the center of sun, the torque of this force on a planet about the center of sun is zero. Therefore, a planet's angular momentum will be conserved and not change with time. This is why the motion of planets occurs in planes. Kepler's second law holds true more generally whenever the torque on a particle vanishes as we show in the next example.

Example 9.4.2. Consider a particle that has zero torque about the origin. The position vector of the particle sweeps out an area with time. Show that the rate of the area swept is constant.

Solution. Since $\vec{L}$ is fixed in time due to zero torque, the motion of the particle will be in the plane of $\vec{r}$ and $\vec{p}$ of the particle. To be concrete, let the a particle be confined to the xy-plane. This will make $\vec{L} = L_z \hat{u}_z$. That is the constant angular momentum will mean constant L_z.

We will show that the rate of change of the area in xy-plane swept out by the position vector $\vec{r}$ from the origin to the position of the particle is proportional to the z-component of the constant angular momentum.

In this problem we work in the polar coordinate to make use of the radial distance of the position vector from the origin. Let the position of the particle at t and $t + \Delta t$ in the polar coordinates be (r, θ) and $(r + \Delta r, \theta + \Delta \theta)$, respectively. The area ΔA swept out by the position vector $\vec{r}$ during this time interval can be approximated by the area of the triangle of base r and height $r\Delta \theta$ as shown in the Fig. 9.36.

$$\Delta A = \frac{1}{2}(r)(r\Delta\theta) = \frac{1}{2}r^2 \Delta\theta.$$

To obtain the rate of change of the area we divide this result by Δt and take the $\Delta t \to 0$ limit. In this limit, the second term goes to zero, and we obtain.

$$\frac{dA}{dt} = \frac{1}{2}r^2 \frac{d\theta}{dt}. \tag{9.79}$$

This can be shown to be proportional to the z-component of the angular momentum as follows. Working in the polar coordinates and using the polar unit vectors $\hat{u}_r$ and $\hat{u}_\theta$ in the xy-plane, the angular momentum of a particle confined to the xy-plane be written using the polar coordinates representation of vectors $\vec{r}$ and $\vec{v}$ given in section 3.7.

$$\begin{aligned} \vec{L} &= \vec{r} \times \vec{p} \\ &= r\hat{u}_r \times \left[m \left(\frac{dr}{dt} \hat{u}_r + r \frac{d\theta}{dt} \hat{u}_\theta \right) \right] \\ &= mr^2 \frac{d\theta}{dt} \hat{u}_z \equiv L_z \hat{u}_z. \end{aligned}$$

Therefore,
$$\frac{dA}{dt} = \frac{1}{2m} L_z. \tag{9.80}$$

Since L_z is constant, the rate of area covered by the position vector $\vec{r}$ is also constant. Therefore, if the torque on a particle about some point O is zero, then the trajectory of the vector from the point O to the position of the particle sweeps out equal area in equal time.

Since a point particle does not have structure, the consequences of the conservation of angular momentum are somewhat limited. We now examine the consequences of vanishing of net torque on an extended object, both rigid and non-rigid.

9.4.2 Conservation of Angular Momentum For Extended Bodies

It was shown above that the rate of change of the total angular momentum of an extended body depends on the net <u>external</u> torque on the body.

$$\frac{d\vec{L}}{dt} = \vec{\tau}_{\text{net}}^{\text{ext}}. \tag{9.81}$$

Therefore, if the external torque is zero, then the angular momentum of the body cannot change, i.e. angular momentum will be conserved.

$$\boxed{\text{If } \vec{\tau}_{\text{net}}^{\text{ext}} = 0, \text{ then } \frac{d\vec{L}}{dt} = 0.} \tag{9.82}$$

This equation is true of all extended bodies, whether a body is rigid or deformable. Since this equation is a vector equation, it is also true for each component separately.

$$\text{If } \tau_{\text{net,x}}^{\text{ext}} = 0, \text{ then } \frac{dL_x}{dt} = 0.$$
$$\text{If } \tau_{\text{net,y}}^{\text{ext}} = 0, \text{ then } \frac{dL_y}{dt} = 0.$$
$$\text{If } \tau_{\text{net,y}}^{\text{ext}} = 0, \text{ then } \frac{dL_z}{dt} = 0.$$

Suppose we choose the positive z-axis direction to be the direction of the constant $\vec{L}$, then for a system of conserved angular momentum, we will have only the z-component non-zero.

$$L_x = 0$$
$$L_y = 0$$
$$L_z \text{ Non-zero, constant.}$$

Let us write the condition for the conservation of angular momentum using this coordinate choice. For teh fixed-axis rotation, we have found that the z-component of angular momentum is equal to the product of the zz component of the moment of inertia and the z-component of the angular velocity. Therefore Eq. 9.82 takes the following form.

$$\boxed{\text{If } \tau^{\text{ext}}_{\text{net},z} = 0, \text{ then } \frac{d}{dt}(I_{zz}\omega_z) = 0,} \qquad (9.83)$$

which says that the product $I_{zz}\omega_z$ is fixed in time when the z-component of the net external torque vanishes. If the extended body is a rigid body, then I_{zz} will not change with time. In that case, Eq. 9.83 says that the z-component of the angular velocity will not change with time. Let $\omega_z^{(1)}$ and $\omega_z^{(2)}$ be the z-component of angular velocity at times t_1 and t_2, respectively, then

$$\text{Net external torque zero, rigid body:} \quad \omega_z^{(1)} = \omega_z^{(2)}. \qquad (9.84)$$

This statement is taken to mean that the rigid body with no external torque will rotate for-ever at the same angular velocity. This tendency of a rigid body is called **rotational inertia**.

On the other hand, if the extended body can change shape during some time without an application of external forces, i.e., purely by internal forces, then I_{zz} could be different at different times, which has important consequences for the motion of the body. If I_{zz} changes, ω_z must change also even when external torque is zero in order for the product $I_{zz}\omega_z$ to remain unchanged.

Let $L_z^{(1)}$ and $L_z^{(2)}$ be the z-component of the angular momentum at times t_1 and t_2. Let $I_{zz}^{(1)}$ and $\omega_z^{(1)}$ be the zz component of moment of inertia and the z-component of the angular velocity at the time t_1 and $I_{zz}^{(2)}$ and $\omega_z^{(2)}$ be the corresponding quantities at the time t_2, then

$$\text{Net external torque zero, general: } L_z^{(1)} = L_z^{(2)}$$
$$\implies \boxed{I_{zz}^{(1)}\omega_z^{(1)} = I_{zz}^{(2)}\omega_z^{(2)}.} \qquad (9.85)$$

Thus, even though the angular momentum of a system cannot change under a condition of the zero external torque, the angular velocity of the body can change if the body is not a rigid body with a change in the moment of inertia of the body: if I_{zz} increases, then ω_z would decrease, and vice-versa.

This is what happens in figure skating. The ice skater has a smaller moment of inertia when the arms are closer to the body and so the skater

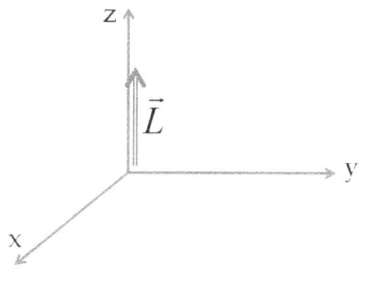

Figure 9.37: Choose z-axis to point in the direction of the conserved angular momentum.

Memory tool:
$I^{(1)}\omega^{(1)} = I^{(2)}\omega^{(2)}$
if $\tau^{\text{ext}} = 0$.

9.4. CONSERVATION OF ANGULAR MOMENTUM

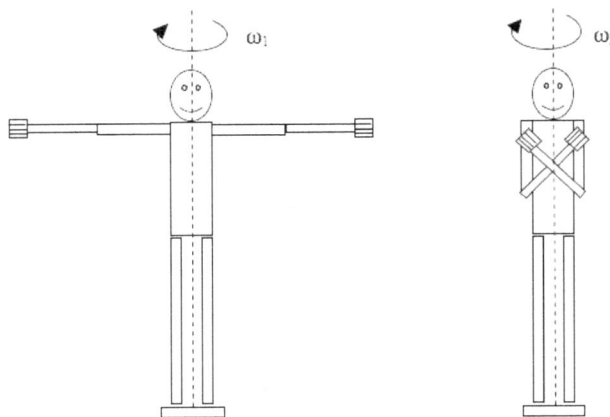

Figure 9.38: With negligible external torque on the skater the angular momentum of the skater about the vertical axis (z-axis) is conserved. The zz component of moment of inertia is larger when the arms are stretched out than when the arms are closer to the axis. Since $I_1 > I_2$, $\omega_1 < \omega_2$. The skater spins faster with arms closer to the body.

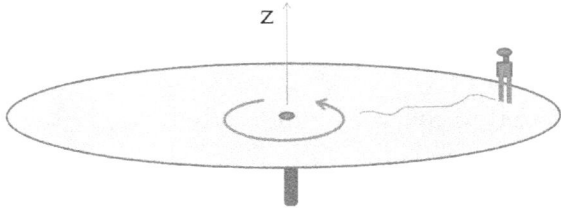

Figure 9.39: Example 9.4.3.

spins faster. As the arm is moved outward, away from the body, the moment of inertia increases, which is accompanied by slower rotation. This is due to the fact that external torques on the skater from the ice and air resistance are negligible, and therefore, the product of the moment of inertia and the angular velocity about the axis must be conserved (Fig. 9.38). To stop the rotation, the skater needs an external torque which is supplied by the reaction force from the ice when she pushes into the ice by her skate.

Example 9.4.3. Walking on a rotating platform. Consider a platform of mass M and radius R with a person of mass m standing at a distance a from the center. The platform is initially rotating at angular speed ω. When the person on the platform walks to another place on the platform with radial distance $b < a$, the platform rotates faster. Find the new rotation speed.

Solution. We consider the person and the platform as one system. This system is not rigid since the relative positions of the person and the platform can change.

Notice that during the walk there is no external torque. There is only torque from the person on the platform and vice-versa. Therefore, the net angular momentum of the combined system is conserved.

We choose z-axis to pass through the center of the platform in the direction of the axis of rotation. The angular speed is ω when the person is at a radial distance a from the center of the platform, and let ω' be the angular speed when the person is at radial distance b.

We will assume that distances a and b are much greater than the diameter of the person's body so that all the mass of the person is at a distance a or b from the center in the two situations. Therefore, I_{zz} of the person will be ma^2 when he is a distance a from the axis and mb^2 when he is a distance b from the axis. As for the platform, we will assume that the density is uniform so that I_{zz} of the platform is $\frac{1}{2}MR^2$, where R is the radius of the platform.

Now, we can write the z-component of the angular momentum in the two situations, before and after the walk, as follows.

Before walk: $\quad L_z^{(1)} = \left(L_z^{(1)}\right)_{\text{platform}} + \left(L_z^{(1)}\right)_{\text{person}} = \frac{1}{2}MR^2\omega + ma^2\omega$

After walk: $\quad L_z^{(2)} = \left(L_z^{(2)}\right)_{\text{platform}} + \left(L_z^{(2)}\right)_{\text{person}} = \frac{1}{2}MR^2\omega' + mb^2\omega'$

Equating the angular momenta at two instants gives us the following relation.
$$\frac{1}{2}MR^2\omega + ma^2\omega = \frac{1}{2}MR^2\omega' + mb^2\omega',$$
which can be solved for the rotation speed after the person has reached the second spot.
$$\omega' = \left(\frac{MR^2 + 2ma^2}{MR^2 + 2mb^2}\right)\omega.$$

9.5 ROTATIONAL WORK AND KINETIC ENERGY

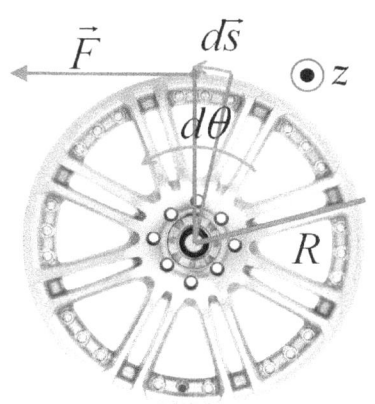

Figure 9.40: Rotational work by the force applied at the rim is $dW = \vec{F} \cdot d\vec{s}$.

Consider rotating a rigid wheel about its axle by applying a force $\vec{F}$ at the rim as shown in Fig. 9.40. In time dt, the force acts along the rim for a displacement ds. Therefore, the work done by the force will be

$$dW = Fds. \tag{9.86}$$

This work can be written in terms of the torque by this force about the axle and the angle $d\theta$ of rotation. If the radius of the wheel is R then, we will have $ds = Rd\theta$, and the work will be given as

$$dW = Fds = FRd\theta. \tag{9.87}$$

The torque by the force applied at the rim is $\tau = FR$. Therefore, the work by the force in rotating the wheel by and angle $d\theta$ is

$$dW = \tau d\theta. \tag{9.88}$$

This work is called rotational work. We will write this work using vector notation so that the result can be applied more generally. We also add "rot" to the symbol of work to indicate that we are talking about rotational work.

$$dW^{\text{rot}} = \vec{\tau} \cdot d\vec{\theta}. \tag{9.89}$$

We can deduce an analogue of the work-energy theorem and discover the expression for the rotational kinetic energy by examining the work done over an interval. The work by all torques will be obtained by integrating $\vec{\tau} \cdot d\vec{\theta}$ from the initial time to the final time. Since we are dealing here with only a fixed-axis rotation about the z-axis, we can write the result as integral over the z-component of the net torque.

$$W_{fi}^{\text{rot,net}} = \int_i^f \tau_z^{\text{net}} d\theta_z \tag{9.90}$$

Using the equation of motion for rotation, we can replace the torque by the rate of change of angular momentum.

$$\tau_z^{\text{net}} = I_{zz} \frac{d\omega_z}{dt} \quad (\text{rigid body}) \tag{9.91}$$

where we have assumed that the moment of inertia component I_{zz} is not a function of time as would be the case with a rigid body. Putting Eq. 9.91 in Eq. 9.90, and using the definition $\omega_z = d\theta_z/dt$, we find that the net rotational work on a body between two instances in time obeys the following relation.

$$W_{fi}^{\text{rot,net}} = \left(\frac{1}{2} I_{zz} \omega_z^2\right)_f - \left(\frac{1}{2} I_{zz} \omega_z^2\right)_i. \tag{9.92}$$

The quantity $\frac{1}{2} I_{zz} \omega_z^2$ is rotational kinetic energy of the body rotating around a fixed axis along the z-axis.

$$\text{Fixed axis rotation about } z\text{-axis:} \quad K_{\text{rot}} = \frac{1}{2} I_{zz} \omega_z^2, \tag{9.93}$$

which is written more compactly as $K_{\text{rot}} = \frac{1}{2} I \omega^2$. Thus, a rotational work gives rise to a change in the rotational kinetic energy. This is analogous to the work-energy theorem we have worked out before. The relation between the rotational work and the resulting change in rotational kinetic energy, often called the **rotational work-energy theorem**, is written as:

$$\boxed{W_{fi}^{\text{rot,net}} = (K_{\text{rot}})_f - (K_{\text{rot}})_i.} \tag{9.94}$$

9.6 ROLLING MOTION IN A STRAIGHT LINE

Rolling motion in a straight line is an example of motion in which both translation and rotation take place at the same time but the axis of rotation is always pointed in the same direction. The rolling motion is different from a pure rotation since, in a rolling motion, the axis of rotation moves with the body. However, if the rolling motion occurs in a straight line, then the direction of the axis remains unchanged. It is possible to separate the translational motion of the center of mass (CM) from the rotational motion about an axis through CM as we will show now.

Separation of angular momentum

Consider a disk rolling on a plane surface as shown in Fig. 9.41. To be concrete, we choose the z-axis of a coordinate system that is fixed to the plane surface along the axis of rotation. Then, the z-component of

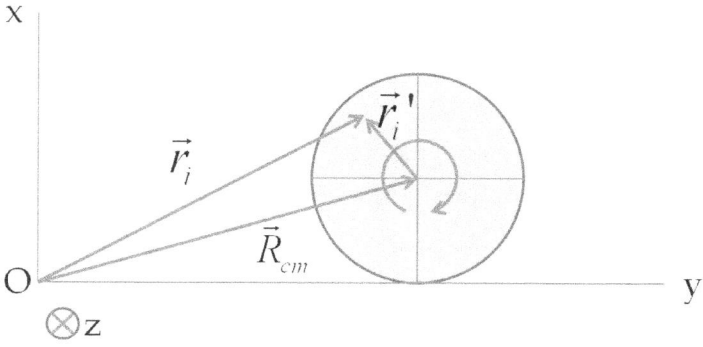

Figure 9.41: Rolling motion. Here z-axis is pointed in-the-page, indicated by x in a circle.

the angular momentum of a rolling body about the origin O of the fixed coordinate system can be obtained by summing over the angular momenta of the elements of the disk as we have done before. We can prove that the z-component of the angular momentum in the rolling motion separates into two terms.

$$L_z^{\text{about O}} = I_{zz}^{\text{CM}}\omega_z + \left(\vec{R}_{\text{CM}} \times M\vec{V}_{\text{CM}}\right)_z \qquad (9.95)$$

where I_{zz}^{CM} is the moment of inertia about an axis through the CM that is parallel to the z-axis through O.

Proof:

Consider the rolling object as collection of masses m_i. Then the angular momentum about O is the sum of angular momentum of each mass.

$$L_z^{\text{about O}} = \sum_i (\vec{r}_i \times m_i \vec{v}_i)_z = \sum_i \left(\vec{r}_i \times m_i \frac{d\vec{r}_i}{dt}\right)_z \qquad (9.96)$$

The position vector of m_i with respect to O, labeled $\vec{r}_i$, differs from its position $\vec{r}_i{'}$ with respect to the CM.

$$\vec{r}_i = \vec{R}_{\text{CM}} + \vec{r}_i{'} \qquad (9.97)$$

Substituting this equation into Eq. 9.96, and then expanding the terms in the sum, we find the angular momentum to be

$$L_z^{\text{about O}} = \left[\vec{R}_{\text{CM}} \times \sum_i m_i \frac{d\vec{R}_{\text{CM}}}{dt} + \left(\sum_i m_i \vec{r}_i' \right) \times \frac{d\vec{R}_{\text{CM}}}{dt} \right.$$
$$\left. + \vec{R}_{\text{CM}} \times \left(\sum_i m_i \frac{d\vec{r}_i'}{dt} \right) + \sum_i \left(\vec{r}_i' \times m_i \frac{d\vec{r}_i'}{dt} \right) \right]_z \quad (9.98)$$

The middle two terms can be shown to be zero as follows. The second term becomes zero using the definition of center of mass,

$$\sum_i m_i \vec{r}_i' = \sum_i m_i \left(\vec{r}_i - \vec{R}_{\text{CM}} \right) = M\vec{R}_{\text{CM}} - M\vec{R}_{\text{CM}} = 0,$$

which makes the third term in Eq. 9.98 zero since it implies

$$\sum_i m_i \frac{d\vec{r}_i'}{dt} = 0.$$

The first term in Eq. 9.98 simplifies to $\left(\vec{R}_{\text{CM}} \times M\vec{V}_{\text{CM}} \right)_z$. The fourth term gives motion about the CM and an be shown to be an angular momentum.

$$\left(\sum_i m_i \vec{r}_i' \times \frac{d\vec{r}_i'}{dt} \right)_z = \left(\vec{r}_i' \times \vec{p}_i' \right)_z,$$

where $\vec{r}_i'$ and $\vec{p}_i'$ are the position and momentum of m_i with respect to the CM. Therefore, this term gives the z-component of angular momentum with respect to the CM.

Fourth term in Eq. 9.98 = $I_{zz}^{\text{CM}} \omega_z$.

Hence,

$$\boxed{L_z^{\text{about O}} = I_{zz}^{\text{CM}} \omega_z + \left(\vec{R}_{\text{CM}} \times M\vec{V}_{\text{CM}} \right)_z.}$$

Separation of torque

There is a similar separation of torque on the body calculated about the z-axis through O. For each force, the torque separates into a torque about the CM and a torque of the force if the force were acting at the CM rather than where it is actually acting. Let $\vec{F}_i$ be an external force on the body acting at a position vector $\vec{r}_i$. Then, the torque from this force about O is

$$\tau_{i,z}^{\text{about O}} = \left(\vec{r}_i \times \vec{F}_i \right)_z$$
$$= \left(\vec{r}_i' \times \vec{F}_i \right)_z + \left(\vec{R}_{\text{CM}} \times \vec{F}_i \right)_z$$
$$= \text{Torque about CM} + \text{Torque about O as if } \vec{F}_i \text{ was acting at CM}.$$

Now, adding torques of all forces on the body we find

$$\tau_{\text{net},z}^{\text{about O}} = \sum_i \left(\vec{r}_i' \times \vec{F}_i \right)_z + \left(\vec{R}_{\text{CM}} \times \vec{F}_{\text{net}} \right)_z$$
$$= \tau_{\text{net},z}^{\text{about CM}} + \left(\text{Torque about O as if } \vec{F}_{\text{net}} \text{ was acting at CM} \right)_z.$$

Separation of the equations of motion

Taking the time derivative of the angular momentum, and then equating the result to the torque, we obtain the equation for rotational dynamics for a fixed axis rotation.

$$\tau_z^{\text{about O}} = \frac{dL_z^{\text{about O}}}{dt}.$$

This relation can be expanded in motion of the CM and motion about the CM as follows. For a rigid body we obtain

$$\left(\vec{r}_i{\,'} \times \vec{F}_i\right)_z + \left(\vec{R}_{\text{CM}} \times \vec{F}_{\text{net}}\right)_z = I_{zz}^{\text{CM}} \alpha_z + \left(\vec{R}_{\text{CM}} \times M\vec{A}_{\text{CM}}\right)_z, \quad (9.99)$$

where α_z is the z-component of the angular acceleration and $\vec{A}_{\text{CM}}$ is the CM acceleration. We know from an earlier chapter that the net external force on a body is equal to the total mass times acceleration of the CM.

$$\vec{F}_{\text{net}}^{\text{ext}} = M\vec{A}_{\text{CM}}$$

Putting this in Eq. 9.99, we can separate out the dynamics of the rotation about the CM from the translation of the CM.

$$\boxed{\text{Rotation about CM:} \quad \tau_z^{\text{about CM}} = I_{zz}^{\text{CM}} \alpha_z} \quad (9.100)$$

$$\boxed{\text{Translation of CM:} \quad \vec{F}_{\text{net}}^{\text{ext}} = M\vec{A}_{\text{CM}}} \quad (9.101)$$

Separation of kinetic energy

Kinetic energy for a rolling motion also separates into the kinetic energy of translation of CM and the kinetic energy of rotation about CM.

$$\boxed{\text{Kinetic Energy} = \frac{1}{2} I_{zz}^{\text{CM}} \omega_z^2 + \frac{1}{2} M V_{\text{CM}}^2.} \quad (9.102)$$

In Table 9.2 the dynamical equations for pure rotations and rolling motion are summarized for a quick reference.

Table 9.2: Pure rotation and rolling motion of a rigid body

Quantity	Pure Rotation	Rolling Motion
Angular momentum	$L_z = I_{zz}\omega_z$	$L_z^{\text{about O}} = I_{zz}\omega_z + \left(\vec{R}_{\text{CM}} \times M\vec{V}_{\text{CM}}\right)_z$
Torque	$\tau_z = \sum \left(\vec{r} \times \vec{F}\right)_z$	$\tau_z = \sum \left(\vec{r} \times \vec{F}\right)_z^{\text{about O}}$ $\tau_z = \left(\vec{R}_{cm} \times \vec{F}_{\text{ext}}\right)_z + \sum \left(\vec{r}\,' \times \vec{F}\right)_z^{\text{about CM}}$
Equations of motion	$\sum \left(\vec{r} \times \vec{F}\right)_z = \frac{d}{dt}(I_{zz}\omega_z)$	Rotation about the CM: $\tau_{z,net}^{\text{about CM}} = \frac{d}{dt}(I_{zz}\omega_z)$ Translation of the CM: $\vec{F}_{\text{ext}} = M\vec{A}_{\text{CM}}$
Kinetic energy	$K = \frac{1}{2} I_{zz}\omega_z^2$	$K = \frac{1}{2} I_{zz}\omega_z^2 + \frac{1}{2} M V_{\text{CM}}^2$

9.6. ROLLING MOTION IN A STRAIGHT LINE

Example 9.6.1. Rolling a drum.

Find the acceleration of a drum rolling downhill without slipping on a slope inclined at angle ϕ.

Solution. First we draw a figure and identify the relevant dynamical variables and forces as shown in Fig. 9.42. For the translation of the CM,

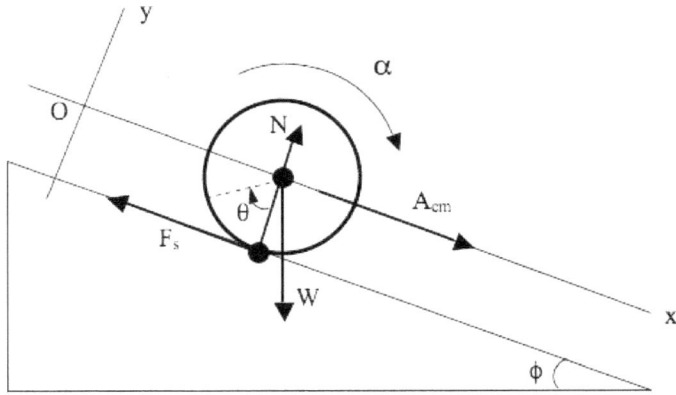

Figure 9.42: Rolling drum down an incline.

it is best to use a Cartesian coordinates with x-axis along the incline as shown in the figure. For the rotation, we can use the angle between a marked line, shown dashed line on the cross-section of the drum, and the vertical to the incline, not the vertical to the ground. From the figure, we find that the angle rotated by the marked line and the distance traveled by the CM have the following relation if the drum rolls without slipping.

$$X_{\text{CM}} = R\theta_z$$

Taking derivative of both sides with respect to time t we find

$$V_x^{\text{CM}} = R\omega_z$$

The x-component of acceleration of the CM is, therefore, related to the z-component of the angular acceleration α_z.

$$A_x^{\text{CM}} = R\alpha_z.$$

The forces acting on the drum are its weight with magnitude $W(=Mg)$, normal force N (unknown) and the static friction F_s (unknown). Now, we will set up the dynamical equations for the translation of the CM and the rotation about the CM. These equations together with the relation between the x-component of the acceleration of the CM and the z-component of the angular acceleration will be sufficient to solve for the acceleration of the CM.

Translation of the CM:

$$\text{x-component: } W\sin\phi - F_s = MA_x^{\text{CM}}$$
$$\text{y-component: } N - W\cos\phi = 0$$

Rotation about the CM:

$$F_s R = \frac{1}{2} M R^2 \alpha_z$$

Replace α_z by A_x^{cm}/R, and solving the three equations simultaneously, we obtain the desired result.

$$A_x^{\text{CM}} = \frac{2}{3} g \sin \phi.$$

Table 9.3: Formulas for moment of inertia components

Shape	Axis	Moment of Inertia	Radius of Gyration
Rod	(through center, perpendicular)	$\frac{1}{12} ML^2$	$\frac{1}{\sqrt{12}} L$
Rod	(through end, perpendicular)	$\frac{1}{3} ML^2$	$\frac{1}{\sqrt{3}} L$
Ring	(through center, perpendicular to plane)	MR^2	R
Ring	(through diameter)	$\frac{1}{2} MR^2$	$\frac{1}{\sqrt{2}} R$
Disk	(through center, perpendicular to plane)	$\frac{1}{2} MR^2$	$\frac{1}{\sqrt{2}} R$
Disk	(through diameter)	$\frac{1}{4} MR^2$	$\frac{1}{2} R$
Cylinder	(through central axis)	$\frac{1}{2} MR^2$	$\frac{1}{\sqrt{2}} R$
Cylinder	(through center, perpendicular to axis)	$\frac{1}{4} MR^2 + \frac{1}{12} ML^2$	$\sqrt{\frac{1}{4} R^2 + \frac{1}{12} L^2}$
Sphere	(through diameter)	$\frac{2}{5} MR^2$	$\sqrt{\frac{2}{5}} R$

9.7 EXERCISES

Rotational Kinematics

Ex 9.7.1. A CD-ROM disk starts to rotate about its axis from rest at a constant angular acceleration of magnitude 4 rad/sec^2. (a) Set up a coordinate system so that one of the Cartesian axis is pointed in the direction of the angular acceleration and use the kinematics of the constant acceleration to find the angle in radians the disk rotates in 15 sec. (b) Find the angular velocity vector at the 15-sec mark.

Ans: (a)450 radians, (b) 60 rad/sec.

Ex 9.7.2. Starting from rest a wheel rotates an angle of 200 radians in 20 seconds at a constant angular acceleration. The axis of rotation is pointed towards the North. (a) Find the magnitude and direction of angular acceleration. (b) Find the magnitude and direction of the angular velocity at $t = 20$ sec.

Ans: (a) Magnitude: 1 rad/s^2, (b) Magnitude: 20 rad/s.

Ex 9.7.3. At $t = 0$ a wheel is rotating at 3 rad per sec with the axis pointed up. You give it a steady torque so that the rotation of the wheel picks up angular speed at a steady rate. After some time of constant angular acceleration, you find that the rotational speed of the wheel is 30 rad per sec. During this time the wheel had rotated by a total angle of 20 radians. Find the magnitude and direction of the angular acceleration.

Ans: Magnitude: 7.1rad/s^2.

Ex 9.7.4. A tire of radius 25 cm rolls on a flat road without slipping such that the axis of rotation is always pointed towards the East and the tire is rolling towards the North. (a) Find the angle rotated when the tire moves a distance of 200 cm in 10 sec. (b) Set up a Cartesian coordinate system so that one axis is towards the direction of motion of the center of mass and another axis is pointed towards the axis of rotation. (c) Find the relation between the components of the velocity of the center of the wheel and the angular velocity of the wheel.

Ans: (a) 8 rad, (c) $\frac{4}{5}$ rad/s.

Ex 9.7.5. A circular groove of diameter 50 cm is made in a horizontal table. A penny (diameter 1 cm) is rolled in the groove. Suppose, the penny rolls at right angle without slipping. (a) Find the total angle the penny rotates about an axis through the center as it goes around the groove once. (b) When the penny rotates at a steady rate, and it takes 10 sec to go around the circle. What is its instantaneous rotation velocity of the penny? (c) Going around the circle can be considered as a revolution of the CM of the penny about an axis that goes through the point of the circle that is momentarily in contact with the groove. What is the angular velocity vector of the CM about this axis?

Ans: (a) 100 π radians; (b) 10 π rad/sec; (c) $\pi/5$ rad/sec.

Angular Momentum of Particles

Ex 9.7.6. A ball of mass 200 grams is moving in a circle of radius 25 cm with a uniform speed of 10 m/s. When observed from above the motion appears counterclockwise. Find the angular momentum about the center of the circle treating the ball as a point particle of mass m. Give both the magnitude and the direction of the angular momentum.

Ans: $0.5\ kg.m^2/s$, up.

Ex 9.7.7. The magnitude of the angular momentum of a steel ball of mass 400 grams moving in a circle in a horizontal plane of radius 50 cm is 3 kg. m^2/sec about the center and the direction of the angular momentum vector is towards the ground. Find the speed of the ball and the clockwise or counterclockwise sense of its motion in the circle as observed from above the circle. Treat the ball as a point particle.

Ans: 15 m/s, clockwise.

Ex 9.7.8. Two point particles of masses m_1 and m_2 are moving uniformly in circles of radii r_1 and r_2 with the same center in the same direction but on the opposite sides such that the total angular momentum of the two has a constant magnitude l_0. The speeds of the two particles are v_1 and v_2 respectively. Set up a coordinate system so that the particles move in the xy-plane and write out the components of the angular momenta of the two masses about the center, and show how the two speeds and radii are related.

Ans: $m_1 v_1 r_1 + m_2 v_2 r_2 = l_0$.

Ex 9.7.9. Let the position vector of a particle of mass 50 g be the vector $(x=2, y=t^2, z=0)$, where t is in sec and coordinates in m. What is the angular momentum at $t = 30$ sec?

Ans: $6 kg.m^2/s$, along z-axis.

Ex 9.7.10. A baton of length b has two balls of mass m at the ends. The mass of the connecting rod is negligible compared to m. The baton is spinning at a constant angular speed ω keeping the mid-point of rod fixed while the masses move in a plane. Let the coordinate system $Oxyz$ be chosen so that the origin is at the middle of baton, and the xy-plane is the plane in which the balls move, and the balls are on the x-axis at time $t = 0$. Write expressions for for an arbitrary time t: (a) the position vectors of the two masses, (b) the momentum of the two masses, (c) the total momentum, (d) the angular momentum of the individual masses, and (e) the total angular momentum. [You should find in this example that even when total momentum is zero, the total angular momentum is not zero!].

Ans: (e) $\frac{1}{2}mb^2\omega \hat{u}_z$.

Ex 9.7.11. Same baton, same mass as above, except this time the baton is spinning so that the plane containing the balls changing with time as shown in Fig. 9.43. Let the connecting rod make an angle θ with the z-axis. Find the same things.

Ans: (e) Answer key: At $t = 0$, $\vec{L} = mR^2\omega\left(-\frac{b}{R}\cos\theta\hat{u}_x + 2\hat{u}_z\right)$.

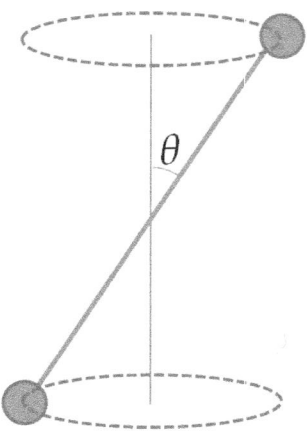

Figure 9.43: Exercise 9.7.11.

Moment of Inertia and Angular Momentum

Ex 9.7.12. A point particle is located at a point P with the coordinates (x,y,z) with respect to a Cartesian coordinate system. Find the moment of inertia of the particle about the x, y and z-axes, i.e. find the formulas for I_{xx}, I_{yy}, and I_{zz}.

Ans: $I_{xx} = m\left(y^2 + z^2\right)$, $I_{yy} = m\left(x^2 + z^2\right)$, $I_{zz} = m\left(x^2 + y^2\right)$.

Ex 9.7.13. Two particles of masses m_1 and m_2 are located at the ends of a light rod of length D. When the rod is placed on a Cartesian coordinate system, the coordinates of the two masses are $(-\frac{D}{2}, 0, 0)$ and $(\frac{D}{2}, 0, 0)$ respectively. Find the moment of inertia of the system containing both the particles about the x, y and z-axes, i.e. find the formulas for I_{xx}, I_{yy}, and I_{zz}. Neglect the mass of the rod compared to m_1 and m_2.

Ans: $I_{xx} = 0$, $I_{yy} = (m_1 + m_2)D^2/4 = I_{zz}$.

Ex 9.7.14. Three particles of masses m_1, m_2 and m_3 are rotating with angular speed ω about the given axis in Fig. 9.44. Find the magnitude and direction of the total angular momentum in two different ways: (a) from a sum of the angular momentum of each particle and (b) first finding the moment of inertia of the three as a system about the axis of rotation, which you can take to be the z-axis, and then multiplying the moment of inertial about the axis with the angular velocity about that axis.

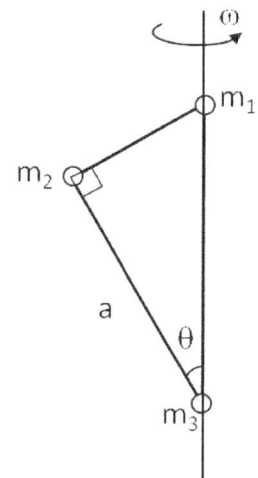

Figure 9.44: Exercise 9.7.14.

Ans: (a) and (b) $m_2(a\sin\theta)^2\omega$.

Ex 9.7.15. Three masses m_1, m_2 and m_3 are rotating about axis with angular speed ω as shown in Fig. 9.45. Find the magnitude and direction of the total angular momentum. Ans: $\left[m_1(a\tan\theta\sin\theta)^2 + m_3(a\cos\theta)^2\right]\omega$.

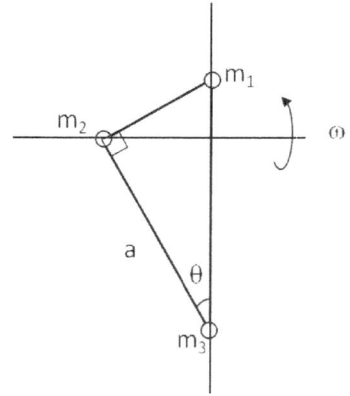

Figure 9.45: Exercise 9.7.15.

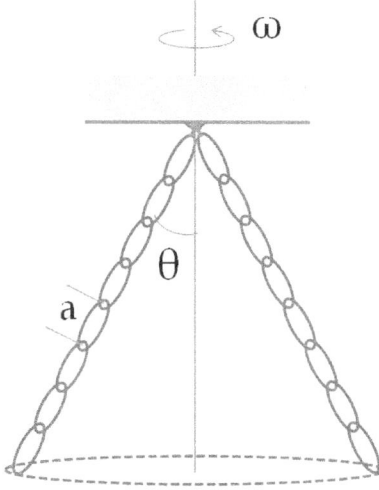

Figure 9.46: Exercise 9.7.16.

Ex 9.7.16. 2N rings, each of mass m, are linked together into a chain and hung from the middle (Fig. 9.46). The chain is then rotated at an angular speed of ω as shown in figure on the right. Model the chain as a sequence of masses separated by distance a and find the magnitude and direction of angular momentum. Ans: $L_z = \frac{ma^2 \sin^2 \theta}{6} N(1+N)(1+2N)\,\omega$.

Ex 9.7.17. Let the z-axis of a Cartesian coordinate system be perpendicular to a thin disk shown in Fig. 9.47. Let the x-axis be parallel to the axis shown in Fig. 9.47(b) but passing through the center. Find the moments of inertia about the axes shown in the figures. Ans: $I_{\text{axis, (a)}} = \frac{3}{2}MR^2$, $I_{\text{axis, (b)}} = \frac{13}{16}MR^2$.

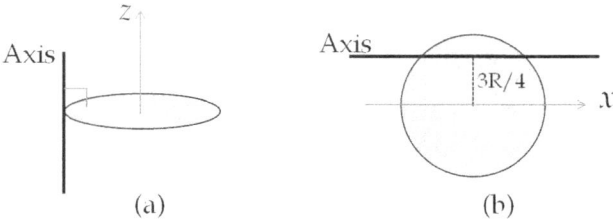

Figure 9.47: Exercise 9.7.17.

Ex 9.7.18. A circular hole of radius a is cut out from the center of a disk of mass M and radius R. Let the disk with the hole be in the xy-plane of a Cartesian coordinate system with the origin of the axes at the center of the disk. Find I_{xx}, I_{yy} and I_{zz}. Ans: $I_{zz} = \frac{1}{2}M\left(R^2 + a^2\right)$, $I_{xx} = I_{yy} = \frac{1}{4}M\left(R^2 + a^2\right)$.

Dynamics of Fixed-Axis Rotation

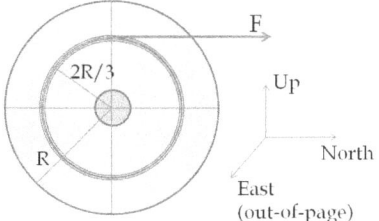

Figure 9.48: Exercise 9.7.19.

Ex 9.7.19. A wheel of mass M and radius R is mounted vertically on an axle so that the axle is parallel to a East-West line. A string is wound on the wheel. When the string is pulled, it unwinds at a distance $\frac{2}{3}R$ from the center of the wheel without slipping as shown in Fig. 9.48. (a) What is the angular acceleration of the wheel if the string is pulled steadily at a constant speed? (b) Suppose the tension in the string has a magnitude T at some instant in time, what will be the angular acceleration of the wheel at instant if the effects of friction at the axle can be neglected? (c) Suppose you find that a tension of magnitude T_0 is needed to pull the string at a constant speed, what friction must be acting at the axle if friction acts at a distance r from the center of the wheel, which would be the radius of the axle? Ans: (a) $\alpha = 0$. (b) $\alpha_z = \frac{4}{3}\frac{T}{MR}$. (c) $\frac{2R}{3r}T_0$.

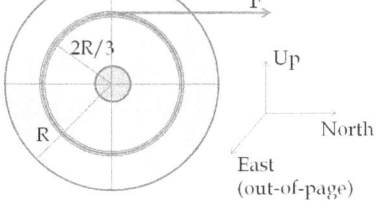

Figure 9.49: Exercise 9.7.20.

Ex 9.7.20. A plank of mass M and length L is pivoted at one end (Fig. 9.49). Initially, the plank is supported at the other end so that the plank is horizontal. When the plank is released, the imbalance of the torque about the pivoted end leads to an angular acceleration of the plank. Find the angular acceleration of the plank at two instants (a) immediately after the plank is released such that the plank is horizontal but not supported and (b) when the plank makes an angle θ with the horizontal. Note: For the purpose of computing the torque, the entire weight may be assumed to act at the center of mass of the body. This does not mean that gravity acts at the center of mass. The force of gravity acts at individual particles of the body at different places. Ans: With the positive z axis out of page, (a) $\alpha_z = -\frac{3}{2}\frac{g}{L}$. (b) $\alpha_z = -\frac{3}{2}\frac{g\cos\theta}{L}$.

Ex 9.7.21. This is a long problem and covers a number of important concepts. Two children of mass m_1 and m_2 are sitting at r_1 and r_2 from the center on the two sides of a plank of mass M and length L that is pivoted at the center as shown in Fig. 9.50. Consider a particular instant when the plank makes an angle θ with the horizontal direction and is rotating counterclockwise. Let the origin of Cartesian coordinate system be at the pivot point, which is the point where the plank touches the support, the z-axis be perpendicular to the drawing shown in Fig. 9.50, and the positive z direction corresponds to the "coming out-of-page" direction. Assume there is enough static friction between the children and the plank so that the children do not slide on the plank.

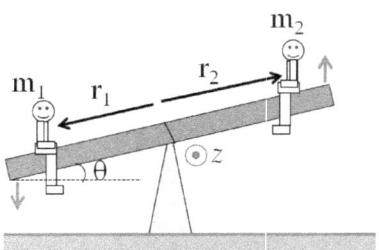

Figure 9.50: Exercise 9.7.21.

(a) Draw free-body diagrams of the forces on each child, choose a coordinate system, and write Newton's second law equations for the two children in the component form.

(b) Draw a diagram showing all forces on the plank and where they act, choose a coordinate system with origin at the pivot point, and find the components of the net torque on the plank about the pivot point.

(c) Draw a free-body diagram of the forces on the plank, and determine the components of the net force on the plank.

(d) Write rotational equation of motion for the plank in the component form. Note that the force by children on the plank will not be mg of the children but the normal and frictional forces between the children and the plank. Note: the force mg on a child acts on that child, not on the plank. The force between the plank and the child consists of normal and frictional forces between the two.

(e) Write any relations between components of the acceleration and the angular acceleration of the plank.

(f) Solve the equations you have generated for (i) acceleration of the children at the instant under consideration, (ii) the angular acceleration of the plank, (iii) the forces between the children and the plank, and (iv) the force of the support on the plank at the pivot point. Ans: $\alpha = \frac{m_1 g r_1 \cos\theta - m_2 g r_2 \cos\theta}{\frac{1}{12}ML^2 + m_1 r_1^2 + m_2 r_2^2}$.

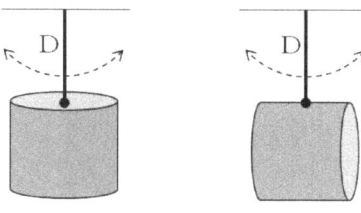

Figure 9.51: Exercise 9.7.22.

Ex 9.7.22. A cylinder of mass M, radius R and length $2R$ is attached to massless rods of length D in two way to form two different physical pendula as shown in Fig. 9.51. Here D is not necessarily large compared to R. Consider an arbitrary instant when these pendula make an angle θ with respect to the vertical line pointed down from the point of suspension.

(a) Find the moment of inertia of the cylinder about an axis through the point of suspension and perpendicular to the drawing, i.e. coming out-of-page.

(b) Draw diagrams of forces, with information regarding where they act. Then, find the net torques on the bodies about the same axis, i.e. if the axis is the z-axis, then find τ_z.

(c) Write the rotational equations of motion in the component form for the two bodies. Make sure you have the signs of various components correctly identified, which follows from your choice of the coordinate system.

(d) Make the small angle approximation and then read off the periods or frequencies of the two physical pendula from their equations of motion. What is the ratio of the frequencies of small oscillations of the two pendula?

Ans: (a) $I_{\text{axis}} = \frac{5}{4}MR^2 + 2MDR$. (d) $f = \frac{1}{2}\sqrt{\frac{Mg(D+\frac{H}{2})}{I_{\text{axis}}}}$.

Conservation of Angular Momentum

Ex 9.7.23. A disk of mass M and radius R is rotating at an angular speed ω about an axle through its center and vertical to its plane. Another disk of mass M' and radius R' falls on the axle and gets stuck. The two rotate at a lower speed ω' about the same axis passing through their common center. Show your choice of a coordinate system and implement the conservation of angular momentum in the component form to find ω' in terms of the other quantities given. Ans: $\omega' = \left[\frac{MR^2}{MR^2+M'R'^2}\right]\omega$.

Ex 9.7.24. A projectile of mass $1,000,000$ kg is fired from Earth horizontally towards the East at the equator with speed 10^6 m/s. This blast will actually change the angular speed of the rotation of the Earth. Find the new time for one full rotation of the Earth (i.e. the duration for new one-day) afterwards. Assume Earth to be a sphere of mass 6.0×10^{24} kg and radius 6.4×10^6 m and the axis of rotation to be perpendicular to the cross-section of earth at the equator. Note: you must use an inertial coordinate system fixed at the center of the Earth for the law of conservation of angular momentum to be applicable. Ans: 7.8×10^{-11} sec.

Ex 9.7.25. A boy of mass 50 kg is standing 7 m from the center of a large circular disk of mass 300 kg and radius 10 m rotating at an angular speed of 0.6 rad/sec. When the boy walks on the platform, the rotation speed changes. Show your choice of a coordinate system and implement the conservation of angular momentum in the component form to find the rotation speed when the boy is 3 m from the center. Ans: 0.68 rad/sec.

Ex 9.7.26. A wheel of mass $M = 10$ kg and radius $R = 0.5$ m is at rest (Fig. 9.52). A putty of mass $m = 0.5$ kg flies at the rim of the wheel striking it at a speed of $v = 15$ m/s. The putty gets stuck after striking the wheel and as a result the wheel and putty start to rotate together. Find the magnitude and direction of angular velocity of the combination in terms of M, R, m and v before putting in the numbers to get a numerical value for the magnitude. Ignore any resistance at the axle of the wheel and assume all mass in the wheel to be concentrated at the rim. Hint: The flying putty has non-zero angular momentum about the center of the disk. Ans: 1.4 rad/sec.

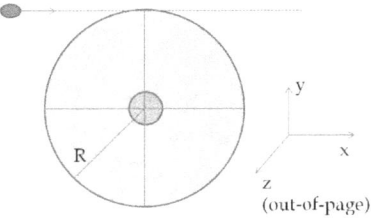

Figure 9.52: Exercise 9.7.26. A coordinate system is shown for convenience.

Rotational Work and Kinetic Energy

Ex 9.7.27. (a) A 12-in wrench is used to tighten a bolt by applying a steady force of 5 N at the end of the handle which is approximately 10 inches from the center of the bolt. How much work is done for each quarter turn?

(b) Suppose, you use a 6-in wrench, and apply the same force, except that the force now would act at a distance of about 5 inches from the center of the bolt. How much work would now be done for each quarter turn?

Ans: (a) 2.0 N.m. (b) 1.0 N.m.

Ex 9.7.28. A uniform density disk of mass 50 kg and radius 12 cm is set on an axle through its center and rotated from rest. (a) How much work is required to bring the angular speed from zero to 10 revolutions per second? (b) How much work is required to change the angular speed from 10 rev/sec to 20 rev/sec? (c) If an average force of 20 N acts at the rim in the direction of the tangent to the rim of the disk continuously, how many rotations would it take to reach 10 rev/sec from rest? (d) If you want to get to 20 rev/sec from 10 rev/sec in half as many rotations that it took you to get from rest to 10 rev/sec using the average 20 N force, what average force would you need to act along the tangent of the rim?

Ans: (a) 4,400 J, (b) 13,00 J, (c) 47, (d) 120 N.

Ex 9.7.29. A steel ball of radius R is attached at the end of a cylindrical steel rod of radius r and length L as shown in Fig. 9.53. The other end of the rod is pivoted about a shaft, and the assembly is rotated. (a) How much work is required to bring the angular speed to ω starting from rest? (b) How much work is required to change its angular speed from ω to 2ω? Use the symbol ρ for the density of steel.

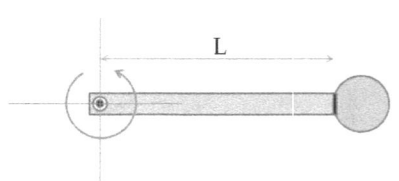

Figure 9.53: Exercise 9.7.29. Ignore the mass to the left of the hole in the rod.

Ex 9.7.30. A ceiling fan rotates at a steady speed of ω. Let the magnitude of average torque applied by the motor be $\bar{\tau}$. (a) Find the rotational work done by the motor in N revolutions. (b) If the moment of inertia of the fan about the axis of rotation is I_0, what is the change in rotational energy of the fans in each turn when the fans are rotating at a steady speed? (c) What happens to the work done by the motor? Explain.

Ans: (a) $2N\pi\bar{\tau}$, (b) No change, (c) Goes towards the work by drag.

Rolling Motion

Ex 9.7.31. A bicycle moving in a straight line speeds from 5 m/s to 15 m/s over a distance of 200 meters at a constant acceleration without slipping. Each wheel has a radius of 25 cm and a mass of 250 grams with all masses assumed to be at the rim. Find the following quantities: (a) the angular speed of the wheel at the initial and the final instants, (b) the magnitude of angular rotation of each wheel over the time interval, (c) the angular acceleration during the interval, (d) the net torque on each wheel about the center of the wheel, and (e) the net force on the bicycle, assuming the mass of the bike is 1.2 kg.

Ans: (a) 20 rad/s and 60 rad/s, (b) 800 radians, (c) 2 rad/s2, (d) 0.03N.m.

Ex 9.7.32. A yoyo of radius R and mass M has a massless thread wound on the inside wheel of radius r (Fig. 9.54). The thread is then held and the yoyo is allowed to fall. As the yoyo falls the thread unwinds smoothly. Find the tension in the thread and the speed of fall of the yoyo when it has fallen a height h vertically. For purposes of the moment of inertia, assume the yoyo to be a uniform thin disk of of mass M and radius R. You may use $R = 3$ cm, $M = 300$ g, $r = 2$ cm, and $h = 0.8$ m if you prefer numerical problem.

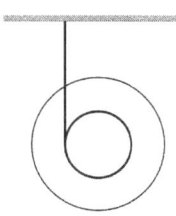

Figure 9.54: Exercise 9.7.32.

Ans: 1.56 N, 2.7 m/s.

Ex 9.7.33. A spherical marble of radius R and mass M rolls down in a straight line on an inclined plane with the angle of inclination θ without slipping. What will be the speed of the marble when it has rolled down a distance D without slipping as measured on the incline?

Ans: $\sqrt{\frac{10}{7}gD\sin\theta}$.

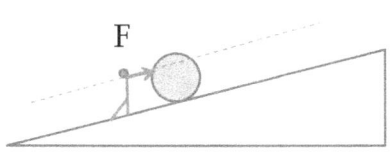

Figure 9.55: Exercise 9.7.34.

Ex 9.7.34. A heavy drum of mass M and radius R is being pushed up an incline of angle of inclination θ by a man who is applying a force of magnitude F horizontally with the incline at a height of $\frac{3}{2}R$ from the incline (Fig. 9.55). (a) What would be the acceleration of the center of mass of the drum if the coefficient of rolling friction is μ_r? (b) How long will it take to move the drum a distance L on the incline?

Ans: (a) $a = \dfrac{F}{m} - g(\sin\theta - \mu_r\cos\theta)$, (b) $t = \sqrt{\dfrac{2\Delta x}{a}}$.

9.8 PROBLEMS

Problem 9.8.1. Consider two blocks of masses m_1 and m_2 attached through a massless string which goes over a pulley of mass M and radius R as shown in Fig. 9.56. Block m_1 slides on a frictionless table while block m_2 moves in the vertical direction. Assume the pulley to be a disk of uniform mass density with no friction at the axle. Note that the pulley here is not an "ideal" pulley. Find the acceleration of the masses and the tension in the string at various points. You can ignore the mass of the connecting strings compared to all other masses in the problem.

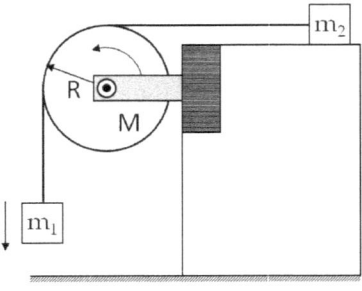

Figure 9.56: Problem 9.8.1.

Ans check: If $m_1 = m_2 = M$ then $a = 2g/5$, $T_1 = 3mg/5$, $T_2 = 2mg/5$.

Problem 9.8.2. In the problem above, suppose an average frictional torque from the axle on the pulley has the magnitude τ_f and the radius of the axle is r. How would your answer for the acceleration and tensions be modified? Assume, block m_2 still slides frictionlessly.

Ans: $a = \frac{m_1 g - \frac{r}{R} F_k}{m_1 + m_2 + \frac{1}{2}M}$. Work out T_1 and T_2 as well.

Problem 9.8.3. A marble of mass m rolls down a circular track without slipping starting from rest at a height h and hits another marble of mass M horizontally (Fig. 9.57). As a result of the collision, the marble m comes to rest while the marble M leaves the track with a horizontal velocity, which you need to determine as an intermediate step towards answering the following question. Where does the marble with mass M land on the floor a height H below?

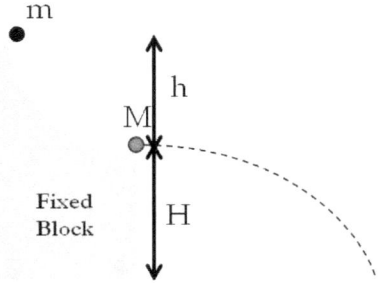

Figure 9.57: Problem 9.8.3.

Ans: $D = \frac{m}{M}\sqrt{\frac{20}{7}hH}$.

Problem 9.8.4. A car rounds a circular bend at a constant speed of 10 m/s. The radius of the bend from the CM of the car is 40 meters. The road can be assumed to be flat. There is enough friction so that car does not slide away. The distance between the left and right wheels is 1.5 meters and between the front and back wheels is 2.5 meters. The CM of the car is in the middle at a height of 30 cm from the ground. Find the value of the magnitudes of the normal forces on the four wheels of the car. Ans: $F_{\text{left wheels}} = 16{,}000$ N, $F_{\text{right wheels}} = 13{,}000$ N.

Problem 9.8.5. A putty of mass m is tossed at a rod of mass M and length l that is free to rotate frictionlessly about an axle through the center (Fig. 9.58). The putty strikes the rod horizontally with a speed v_0 at $\frac{1}{4}l$ from the center. (a) Find the angular speed of the rod if the putty sticks to the rod. (b) Find the angular speed of the rod if the putty bounces straight back. You can assume the putty is small enough that we can treat it as a point mass. Ans: (a) $\omega = \frac{12mv_0}{3ml + 4Ml}$, (b) $\omega' = \frac{6mv_0}{Ml}$.

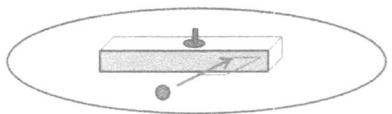

Figure 9.58: Problem 9.8.5.

Problem 9.8.6. A spherical ball of mass M and radius R is attached to a thick rod of mass m and length l which is pivoted at the other end (Fig. 9.59). Treating the system as a physical pendulum find the frequency of small oscillation. Ans Key: If $m = M$ and $l = R$ the $f = \frac{1}{2\pi}\sqrt{\frac{25g}{142l}}$.

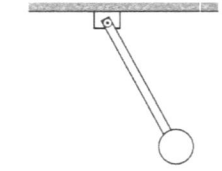

Figure 9.59: Problem 9.8.6.

Figure 9.60: Problem 9.8.7.

Problem 9.8.7. When a string is twisted, it applies a torque at the ends. The magnitude of the torque depends on the twist angle similar to the spring force. For small angles of twist, Hooke's law can be applied to express the torque as a linear function of the angular displacement, i.e the twist angle, θ.

$$\tau = \kappa \theta,$$

where κ is a constant. Now, consider tying a string to the center of a disk of mass M and radius R, and hanging the mass from a fixed support as shown in Fig. 9.60. The string is then given a twist by a small amount and released. The disk rotates back and forth with a definite frequency. Find the frequency of these oscillations. Hint: Write the equation of motion for the rotation of the disk. Ans: $f = \frac{1}{2\pi}\sqrt{\frac{2\kappa}{MR^2}}$.

Problem 9.8.8. A meter stick of length l and mass m is held against a frictionless wall and a frictionless floor (Fig. 9.61). When the stick is let go from rest, it starts to slide down the wall keeping in contact with the wall till it gains enough speed and then the stick leaves the wall at a particular point. Find the point at which the meter stick leaves the wall. Ans: $h = \frac{2}{3}h_0$ where h_0 is the initial height.

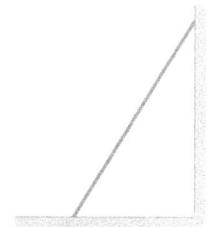

Figure 9.61: Problem 9.8.8.

Problem 9.8.9. A meter-stick is set almost vertically on a frictionless table. When the stick is let go at rest from the nearly vertical orientation it falls to the table (Fig. 9.62). Since there is no horizontal force on the stick and it starts out at rest, the center of mass of the stick falls straight down. Find the speed of the CM as a function of the position of the CM with respect to the table. Hint: Let the y-axis be the vertical direction and the z-axis into-the-page, and write the y-component of equation of motion for the translation of CM and the z-component of the rotational motion of the stick about the CM.

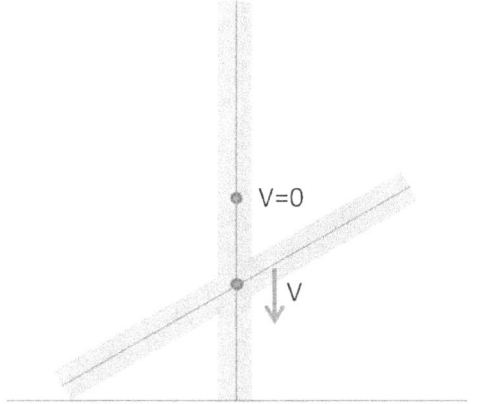

Figure 9.62: Problem 9.8.9.

9.8. PROBLEMS

Problem 9.8.10. A "massless" tape is wound on one disk of mass m_1 and radius R_1. The tape then goes over a pulley of mass M and radius R and the other end of the tape is wound on another disk of mass m_2 and radius R_2 as shown in Fig. 9.63. The disks are then released from rest with tapes taut on each side. Assume the tape unwinds smoothly on each side. Determine the accelerations of each mass and the angular acceleration of the pulley.

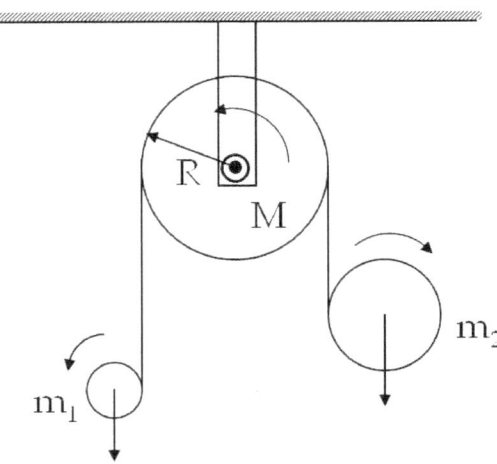

Figure 9.63: Problem 9.8.10.

Chapter 10

NONINERTIAL FRAMES

Contents

10.1 ACCELERATING FRAME	**416**
10.1.1 Kinematics in Accelerating Frame	416
10.1.2 Newton's Second Law in Accelerating Frame .	417
10.2 ROTATING FRAME	**420**
10.2.1 Kinematics in a Uniformly Rotating Frame . .	420
10.2.2 Newton's Second Law in Uniformly Rotating Frame	425
10.2.3 Newton's Second Law in Earth's Frame	425
10.3 CORIOLIS FORCE	**430**
10.4 EXERCISES	**432**
10.5 PROBLEMS	**436**

So far we have studied motion with respect to inertial observers. For these observers, Newton's second law of motion takes the familiar form, "the rate of change of momentum equals the net force", or $\vec{F} = m\vec{a}$ for a constant mass particle.

Observers that have a non-zero acceleration with respect to an inertial observer are called non-inertial. When the second law of motion is written in a non-inertial frame, the equation of motion is modified. The new equations of motion contain forces not present in the inertial frame are called fictitious or inertial forces. We will study these modifications in this chapter.

A particularly important application of these modified equations is to the frames fixed to the Earth, also called Earth-based frames. Since Earth is rotating about its axis, all observers on the Earth are accelerating with respect to any fixed inertial frame. This makes all Earth-based frames non-inertial.

Newton's second law of motion in rotating frames that have a constant angular velocity with respect to an inertial frame, two inertial forces arise. These inertial forces are called centrifugal and Coriolis forces. We will study their implications in some detail in this chapter. But, first we will study a simple situation of a frame accelerating in a straight line with respect to an inertial frame.

10.1 ACCELERATING FRAME

10.1.1 Kinematics in Accelerating Frame

The position of a particle in an accelerating frame is defined in the same way as in any other frame. The position vector of a point particle is the displacement vector from the origin to the current position of the particle. The relation with a non-accelerating frame can be established by drawing the two frames and noting the triangle of vectors formed by the position of the particle $\vec{r}$ and $\vec{r}\,'$ in the non-accelerating and accelerating frames respectively and the vector $\vec{R}$ from the origin O of the non-accelerating frame to the origin O' of the accelerating frame (Fig 10.1).

$$\vec{r} = \vec{R} + \vec{r}\,' \tag{10.1}$$

The velocity $\vec{V}$ and acceleration $\vec{A}$ of the accelerating frame with respect to the non-accelerating frame are simply the velocity and acceleration of origin O' of the accelerating frame, and can be obtained by successively taking time derivatives of vector $\vec{R}$.

$$\vec{V} = \frac{d\vec{R}}{dt} \tag{10.2}$$

$$\vec{A} = \frac{d\vec{V}}{dt} = \frac{d^2\vec{R}}{dt^2} \tag{10.3}$$

10.1. ACCELERATING FRAME

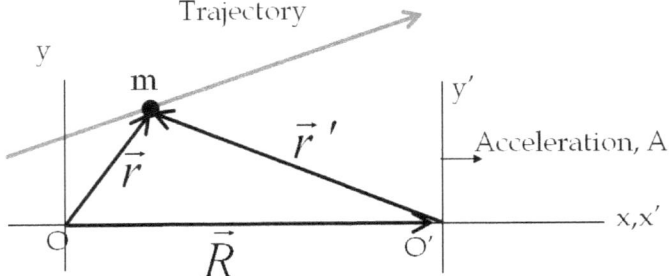

Figure 10.1: Position vectors of a point mass in two frames.

The velocity $\vec{v}'$ of the particle with respect to the accelerating frame is simply the rate at which position $\vec{r}'$ of the particle changes with time.

$$\vec{v}\,' = \frac{d\vec{r}\,'}{dt} \qquad (10.4)$$

Similarly, the acceleration $\vec{a}\,'$ of the particle with respect to the accelerating frame is simply the rate at which velocity $\vec{v}\,'$ of the particle with respect to the same frame changes with time.

$$\vec{a}\,' = \frac{d\vec{v}\,'}{dt} \qquad (10.5)$$

Now, by taking successive derivatives of both sides of Eq. 10.1 with respect to time gives us the relation between velocity and acceleration of a particle in an in an inertial and a non-inertial frame.

$$\boxed{\vec{v} = \vec{V} + \vec{v}\,'} \qquad (10.6)$$

$$\boxed{\vec{a} = \vec{A} + \vec{a}\,'} \qquad (10.7)$$

10.1.2 Newton's Second Law in Accelerating Frame

Newton's second law of a point particle of mass m in an inertial frame is given by

$$m\vec{a} = \vec{F} \qquad (10.8)$$

where $\vec{F}$ is the net force on the particle and $\vec{a}$ the acceleration. This equation remains the same in all inertial frames, which are frames that have constant velocity with respect to each other. Now, we wish to find the corresponding equation of motion if the particle's motion is observed from an accelerating frame. Let $\vec{a}\,'$ be the acceleration of the particle as observed from an accelerating frame. Let the acceleration of the accelerating frame with respect to the inertial frame be $\vec{A}$. By substituting Eq. 10.7 from the last section, we find that the following relation holds.

$$m\vec{a}\,' = \vec{F} - m\vec{A}. \qquad (10.9)$$

Therefore, mass times acceleration of a particle with respect to an accelerating frame is not equal to the net force, but to the net force minus mass of the particle times acceleration of the frame itself. Thus, in an accelerating frame one needs to add $(-m\vec{A})$ to the real net force $\vec{F}$ to equate the "corrected" force to mass times acceleration.

The quantity $(-m\vec{A})$ acts like an additional force on the mass m, and is called a **"fictitious force"** or **"inertial force"** to distinguish it from the "real force" $\vec{F}$. In the accelerating frame the inertial forces are felt the same way as real forces except there is no agent that applies them. The inertial force in a uniformly accelerating frame acts just like the gravitational force since it is proportional to mass. This observation led Albert Einstein to develop an alternate theory of gravitation called the general theory of relativity.

Example 10.1.1. Apparent Weight in an Elevator. As an example of an accelerating frame, consider observation made by a person in an elevator which accelerates with respect to the ground. We treat the frame of a ground-based observer as an inertial frame for purposes of this example. Both the ground-based observer and the observer in the elevator have access to a reading of a weighing scale fixed to the elevator. A person stands on the weighing scale in the elevator. What will be the reading in the weighing scale as recorded by the two observers?

Solution. First, we note that the reading in the scale is that of the normal force between the scale and the person. Since the force is a real force, the two frames will have the same values for the normal force, although the calculations in the two frames will differ. In the following we present the two calculations.

Ground-based frame.

This is the non-accelerating frame, therefore there will be no inertial forces in this frame. The person on the scale has acceleration $\vec{A}$ with respect to this frame since the acceleration of the person is same as that of the elevator. The forces on the person are (1) gravity (Mg, pointed down) and (2) normal (unknown, pointed up). The free-body diagram is shown in Fig. 10.2.

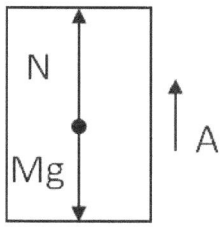

Figure 10.2: The free-body diagram of forces on a person in an accelerating elevator drawn from the perspective of the ground-based frame. The person accelerates with the elevator.

Let y-axis be pointed up. Now, we write down the y-component of Newton's second law in this inertial frame.

$$N - Mg = MA$$

Therefore, the reading on the scale will be:

$$N = M(g + A).$$

This result says that, when the accelerator is accelerating up (meaning the direction of the acceleration is pointed up), the scale will give a higher reading than Mg. If the acceleration of the elevator is pointed down, then A will be negative. In that case the scale will read less than Mg. Note that the reading on the scale does not depend upon the direction of the motion

10.1. ACCELERATING FRAME

of the elevator but rather the direction of the acceleration. Therefore the scale reading is more than Mg if the acceleration is pointed up regardless of whether the elevator is going up or going down.

Elevator frame.

This is non-inertial frame when the elevator has non-zero acceleration $\vec{A}$ with respect to the ground. In this frame the real forces on the person are the same two force, viz. the forces of gravity (Mg, down) and normal (N unknown, up). But the statement of the second law is different. We need to subtract $M\vec{A}$, from the real forces to get the net force on the person as shown in the free-body diagram of the person in the elevator frame given in Fig. 10.3.

The person on the scale does not have any acceleration with respect to this frame since the person moves with the elevator.

$$\vec{a}\,' = 0.$$

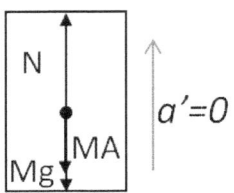

Figure 10.3: The free-body diagram of forces on a person in an accelerating elevator drawn from the perspective of the elevator frame. Since the person accelerates with the elevator, his acceleration with respect to the elevator is zero.

Taking the vertically up direction as the positive y-axis, the y-component of the modified equation of motion is

$$N - Mg - MA = 0$$

Therefore, we find $N = M(g+A)$, which is the same conclusion for the reading on the scale as we found by working in the ground-based inertial frame.

Example 10.1.2. Pendulum In An Accelerating Train.

As a second example, consider a pendulum of mass M hanging from the ceiling of a train. When the train is at rest or coasting at a constant velocity, the pendulum hangs vertically. But when the train is accelerating the bob hangs at an angle θ to the vertical. We wish to find this angle when the train's acceleration is $\vec{A}$. We will do this problem in two frames to illustrate a non-accelerating frame and an accelerating frame.

Equations of motion in the ground-based frame ($\vec{a} \neq 0$):

$$\text{x component: } T\sin\theta = MA$$
$$\text{y component: } T\cos\theta - Mg = 0$$

Equations of motion in the train-based frame ($\vec{a}' = 0$):

$$\text{x component: } T\sin\theta - MA = 0$$
$$\text{y component: } T\cos\theta - Mg = 0$$

Note that the two frames yield identical relations for the equations of motion. Both frames will yield the same value for the real forces such as the tension in the string.

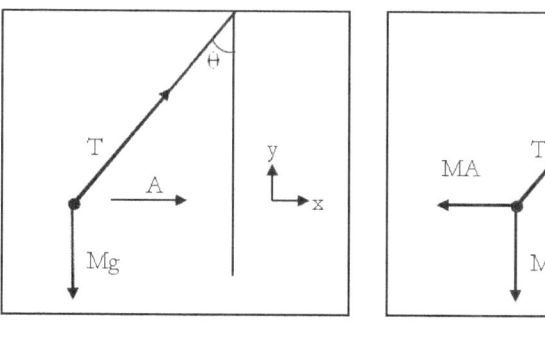

(a) Inertial Ground Frame   (b) Accelerating Train Frame

Figure 10.4: Example 10.1.2. Free-body diagrams in (a) inertial and (b) non-inertial frames. Note the additional inertial force on the bob in the accelerating frame.

10.2 ROTATING FRAME

10.2.1 Kinematics in a Uniformly Rotating Frame

Since a rotating frame is not an inertial frame, the form of the second law, $\vec{F} = m\vec{a}$, is modified to $\vec{F} - m\vec{A} = m\vec{a}\,'$, where $\vec{A}$ is the acceleration of the non-inertial frame with respect to the inertial frame. In this section, we will work out the implications of the acceleration of the frame arising from the rotation of the frame with respect to an inertial frame.

Let us denote the kinematic variables in the accelerating frame with a prime as above. Let $\vec{r}\,'$, $\vec{v}\,'$, and $\vec{a}\,'$ be the position, velocity, and acceleration of the particle as observed in the rotating frame. Their definitions are the same in the rotating frame as in any other frame, viz., the velocity being the rate of change of the position and the acceleration the rate of change of the velocity.

To relate the position, velocity and acceleration vectors in the rotating frame to the corresponding quantities in the non-rotating inertial frame, it is helpful to examine these quantities for a point particle in two different frames that share a common origin but one rotating with respect to the other about the common z-axis as shown in Fig. 10.5.

For simplicity we pick the zero of time at the instant when the two frames have their axes lined up in the same directions so that all the coordinates of the same particle in the two frames have the same values at $t = 0$. In the following, we will consider two situations. In the first situation, we will choose a particle at rest in the rotating frame, and in the second situation, the particle will be allowed to move in a plane perpendicular to the rotation axis of the rotating frame.

10.2. ROTATING FRAME

A particle fixed at the x-axis of a uniformly rotating frame

Consider a particle fixed at $(x', 0, 0)$ as shown in Fig. 10.5. Although this particle's position in the rotating frame is fixed, its position in the fixed frame is changing. As a matter of fact, as far as the fixed frame is concerned, the particle moving in xy-plane in a circle of radius equal to x' with angular speed Ω.

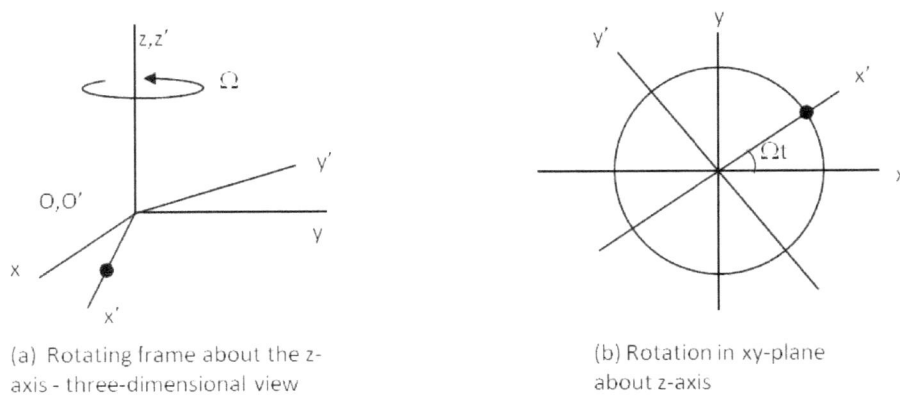

(a) Rotating frame about the z-axis - three-dimensional view

(b) Rotation in xy-plane about z-axis

Figure 10.5: Looking at a particle at rest in a rotating frame.

The position of the particle in the two frames can be obtained by simple trigonometry of the situation in the xy-plane. We will omit the z-components since they are all zero in the two frames.

$$\text{Rotating frame:} \quad x' = \text{fixed}, \; y' = 0$$
$$\text{Inertial frame:} \quad x = x'\cos\Omega t, \; y = x'\sin\Omega t$$

Therefore, the velocities of the particle in the two frames, obtained by taking derivatives of the positions, are as follows.

$$\text{Rotating frame:} \quad v'_x = \frac{dx'}{dt} = 0, \; v'_y = \frac{dy'}{dt} = 0$$
$$\text{Inertial frame:} \quad v_x = \frac{dx}{dt} = -\Omega x'\sin\Omega t, \; v_y = \frac{dy}{dt} = \Omega x'\cos\Omega t$$

The acceleration of the particle in the two frames can be obtained by taking derivatives of the corresponding velocities.

$$\text{Rotating frame:} \quad a'_x = \frac{dv'_x}{dt} = 0, \; a'_y = \frac{dv'_y}{dt} = 0$$
$$\text{Inertial frame:} \quad a_x = \frac{dv_x}{dt} = -\Omega^2 x'\cos\Omega t, \; a_y = \frac{dv_y}{dt} = -\Omega^2 x'\sin\Omega t$$

The x and y-components of acceleration in the inertial frame are equivalent to the radially inward acceleration, i.e. the centripetal acceleration.

Inertial frame:

$$\vec{a} = \begin{cases} \text{Magnitude} = \Omega^2 x' \\ \text{Direction: Towards the origin.} \end{cases} \tag{10.10}$$

This makes sense, because from the perspective of the fixed frame, the particle moves in a uniform circular motion of radius equal to x' at constant angular speed Ω. Therefore, the acceleration in the fixed frame must be pointed radially inwards.

Vector Notation

It is instructive to write the kinematic quantities in the fixed and rotating frames given above in a vector notation. The example particle does not move with respect to the rotating frame. This particle will appear to rotate in a circle about the z-axis when observed from the fixed frame. Therefore, in the fixed frame the angular velocity vector will be pointed along the z-axis.

$$\vec{\Omega} = \Omega \hat{u}_z. \tag{10.11}$$

The velocity of the particle in the fixed frame is tangential to the circle in the xy-plane and is given by the cross product of the angular velocity vector and the position vector.

$$\vec{v} = \frac{d\vec{r}}{dt} = \vec{\Omega} \times \vec{r}. \tag{10.12}$$

Since acceleration is equal to the time-derivative of velocity, we obtain the following for the acceleration in the fixed frame.

$$\vec{a} = \frac{d\vec{v}}{dt} == \vec{\Omega} \times \frac{d\vec{r}}{dt}$$
$$\implies \vec{a} = \vec{\Omega} \times \left(\vec{\Omega} \times \vec{r}\right) \quad (\text{Const } \vec{\Omega})$$

We find that, although the particle is not moving in the rotating frame, the particle has an acceleration in the fixed frame.

A particle moving in the xy-plane of a uniformly rotating frame

Let us consider a particle that is not fixed but moves in the xy-plane of a uniformly rotating frame $Ox'y'z'$ which is rotating with respect to a fixed frame $Oxyz$ about their common z-axis. Since the frame is rotating about the z-axis, the particles moves in the xy-plane of the fixed frame as well. The situation between time t and $t + \Delta t$ is shown in Fig. 10.6. For simplicity, let the axes of the two frames be coincident at time t.

Let (x, y) and (x', y') be the coordinates of the particle at some time t in the $Oxyz$ and $Ox'y'z'$ frames respectively. The change of the coordinates over the interval from t to $t+\Delta t$ are $(\Delta x, \Delta y)$ and $(\Delta x', \Delta y')$ respectively. What are their relations? The relation between the displacements in the two frames is more conveniently written in the vector notation. Therefore, we will work with vector notation here.

At time t, let the particle be at point P along momentarily coincident x and x'-axes (Fig. 10.7). The non-rotating frame marks the location with a marker at P at its x-axis and the rotating frame marks the same

10.2. ROTATING FRAME

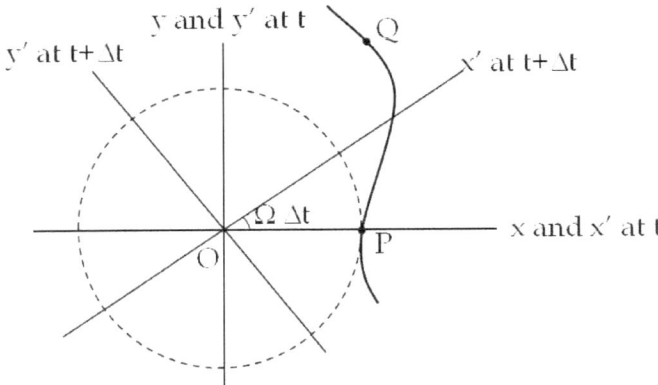

Figure 10.6: The physical situations at times t and $t+\Delta t$. In duration t to $t+\Delta t$, the particle moves from P to Q, and the rotating frame rotates by an angle Ωt. The common z-axis pointed out-of-page.

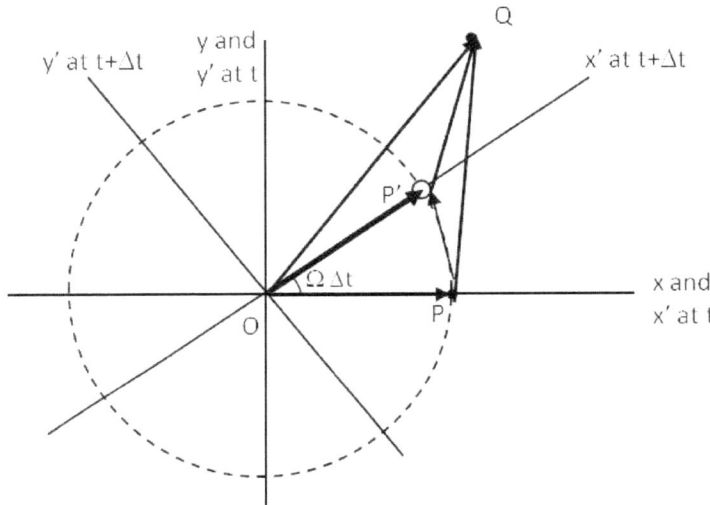

Figure 10.7: Position of a point particle P from rotating frame.

location in its frame as P' at its x-axis. As time passes, the particle moves to another location shown as Q at time $t + \Delta t$. Meanwhile the rotating frame has also moved.

Therefore, from the perspective of the rotating frame, the displacement of the particle between t and $t + \Delta t$ is the vector from P' on its x-axis where the particle was at t to the space point Q. The same displacement of the particle from the perspective of the non-rotating frame is PQ vector in space. From $\triangle PP'Q$ we obtain the following relation in space.

$$\overrightarrow{PQ} = \overrightarrow{PP'} + \overrightarrow{P'Q}. \tag{10.13}$$

Now, $\overrightarrow{PP'}$ is on the arc of a circle of radius $|\vec{r}|$ with the arc angle $\Omega \Delta t$, therefore

$$\overrightarrow{PP'} = \left(\vec{\Omega}\Delta t\right) \times \vec{r}.$$

The vector $\overrightarrow{PQ}$ is the displacement $\Delta \vec{r}$ of the particle in the fixed frame and $\overrightarrow{P'Q}$ is the displacement $\Delta\vec{r}\,'$ of the particle in the rotating frame. Putting these quantities in Eq. 10.13 we have the following relation among displacements in the time interval Δt.

$$\Delta\vec{r} = \left(\vec{\Omega}\Delta t\right) \times \vec{r} + \Delta\vec{r}\,'.$$

Dividing by Δt and taking the $\Delta t \to 0$ limit, we find that the velocity in the inertial frame is equal to the sum of the velocity of the particle in the rotating frame and an additional term resulting from the rotation of the frame.

$$\vec{v} = \vec{v}\,' + \vec{\Omega} \times \vec{r}. \tag{10.14}$$

Rather than use prime for the quantities in the rotation frame, sometimes a different notation is use in which we attach a subscript "in" and "rot" for quantities in the fixed frame and the rotating frame respectively.

$$\vec{v}_{in} = \vec{v}_{rot} + \vec{\Omega} \times \vec{r}. \tag{10.15}$$

There is no need to put a subscript to $\vec{r}$ since $\vec{r}$ is a vector from the origin to the position of the particle and the origins of the two frames are at the same point. The result of relation between the velocity vectors is also often written as

$$\left(\frac{d\vec{r}}{dt}\right)_{in} = \left(\frac{d\vec{r}}{dt}\right)_{rot} + \vec{\Omega} \times \vec{r}. \tag{10.16}$$

where the subscript "in" denotes the quantity in the inertial or non-rotating frame and the subscript "rot" for a quantity with respect to the rotating frame.

Note that to derive this relation, we only used the geometric property of the position vector. A similar argument can be made about the rate of change of any arbitrary vector $\vec{w}$ in the two frames.

$$\boxed{\left(\frac{d\vec{w}}{dt}\right)_{in} = \left(\frac{d\vec{w}}{dt}\right)_{rot} + \vec{\Omega} \times \vec{w}.} \tag{10.17}$$

For instance, we can obtain the time-derivative of the velocity vector $\vec{v}_{in}$ in the inertial frame by setting $\vec{w} = \vec{v}_{in}$ in this equation.

$$\boxed{\left(\frac{d\vec{v}_{in}}{dt}\right)_{in} = \left(\frac{d\vec{v}_{in}}{dt}\right)_{rot} + \vec{\Omega} \times \vec{v}_{in}.} \tag{10.18}$$

The left hand side of this equation is the acceleration of the particle in the inertial frame since it is the rate of change of the corresponding velocity in the same frame. The first term on the right side of this equation is not acceleration of anything since this term mixes information from the two frames - this term is the rate of change of the velocity in the inertial frame as observed from the rotating frame.

Finally, by substituting $\vec{v}_{in}$ in the right-side of Eq.rot-inert-eq-2 in terms of $\vec{v}_{rot}$ as given in Eq. 10.15 we can work out the relation between

10.2. ROTATING FRAME

accelerations in the two frames.

$$\begin{aligned}\left(\frac{d\vec{v}_{in}}{dt}\right)_{in} &= \left[\frac{d}{dt}\left(\vec{v}_{rot} + \vec{\Omega} \times \vec{r}_{in}\right)\right]_{rot} + \vec{\Omega} \times \left(\vec{v}_{rot} \times \vec{\Omega} \times \vec{r}_{in}\right) \\ &= \vec{a}_{rot} + 2\vec{\Omega} \times \vec{v}_{rot} + \vec{\Omega} \times \left(\vec{\Omega} \times \vec{r}\right)\end{aligned}$$

Therefore, the accelerations of the particle in the two frames are related as follows.

$$\boxed{\vec{a}_{in} = \vec{a}_{rot} + 2\vec{\Omega} \times \vec{v}_{rot} + \vec{\Omega} \times \left(\vec{\Omega} \times \vec{r}\right) \quad \text{(Constant } \vec{\Omega}\text{)}.} \quad (10.19)$$

10.2.2 Newton's Second Law in Uniformly Rotating Frame

Newton's second law for a particle of fixed mass m is written in inertial frame as $m\vec{a}_{in} = \vec{F}$, where $\vec{F}$ is the net real force. Substituting $\vec{a}_{in}$ from Eq. 10.19 into the second law we find that mass times acceleration in a rotating frame has the following form.

$$\boxed{m\vec{a}_{rot} = \vec{F} - m\left[2\vec{\Omega} \times \vec{v}_{rot} + \vec{\Omega} \times \left(\vec{\Omega} \times \vec{r}\right)\right] \quad \text{(Constant } \vec{\Omega}\text{)}} \quad (10.20)$$

Therefore, in a rotating frame, mass times acceleration is not equal to the net real force on the particle. In addition to the net real force $\vec{F}$, there are terms that have units of force but are dependent on the rotation of the frame. These are the "**fictitious forces**" or "**inertial forces**". In a rotating frame, the "real" and "fictitious" forces are indistinguishable in their effect on the acceleration except that there appears to be no agent(s) for the fictitious forces as far as the rotating frame is concerned.

The first term $\left(-2m\vec{\Omega} \times \vec{v}_{rot}\right)$ is called the **Coriolis force**, and the second term $\left[-m\vec{\Omega} \times \left(\vec{\Omega} \times \vec{r}\right)\right]$ the **centrifugal force**. While the direction of the Coriolis force is perpendicular to the axis of rotation and the velocity of the particle, the centrifugal force is always pointed away from the axis of rotation and remains perpendicular to the axis.

10.2.3 Newton's Second Law in Earth's Frame

An Earth-based frame is a rotating frame. Therefore, the equation of motion of a particle in an Earth-based frame will have Coriolis force and centrifugal force. The centrifugal force is included in mg, therefore, the equation of motion in the rotating frame of the Earth has Coriolis force only.

$$\boxed{m\vec{a}_{rot} = m\vec{g} + \vec{F}_{\text{other}} - 2m\vec{\Omega} \times \vec{v}_{rot},} \quad (10.21)$$

where $\vec{F}_{\text{other}}$ are forces other than gravity and $\vec{g}$ is the acceleration due to gravity written as a vector to include the direction information in the equation. We will use Eq. 10.21 to study the motion the motion of particles near the surface of the Earth in an Earth-based frame.

Example 10.2.1. Freely Falling Particle In An Earth-Based Frame.
The rotation of earth has observable effects on a freely falling object. If a particle of mass m is released at rest from a height h above the surface of the Earth, it will fall towards the center of the Earth if the Earth were not rotating. But, because the Earth is rotating, the path of the particle will deviate from this direction in an Earth-based frame. In this example we wish to determine the deviation at the equator.

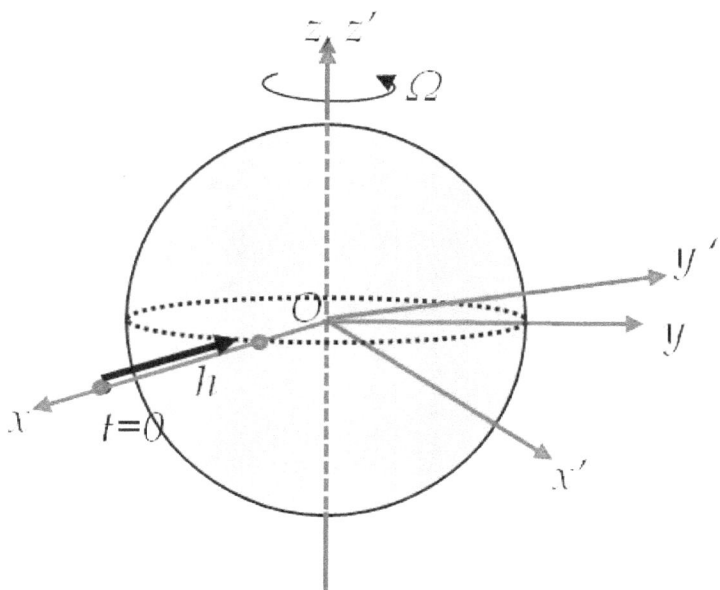

Figure 10.8: Example 10.2.1. The rotating frame $O'x'y'z'$ and the fixed frame $Oxyz$ coincide at $t = 0$. That is, Ox' and Ox are in the same direction at $t = 0$ when the particle is released on the x-axis. The particle falls down the x-axis of the fixed coordinate system.

Solution. Consider two coordinate frames $Oxyz$ and $Ox'y'z'$ with their origins at the center of Earth that have the z-axis pointed in the direction of the axis of rotation of the Earth. Let the frame $Ox'y'z'$ rotate with respect to the fixed inertial frame $Oxyz$ at angular speed Ω, and let the axes of the two systems be coincident at $t = 0$. We will use M for the mass of the Earth and assume Earth to be a sphere of radius R.

Suppose now a particle of mass m is released at $t = 0$ from rest at $x = x' = h + R$. In the inertial frame the particle moves straight down the x-axis and reaches $x = R$ at some time Δt. During Δt the x' and y' axes of the rotating frame move out to another direction. Since the rotation is about the z-axis, the particle will fall in the $x'y'$ plane of the rotating frame. We wish to determine the y' component of the displacement when the particle has dropped a distance of h in the inertial frame. We start by

10.2. ROTATING FRAME

writing the components of the equation in the rotating frame.

$$\frac{dv'_x}{dt} = -g - \left(2\vec{\Omega} \times \vec{v}_{rot}\right)_{x'} \quad (10.22)$$

$$\frac{dv'_y}{dt} = 0 - \left(2\vec{\Omega} \times \vec{v}_{rot}\right)_{y'} \quad (10.23)$$

$$\frac{dv'_z}{dt} = 0 - \left(2\vec{\Omega} \times \vec{v}_{rot}\right)_{z'} \quad (10.24)$$

The x' and y' components of the Coriolis and centrifugal terms are as follows.

$$\vec{\Omega} \times \vec{v}_{rot} = \begin{vmatrix} \hat{u}_{x'} & \hat{u}_{y'} & \hat{u}_{z'} \\ 0 & 0 & \Omega \\ v_{x'} & v_{y'} & v_{z'} \end{vmatrix} = -\Omega v_{x'} \hat{u}_{y'} + \Omega v_{x'} \hat{u}_{z'}.$$

Hence the equations of motion of the particle are:

$$\frac{dv_{x'}}{dt} = -g - 2\Omega v_{y'} \quad (10.25)$$

$$\frac{dv_{y'}}{dt} = 2\Omega v_{x'} \quad (10.26)$$

$$\frac{dv_{z'}}{dt} = 0 \quad (10.27)$$

Solving these equations with the initial condition $x' = h + R$, $y' = 0$, $z' = 0$, and $v_{0x'} = v_{0y'} = v_{0z'} = 0$ will give us the trajectory of the particle from the perspective of the rotating frame, i.e., from someone observing the particle from Earth-based frame. Since there is no motion along the z-axis, we will work out the solution for x' and y' components only.

Approximate solution: Observe that the particle will pick up velocity more along the vertical direction than along the horizontal direction. Therefore, we can assert that in time t, $|v'_x| \approx gt$. Using this in v'_x equation, we find the following for v'_y.

$$\frac{dv_{y'}}{dt} \approx -2\Omega gt \implies v_{y'} = -\Omega gt^2,$$

where we have used the initial condition on $v_{y'}$. Integrating $v_{x'}$ and $v_{y'}$ we obtain the following for x' and y' coordinates after the particle is released at $(x' = h + R, y' = 0)$ at $t = 0$.

$$x' = h + R - \frac{1}{2}gt^2 \quad (10.28)$$

$$y' = -\frac{1}{3}\Omega gt^3 \quad (10.29)$$

From the x' equation we can determine the time T for the particle to the surface of Earth, which has $x' = R$. Using this time in the y' equation we find the horizontal deviation y'.

$$\boxed{\Delta y' = -\frac{1}{3}\Omega g \left(\frac{2h}{g}\right)^{3/2}}$$

For a 100 meter drop we will find the deviation to be

$$\Delta y' = -\frac{1}{3}\frac{2\pi}{24\times 3600\ s}\times 9.81\ \text{m/s}^2 \times \left(\frac{2\times 100\ \text{m}}{9.81\ \text{m/s}^2}\right)^{3/2}$$
$$= -2.2\times 10^{-2}\ \text{m}.$$

Since the rotation axis is towards the North, the y' axis is towards the East at the surface of the Earth. Therefore, a particle dropped from 100 m above the ground will land approximately 2.2 cm to the West of the line connecting the original position to the center of the Earth.

Figure 10.9: Foucault pendulum. The rotation of the Earth causes the plane of oscillation of the pendulum to change over time. With changing plane of oscillation, the pins at different positions are knocked down at different times. Photo credit: Ciudad de las Artes y de las Ciencias de Valencia by Daniel Sancho, Wikicommons.

Example 10.2.2. Surface of a Rotating Fluid in a Bucket.

A bucket of water is rotated with a uniform rotational speed Ω. It is found the surface assumes a steady shape. Determine the shape. Ignore the rotation of earth.

Solution. Due to the symmetry in the problem, it is sufficient to work in a plane containing the axis of rotation and a horizontal direction as shown in Fig. 10.10. We will call the axis of rotation the z-axis and the horizontal direction will be taken to be the x-axis. Therefore, to find the equation of the surface, we need to work out the function $z(x)$. We will make use of steady condition on a mass element at the surface.

Here, notice that it is easier to work in the rotating frame of the bucket since in this frame, liquid in the bucket will not be moving and will have zero acceleration. That is, in this frame, real forces will be balanced by inertial force(s). Let us figure out the real and inertial forces on a mass element at the surface.

10.2. ROTATING FRAME

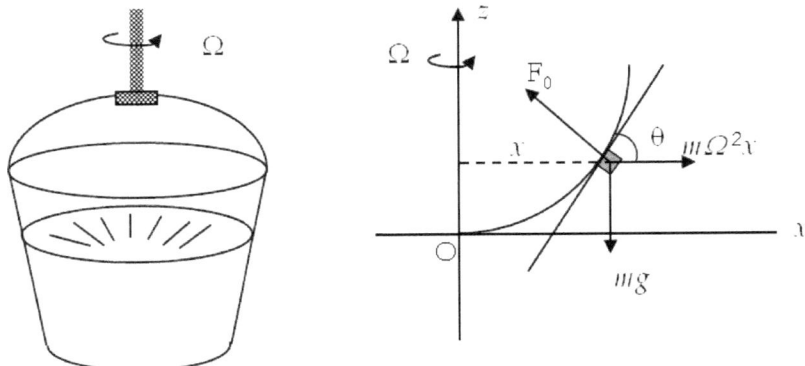

Figure 10.10: Example 10.2.2.

In the rotating frame once the steady state has reached, water would not be moving any more. Therefore the Coriolis force will be zero. Thus, the only inertial force on particles of water will be the centrifugal force. The only other force is the weight.

Consider a mass element of mass m at the surface of the steady fluid. The weight has magnitude mg and acts straight down. The centrifugal forces on various water particles and the weight of the water molecules will press on the layers of water in contact. The water molecules on the surface will press on the molecules just below the surface. The reaction force from the layer just below the surface will be in the normal direction of the surface as shown by F_0 in the figure.

Therefore, the x and z-components of the equations of motion of an element at the surface of water in the rotating frame for a mass at the surface are:

$$x\text{-component:} \quad m\Omega^2 x - F_0 \sin\theta = 0$$
$$z\text{-component:} \quad F_0 \cos\theta - mg = 0$$

Therefore,

$$\tan\theta = \frac{\Omega^2}{g} x$$

But this tangent must equal the slope of the tangent to the curve $z(x)$ at the point.

$$\frac{dz}{dx} = \tan\theta.$$

Therefore, we obtain the following equation for $z(x)$.

$$\frac{dz}{dx} = \frac{\Omega^2}{g} x,$$

which can be immediately integrated to yield

$$z = \frac{\Omega^2}{2g} x^2 + C,$$

where C is the constant of integration. From the figure, $x = 0$ corresponds to $z = 0$ on the surface, therefore $C = 0$. Hence, the equation for the surface is

$$z = \frac{\Omega^2}{2g}x^2.$$

The situation is symmetric in the xy-plane and there is nothing special about x-axis. To obtain the equation for the entire surface and not just a slice of the surface, all we need to do is to replace x by the radial distance r of polar coordinates.

$$z = \frac{\Omega^2}{2g}r^2. \tag{10.30}$$

Hence, the surface is a paraboloid of revolution about the axis of rotation.

10.3 CORIOLIS FORCE

In the Earth-based frame, there are two inertial forces, one directed away from the center of earth, and the other pointed in the plane perpendicular to the axis of rotation. The centrifugal force subtracts from the centrally directed force of gravitation, and is absorbed in the value for the acceleration due to gravity g. Therefore, the inertial force that leads to unexpected effects in the Earth-based frame is the Coriolis force $\vec{F}_{cor}$.

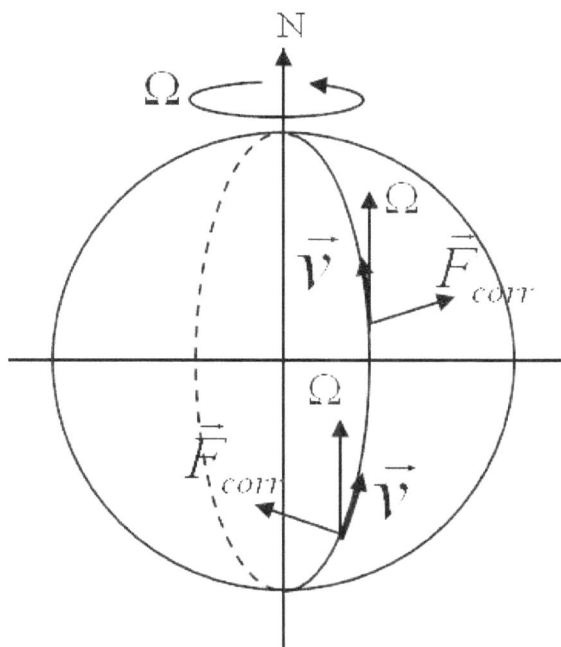

Figure 10.11: Coriolis force in northern and southern hemisphere.

$$\boxed{\vec{F}_{cor} = -2m\vec{\Omega} \times \vec{v}.} \tag{10.31}$$

Note that the Coriolis force is perpendicular to the direction of velocity of the particle, as evident from the cross product. Therefore, a moving

10.3. CORIOLIS FORCE

particle is deflected perpendicular to its direction of motion by the Coriolis force. As a result, northward moving particles in the northern hemisphere are deflected to the East while northward moving particles in the southern hemisphere are deflected to the West as illustrated in Fig. 10.11.

The atmosphere of Earth is usually modeled as particles in motion. The forces on particles of atmosphere are from gravity and pressure differences. If there is a pressure gradient, then there is a force from high pressure region towards the low pressure region. Due to the Coriolis force, when a particle has nonzero velocity towards the lower pressure region, the path of the particle is changed towards the perpendicular direction. As a result the winds circulates differently in the northern and souther hemispheres, being in the counterclockwise sense in the northern hemisphere and in the clockwise sense in the southern hemisphere as shown in Figs. 10.12 and 10.13 respectively.

Figure 10.12: Hurricane Katrina in the Gulf of Mexico rotating counterclockwise on August 28, 2005. Hurricane Katrina later devastated the city of New Orleans in the state of Louisiana causing massive flooding. Credit: National Oceanic and Atmospheric Administration, USA.

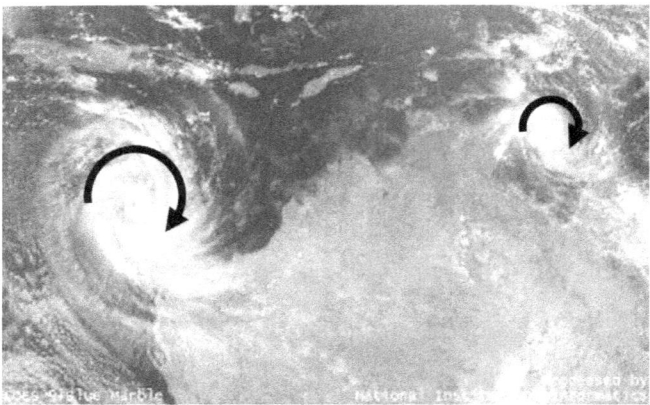

Figure 10.13: Rotation of winds in southern hemisphere. Cyclones Willy (left) and Ingrid (right) off the northern coast of Australia on 11th March 2005. (Credits: National Institute of Informatics, Japan).

10.4 EXERCISES

Accelerating Frame

Ex 10.4.1. Two friends John and Jane observe the location of a building from their own frames which are accelerating with respect to each other (Fig. 10.14). In Jane's frame John has a constant acceleration $\vec{A}$ whose direction makes an angle θ with respect to the direction of the building. Let the building be at a distance R from Jane who is fixed to the Earth.

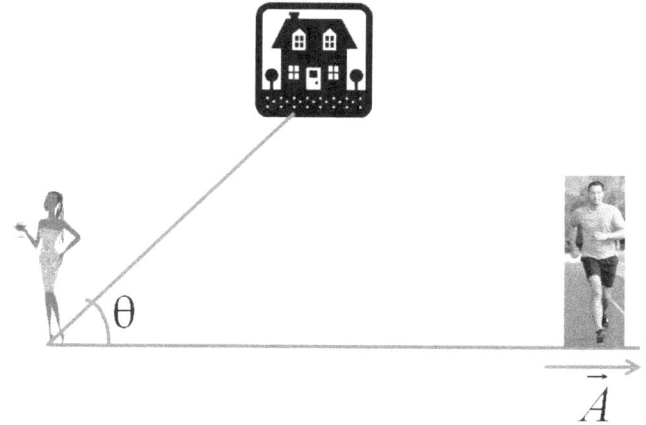

Figure 10.14: Exercise 10.4.1.

Suppose at $t = 0$, John and Jane were at the same location and their relative velocity was zero. (a) Draw a figure in Jane's frame showing a choice of coordinates for both John and Jane supposing that the position vector of the building and acceleration vector $\vec{A}$ are in the xy-plane. You can also take the x-axis to be the direction of vector $\vec{A}$. (b) Deduce the relation between the coordinates of the building in the two frame at an arbitrary time? (c) Describe the motion of the building in the two frames.

Ex 10.4.2. A box sits on a rubber floor in a truck. The driver starts the truck and increases the acceleration steadily from zero to 5 m/s^2 in 2 seconds. If the coefficient of static friction between the box and the rubber floor is 0.2, determine if the box will slide, and if so, at what acceleration?

Ans: The box will not slide.

Ex 10.4.3. The coordinates of a particle in space from two frames $Oxyz$ and $O'x'y'z'$ are related as follows.

$$x' = x + (5 \text{ m/s})\, t$$
$$y' = y + (5 \text{ m/s})\, t + (2 \text{ m/s}^2)\, t^2$$
$$z' = z$$

(a) Describe the relative motion of the two frames with respect to each other. (b) If a particle has a constant velocity with respect to $Oxyz$ with components $(2 \text{ m/s}, 0, 0)$, what are the velocity and acceleration of this particle with respect to $O'x'y'z'$? (c) If a particle has a constant velocity with respect to $O'x'y'z'$ with components $(2 \text{ m/s}, 0, 0)$, what are the velocity and acceleration of this particle with respect to $Oxyz$?

Ans: (c) $v_x = -3$ m/s, $v_y = -3$ m/s $-(4 \text{ m/s}^2) \, t$, $v_z = 0$; $a_x = 0$, $a_y = -4$ m/s^2, $a_z = 0$.

Ex 10.4.4. A box is sliding on a floor so that its coordinates with respect to a frame $Oxyz$ fixed with respect to the floor are given as

$$x = (1 \text{ m}) + (2 \text{ m/s}) \, t + (3 \text{ m/s}^2) \, t^2$$
$$y = z = 0$$

What would be the coordinates of this box when observed with respect to another frame $O'x'y'z'$ that has the following acceleration with respect to $Oxyz$?

$$A_x = 2 \text{ m/s}^2$$
$$A_y = A_z = 0$$

Assume origins O and O' coincide at $t = 0$ and the axes of the two frames are parallel with each other.

Ans: $x' = 1$ m $+ (2 \text{ m/s}) \, t + (1 \text{ m/s}^2) \, t^2$, $y' = 0$, $z' = 0$.

Ex 10.4.5. A box is sliding on a floor so that its coordinates with respect to a frame $Oxyz$ fixed with respect to the floor are given as

$$x = (1 \text{ m}) + (2 \text{ m/s}) \, t + (3 \text{ m/s}^2) \, t^2$$
$$y = 0$$
$$z = 0$$

What would be the coordinates of this box when observed with respect to another frame $O'x'y'z'$ that has the following acceleration with respect to $Oxyz$?

$$A_x = 0$$
$$A_y = 2 \text{ m/s}^2$$
$$A_z = 0$$

Assume the origins O and O' coincide at $t = 0$ and the axes of the two frames are parallel with each other.

Ans: $x' = (1 \text{ m}) + (2 \text{ m/s})t + (3 \text{ m/s}^2)t^2$, $y' = -(1 \text{ m/s}^2)t^2$, $z' = 0$.

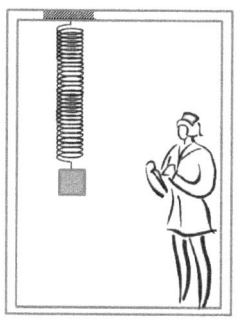

Figure 10.15: Exercise 10.4.6.

Ex 10.4.6. A block of mass m is hung from the ceiling of an elevator using a spring of spring constant k (Fig. 10.15). Find the change in the length of the spring under the following situations. (a) Elevator going up with constant velocity, $\vec{v}_1$. (b) Elevator going down with constant velocity, $\vec{v}_2$. (c) Elevator going up with acceleration $\vec{a}_1$ point up. (d) elevator going up with acceleration $\vec{a}_2$ pointed down. (e) elevator going down with acceleration $\vec{a}_3$ pointed up. (f) elevator going down with acceleration $\vec{a}_4$ pointed down.

Ans: (a) $\Delta l = mg/k$. (b) $\Delta l = mg/k$. (c) $\Delta l = \frac{m}{k}(g + a_1)$. (d) $\Delta l = \frac{g}{k}(g - a_2)$. (e) $\Delta l = \frac{g}{k}(g + a_3)$. (f) $\Delta l = \frac{g}{k}(g - a_4)$.

Ex 10.4.7. A ball is dropped from a 50-m building (Fig. 10.16). In a frame fixed to Earth, the ball drops vertically with a constant acceleration g pointed down. The same ball is observed from a car that is accelerating with respect to the fixed frame. The acceleration of the car is pointed horizontally and away from the building.

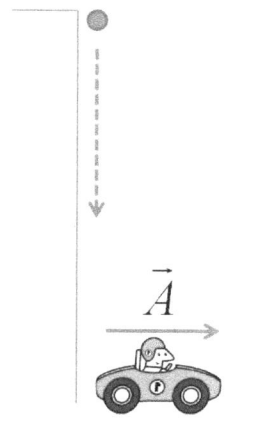

Figure 10.16: Exercise 10.4.7.

Suppose the fixed frame has a coordinate system $Oxyz$ whose y-axis is pointed up and the x-axis is pointed in the direction of the acceleration of the car. Let $O'x'y'z'$ be the coordinate system whose origin is fixed to the car. The coordinate axes of the two frame are parallel to each other. The time $t = 0$ is chosen when the two origins coincide and the velocity of the car is zero with respect to the ground. (a) Write the coordinates of the ball in $Oxyz$ frame during the flight, i.e., before the ball hits the ground. (b) Write the coordinates of the ball in $O'x'y'z'$ frame during the flight, i.e., before the ball hits the ground. (c) What are the trajectories of the ball in the two frames? Do the trajectory equations make sense, i.e. does the ball fall vertically in both frames?

Ans Key: (c) straight down in one frame, slanted in another frame.

Ex 10.4.8. A freely falling sky diver looks at another diver directly above him (Fig. 10.17) that has his parachute on and finds his position in a coordinate system that has the y-axis vertically up as follows.

$$y = 4 + 5t + 3t^2.$$

Figure 10.17: Exercise 10.4.8.

At $t = 0$, the freely falling person was 100 meters from the ground and had zero speed. Find the position of the person with his parachute on as a function of time when observed from the ground?

Ans: $x' = 0$, $y' = (104) + 4$ t $- \frac{1}{2}(g - 3)t^2$, $z' = 0$.

Rotating Frame

Ex 10.4.9. A large platform is rotating uniformly with an angular speed Ω about the z-axis of a fixed frame $Oxyz$ with the origin at the center of the platform. A man is on the platform at a distance R from the center. (a) What is the motion of the man in the fixed frame? (b) What is the motion of the man with respect to a frame $O'x'y'z'$ that has the same origin and z-axis as the fixed frame but whose x and y-axes rotate with the platform? (c) Let the man be along the x' axis at $t = 0$ when he starts to walk directly towards the center at a constant speed v'. That is, he is walking along the x' axis towards origin with speed v' with respect to the origin O'. What is the motion of the person as seen from the fixed frame?

Ex 10.4.10. A car in Los Angeles, California, USA, is moving with a velocity of 30 m/s towards North with respect to an observer on the ground. Find the magnitude and direction of the Coriolis force on the car. Los Angeles, USA has latitude 34.0522° N, and longitude 118.2428° W. Ans: $F_c = 7.3$ N towards the local East.

Ex 10.4.11. A car of mass 3000 kg in Canberra, Australia, is moving with a velocity of 30 m/s towards North with respect to an observer on the ground. Find the magnitude and direction of the Coriolis force on the car. Canberra, Australia has latitude 35.2828° S, and longitude 149.1314° E. Ans: Magnitude $F_c = 7.56$ N. Hint: Figure out the direction using the cross product.

Ex 10.4.12. In exercise, Ex 10.4.9, write the equation of motion of the CM of the man with respect to (a) the fixed frame and (b) the rotating frame. Ans: (a) In the fixed frame, Centripetal: $F_s = m\Omega^2 R$; Vertical: $F_N - mg = 0$.

Ex 10.4.13. A satellite is in the Geocentric orbit about Earth at a distance R from the center of the Earth. The satellite revolves in a circular motion with the angular speed Ω, which is equal to the angular speed of the Earth about the same axis, as observed from a fixed frame. (a) Describe the motion of the satellite with respect to a frame with the origin at the center of Earth and which rotates with the Earth. (b) Write the equation of motion of the satellite in the two frames. Ans: (b) Fixed Frame: $G_N \frac{Mm}{R^2} = m\frac{v^2}{R}$

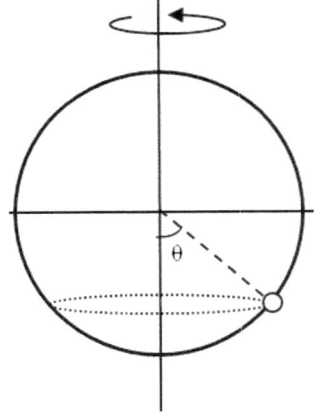

Figure 10.18: Exercise 10.4.14.

Ex 10.4.14. A bead of mass m can slide frictionlessly on a rotating ring of mass M and radius R which rotates at an angular speed of ω about a vertical axis through its center (Fig. 10.18). When the bead is at a particular location, it does not slide. (a) Find this special position of the bead using calculations in a rotating frame. (b) Repeat the calculation in an inertial frame. Ans: (a) $\theta = \cos^{-1}\left(\frac{g}{\omega^2 R}\right)$ with $\omega^2 R > g$.

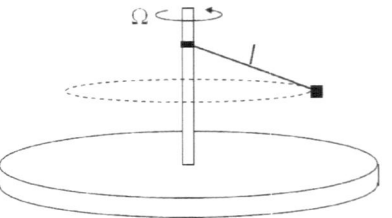

Ex 10.4.15. A block of mass M is attached to a string of length l. The other end of the string is attached to a post at the center of a rotating platform as shown in Fig. 10.19. When the platform is rotating at a steady angular speed Ω the block moves in a circular motion. Find the radius of the circular motion of the block performing the calculations in a rotating frame. Ans: $\frac{l}{\omega^2}\sqrt{\omega^4 l^2 - g^2}$.

Figure 10.19: Exercise 10.4.15.

10.5 PROBLEMS

Problem 10.5.1. A rocket of mass M is moving at constant velocity $\vec{v}$ in zero gravity environment. It ejects fuel from the back at a steady rate of $(dm/dt = \alpha)$ at speed u with respect to the rocket. What is the increase in speed of the rocket when a mass m_f of fuel has been ejected? Do this problem in the frame of the rocket.

Ans: $v_x(T) - v_0 = u \ln \left(\frac{M}{M - m_f} \right)$.

Problem 10.5.2. A stone is tied to a string and rotated in a vertical circle at constant speed v inside an accelerating train. The circle of rotation is perpendicular to the train's velocity as observed from a frame outside the train. The train has a constant acceleration of magnitude A and direction East with respect to an observer outside the train. The train was at rest at $t = 0$. Find the tension in the string at four instances in the vertical circle of the motion of the stone: (a) when the stone is at the top, (b) when the stone is at the bottom, (c) when the stone is horizontal left, and (d) when the stone is horizontal right.

Ans: $T_x = mA$, $T_y = mg - m\frac{v^2}{R}$.

Problem 10.5.3. Calculate the percentage mistake made in evaluating the acceleration due to gravity when ignoring the rotation of earth?

Ans: 0.35%.

Problem 10.5.4. Calculate the percentage mistake made in evaluating acceleration due to gravity when ignoring the revolution of earth around the sun?

Ans: 0.06% of g.

Problem 10.5.5. You are inside a large enclosed container and everything in the entire container, including you, is rotating at a constant rate about some fixed axis, but you do not know the rate of rotation. To find the rate of rotation you place a penny at different places on a frictionless floor, and discover that the penny accelerates everywhere you place the penny. The directions of the acceleration of the penny at various places on the flat floor meet at a point X. (a) If the penny accelerates at 30 m/s^2 at a distance of 10 meters from point X, what is the rate of rotation of the entire container? (b) What is the acceleration of the penny when it is at a point 20 meters from point X?

Ans: (a) $\Omega = \sqrt{3}$ rad/s. (b) 60 m/s^2.

Problem 10.5.6. A pendulum of length L and mass m in an accelerating frame is in equilibrium at an angle θ_0. (a) What is the acceleration of the accelerating frame with respect to an inertial frame? (b) When the pendulum is disturbed from the equilibrium angle by a small angle θ about θ_0, the pendulum oscillates. Determine the frequency of small oscillations in the accelerating frame.

Ans: (a) $A = g\tan\theta_0$. (b) $\omega = \sqrt{\dfrac{A^2 + g^2}{l}}$.

Problem 10.5.7. A stone is dropped from rest from a height of 200 meters in a place at a latitude of 60 degrees and longitude 30 degrees. Where will it land on the surface of the Earth?

Ans: Local directions: x vertically up, y towards East, z towards North. $x = 0$, $y = -\frac{1}{3}\Omega_z g T^3$, $z = \frac{1}{6}\Omega_x \Omega_z g T^4$. Here $\Omega_x = \Omega\sin\lambda$ and $\Omega_z = \Omega\cos\lambda$, Ω is the rotation speed of Earth, λ is the latitude at the point on the surface of the Earth, $T = \sqrt{2h/g}$, h is the height of fall.

Problem 10.5.8. A stone is dropped from rest from a height of 200 meters in a place at a latitude of 30 degrees and longitude 60 degrees. Where will it land on the surface of the Earth?

Index

Accelerating frame, 416
Acceleration, 88
Acceleration, Analytic, 90
Acceleration, Constant, 125
Acceleration, Geometric, 90
Acceleration, polar coords, 101
Acceleration, Variable, 145
Accuracy, 16
Adding vectors, 44
Addition of vectors, 29
Analytic method, 37
Apparent weight, 418
Atomic Clock, 7
Average speed, 73
Average velocity, 69

Base vectors, 37
Burn rate, 297

Cantilever, 204
Center of Mass, 273
Centrifugal force, 425
Centripetal acceleration, 104
CGS system, 9
Circular motion, 102
Circular motion, Uniform, 105
Classical physics, 2
CM, 273
CM frame, 291
Coefficient of kinetic friction, 187
Collision, 286
Common forces, 174
Components of velocity, 70
Conical pendulum, 365
Conservation of momentum, 231, 283
Conservative forces, 325
Constant acceleration, 125
Constant speed, 125
Constant velocity, 124

Coriolis force, 425
Couple, 213
Cross product, 35

Derivatives from tangent, 79
Derived Units, 10
Dimensional analysis, 20
Displacement, 26, 61
Dot product, 33, 46

Empirical science, 2
Enternal forces, 196
Error
    Absolute, 13
    Relative, 13
Escape speed, 335

Fictitious force, 418
First law, 221
Force center, 346
Frame, 60, 150
Free fall, 131
free-body diagram, 239
Fundamental forces, 174
Fundamental Units, 4

g, 132
Galilean relativity, 150
Galileo Galilei, 6
Gravitation force, 175
Gravitational mass, 223
Gravitational potential energy, 329
Guessing physics, 20

Hooke's law, 177
Horsepowe, 339

Imperial units, 8
Impulse, 264
Inertia, 221
Inertial force, 418

Inertial frame, 222
Inertial mass, 223
Instantaneous speed, 87
Instantaneous velocity, 83
Internal forces, 196
Isolated system, 282

Kilogram, 8
Kinematics, 60
Kinetic energy, 320
Kinetic friction force, 186
Kinetic friction, Coefficient, 187

LAB frame, 292
Laws of physics, 3
Length, 4
Lever arm, 190
Linear density, 278

Mass, 7
Mass, Gravitational, 223
Mass, Inertial, 223
Measurements, 12
Meter, 5
Moment, 189
Momentum, 223
MSK system, 9
Multiparticle system, 271
Multiplication of vectors, 32

Net force, 171
Newton's first law, 221
Newton's second law, 225
Newton's third law, 230
Newton, Principia, 168
Newton, Third Law, 168
Nonconservative force, 325
Normal force, 179

One-dimensional motion, 129
Open system, 294
Orders of magnitude, 19

Parallelogram law of addition, 27
Planar motion, 126
Point particle, 60
Polar coordinates, 98
Position, 61

Potential energy, 326
Potential Energy, Gravitational, 329
Pound, 9
Power, 339
Precision, 12
Principia, 168
Projectile motion, 137

Recoil force, 306
Relative Motion, 149
Resultant force, 171
Right Hand Rule, 36
Rocket equation, 297
Rotational inertia, 394
Rotational Work-Energy Theorem, 397

Scalar product, 33, 46
Scattering angle, 293
Scientific notation, 15
Second, 7
Sense of rotation, 191
SI prefixes, 9
SI units, 10
Significant figures, 13
Sliding friction force, 186
Speed, 87
Speed, Average, 73
Speed, General, 87
Spring balance, 167
Spring force, 178
Static friction force, maximum, 185
Static frictional force, 184

Tangential acceleration, 104
Theory, 2
Thrust, 295
Time, 5
Tip to tail method, 30
Torque, 189
Two-particle system, 271

Uncertainties - Propagation, 16
Uncertainty, 12
    Absolute, 13
    Relative, 13
Uniform circular motion, 105

INDEX

Unit vector, 29
Unit vectors, polar, 99
Units, 4
Units conversion, 10

Vector addition, 29
Vector equations, 32
Vector product, 35, 50
Vectors, 27
Vectors in a polygon, 32
Velocity, 1-Dimensional, 74
Velocity, Analytic, 83
Velocity, Average, 69
Velocity, Components, 70, 77
Velocity, General Discussion, 82
Velocity, Geometric, 82
Velocity, Instantaneous, 83
Velocity, polar coords, 101
Viscous drag, 188

Watt Balance, 8
Weight, 166
Work-Energy Theorem, 321

FUNDAMENTAL PHYSICAL CONSTANTS

http://www.physics.nist.gov/cuu/Constants/index.html

QUANTITY	SYMBOL	VALUE
UNIVERSAL		
Speed of light in vacuum	c	2.99792458×10^8 m/s (Exact)
Magnetic permeability of vacuum	μ_0	$4\pi \times 10^{-7}$ N/A^2 (Exact)
Electric permittivity of vacuum	$\epsilon_0 = 1/c^2\mu_0$	$8.854187817... \times 10^{-12}$ F/m (Exact)
Newtonian Constant	G	$6.6742(10) \times 10^{-11}$ N.m^2/kg^2
Planck Constant	h	$6.6260755(40) \times 10^{-34}$ J.s
h-bar or $\hbar$	$h/2\pi$	$\sim 1.05457 \times 10^{-34}$ J.s
ELECTROMAGNETIC		
Bohr magneton	μ_B	$9.27400949(80) \times 10^{-24}$ J/T
Elementary charge	e	$1.60217653(14) \times 10^{-19}$ C
Josephson constant	K_J	$4.83597876(41) \times 10^{14}$ Hz/V
Magnetic quantum flux	Φ_0	$2.06783372(18) \times 10^{-15}$ Wb
Nuclear magneton	μ_N	$5.05078343(43) \times 10^{-27}$ J/T
Von Klitzing constant	R_K	$2.5812807449(86) \times 10^4$ Ω
ATOMIC		
Electron mass	m_e	$9.1093826(16) \times 10^{-31}$ kg
Proton mass	m_p	$1.67262171(29) \times 10^{-27}$ kg
Neutron mass	m_n	$1.67492728(29) \times 10^{-27}$ kg
Bohr radius	a_0	$0.5291772108(18) \times 10^{-10}$ m
Compton wavelength	$\lambda_c = \frac{h}{m_e}$	$2.426310238(16) \times 10^{-12}$ m
Electron gyromagnetic ratio	g_e	$1.76085974(15) \times 10^{11}$ 1/T.s
Proton magnetic moment	μ_p	$1.41060671(12) \times 10^{-26}$ J/T
Rydberg constant	R_H	$1.0973731568525(73) \times 10^7$ 1/m
PHYSICOCHEMICAL		
Atomic mass	u or m_u	$1.66053886(28) \times 10^{-27}$ kg
Avogadro number	N_A	$6.0221415(10) \times 10^{23}$ 1/mol
Boltzman constant	k or k_B	$1.3806505(24) \times 10^{-23}$ J/K
Faraday constant	F	$96485.3383(83)$ C/mol
Molar gas constant	R	$8.314472(15)$ J/K.mol
Stephan-Boltzman constant	σ	$5.670400(40) \times 10^{-8}$ W/m^2.K^4

USEFUL CONVERSION FACTORS

Length
 1 in = 2.54 cm
 1 ft = 12 in
 1 mi = 5280 ft $\approx$ 8/5 km
 1 nautical mile $\approx$ 1.1584 mi $\approx$ 1.852 km
 1 astronomical unit (au) $\approx 1.5 \times 10^{11}$ m
 1 light year (ly) $\approx 9.46 \times 10^{15}$ m
 1 parsec $\approx 3.1 \times 10^{16}$ m
 1 fermi (fm) $= 10^{-15}$ m
 1 angstrom () = 0.1 nm = 10^{-10} m

Mass
1 pound (lb) = 453.59237 g
1 kg = 0.0685 slug
1 ounce (oz) $\approx$ 28.35 g
1 (metric) ton = 1000 kg
1 u $\approx 9.1 \times 10^{-31}$ kg

Volume
1 liter (L) = 1000 cm^3 (cc) = 10^{-3} m^3
1 US-quart (qt) = 32 fluid ounce (oz)= 0.946 L
1 US-gallon (gal) = 4 qt = 0.83 gal (Imperial)

Energy
1 Joule (J) = 1 Volt-Coulomb
= 10^7 ergs
$\approx$ 0.74 lb-force.ft
$\approx 9.5 \times 10^{-4}$ British thermal unit (Btu)
$\approx 2.8 \times 10^{-5}$ kiloWatt-hour (kWh)
1 calorie (cal) $\approx$ 4.186 J $\approx 4 \times 10^{-3}$ Btu
1 electron volt (eV) = 1.602×10^{-19} J
1 liter.atm = 101.3 J

METRIC MULTIPLIERS

Factor	Prefix	Symbol	Factor	Prefix	Symbol
10^1	deka-	da	10^{-1}	deci-	d
10^2	hecto-	h	10^{-2}	centi-	c
10^3	kilo-	k	10^{-3}	milli-	m
10^6	mega-	M	10^{-6}	micro-	μ
10^9	giga-	G	10^{-9}	nano-	n
10^{12}	tera-	T	10^{-12}	pico-	p
10^{15}	peta-	P	10^{-15}	femto-	f
10^{18}	exa-	E	10^{-18}	atto-	a
10^{21}	zetta-	Z	10^{-21}	zepto-	z
10^{24}	yotta-	Y	10^{-24}	yocto-	y

www.ingramcontent.com/pod-product-compliance
Lightning Source LLC
Chambersburg PA
CBHW062211220526
45471CB00009B/3157